Small Country Innovation Systems

Small Country Innovation Systems
Globalization, Change and Policy in Asia and Europe

Edited by

Charles Edquist

CIRCLE *(Centre for Innovation, Research and Competence in the Learning Economy), Lund University, Lund, Sweden*

and

Leif Hommen

CIRCLE *(Centre for Innovation, Research and Competence in the Learning Economy), Lund University, Lund, Sweden*

Edward Elgar
Cheltenham, UK • Northampton, MA, USA

© Charles Edquist and Leif Hommen, 2008

All rights reserved. No part of this publication may be reproduced, stored in a retrieval system or transmitted in any form or by any means, electronic, mechanical or photocopying, recording, or otherwise without the prior permission of the publisher.

Published by
Edward Elgar Publishing Limited
Glensanda House
Montpellier Parade
Cheltenham
Glos GL50 1UA
UK

Edward Elgar Publishing, Inc.
William Pratt House
9 Dewey Court
Northampton
Massachusetts 01060
USA

A catalogue record for this book
is available from the British Library

Library of Congress Control Number: 2007943826

ISBN 978 1 84542 584 5 (cased)

Printed and bound in Great Britain by MPG Books Ltd, Bodmin, Cornwall

Contents

List of contributors	vii
Preface	ix
List of abbreviations	xi

1 Comparing national systems of innovation in Asia and Europe: theory and comparative framework 1
   Charles Edquist and Leif Hommen

PART I   FAST GROWTH COUNTRIES

2 The rise and growth of a policy-driven economy: Taiwan 31
   Antonio Balaguer, Yu-Ling Luo, Min-Hua Tsai, Shih-Chang Hung, Yee-Yeen Chu, Feng-Shang Wu, Mu-Yen Hsu and Kung Wang

3 From technology adopter to innovator: Singapore 71
   Poh Kam Wong and Annette Singh

4 Towards knowledge generation with bipolarized NSI: Korea 113
   Chaisung Lim

5 High growth and innovation with low R&D: Ireland 156
   Eoin O'Malley, Nola Hewitt-Dundas and Stephen Roper

6 From trade hub to innovation hub: Hong Kong 194
   Naubahar Sharif and Erik Baark

PART II   SLOW GROWTH COUNTRIES

7 Reconsidering the paradox of high R&D input and low innovation: Sweden 237
   Pierre Bitard, Charles Edquist, Leif Hommen and Annika Rickne

8 Low innovation intensity, high growth and specialized trajectories: Norway 281
   Terje Grønning, Svein Erik Moen and Dorothy Sutherland Olsen

9	Challenged leadership or renewed vitality? The Netherlands *Bart Verspagen*	319
10	Not just Nokia: Finland *Ville Kaitila and Markku Kotilainen*	355
11	An NSI in transition? Denmark *Jesper Lindgaard Christensen, Birgitte Gregersen, Björn Johnson, Bengt-Åke Lundvall and Mark Tomlinson*	403
12	Globalization and innovation policy *Leif Hommen and Charles Edquist*	442

Appendix: statistical bases of comparison for ten 'small country' NSI 485
Pierre Bitard, Leif Hommen and Jekaterina Novikova

Index 531

Contributors

Erik Baark, Division of Social Science, The Hong Kong University of Science and Technology, Hong Kong

Antonio Balaguer, Science & Technology Policy Research and Information Centre, National Applied Research Laboratories, Taipei, Taiwan

Pierre Bitard, Innovation Intelligence, Europe Department, Paris

Yee-Yeen Chu, Industrial Engineering and Engineering Management Department, National Tsing Hua University, Hsinchu, Taiwan

Charles Edquist, CIRCLE (Centre for Innovation, Research and Competence in the Learning Economy), Lund University, Lund, Sweden

Birgitte Gregersen, Department of Business Studies, Aalborg University, Aalborg, Denmark

Terje Grønning, Institute for Educational Research, University of Oslo, Oslo, Norway

Nola Hewitt-Dundas, School of Management and Economics, Queen's University Belfast, Belfast, Northern Ireland

Leif Hommen, CIRCLE (Centre for Innovation, Research and Competence in the Learning Economy), Lund University, Lund, Sweden

Mu-Yen Hsu, Graduate Institute of Technology and Innovation Management, National Chengchi University, Taipei, Taiwan

Shih-Chang Hung, Institute of Technology Management, National Tsing Hua University, Hsinchu, Taiwan

Björn Johnson, Department of Business Studies, Aalborg University, Aalborg, Denmark

Ville Kaitila, The Research Institute of the Finnish Economy, Helsinki, Finland

Markku Kotilainen, The Research Institute of the Finnish Economy, Helsinki, Finland

Chaisung Lim, School of Business, Konkuk University, Seoul, Korea

Jesper Lindgaard Christensen, Department of Business Studies, Aalborg University, Aalborg, Denmark

Bengt-Åke Lundvall, Department of Business Studies, Aalborg University, Aalborg, Denmark

Yu-Ling Luo, Science & Technology Policy Research and Information Centre, National Applied Research Laboratories, Taipei, Taiwan

Svein Erik Moen, Centre for Technology, Innovation and Culture, University of Oslo, Oslo, Norway

Jekaterina Novikova, European Commission, DG Regional Policy, Brussels, Belgium

Dorothy Sutherland Olsen, Institute for Educational Research, University of Oslo, Oslo, Norway

Eoin O'Malley, The Economic and Social Research Institute, Dublin, Ireland

Annika Rickne, CIRCLE (Centre for Innovation, Research and Competence in the Learning Economy), Lund University, Lund, Sweden and The Dahmén Institute, Örebro, Sweden

Stephen Roper, Economy and Strategy Group, Aston University, Birmingham, UK

Naubahar Sharif, Division of Social Science, The Hong Kong University of Science and Technology, Hong Kong

Annette Singh, NUS Entrepreneurship Centre, National University of Singapore, Singapore

Mark Tomlinson, Department of Social Policy and Social Work, University of Oxford, Oxford, UK

Min-Hua Tsai, Science & Technology Policy Research and Information Centre, National Applied Research Laboratories, Taipei, Taiwan

Bart Verspagen, Eindhoven Centre for Innovation Studies, Eindhoven University of Technology, Eindhoven, The Netherlands

Kung Wang, School of Management, National Central University, Jonghli City, Taiwan

Poh Kam Wong, NUS Entrepreneurship Centre, National University of Singapore, Singapore

Feng-Shang Wu, Graduate Institute of Technology and Innovation Management, National Chengchi University, Taipei, Taiwan

Preface

This book is the outcome of a comparative study of ten small country national systems of innovation (NSI) in Europe and Asia, conducted from 2002 to 2007 under the auspices of the European Science Foundation (ESF, website: www.esf.org) and financed by participating national research councils. The funding source for each of the separate national case studies was one or more national research councils in the country concerned, as will be detailed in each case study. For Ireland, the Nordic countries and the Netherlands, the funding was coordinated by the ESF. The coordination of the whole project and the writing of this introductory chapter as well as the concluding chapter were financed by the Swedish Agency for Innovation Systems (VINNOVA) and the Swedish Institute for Growth Policy Studies (ITPS). We wish to thank all of these organizations for their generous support.

Contributors to the project included national research teams in Denmark, Finland, Hong Kong, Ireland, (South) Korea, the Netherlands, Norway, Singapore, Sweden and Taiwan. NSI in these countries were the objects of study, and participants in the research project eventually came to refer to it as the 'ten countries project'. The total number of researchers involved was about 35. Within the project, several workshops were held in Copenhagen, Oslo, Taipei, Lund and Seoul, and organized by the relevant national research teams. These events not only marked progress in the research process but were also very important for driving the project forward. In this and other respects, the 'ten countries project' proceeded in a truly evolutionary manner – theoretically, methodologically and empirically (as described in the introductory chapter to this book). We are therefore very grateful to the hosts for the workshops, whose efforts and achievements made these meetings so successful. Finally, we want to express our deep gratitude to all the participants in the project. (They are too many to name here, but please see the list of participants!) This project owes its coherence and successful completion to their knowledge, abilities and willingness to collaborate.

In preparing this book for publication we have also incurred several additional debts of gratitude. In particular, we wish to thank the management and staff of Edward Elgar Publishing Ltd, not only for having agreed to publish the work but also for their patience, understanding, and kind assistance with preparing the manuscript for publication. We are also very grateful to Emelie Stenborg for her vital contributions to this process.

Charles Edquist and Leif Hommen
Lund, 2007

Abbreviations

A*STAR	Agency for Science, Technology and Research (Singapore)
ACE	Action Community for Entrepreneurship (Singapore)
ADSL	asymmetrical digital subscriber line
ARF	Applied Research Fund (Hong Kong)
ASTRI	Applied Science and Technology Research Institute (Hong Kong)
BERD	business expenditure on research and development
BIT	efficient use of ICT and e-business programme
BMRC	Bio-Medical Research Council (Singapore)
BUNT	Business Development Using New Technology (Norway)
CDMA	code division multiple access
CEPD	Council for Economic Planning and Development (Taiwan)
CIS(1/2/3)	Community Innovation Survey (1, 2 or 3)
CITB	Commerce, Industry and Technology Bureau (Hong Kong)
CMEs	coordinated market economies
CVU	centres of tertiary education (Denmark)
DEMO 2000	Technology development in the petroleum sector programme (Norway)
DETE	Department of Enterprise, Trade and Employment (Ireland)
DISKO	survey on the Danish NSI
DK	Denmark
DUI learning	learning by doing, using and interacting
EC-BIC	European Business Innovation Centre
EDB	Economic Development Board (Singapore)
EFTA	European Free Trade Association
EISC	Entrepreneurship and Internationalization Sub-Committee (Singapore)
EPA	Environmental Protection Agency (Hong Kong)
EPD	Environmental Protection Department (Hong Kong)
EPO	European Patent Office

ERC	Economic Review Committee (Singapore)
ESD	electronic service delivery
ESF	European Science Foundation
FDI	foreign direct investment
FI	Finland
FFI	Foundation for Finnish Inventions
FII	Finnish Industry Investment
FR	France
FRAM	programme for leadership and strategy development (Norway)
FRI	Fund for Research and Innovation (Norway)
FSP	facility-based service provider
FTE	full-time equivalent
GATT	General Agreement on Tariffs and Trade
GDP	gross domestic product
GEM	Growth Enterprise Management
GER	Germany
GERD	gross expenditure on research and development
GLC	government-linked company
GNI	gross national income
GNP	gross national product
GRI	government research institute
GTS institutes	Approved Technological Service Institutes (Denmark)
HCI	heavy and chemical industry
HE	higher education
HEA	Higher Education Authority (Ireland)
HEI	higher education institutions
HERD	higher education R&D
HK	Hong Kong
HKIB	Hong Kong Institute of Biotechnology Limited
HKITCC	Hong Kong Industrial Technology Centre Corporation
HKPC	Hong Kong Productivity Council
HKSAR	Hong Kong Special Administrative Region
HKSI	Hong Kong Safety Institute Ltd
HKSTP	Hong Kong Science and Technology Parks Corporation
HKUST	Hong Kong University of Science and Technology
HOTSpots	Hub Of Technopreneurs (Singapore)
HSIP	Hsinchu Science-based Industrial Park (Taiwan)
IC	integrated circuit(s)
ICSTI	Irish Council for Science, Technology and Innovation
ICT	information and communication technology

IDA	Industrial Development Agency (Ireland)
IDA	Infocomm Development Authority (Singapore)
IDB	Industrial Development Bureau (Taiwan)
IE	International Enterprise (Singapore)
III	Institute for Information Industry (Taiwan)
ILO	International Labour Organization
ININT2K	innovation intensity in the year 2000
INNO	innovation activities
INNU	innovating firms
INPCS	process innovation
INPDT	product innovation
INTSOK	The Norwegian Oil and Gas Partners
IP	intellectual property
IPD	Intellectual Property Department (Hong Kong)
IPR	intellectual property rights
IR	Ireland
ISF	industrial support fund
ISO	International Organization for Standardization
ITPS	Institute for Growth Policy Studies (Sweden)
IT	information technology
ITC	Innovation and Technology Commission (Hong Kong)
ITF	Innovation and Technology Fund (Hong Kong)
ITRI	Industrial Technology Research Institute (Taiwan)
IVE	Institute of Vocational Education (Hong Kong)
JPN	Japan
KFTC	Korea Fair Trade Commission
KIBS	knowledge-intensive business services
KIS	knowledge-intensive services
KITA	Korea International Trade Association
KMT	Kuomintang (regime)
KOITA	Korea Industrial Technology Association
KR	South Korea
LCD	liquid crystal display
LEs	large enterprises
LMEs	liberal market economies
LOK	Management, Organization and Competence (Denmark)
MBA	Master's of Business Administration Degree
MICA	Ministry of Information, Communications and the Arts (Singapore)
MNCs	multinational corporations
MNEs	multinational enterprises

MOCIE	Ministry of Commerce, Industry and Energy (Korea)
MOEA	Ministry of Economic Affairs (Taiwan)
MOFE	Ministry of Finance and Economy (Korea)
MOST	Ministry of Science and Technology (Korea)
MTI	Ministry of Trade and Industry (Finland)
MTI	Ministry of Trade and Industry (Singapore)
NICs	newly industrializing countries
NL	Netherlands
NMT	Nordic Mobile Telephony
NO	Norway
NPR	non-profit research organization
NRF	National Research Foundation (Singapore)
NSC	National Science Council (Taiwan)
NSE	natural scientists and engineers
NSF	National Science Foundation (Singapore; USA)
NSI	national system(s) of innovation
NSTB	National Science and Technology Board (Singapore)
NSTP	National Science and Technology Plan (Singapore)
NTBFs	new technology-based firms
NTD	new Taiwan dollar
NTP	National Technology Plan (Singapore)
NUI	New University of Ireland
NUS	National University of Singapore
Nutek	Swedish Agency for Economic and Regional Growth
NVCA	Norwegian Venture Capital Association
NWO	Netherlands Organization for Scientific Research
OBM	own brand manufacturing
ODM	original design manufacturing
OECD	Organisation for Economic Co-operation and Development
OEM	original equipment manufacturing
OG21	Oil and Gas in the 21st Century Programme (Norway)
OST	Office of Science and Technology (Ireland)
OSTI	Office of Science and Technology Innovation
OTC	over the counter
PATs	Programmes for Advanced Technologies (Ireland)
PCCW	Pacific Century CyberWorks
PISA	Programme for International Student Assessment
PPP	purchasing power parity
PPP	public–private partnership
PRC	People's Republic of China
PRD	Pearl River Delta

PRIC	Public Research and Development Institutes/Centres (Singapore)
PRTLI	Programme for Research in Third Level Institutions (Ireland)
PTP	public technology procurement
R&D	research and development
RCA	revealed comparative advantage
RDIN2K	research and development intensity in the year 2000
RIEC	Research, Innovation and Enterprise Council (Singapore)
RIVM	National Intitute for Health and the Environment (Netherlands)
RSE	research scientists and engineers
RTDI	research, technological development and innovation
S&E	science and engineering
S&T	science and technology
SAR	Special Administrative Region
SBI	School of Business and Information Systems (Hong Kong)
SBIR	Small Business Innovation Research programme
SE	Sweden
SEEDS	Startup EnterprisE Development Scheme (Singapore)
SEP	Strategic Economic Plan (Singapore)
SERC	Science and Engineering Research Council (Singapore)
SFI	Science Foundation Ireland
SG	Singapore
SI	system(s) of innovation
Sitra	Finnish National Fund for Research and Development
SIVA	Industrial Development Corporation (Norway)
SMEs	small and medium-sized enterprises
SSI	sectoral system(s) of innovation
Stakes	National Research and Development Centre for Welfare and Health (Finland)
STIAC	Science, Technology and Innovation Advisory Council (Ireland)
STPC	Science and Technology Policy Council (Finland)
STW	Dutch Technology Foundation
T&E Centres	Employment and Economic Development Centres (Finland)
T21	Technopreneurship 21 (Singapore)
TEA	total entrepreneurial activity
Tekes	National Technology Agency (Finland)

TFP	total factor productivity
TIF	Technopreneurship Investment Fund (Singapore)
TIMSS	Trends in International Mathematics and Science Study
TNO	Netherlands Organization for Applied Scientific Research
TRIPS	trade-related aspects of intellectual property rights
TSE	Taiwan Stock Exchange
TTIs	Technology Top Institutes (Netherlands)
TTIS	Taiwan Technology Innovation Survey
TUP	Technology Development Programme (Denmark)
TW	Taiwan
UCG	University College Galway (Ireland)
UGC	University Grants Committee (Hong Kong)
UICP	University–Industry Collaboration Programme (Hong Kong)
UL	University of Limerick (Ireland)
USPTO	US Patent and Trademark Office
VC	venture capital
VINNOVA	Swedish Agency for Innovation Systems
VoC	varieties of capitalism
WCY	*World Competitiveness Yearbook*
VTC	Vocational Training Council (Hong Kong)
WTO	World Trade Organization
VTT	Technical Research Centre (Finland)
WUR	Wageningen University and Research Institute (Netherlands)

1. Comparing national systems of innovation in Asia and Europe: theory and comparative framework

Charles Edquist and Leif Hommen

1 INTRODUCTION

The concept of national systems of innovation (NSI) emerged in the late 1980s and started to diffuse more rapidly in the early 1990s with the seminal contributions of Lundvall (1992) and Nelson (1993). It has attracted the attention of many innovation researchers and policy makers (e.g. Amable, 2000; Edquist, 1997, 2005; Freeman, 1997, 2002; Lundvall, 1988; Mytelka and Smith, 2002; OECD, 1997, 2002; Saviotti, 1996) and has rapidly achieved broad international diffusion in both developed and developing countries (e.g. Correa, 1998; Kaiser and Prange, 2004; Liu and White, 2001; Niosi, 1991).[1] However, progress in refining the NSI concept has been uneven and difficult to assess, given that 'no single definition has yet imposed itself on NSI research' (Niosi, 2002, p. 291) and many of the key terms are used in an ambiguous way. As argued previously (Edquist, 2005, pp. 201–3), there is therefore a need for theoretically based empirical research to 'straighten up' the approach and make it more 'theory-like'. A comparative research project on varieties of NSI, as well as determinants of innovation processes within them, may make particularly valuable contributions to such an effort.

The 'ten countries' research project addressed in this volume – so called because it compared ten 'small economy' SI – started operating in a practical sense in the latter half of 2002. However, it also had a lengthy 'pre-history', in which different versions of the project description were discussed by various constellations of researchers from some of the countries that were finally involved.[2] Eventually, the project started up in 2002–3, some ten years after the publication of Lundvall's and Nelson's landmark anthologies on NSI.

As a consequence of the long build-up to this project, the ground was quite well prepared by the time that financing arrangements for the project had been finalized, and the first project meeting was held in Copenhagen in August 2002. At that meeting we could collectively define the project objectives listed below by selecting from a broader range of objectives discussed during the 'pre-history'. The following objectives were agreed upon:

1. To further refine, elaborate and operationalize the SI approach. This means making the approach more 'theory-like'.[3] Moreover, 'straightening up' the approach theoretically should go hand in hand with increasing the usability of the SI approach for empirical studies, by:
    - developing concepts and methodologies suitable for empirical analysis;
    - translating its key concepts into empirical 'correspondents', i.e. variables reflecting concepts, indicators measuring variables, and using comparable sources (e.g. databases) in quantitative work;
    - developing a 'framework' for empirical studies of NSI that includes both quantitative and qualitative elements. Alternatively this might be called a 'methodology' for analysing different NSI in a comparative perspective.
2. To use the SI approach by actually carrying out (quantitative and qualitative) empirical and comparative studies of different NSI.
3. To draw policy conclusions. This means studying earlier and current innovation policies that have been or are being pursued in the ten countries. It also means identifying 'problems' and opportunities that should be subject to future innovation policy in the ten NSI, based on an analysis of strengths, weaknesses and challenges in these systems.

In order to achieve these objectives, conceptual, theoretical and methodological work was conducted partly outside the ten countries project, and published in Edquist (2005). Some of the main results are summarized in Section 2 below. Within the project, we devoted much effort to transforming key concepts into empirical correspondents by developing quantitative indicators of relevance to NSI. We discussed the so-called indicator work at workshops in Oslo in March 2003 and in Taipei in November 2003.[4]

At these early workshops, collective demand emerged among the researchers for a joint conceptual and comparative framework. To make the project truly comparative, it was agreed that we should develop a common framework that could be used for all the case studies of NSI. In Taipei, a number of people argued for a very standardized, detailed and rigid framework, but others wanted more degrees of freedom. The consensus that finally emerged was to carry out work that would facilitate cross-national comparisons of the same elements and activities in all NSI. This would be accomplished by using the same concepts, the same comparative framework and a standardized structure of presentation, rather than simply producing a collection of individual – and idiosyncratic – case studies. At the same time, however, each case study would also point to unique features of the NSI in question. We agreed that we did not only want to tell ten separate stories, but also create a basis for comparative analysis. The underlying conviction was that this approach would increase the quality of all the national studies and – moreover – of the resulting book. The common comparative framework and the streamlined structure of presentation would provide opportunities to highlight diversity as well as similarities among the national systems studied.

The Swedish team was given the responsibility of preparing a draft proposal for developing the comparative framework. A draft of the framework was sent out to the project participants for comments, and, on the basis of many such comments and further revisions, it was finalized and distributed in February 2004 (Edquist and Hommen, 2004).[5]

Since the SI approach is still diffuse and under-theorized (Edquist, 2005), we were unable to come up with a perfect comparative framework. There is simply no such thing as an optimal framework, since the SI approach as such is still evolving. That consideration made it even more important for us to reach a compromise on the comparative framework for this project. There were a number of trade-offs to be made in designing the *ad hoc* comparative framework that we eventually agreed upon.

Formulating a framework to be used by all is also a sensitive issue in the academic world. Researchers are independent creatures and do not want to be too closely governed or managed. We want to be free to design and carry out our analyses in the way we believe is best. To achieve comparability, it was therefore very important that everyone participating in the project could influence the design of the comparative framework. Consequently, the design process required several rounds before consensus could be reached. As editors of this volume, we are extremely grateful that the participating national teams decided to follow the framework, once we had agreed on it. We greatly appreciate their flexibility and generosity!

In addition to concepts and theory, the framework addressed the propensity to innovate in NSI, consequences of innovation, the role of globalization for small NSI, and innovation policy.[6] We agreed that the framework would serve as a basis and a common structure for carrying out the ten empirical studies included in this volume. Drafts of these studies were discussed at the workshop in Lund in September 2004. Subsequently, they were revised, discussed again at the Seoul workshop in March 2005 and then finalized for publication. In this process, we designed and developed the case studies upon the basis of the comparative framework, which, in turn, was 'theory-based'. The framework and the related conceptual and theoretical issues are discussed in Section 2 below.

The remainder of this chapter is structured as follows. This introductory section is immediately followed by an extended conceptual and theoretical discussion. Using the traditional systems of innovation approaches as the point of departure, we specify the most important concepts used in this book. Subsequently, we outline an activities-based framework for studying and comparing NSI. Then we discuss the characteristics of the countries selected for study. Finally we outline the common structure of each of the country[7] studies, including the presentation of a model table of contents.

2 CONCEPTUAL AND THEORETICAL FRAMEWORK

For the reasons explained above, it is highly advantageous for a project dealing with different NSI in a comparative perspective to use the main concepts in the same way in all the national studies. It is also important that the theoretical framework is similar – and explicit. In this section, therefore, we will specify the key concepts used in this book, as well as the theoretical approach agreed upon by the participants in this project.

2.1 The Traditional SI Approaches

When the project started (in 2002), the SI approach had, as discussed above, been well established for more than a decade and had become very widely diffused. The approach had also been developed theoretically thanks to the contributions of a very large group of scholars with different disciplinary backgrounds. However, broad acceptance and further development along a variety of trajectories had naturally led to many ambiguities and inconsistencies in the research literature on SI.

As discussed elsewhere (e.g. in Edquist, 1997), the term NSI was first used in published form by Freeman (1987). He defined an NSI as 'the network of institutions in the public and private sectors whose activities and interactions initiate, import, and diffuse new technologies' (Freeman, 1987, p. 1).[8] Subsequently, Lundvall (1992) and Nelson (1993) both published their major anthologies on NSI, but used different approaches to their studies. Nelson's (1993) book included case studies of the NSI in 15 countries – and is, in that respect, similar to the present volume. The Nelson anthology emphasized empirical case studies more than theory development.[9] These case studies, moreover, were not designed to have the same structure and focus. Some focused narrowly on national research and development (R&D) systems, but others were broader in scope. In contrast, Lundvall's (1992) book was more theoretically oriented and it followed a 'thematic' approach rather than a 'national' one. It placed interactive learning, user–producer interaction and innovation at the centre of the analysis.

Lundvall argued that 'the structure of production' and 'the institutional set-up' are the two most important dimensions that 'jointly define a system of innovation' (Lundvall, 1992, p. 10).[10] In a similar way, Nelson and Rosenberg (1993) singled out organizations supporting R&D – i.e. they emphasized those organizations that promote the creation and dissemination of knowledge as the main sources of innovation. Organizations disseminating knowledge include firms, industrial research laboratories, research universities and government laboratories.[11] Lundvall's broader approach recognized, though, that such organizations are 'embedded in a much wider socio-economic system in which political and cultural influences as well as economic policies help to determine the scale, direction and relative success of all innovative activities' (Freeman, 2002, p. 195). Thus both Nelson and Lundvall defined NSI in terms of determinants of, or factors influencing, innovation processes.[12] However, they specified different determinants in their definitions of the concept, presumably reflecting their judgement about the most important determinants of innovation. In brief, they used the same term but proposed different definitions, thus contributing to the ongoing lack of a generally accepted definition of NSI (Niosi, 2002, p. 291).

As we have seen, Nelson and Lundvall offered definitions of NSI that focused on their constituents (e.g. the main organizations and institutions as well as relations among them). As already noted above, Lundvall (1992) promoted a 'broad' conception of NSI – embedded in a wider socioeconomic system. In contrast, Nelson (1993) advanced a more narrow approach, focusing on national R&D systems and organizations supporting R&D as the main source of innovation. Remarkably, these two approaches have not really confronted each other within the research literature. Instead, researchers have tended to adopt one or the other of these two basic approaches, or to elaborate variants of them, without giving much consideration to the alternative approach. Thus Lundvall et al. (2002, p. 217, n. 2) have discussed the further development of their 'broad' approach to NSI without making many explicit comparisons with the 'narrow' approach, except to comment that their own approach is particularly 'relevant for understanding economic growth and innovation processes in small countries'. Similarly, Larédo and Mustar (2001) have applied the Nelsonian version of the NSI concept in their international survey of research and innovation policies without much consideration of its merits relative to the Lundvallian alternative.[13] It is fair to say that these two versions of the NSI concept have enjoyed a peaceful coexistence, and there has been only a limited dialogue between them in the research literature. We will return to this question in the theoretical discussion that follows in Section 2.3 of this introductory chapter, as well as in the concluding chapter of this book.

A more general definition of an SI includes 'all important economic, social, political, organisational, institutional and other factors that influence the development, diffusion and use of innovations' (Edquist, 1997, p. 14). If an SI definition does not include all factors that influence innovation processes, one has to decide which potential factors should be excluded – and why. This is quite difficult since, at the present state of the art, we do not know the determinants of innovation processes systematically and in detail. Obviously, then, we could miss a great deal by excluding some determinants, since they might prove to be very important once the state of the art has advanced. For example, 25 to 30 years ago, it would have been natural to exclude the interactions between organizations as a determinant of innovation processes. Both the relationships among the factors listed and the actions of both firms and governments are included in the general definition above. This definition, moreover, is fundamental to the 'activities-based' approach to studying SI (Edquist, 2005; Edquist and Chaminade, 2006) discussed in Section 2.3. Obviously, this is a conception of SI that is as broad as Lundvall's, if not broader.

2.2 Main Terms Used in this Book

Everyday language and the scientific literature ('general systems theory') give a common answer to the question 'What is a system?' focusing on three features (Ingelstam, 2002).[14] First, a system consists of two types of constituents: components and relations among them. The components and relations should form a coherent whole, with properties different from those of the constituents. Second, the system has a function – i.e. it is dedicated to performing or achieving something. Third, it must be possible to discriminate between the system and the rest of the world (i.e. the system's 'environment') – that is, it must be possible to identify the boundaries of the system[15] (Edquist, 2005). Obviously, for empirical studies of specific systems, one must know their extension.

Making the SI approach more theory-like – as proposed in Section 1 – does not require specifying all components and all relations among them in detail. At the present state of the art, this would be too ambitious. For the time being, it is not a matter of transforming the SI approach into a 'general theory of innovation', but one of making it clearer and more consistent so that it can serve better as a basis for generating hypotheses about relations between specific variables within SI (which might be rejected or supported through empirical work). Even the much more modest objective of specifying the most important components of the SI, their main function and specific activities influencing the main function and the key relations among the components and the activities, would represent a considerable advance. Used in this way, the SI approach can help to develop theories about relations between specific variables within SI.

There seems to be general agreement in the literature that the main components in SI are institutions and organizations – among which firms are often considered to be the most important organizations. However, particular definitions of SI specify different sets of institutions and organizations and, moreover, set-ups of institutions and organizations vary across systems. Consequently, in a project such as this one, it is important to specify the main terms used. We therefore present, in Box 1.1, a list of specifications of the key terms used in this book.[16]

2.3 An Activities-based Framework for Analysing SI

As indicated in Box 1.1, the main or 'overall' function of SI is to pursue innovation processes: to develop and diffuse innovations. What we call 'activities' in SI from here on are those factors that influence the development and diffusion of innovations.[17] We use the term 'activities' as equivalent to determinants of the innovation process.

The theoretical framework employed in this book includes a central focus on 'activities' within systems of innovation. It is new in the sense that it focuses strongly on what 'happens' in the systems – rather than on their constituents – and that it thus uses a more dynamic perspective.

The traditional SI approaches, such as Lundvall (1992) and Nelson (1993), focused strongly on the components within the systems, i.e. organizations and institutions. Since the late 1990s, some authors have addressed issues related to the issue of specification of activities influencing the overall function of SI (Galli and Teubal, 1997; Johnson and Jacobsson, 2003; Liu and White, 2001; Rickne, 2000).

Clearly, no consensus has yet emerged among innovation researchers as to which activities should be included, and this provides abundant opportunities for further research. In Box 1.2 we present a hypothetical list of ten activities that we have adopted in this book. This list is based on the literature and on our own knowledge about innovation processes and their determinants, as discussed in Edquist (2005) and Edquist and Chaminade (2006).

BOX 1.1 DEFINITIONS OF KEY TERMS

Innovations	New creations of economic significance, primarily carried out by firms (but not in isolation). They include product innovations as well as process innovations.[1]
Product innovations	New – or improved – material goods as well as new intangible services; it is a matter of *what* is produced.[2]
Process innovations	New ways of producing goods and services. They may be technological or organizational; it is a matter of *how* things are produced.[3]
Creation versus diffusion of innovations	This dichotomy is partly based on a distinction between innovations that are 'new to the market' (brand new, or globally new) and innovations that are 'new to the firm' (being adopted by or diffused to additional firms, countries or regions). In other words, 'new-to-the-firm' innovations are (mainly) a measure of the diffusion of innovations. For many small countries diffusion (absorption) is more important than the creation of new innovations.[4]
SI	Determinants of innovation processes – i.e. all important economic, social, political, organizational, institutional and other factors that influence the development and diffusion of innovations.[5]
Components of SI	These include both organizations and institutions.[6]
Constituents of SI	These include both components of SI and relations among them.[7]
Main function of SI	To pursue innovation processes – i.e. to develop and diffuse innovations.[8]
Activities in SI	Those factors that influence the development and diffusion of innovations. The activities in SI are the same as the determinants of the main function. (A list of activities is presented in Box 1.2.) The same activity (e.g. R&D) can be per-

	formed by several categories of organization (universities, public research organizations, firms), and the same kind of organization (e.g. universities) can perform more than one kind of activity (e.g. research and teaching).[9]
Organizations	Formal structures that are consciously created and have an explicit purpose. They are players or actors.[10]
Institutions	Sets of common habits, norms, routines, established practices, rules or laws that regulate the relations and interactions between individuals, groups and organizations. They are the rules of the game.[11]
Innovation policy	Actions by public organizations that influence the development and diffusion of innovations.[12]

Notes and sources:

[1] This definition and those immediately following it – i.e. definitions of different kinds and aspects of innovations – are based on an earlier specification of basic concepts by Edquist et al. (2001, ch. 2). In turn, their taxonomy of innovations is based on Schumpeter (1911).

[2] In this taxonomy, only goods and technological process innovations are material; the other categories are non-material and intangible. Thus, for example, innovations in service products are considered to be non-material or intangible innovations; so too are organizational process innovations.

[3] In this study, only product and process innovations as specified here are considered as innovations. This means that new markets, new research results, new organizations, new institutions etc. are not called 'innovations' here. These phenomena are instead dealt with in terms of 'activities' that influence the development and diffusion of (product and process) innovations (see Section 2.3). Hence we strongly stress the crucial role of new markets, new R&D results, new institutions and new organizations in this project – but we do not call them 'innovations'.

[4] It might also be useful to distinguish between *incremental* and *radical* innovations and between *science-based* and *experience-based innovations*. A general remark is that the notion of innovation in a general sense is so comprehensive and heterogeneous that it is useful to create various taxonomies of innovations and deal with them separately when describing and explaining innovation processes. For a further discussion of taxonomies of innovation, see Edquist et al. (2001), Appendix C.

[5] This definition of an SI is taken from Edquist (1997), p. 14.

[6] See Edquist (1997), pp. 113–15; Edquist (2005), p. 189 and Edquist and Johnson (1997), pp. 60–61.

[7] See Edquist (2005), p. 187.

[8] Ibid., p. 190.

[9] Ibid., pp. 189–91.

[10] See Edquist and Johnson (1997), p. 47.

[11] Ibid.

[12] See Edquist (2001), p. 19.

BOX 1.2 KEY ACTIVITIES IN SYSTEMS OF INNOVATION

I. Provision of knowledge inputs to the innovation process
1. Provision of R&D and, thus, creation of new knowledge, primarily in engineering, medicine and natural sciences.
2. Competence building through educating and training the labour force for innovation and R&D activities.

II. Demand-side activities
3. Formation of new product markets.
4. Articulation of quality requirements emanating from the demand side with regard to new products.

III. Provision of constituents of SI
5. Creating and changing organizations needed for developing new fields of innovation. Examples include enhancing entrepreneurship to create new firms and intrapreneurship to diversify existing firms; and creating new research organizations, policy agencies, etc.
6. Networking through markets and other mechanisms, including interactive learning between different organizations (potentially) involved in the innovation processes. This implies integrating new knowledge elements developed in different spheres of the SI and coming from outside with elements already available in the innovating firms.
7. Creating and changing institutions – e.g. patent laws, tax laws, environment and safety regulations, R&D investment routines etc. – that influence innovating organizations and innovation processes by providing incentives for and removing obstacles to innovation.

IV. Support services for innovating firms
8. Incubation activities such as providing access to facilities and administrative support for innovating efforts.
9. Financing of innovation processes and other activities that can facilitate commercialization of knowledge and its adoption.
10. Provision of consultancy services relevant for innovation processes, e.g. technology transfer, commercial information and legal advice.

Note: Each activity is described in more detail in Edquist and Chaminade (2006).

Source: Edquist (2005).

The activities are not ranked in order of importance, but the list is structured into four thematic categories: (I) provision of knowledge inputs to the innovation process, (II) demand-side activities; (III) provision of constituents of SI and (IV) support services for innovating firms. The activities can each be considered as partial determinants of the development and diffusion of innovations. The list is certainly provisional and will be subject to revision as our knowledge about determinants of innovation processes increases. Public innovation policy is an element of all the ten activities.

In relation to the distinction between 'narrow' and 'broad' conceptions of SI discussed in Section 2.1, the activities-based framework is certainly as broad as Lundvall's. We agreed at the Taipei workshop in November 2003 that we would use activities as the 'point of entry' into the subject matter and as a structuring device for the empirical studies of factors hypothetically influencing innovation processes in the NSI in the ten countries. Thus the activities-based framework forms the basis of the common framework used in all the ten case studies of NSI reported in this book, as discussed in Section 1. This volume is therefore the first comparative study of NSI that has focused on 'activities' in a systematic manner.

Our focus on activities has not meant, however, that we have disregarded or neglected the organizations and institutions that constitute the components of SI. When addressing activities in the NSI studies we have also had to address the organizations (or organizational actors) that carry out these activities and the institutions (institutional rules) that constitute incentives and obstacles affecting the innovation efforts of these organizations. In order to understand innovation processes we need to address the relations among activities and components, as well as among different kinds of components. In addition to basing this approach upon quantitative indicators, the case studies also include a strong emphasis on qualitative aspects – including narrative accounts – related to the creation, change and abolition of organizations and institutions, and to other activities.

To sum up, activities, organizations and institutions are discussed in an integrated manner in the longest section of the NSI studies (Section 4) presented in this book. The activities are used in each national study as the point of entry into the subject matter and serve as a structuring device. Where possible, the various studies also try to determine with what effectiveness and efficiency the activities are performed, i.e. how they have influenced the development and diffusion of innovations.

3 COUNTRIES SELECTED FOR STUDY

As indicated in Section 1, this volume is intended as a contribution to the research literature on NSI. Hence this book is intrinsically based upon both Lundvall (1992) and Nelson (1993), as well as other literature on NSI. As also suggested previously, certain parallels can be drawn between this book and both Nelson's and Lundvall's volumes. On the one hand, this book may be considered to resemble Lundvall's, in so far as it is theoretically oriented and follows a thematic approach. On the other, there are also very clear – and arguably greater – similarities with the Nelson anthology, since both this book and Nelson's are collections of national case studies. Therefore it may be useful to make a somewhat more detailed comparison between these two works, focusing particularly on the countries selected for study.

The 15 countries studied in the Nelson book covered a wide range of national contexts. Geographically, they represented four continents (North and South America, Europe and Asia). The countries selected included both large ones (e.g. the USA) and small ones (e.g. Denmark and Sweden). In addition, the selection of countries also represented different levels of economic development, including both industrialized countries (e.g. Japan) and developing countries (e.g. Brazil). In the book they were clustered into the following groups: 'Large high-income countries', 'Smaller high-income countries' and 'Lower-income countries'. Hence the dimensions of classification were large/small and high/low income. Another, more recent, edited volume on NSI (Lundvall et al., 2006) addresses only systems in Asia, including large ones (such as Japan, India and China) and smaller ones (such as Hong Kong and Singapore).

In this book we address the following NSI located in the two continents of Europe and Asia: Denmark, Finland, Hong Kong, Ireland, the Netherlands, Norway, Singapore, South Korea, Sweden and Taiwan. As noted earlier in this chapter, though, not all these units are countries or nations in a political and cultural–historical sense (see endnote 7).

The title of our book refers to 'small country innovation systems'. As can be seen from the Appendix: Statistical bases of comparison for ten 'small country' NSI, most of the ten economies are indeed small. Seven of them have populations of between four and nine million. The Netherlands (16 million) and Taiwan (23 million) are also fairly small. The main outlier in this respect is Korea, with a population of 49 million. Korea thus approaches some of the larger European countries (e.g. France and the UK) in terms of population size. However, like nearly all European countries, Korea can still be viewed as relatively small when considered in relation to 'giants' such as the USA and Japan (or, for that matter, China and India). Thus, unlike Nelson's anthology, which made a point of including such large economies, the present volume develops an exclusive focus on small ones.

With reference to Nelson's other main set of selection criteria – i.e. high versus low income levels – we have deliberately opted to focus only on relatively high-income economies. As shown in Table A1.2 of the Appendix, the ten small economies represented here were ranked among the top 28 (out of a total of 177) countries on the UNDP Human Development Index for 2004.[18] Moreover, four of these countries (Denmark, Ireland, Norway and Sweden) were ranked among the top ten. According to the same source, the ten small economies included in this volume had gross domestic product per capita rankings among the top 37 countries, with three of them (Denmark, Ireland and Norway) belonging to the top ten. Table A1.2 of the Appendix also indicates that all ten of these small economies have very high levels of individual life-expectancy and very high levels of combined enrolment in primary, secondary and tertiary education.

The combination of relatively small size with relatively high income implies a third quality of the ten small economies selected for this study: dynamism. This quality can be measured in a variety of ways, and we will mention only a very few of them here. To begin with, these economies fare very well in terms of technological development, as measured by, for example, advanced infrastructure development. For instance, when 178 countries were compared with regard to Internet access in 2002, eight of our ten economies (Sweden, Denmark, South Korea, Norway, the Netherlands, Hong Kong, Finland and Taiwan) were ranked among the top nine, intersected only by Iceland at number 3. Further, the ten economies also have high ratings with respect to economic performance. In the 2004 issue of the *Annual Review of Global Competitiveness* published by the World Economic Forum, our ten economies came out very well with regard to growth potential (WEF, 2004). Six of them (Finland, Sweden, Taiwan, Denmark, Norway and Singapore) were ranked among the first seven countries (this series was interrupted only by the USA at number two). In addition, the Netherlands was ranked at number twelve (Balls, 2004).

More generally, these ten economies have all established reputations for success within the context of globalization and the advent of a 'new' economy where competition is increasingly based on innovation. Four of these economies have gained wide renown as the 'four tigers' of Asia, and they have at least one European counterpart in Ireland, which is also known as the 'Celtic tiger'. Our other European economies represent the Nordic countries (sans Iceland) and the Netherlands. These countries have perhaps not attracted the same degree of publicity, but in many respects they have been very high achievers, as noted above. A recent study by Florida and

Tinagli (2004) on 'Europe in the creative age' pointed to a high degree of learning and innovativeness in the contemporary Nordic countries, with Sweden as the leading country, outperforming not only all the other European countries, but also the USA. The other Nordic countries, as well as other small northern European countries, including the Netherlands, Belgium and Ireland, also scored highly on the 'Euro-creativity index'. Although the ten countries are similar on many dimensions (size, economic performance, diffusion of ICT, etc.), they are very different in many other respects – as we will see in the concluding chapter of this book.

As compared to the Nelson book, then, our selection of NSI represents a narrower – or, rather, more focused – sample with respect to the dimensions of size and levels of income. Its geographical coverage is also more restricted, representing only two continents instead of four. Based on these considerations, it might be argued that one advantageous feature of the present volume is that it deals with a set of countries that are – in some respects, at least – more directly comparable with one another than those included in the Nelson anthology.

Notwithstanding these considerations, we should urge some caution with respect to conducting comparative analyses that cover this whole set of cases. It is clear that our 'sample' of NSI cuts across some fundamentally different contexts, which ought to be taken into careful account in any comparative analysis. First, our cases represent two very different regions of the world: the Asia-Pacific and North European regions. Second, these cases also constitute two very different groups defined from an economic-history perspective. On the one hand, the Nordic countries and the Netherlands represent late industrializing countries whose economies reached the 'take-off' point in the early twentieth century. On the other hand, the Asian and Irish economies represent 'newly industrialized countries' that reached this point much later, and under different conditions. These differences, among others, should be addressed in making comparisons across this set of cases – an issue to which we shall return in the concluding chapter.

As stated in Section 1, one of our objectives has been to contribute to the development of comparative studies of NSI. One means of realizing this goal is, of course, to identify and describe comparable cases in a systematic way. Other means of increasing the potential for comparative analysis are to identify common bases of comparison and to elaborate a conceptual framework for this purpose. Thus, as compared to the Nelson book, we have also tried to increase the degree of comparability by using concepts that are standardized or harmonized to a larger extent and by adopting a common theoretical framework. Along the same lines, we have also streamlined the structure of presentation in all the case studies. As noted previously, all the national studies included in this collection use the same model table of contents, which will be discussed in the next section.

4 THE STRUCTURE OF THE CASE STUDIES

In this section, the common structure of the case studies in this book is briefly outlined and explained. The discussion here concludes with the presentation of a model table of contents used in all the case studies.

After a short introduction (Section 1), the case studies outline the main trends in the history of the NSI (Section 2). These opening sections are intended to characterize the NSI, often in relation to a central theme identified in the title of the case study, and point out the most important aspects of the system's development over time.

In Section 3, the propensity to innovate (or innovation intensity) of the NSI is addressed. Ideally, this discussion addresses both the development (creation) and the diffusion of innovations, including product as well as process innovations.[19] To a large extent the Community Innovation Surveys and similar surveys carried out in non-European countries have been used to describe the propensity to innovate in the various countries. To the extent possible, these descriptions of the propensity to innovate are structured in order to facilitate the development of a comparative perspective across the various case studies.[20] Some of the data on which these comparisons are based, as well as additional data, are presented in the Appendix.

As discussed above in Sections 2.2 (including Box 1.1) and 2.3, the main function in systems of innovation is to pursue innovation processes – i.e. to develop and diffuse innovations. Hence the development and diffusion of (different kinds of) innovation is what 'comes out' of the systems. In the case studies, these outcomes are measured and assessed in Section 3's discussion of the 'propensity to innovate' (or 'innovation intensity').

The propensity to innovate is actually what should be explained – if possible – by accounting for the determinants of the development and diffusion of innovations. In the conventional terms of scientific method, the propensity to innovate is *explanandum* and the determinants are the *explanans*. These determinants were referred to as 'activities' in Section 2.3, where we hypothetically listed ten such activities. They were clustered in four main categories (see Box 1.2). In Section 4 of the case studies, these ten activities are discussed in depth. Due to the detailed description of these activities, Section 4 is the longest part of each national case study.

The research question asked in Section 4 is, in effect: 'What are the national characteristics of the factors (or 'activities') that influence (product and process) innovation processes in the NSI?'[21] In this section, the country case studies aim to identify factors that influence – and, in this sense, help to explain – the propensity to innovate.

Systematic identification of such determinants of innovation processes is a surprisingly under-researched area in innovation studies. Partly for this reason, but also because of the very complex nature of innovation processes, as well as the difficulty of developing causal explanations in the social sciences, none of the case studies arrives at a 'complete' causal explanation of the propensity to innovate in their respective NSI. What they do present are structured and illuminating discussions of the factors that influence that propensity. What we have learned in the work of this project is that a much deeper analysis of each of the potential determinants is both necessary and possible – but far beyond the scope of a volume such as this. For that reason, many of the researchers involved in this project have discussed the possibility of writing a whole book on their respective NSI.

As pointed out in Section 2.1 above, the generally accepted definitions of SI do not include the consequences of innovations, and the various systems of innovation approaches do not, as such, deal with the consequences of innovations. However, innovations, developing and diffusing in innovation systems, have extremely important consequences for socioeconomic variables (such as productivity growth, employment and sustainability). Therefore Section 5 in the case studies provides a brief discussion of some consequences of innovations, mainly emphasizing consequences for productivity growth[22] (although some case studies also mention other effects, such as employment and sustainability). This issue could also, of course, have been dealt with at much greater length and depth in specialized analyses than has been possible in one section of a chapter in an edited volume.

This project has also sought to counter the criticism that NSI analyses pay too little attention to 'external' factors by explicitly taking into account processes of globalization and issues raised by this phenomenon. Globalization is dealt with as a basic issue and profoundly integrated into each of the national studies. Thus we have dealt with aspects of globalization when discussing the various activities pursued in the NSI (see the discussion above of Section 4). We have also tried to address the extent to which various activities have been influenced by globalization. Because of the importance of the issue for this project, each national study also includes a section summing up the impact of globalization on the NSI (Section 6). In Section 6, the case studies address questions such as the following:

- What have been the relations between globalization and NSI?
- What does globalization mean for the NSI of small countries?
- How has globalization influenced NSI – positively and negatively?

- How has globalization influenced innovation policies of small countries?
- How has globalization been incorporated into innovation patterns – e.g. in capturing technological opportunities abroad and capturing global market opportunities?
- How have various countries influenced processes of globalization?

We shall return to the issue of globalization in the concluding chapter of this book.

A general definition of innovation policy was presented above in Box 1.1. Just as in the case of globalization, innovation policies are discussed in relation to various activities addressed in Section 4 of the case studies – but we also devote one separate section to innovation policy in each case study. In the context of this book, 'innovation policy' means two things:

1. The policies that have historically been pursued in the ten countries.
2. The policies proposed for the future.

We deal with both of these aspects of innovation policy in the case studies. To analyse national innovation policies, it is necessary to relate them to strengths and weaknesses of the NSI. On this basis, innovation policies are discussed in Section 7 of the case studies. This section addresses both those policies that have been pursued during the last few decades and those proposed for the future. Since policy will be one of the key issues discussed in the concluding chapter, we abstain from going deeper into this issue here.

The above description of the contents of the various sections results in the model table of contents (Box 1.3) that is used in all country chapters. This model is not followed slavishly by all authors. All of them have addressed all the headings outlined above. However, to prevent the model from becoming too much of a straitjacket, many chapters have improved upon the outline by adding various subsections to account for national peculiarities and deal with issues that may not be as relevant or as important in other countries. The length and weight of each of the sections also differ among the chapters. Thus we have ended up with a joint structure pointing out the 'minimum' requirements for what should be addressed and included in all the national studies. This framework is intended to be a common denominator to achieve comparability, without being too inhibiting. Therefore additional issues are covered in the country case studies, and the way that specific issues are covered varies across the chapters. This diversity is in the interest of comparability as well as of creativity!

Another salient feature of both the framework and the flexible manner in which it has been implemented is that the borders between sections are not razor-sharp. For example, institutions (rules of the game) are

> **BOX 1.3   MODEL TABLE OF CONTENTS FOR NSI CASE STUDIES**
>
> 1. Introduction
> 2. Main historical trends
> 3. Innovation intensity
> 4. Activities that influence innovation
>   - 4.1 Knowledge inputs to innovation
>   - 4.1.1 R&D activities
>   - 4.1.2 Competence building
>   - 4.2 Demand-side factors[1]
>   - 4.3 Provision of constituents
>   - 4.3.1 Provision of organizations
>   - 4.3.2 Networking, interactive learning and knowledge integration
>   - 4.3.3 Provision of institutions
>   - 4.4 Support services for innovating firms
>   - 4.4.1 Incubating activities
>   - 4.4.2 Financing
>   - 4.4.3 Provision of consultancy services
>   - 4.5 Summary of the main activities influencing innovation
> 5. Consequences of innovations
> 6. Globalization
> 7. Strengths and weaknesses of the system and innovation policies
>   - 7.1 Strengths and weaknesses
>   - 7.2 Summary and evaluation of the innovation policy pursued
>   - 7.3 Future innovation policy
>
> Notes
> References
>
> *Note:* [1] Originally the demand-side factors were divided into two different activities, but they were merged as a consequence of discussions within the project group.

mentioned in many sections, since they are certainly relevant for demand, for R&D and other knowledge inputs, etc. In addition, however, institutions are also addressed in a separate subsection. A further example concerns the provision of organizations. Like the provision of institutions, this topic is the subject of a separate subsection – but is also addressed under other headings in each national case study. There are also differences in the approaches to and coverage of the same heading across the different country studies. In some chapters, for example, 'provision of organizations' is dealt with exclusively in terms of the birth and death of firms, and other kinds of organizations are dealt with elsewhere. In other chapters, a focus on new public sector organizations is developed under this heading, and firms are addressed elsewhere.

Notwithstanding this kind of flexibility, the fact that the model table of contents is used in all country chapters means that they all address the same issues and activities in similar ways. It also means that comparisons between the various cases are facilitated by the adoption of a common framework, as should be obvious to the readers of this book. To a large extent, however, we shall leave such comparisons to be drawn by the readers themselves. Only a few dimensions will be explicitly compared in the concluding chapter.

Readers will probably form views about which national studies provide a better structure and content under each heading than do others. On this basis, the next attempt – by us or by others – at systematically describing NSI in a comparative manner will provide an alternative which competes with the attempts that are included in this book. The same is true with regard to the framework outlined in Box 1.1. There are certainly other ways to specify the main terms, and other researchers should be encouraged to elaborate them. However, we do hold the view that specification as such is a virtue. There is no advantage to using common terms in ambiguous and unclear ways.

To sum up, we have managed to use a fairly standardized conceptual 'language' and to structure the case studies along similar lines. We believe that using this common 'format' is an achievement; certainly, it has not been done before in a comparative study of several NSI.

5 FINAL REMARKS AND OVERVIEW OF THE CHAPTERS

The remainder of this book is dedicated, for the most part, to the presentation of the national case studies. There are two main groups of chapters: Part I presents the case studies of what we identify as the fast growth countries during the last three decades (Taiwan, Singapore, Korea, Ireland and Hong Kong), and the second presents what we call the slow growth countries (Sweden, Norway, the Netherlands, Finland and Denmark).[23] With regard to these case studies, it should be said at the outset that all authors have been forced to economize strictly with regard to space. And, as noted above, most authors would actually have wanted to devote a whole book to their respective NSI. We are confident that some of them will do so in the future!

5.1 Part I: Fast Growth Countries

5.1.1 Taiwan

Following an NSI approach, this chapter addresses the story of economic transformation in Taiwan. It emphasizes the key role of policy in leading the process of systemic upgrading, which has involved complex processes of co-evolution among actors, institutions, knowledge, technology and markets. Three elements are emphasized here. The first is the role of Taiwan as a latecomer economy, learning to compete in world markets. The second concerns the strategic role of the government in changing the economic base for competition on the part of Taiwanese firms. The third is specialization, referring to a unique capacity to adjust quickly to patterns of change in global demand by upgrading and excelling at the project execution level in original equipment manufacturing (OEM) and original design manufacturing (ODM) production.

5.1.2 Singapore

Singapore has experienced rapid economic and technological development since political independence in 1965. Until the late 1990s, this rapid growth was accomplished largely through heavy reliance on foreign direct investment (FDI), leveraging foreign multinational corporations (MNCs) to transfer and diffuse technology to local companies and employees. The government has played a central role in this development, providing incentives for MNCs to locate in Singapore, developing relevant training programmes and institutions, providing the necessary infrastructure and setting an example by itself being a lead user of new technologies. However, this approach has also produced an imbalance in the NSI, with greater emphasis on the adoption of advanced technologies at the expense of developing indigenous R&D and innovation capabilities. This is reflected in the relatively low innovation intensity and patenting levels prevailing in Singapore up to the late 1990s. Since then, policy efforts aimed at redressing this imbalance have been implemented, particularly over the last two to three years. However, weaknesses remain, particularly in the basic research system and the technology entrepreneurial ecosystem. Future policy will need to address these issues, including mechanisms to fund technology commercialization efforts, programmes to promote R&D cooperation with international partners and, perhaps most fundamentally, policies to foster a change in the cultural mindset in order for the population to embrace entrepreneurship.

5.1.3 Korea

Korea is comparable to Germany and the UK with regard to complexity and diversity of organizations, institutions and industrial structures. Its NSI has been developed through aggressive investment in R&D and innovation activities, led by large firms and the government. However, this has resulted in low innovativeness and productivity relative to the advanced countries, and lower gross domestic product (GDP) per capita than the first-tier Asian 'tigers'. Korea's NSI is characterized by a group of 'strong large firms and weak small firms' – that is, innovative large firms able to exploit technological and market opportunities abroad, and laggard small firms. In terms of public R&D, Korea has a relatively large government research institute sector, but university research activities are comparatively small. Industry networks comprising *chaebol* groups and their affiliated firms are dominant in the major industry sectors, and university–industry–government research institutes networks are at an early stage of development. The education system in Korea has been continuously expanding, but is under increasing pressure to upgrade the quality of education being offered. The financial system, which is predominantly a banking system, has been reformed. Under the liberalized environment, which emerged rapidly after the 1990s financial crisis, coordination and networking of innovative actors and resolution of mismatches in the system of innovation have become urgent issues in Korea's bid to become active in knowledge generation and effective utilization of technology from abroad.

5.1.4 Ireland

High levels of inward investment have helped Ireland to achieve extremely rapid growth over recent years compared to other European countries. Innovation levels have also been high, supported primarily by inward technology transfer and despite historically low levels of both public and private R&D spending and weaknesses in Ireland's NSI. Acknowledgement of these issues in the mid-1990s, and increasing uncertainty over whether Ireland would continue to attract high levels of inward investment, led to a refocusing of policy towards support for domestic R&D, innovation and new technology adoption. Since 2000 in particular, public investment in higher education R&D has increased rapidly, supported by policy innovations such as the introduction of the Programme for Research in Third Level Institutions (i.e. organizations) and Science Foundation Ireland. Efforts to boost levels of business R&D and connectivity have also been intensified, with a particular focus on indigenously owned and smaller firms. Over the same period, Ireland has tightened its intellectual property rights regime, strengthened corporate governance legislation and continued to develop organizations to support business start-up and service sector growth.

5.1.5 Hong Kong

From 1847 to 1997, Hong Kong was a Crown colony of Great Britain. The five decades leading up to 1997 saw Hong Kong becoming a newly industrialized economy and then developing extensive services to become an unrivalled trade hub between the People's Republic of China and the rest of the world. This role has contributed directly to the prosperity and standards of living Hong Kong enjoys today. Since Hong Kong became a Special Administrative Region under Chinese sovereignty in 1997, however, a series of events has created new pressures on Hong Kong to diversify its role as a regional hub. One such event has been the accelerated integration of Hong Kong's production networks into the Chinese mainland, specifically the Pearl River Delta region of Southern China. Another important event was the Asian financial crisis that struck in 1998, initiating a prolonged economic recession out of which Hong Kong has only recently emerged.

In reaction to these events, the Hong Kong government has launched major initiatives to improve innovation in the economy. The low level of R&D investment in industry has been gradually improving, and attempts have been made to generate new technologies through public support on a continued basis – to transform Hong Kong into an innovation hub with global links to and from China.

The point of departure for this chapter is thus that the transition to a new status, from that of a trade hub for China to that of an innovation hub, presents new challenges for Hong Kong's NSI. Hong Kong must leverage its unique position as a gateway that provides high value-added services to global production chains linking China and the world, and in the process upgrade its expertise and knowledge for trade and production chain orchestration into the resources needed to contribute substantially to product and process innovation in China.

5.2 Part II: Slow Growth Countries

5.2.1 Sweden

This chapter takes its point of departure in the so-called Swedish paradox, according to which the Swedish NSI is plagued by low pay-off in relation to very high investments in R&D and innovation efforts. Using new data, we show that this paradox is still in operation, i.e. the productivity or efficiency of the Swedish NSI remains low. We also specify the paradox in several respects. By focusing on nine activities in the NSI, we attempt to explain why and how the paradox operates. The paradox is also related to the moderate growth of labour productivity in Sweden. Further, we show that the paradox is linked to globalization: internationalization of production by

Swedish firms has proceeded further than the internationalization of R&D. On the basis of this analysis, we identify strengths and weaknesses of the Swedish NSI – many of which are related to the Swedish paradox. We take account of the history of innovation policy in Sweden and – on the basis of the analysis as a whole – we identify future policy initiatives that might help to mitigate the Swedish paradox.

5.2.2 Norway
Norway ranks low regarding average innovation outcome, but performs well regarding economic output and standard of living. We provide a description of activities within the NSI, with a focus on technological trajectories. Norway has been blessed with an abundance of natural resources, and this rich endowment partly explains the country's affluence based on resource extraction specializations. But an over-emphasis on overall low innovation intensity is misleading. The specialization in low-technology resource extraction would not have been possible without innovation-intensive technological trajectories working next to resource extraction sectors, such as mechanical engineering, engineering consultancy and suppliers to the aquaculture sectors.

5.2.3 The Netherlands
The Netherlands NSI has deep roots in the history of the country. The industrial structure and the common mode of societal organization (the *poldermodel*) go back to the sixteenth-century history of the Republic of the Netherlands. At the end of the 1960s, the Netherlands was a leading industrial nation, and innovation (especially by a few large firms) was at the heart of this economic success. Since then, innovation intensity has been in relative decline, partly because other nations have successfully caught up. What results is a relatively rich NSI, in which many actors (public and private) play a role, the science and technology infrastructure is well developed, and innovation policy (including policy employing a systems perspective) has a long tradition. But performance is declining, in terms of both innovation and science and technology indicators, as well as in terms of economic indicators such as productivity. The challenge for innovation policy is to overcome this situation, but policy makers have been faced with budget cuts, and, despite well-recognized elements of a systems approach in policy thinking, innovation policy is still very much steered by scoreboard indicators.

5.2.4 Finland
Industrial development in Finland can be divided into three phases: (1) a factor-driven economy from the mid-1800s to the early 1900s; (2) an investment-driven economy from the end of the Second World War to the

1980s; and (3) an innovation-driven economy since the late 1980s. Finland experienced a severe depression in the early 1990s, and the recovery from it was to a large extent due to fast growth in the information and communication technology (ICT) sector. Lately, innovative activity in Finland has been dominated by the electronics industry, as reflected in the success of this sector, and particularly of Nokia. Even though the electronics industry and especially Nokia dominate innovative activity in Finland, there are also other innovative sectors in the country, such as knowledge-intensive business services (KIBS). Many traditional sectors, such as the engineering and paper industries, are also quite innovative by international standards. All in all, Finland ranks among the top countries in innovativeness. The future challenges of the Finnish NSI include strengthening of innovative activities in traditional manufacturing industries and in service sectors. In addition to technical innovations, the role of organizational innovations should be strengthened and technical and organizational innovations should be integrated more than is currently the case.

5.2.5 Denmark

The Danish NSI is characterized by many small and medium-sized enterprises (SMEs) with only a few (in international terms) large firms. In general, Danish firms are innovative (making both product innovations, process innovations and organizational innovations), but their innovations mainly take the form of incremental changes. Such innovations often reflect a practical and experience-based interaction between skilled labour, engineers and marketing people. The firms mainly build up competences by employing experienced labour on a flexible labour market and through intensive interfirm collaboration – especially with domestic and foreign customers and suppliers. However, there are signs indicating that important changes in the traditional Danish mode of innovation may be under way. First, Danish firms – including many SMEs – are increasingly investing in R&D, collaborating more with universities than before and employing more personnel with higher education. Second, ongoing globalization implies on the one hand an outsourcing of low-skilled jobs – for instance within traditional scale-intensive food processing sectors – and on the other an increasing number of high-skilled jobs in high-tech sectors – for instance within biotechnology, ICT and various knowledge-intensive service industries.

5.3 Concluding Chapter

The concluding chapter of this book develops a comparative analysis that deals with only a very few of the many issues addressed empirically by the country case studies. The concluding chapter focuses to some extent on issues related to globalization, but devotes most of its attention to innovation policy. Our concluding chapter is intended as a contribution to the comparative analysis of NSI, conceived in the spirit of 'appreciative theorizing'.

ACKNOWLEDGEMENTS

We are extremely grateful to several colleagues within the project for valuable comments on this chapter.

NOTES

1. By March 2007 'innovation systems' had 792 000 hits on Google and 'systems of innovation' had 224 000. As a comparison 'economics of innovation' had 219 000 and 'neo-classical economics' had 285 000 hits.
2. Previous initiatives were taken by research groups in Aalborg, Denmark (Bengt-Åke Lundvall) and Oslo, Norway (Jan Fagerberg) before Lund, Sweden (Charles Edquist) was invited to take over the coordination.
3. What this may mean is discussed in Section 2.2 below.
4. The Appendix in this book is a result of that work.
5. We received comments from all the national teams, and in some cases from several members of the same team. The framework was revised substantially on the basis of these comments. This introduction is partly based upon the framework, but certainly does not reproduce it entirely. (The framework document was quite specific and the document describing it ran to 112 pages.)
6. We will briefly return to these issues later in this chapter. In addition, globalization and innovation policy will be discussed in more detail in the concluding chapter of this book.
7. We have adopted the common term 'countries' in this introductory chapter and in the concluding chapter. However, Hong Kong is not, properly speaking, a country in the sense of a nation-state. Formerly a British Crown colony, Hong Kong was made a Special Administrative Region of the People's Republic of China by the Sino-British Joint Declaration of 1984, and assumed that status in 1999. However, the 1984 Joint Declaration ensured preservation of Hong Kong's capitalist system and 'way of life' for 50 years, and this principle is reflected in the 'one country–two systems' framework that was subsequently enshrined in the constitution of the Hong Kong SAR. The innovation system of Taiwan covers only the Republic of China, which operates like a country, but is considered as a part of China. In addition, South Korea (the Republic of Korea) is only a part of the Korean peninsula and the case study of Ireland does not include the north-eastern part of the island (i.e. Northern Ireland).
8. Freeman here means 'organizations' in the sense of actors and not 'institutions' in the sense of rules. In addition, we currently often use the term innovations instead of technologies – implying that we also include in this category new creations also of a non-material nature, e.g. service product innovations and organizational process innovations (see specification of key terms in Section 2.2).
9. This emphasis is clear from Nelson and Rosenberg (1993, p. 4): 'the orientation of this project has been to carefully describe and compare, and try to understand, rather than to theorise first and then attempt to prove or calibrate the theory'. In the current project we have tried to do it partly in the opposite way – specifying concepts and theories first and carrying out empirical work in a comparative way thereafter.

10. Lundvall uses the term 'institution' in an ambiguous way. Sometimes, he uses it in the sense of 'rules' only; at other times, he uses it to denote 'organizations' – see Section 2.2.
11. Nelson (and Rosenberg) use the term 'institutions' to denote these organizations.
12. Their definitions of NSI do not include, e.g., consequences of innovation. This does not mean that innovations emerging in SI do not have tremendously important consequences for socioeconomic variables such as productivity growth and employment – on the contrary. Moreover, distinguishing between determinants and consequences does not, of course, exclude feedback mechanisms between them.
13. Curiously enough, these authors see Nelson as the proponent of a 'broad' approach. But here they are drawing a contrast, not with Lundvall, but rather with Bozeman and Dietz (2001), who propose a definition of NSI that is even narrower and more restrictive than Nelson's.
14. Like the SI approach, general systems theory might be considered an approach rather than a theory.
15. Only in exceptional cases is the system closed in the sense that it has nothing to do with the rest of the world (or because it encompasses the whole world).
16. Before going into these definitional issues, we want to stress that definitions and taxonomies are neither right nor wrong; they are more or less useful for certain purposes.
17. The activities in SI are the same as the determinants of the main function. An alternative term for 'activities' could have been 'subfunctions'. We chose 'activities' in order to avoid the connotation with 'functionalism' or 'functional analysis' as practised in sociology, which focuses on the consequences of a phenomenon rather than on its causes, which are the focus here (Edquist, 2005, p. 204, n. 16). In order to avoid all connotations, it would perhaps be better to use term 'x' to denote the concept – but this might seem too radical for some social scientists.
18. The Human Development Index is a composite index that measures the average achievements in a country in three basic dimensions: life expectancy at birth, adult literacy and the combined gross enrolment ratio for education at all levels, and GDP per capita.
19. See Box 1.1 for concept specifications and how 'development' and 'diffusion' of innovations relate to 'new-to-the-market' and 'new-to-the-firm' innovations.
20. The reason for this is that it is not possible to say that innovation intensity is high or low in a certain system if there is no comparison with innovation intensity in other systems. This has to do with the fact that we can not identify an 'optimal' or 'ideal' innovation intensity. The notion of optimality will be discussed related to policy issues in the concluding chapter of this book.
21. The innovation policies pursued during recent decades and relevant aspects of globalization are also discussed in the context of the factors influencing innovation processes. Since they are considered crucial issues in this project, they are also addressed in separate sections. They will also be discussed in the concluding chapter of the book.
22. However, dealing with consequences for productivity growth is done in a very different way in each case study.
23. Within these groups we simply present the chapters in reverse alphabetical order. The identification of fast and slow growth countries respectively is done in the concluding chapter.

REFERENCES

Amable, B. (2000), 'Institutional complementarity and diversity of social systems of innovation and production', *Review of International Political Economy*, 7(4), 645–87.

Balls, A. (2004), 'Nordic nations "top for growth potential"', *Financial Times*, 14 October, p. 6.

Bozeman, B. and J.S. Dietz (2001), 'Research policy trends in the United States: civilian technology programs, defence technology, and the development of the national laboratories', in P. Larédo and P. Mustar (eds), *Research and Innovation Policies in the New Global Economy: An International Comparative Analysis*, Cheltenham, UK and Northampton, MA, USA: Edward Elgar, pp. 47–78.

Correa, C. (1998), 'Argentina's national system of innovation', *International Journal of Technology Management*, **15**(6–7), 721–60.

Edquist, C. (1997), 'Systems of innovation approaches – their emergence and characteristics', in C. Edquist (ed.), *Systems of Innovation: Technologies, Institutions and Organizations*, London: Pinter, pp. 1–35.

Edquist, C. (2001), 'Innovation policy – a systemic approach', in D. Archibugi and B.-Å. Lundvall (eds), *The Globalizing Learning Economy*, Oxford: Oxford University Press, pp. 219–38.

Edquist, C. (2005), 'Systems of innovation – perspectives and challenges', in J. Fagerberg, D.C. Mowery and R.R. Nelson (eds), *The Oxford Handbook of Innovation*, Oxford: Oxford University Press, pp. 181–208.

Edquist, C. and C. Chaminade (2006), 'Industrial policy from a systems-of-innovation perspective', *European Investment Bank Papers*, **11**(1), 108–33.

Edquist, C. and L. Hommen (2004), 'Comparative framework for and proposed structure of the studies of national innovation systems in ten small countries', Lund, Sweden: CIRCLE, Lund University, mimeo (unpublished, but available from the authors).

Edquist, C., L. Hommen and M. McKelvey (2001), *Innovation and Employment: Product versus Process Innovation*, Cheltenham, UK and Northampton, MA, USA: Edward Elgar.

Edquist, C. and B. Johnson (1997), 'Institutions and organisations in systems of innovation', in C. Edquist (ed.), *Systems of Innovation: Technologies, Institutions and Organizations*, London: Pinter, pp. 41–63.

Florida, R. and I. Tinagli (2004), *Europe in the Creative Age*, Pittsburgh, PA: Software Industry Center, Carnegie Mellon University (co-published in Europe with DEMOS).

Freeman, C. (1987), *Technology Policy and Economic Performance: Lessons from Japan*, London: Pinter.

Freeman, C. (1997), 'The "national system of innovation" in historical perspective', in D.J. Archibugi and J. Michie (eds), *Technology, Globalization and Economic Performance*, Cambridge: Cambridge University Press, pp. 24–46.

Freeman, C. (2002), 'Continental, national and sub-national innovation systems – complementarity and economic growth', *Research Policy*, **31**(2), 192–211.

Galli, R. and M. Teubal (1997), 'Paradigmatic shifts in national innovation systems', in C. Edquist (ed.), *Systems of Innovation: Technologies, Institutions and Organizations*, London: Pinter, pp. 342–70.

Ingelstam, L. (2002), *System: Att tänka över samhälle och teknik* (*Systems: To Reflect over Society and Technology* – in Swedish), Stockholm: Energimyndighetens förlag.

Johnson, A. and S. Jacobsson (2003), 'The emergence of a growth industry: a comparative analysis of the German, Dutch and Swedish wind turbine industries', in S. Metcalfe and U. Cantner (eds), *Transformation and Development: Schumpeterian Perspectives*, Heidelberg: Physica/Springer, pp. 64–88.

Kaiser, R. and H. Prange (2004), 'The reconfiguration of national innovation systems – the example of German biotechnology', *Research Policy*, **33**, 395–408.

Larédo, P. and P. Mustar (2001), *Research and Innovation Policies in the New Global Economy: An International Comparative Analysis*, Cheltenham, UK and Northampton, MA, USA: Edward Elgar.

Liu, X. and S. White (2001), 'Comparing innovation systems: a framework and application to China's transitional context', *Research Policy*, **30**, 1091–114.

Lundvall, B.-Å. (1988), 'Innovation as an interactive process: from user–producer interaction to the national system of innovation', in G. Dosi, C. Freeman R. Nelson, G. Silverberg and L. Soete (eds) (1988), *Technical Change and Economic Theory*, London and New York: Pinter, pp. 349–69.

Lundvall, B.-Å. (1992), *National Systems of Innovation: Towards a Theory of Innovation and Interactive Learning*, London: Pinter.

Lundvall, B.-Å., P. Intarakumnerd and J. Vang (2006), *Asia's Innovation Systems in Transition*, Cheltenham, UK and Northampton, MA, USA: Edward Elgar.

Lundvall, B.-Å., B. Johnsson, E.S. Andersen and B. Dalum (2002), 'National systems of production, innovation and competence building', *Research Policy*, **31**(2), 213–31.

Mytelka, L.K. and K. Smith (2002), 'Policy learning and innovation theory: an interactive and co-evolving process', *Research Policy*, **31**, 1467–79.

Nelson, R.R. (1993), *National Systems of Innovation: A Comparative Study*, Oxford: Oxford University Press.

Nelson, R.R. and N. Rosenberg (1993), 'Technical innovation and national systems', in R.R. Nelson (ed.), *National Systems of Innovation: A Comparative Study*, Oxford: Oxford University Press, pp. 3–21.

Niosi, J. (1991), 'Canada's national system of innovation', *Science and Public Policy*, **18**(2), 83–93.

Niosi, J. (2002), 'National systems of innovation are "x-efficient" (and x-effective): why some are slow learners', *Research Policy*, **31**(2), 291–302.

OECD (1997), *National Innovation Systems*, Paris: OECD.

OECD (2002), *Dynamizing National Innovation Systems*, Paris: OECD.

Rickne, A. (2000), *New Technology-based Firms and Industrial Dynamics: Evidence from the Technological Systems of Biomaterials in Sweden, Ohio and Massachusetts*, Gothenburg, Sweden: Department of Industrial Dynamics, Chalmers University of Technology.

Saviotti, P.P. (1996), *Technological Evolution, Variety and the Economy: Planning in an Era of Change*, Cheltenham, UK and Brookfield, USA: Edward Elgar.

Schumpeter, J.A. (1911), *The Theory of Economic Development* (R. Opie, trans.) (1934, English edn), Cambridge, MA: Harvard University Press.

WEF (2004), *Annual Review of Global Competitiveness*, Geneva: World Economic Forum.

PART I

Fast Growth Countries

2. The rise and growth of a policy-driven economy: Taiwan

Antonio Balaguer, Yu-Ling Luo, Min-Hua Tsai, Shih-Chang Hung, Yee-Yeen Chu, Feng-Shang Wu, Mu-Yen Hsu and Kung Wang

1 INTRODUCTION

Over the last 40 years, Taiwan has been an example of sustained economic growth. The empirical richness of Taiwan's economic and social transformation provides a good source for testing theories, hypotheses and models of growth. 'Systems of innovation' approaches and concepts are useful in capturing some of the key elements of the 'Taiwanese story', which we have characterized as a 'policy-led systemic upgrading', involving a complex process of co-evolution among actors, institutions, knowledge, technology and markets.

The foremost aspect in the 'Taiwanese story' is that Taiwan was a latecomer in terms of its industrial development and learning how to compete in world markets. A second and related aspect is the role of government and public policy. Here, the crucial point is not simply that the government intervened in development, but that it played a key strategic role by changing the economic base to produce new comparative advantages and creating new market segments in which Taiwanese firms could compete (Amsden and Chu, 2003). A third aspect of the 'Taiwanese story' is specialization. Rather than sectoral specialization, we refer to a unique capacity to adjust quickly to patterns of changes in global demand by upgrading and excelling at the project execution level in original equipment manufacturing (OEM) and original design manufacturing (ODM). Specialization in these areas means that Taiwanese firms occupy a particular role in the international division of labour, creating the basis for relationships with multinational corporations (MNCs), which have become dependent on Taiwanese manufacturing and process innovation skills.

2 MAIN HISTORICAL TRENDS

In Taiwan's short history, many watershed events have marked the development of its NSI. The influence of Japanese colonial rule from 1895 to 1945 can be regarded as the starting point for this development. One important feature of this period was the dispersion of new infrastructure throughout the island. This helped to avoid city-centred development in Taiwan, a common characteristic of other colonies (Gold, 1988, p. 116). Lin (1973, pp. 13–27) argued that the years of Japanese rule were very important in the process of forsaking the tradition-bound static life of Chinese society and preparing the conditions for Taiwan's incorporation into the modern industrial life of the postwar years.

Between 1945 and 1949, the economic development of Taiwan was slow. Taiwan suffered severe damage after allied bombing, but the deterioration of the economic and social situation on the mainland left no room for Taiwan's reconstruction. From 1949, the situation changed dramatically. After the defeat of the Nationalists (Kuomintang, or KMT) and the 'relocation' of the Republic of China to the island of Taiwan, Taiwan began a new era of growth. Economically speaking, this period was characterized by three major successful polices that had a big impact on Taiwan's further development: land reform (followed by impressive agricultural development), industrialization based on import substitution, and support for state enterprises. In this period, Taiwan experienced unprecedented growth. Between 1958 and 1973, real gross national product (GNP) grew at an annual average of 9.8 per cent, and the rate of industrial production grew at 16.3 per cent. Exports increased from US$156 million in 1958 to US$4483 million in 1973.

From the late 1970s, Taiwan launched a major initiative to promote capital-intensive and high-technology industries. In addition, research and development (R&D) began to be considered an important issue in the Taiwanese policy makers' agenda. Local content policies and moderate import tariffs were applied in some industries, such as the automobile and electronics industries, but subsidization, tax incentives, strategic partnership, foreign direct investment (FDI), joint ventures and other mechanisms for technology transfer from overseas were common.

After the second oil shock in the late 1970s, the government took a much more decisive approach to promoting strategic high-technology and value-added industries. During the mid-1970s, notable advances were made in building a scientific and technological infrastructure. In 1973, the Ministry for Economic Affairs established the Industrial Technology Research Institute (ITRI) by merging three existing R&D institutes. In addition, the Hsinchu Science-based Industrial Park (HSIP) was established in the latter half of the 1970s to promote high-tech firms in information industries, new materials and biotechnology.

Research and development became a priority from the 1980s, when Taiwan embarked on its high-tech period. To encourage technological upgrading, the government used mechanisms such as exempting R&D equipment and instruments from import duties as well as tax refunds and regulations to induce R&D spending. This process of technological upgrading was accompanied by the development of new industries, notably integrated circuits (IC) (Tung, 2001).

The 1980s was a period of OEM/ODM sophistication, but it also saw the internationalization of some Taiwanese own brand manufacturing (OBM) companies (Hung and Whittington, 1997). In electronics, the 1980s witnessed considerable advances as Taiwanese firms attempted to catch up to the world's technological frontier (Dedrick and Kraemer, 1998; Hobday, 1995). From the 1990s, industrial restructuring accelerated in Taiwan despite the fact that the government took a much more neutral approach to the selection of strategic industries and prioritized policies such as the privatization of government enterprises. However, the government did select the so-called group of ten important 'technology-based' industries that received special treatment because of their technological intensiveness.[1] Between 1986 and 1996, technology-intensive exports grew from 18.4 per cent to 39.6 per cent of Taiwan's exports. By the latter date, Taiwan had achieved an impressive share of the world market in a number of information technology products; for example, it sold 53.4 per cent of the world's monitors, 74.2 per cent of motherboards, 95 per cent of handy scanners, 52 per cent of desktop scanners, 61 per cent of keyboards, and 65 per cent of computer mice (CEPD, 1997).

From July 1997 to mid-1998, most Asian economies were deeply shaken by the Asian crisis, and their currencies and stock markets dropped dramatically. In contrast to most Asian economies, Taiwan maintained most of its economic growth, despite the fact that its stock market and currency dropped significantly. In 1998, GDP (gross domestic product) grew at 5.2 per cent (Asian Development Bank, 1998). Healthy macroeconomic conditions, including a budget surplus, positive trade accounts and a relatively sound financial system, were instrumental in curbing speculative flows and maintaining stability.

The economic interdependence between China and Taiwan accelerated dramatically between the late 1990s and early 2000. Since 2002, the bilateral trade between Taiwan and Mainland China has surpassed that between Taiwan and the USA. The situation is complex, as both threats and opportunities are present. At present, the 'hollowing out' caused by the massive migration of Taiwanese firms to China has created an enormous pressure for the government to ensure that investment remains in Taiwan. Thus one of the most important recent Taiwanese government policies has been to encourage both domestic and foreign firms to establish their R&D centres and headquarters in Taiwan.

3   INNOVATION INTENSITY

Based on the first Taiwan Technology Innovation Survey (TTIS), we observed that in both input and output indicators the manufacturing sector shows a considerably higher propensity to innovate than the rest of Taiwan's firms (see Table 2.1). The innovation intensity of the manufacturing sector is 1.5 times higher than Taiwan's average and the percentage of turnover due to new-to-the-firm products is 2.6 times higher for manufacturing firms than for service firms. In addition, the focus of the innovative effort differs between the manufacturing and services sectors. The NSC (National Science Council) (2003) indicates that a large share of innovation expenditure for the manufacturing sector goes to in-house R&D, whereas in the service sector it is mainly related to technology acquisition.

Among the manufacturing sectors included in the first TTIS,[2] the most innovative is the machinery, equipment and instruments sector, which includes the computer, communication and semiconductor industries – the backbone of Taiwan's high-tech growth since the 1990s. In addition, this sector has the highest percentage of firms undertaking process and product innovation. In 2000, innovation intensity (ININT2K) and R&D intensity (RDIN2K) in machinery and equipment were 5.20 per cent and 2.06 per cent, respectively. Production efficiency, quality control and technology adoption are the main objectives for R&D spending in local firms of the ICT (information and communication technology) sector (see Table 2.2).

The group including petroleum refining, nuclear fuels, chemicals and plastics provided the foundation of Taiwan's industrialization. For many years, these industries were strongly supported (and partially owned) by the government. Although most of the firms in these industries rely on foreign technology, their key role as suppliers of materials for export-oriented downstream industries, such as electronics, has forced them to upgrade production. The plastic materials industry hosts some of the world's largest manufacturers of engineering and commodity plastics, which have used its indigenous R&D process capability to adopt and improve foreign technologies. A key driver of innovative efforts in these industries has been the need to increase the scale of production in order to compete. Some government-led R&D collaborative initiatives with national R&D

Table 2.1 Selected TTIS-1 indicators: Taiwan

Indicators[1]	Input indicators (%)		Output indicators (%)					
	ININT2K	RDINT2K[2]	TURNIN	TURNMAR	INPDT	NTTM	INPCS	INNO
Taiwan	2.8	1.3	11.7	6.0	20.0	n.a.	31.0	40.0
Taiwan manufacturing	4.1	1.2	17.9	7.0	25.0	n.a.	35.0	45.0
Taiwan services	1.8	n.a.	6.9	5.4	17.0	n.a.	28.0	37.0

Notes:
[1] ININT2K: innovation intensity in year 2000; RDINT2K: R&D intensity in 2000; TURNIN: turnover due to 'new to the firm' products; TURNMAR: turnover due to 'new-to-the-market' products; INPDT: introduction of 'new-to-the-firm' products; NTTM: introduction of 'new-to-the-market' products; INPCS: introduction of new processes; INNO: innovating firms.
[2] *Source:* NSC (2001).

Source: NSC (2003).

Table 2.2 Selected CIS indicators of the manufacturing and service sectors in Taiwan

	Input indicators (%)		Output indicators (%)		
Indicators[1]	ININT2K	RDIN2K[2]	INPDT	INPCS	INNO
Manufacturing industry	4.1	1.23	25.0	35.0	45.0
Food, beverages & tobacco	2.2	0.36	28.0	46.0	50.0
Textiles, wearing apparel, fur & leather	4.8	0.40	18.0	34.0	41.0
Wood, paper, printing, publishing	4.3	0.13	12.0	33.0	39.0
Petroleum, nuclear fuel, chemicals, rubber & plastics	2.5	0.71	30.0	37.0	48.0
Non-metallic mineral products	2.5	0.24	20.0	35.0	40.0
Basic metals	1.9	0.17	12.0	32.0	38.0
Fabricated metal products	3.3	0.22	25.0	32.0	43.0
Machinery equipment, instruments & transport equipment	5.2	2.06	36.0	37.0	52.0
Furniture, other manufacturing nec (not elsewhere classified)	5.7	0.47	25.0	33.0	43.0
Service industry	1.8	n.a.	17.0	28.0	37.0
Wholesale, retail trade, motor vehicle repair etc	1.3	n.a.	12.0	29.0	32.0
Hotel & restaurants	1.8	n.a.	18.0	29.0	31.0
Transport & storage	1.0	n.a.	15.0	26.0	35.0
Communications	6.6	n.a.	43.0	29.0	43.0
Financial intermediation (incl. insurance)	1.9	n.a.	36.0	39.0	53.0
Real estate, renting & business activities	4.0	n.a.	20.0	33.0	41.0
Commercial, social & personal services activities, etc	1.1	n.a.	36.0	50.0	61.0
Total	2.8	1.25	20.0	31.0	40.0

Notes:
[1] ININT2K: innovation intensity in 2000; INPDT: introduction of 'new-to-the-firm' products; INPCS: introduction of new processes; INNO: innovating firms.
[2] *Source:* NSC (2001).

institutes have been important within this group – for example, in the upgrading of the polyester industry.

Table 2.1 shows that process innovation is a more common activity than product innovation in both manufacturing and services. Taiwanese specialization in OEM manufacturing has been an important cause of process innovation as OEM arrangements governed more than half of the data-processing equipment production in Taiwan. Under an OEM agreement, a Taiwanese firm produced a finished product by following the precise specifications of multinational corporations, which meant that profitability for the Taiwanese firm was directly linked to its capacity to reduce manufacturing costs. On the other hand, the intense competitive environment among OEM firms in the local market created significant pressure for the development of efficient and flexible processes that could satisfy multinational corporations' strict specifications and standards.

Indicators of the propensity to innovate differ depending on the size of the firms involved. As in most countries and industries, the extent of innovation is considerably higher in Taiwan's large enterprises than in its small and medium-sized enterprises (SMEs). Innovation activities (INNO) are undertaken by 77 per cent of large firms, a figure about 20 per cent higher than that for medium-sized firms. The percentage of medium-sized firms engaging in product innovation (INPDT) or process innovation (INPCS) exceeds the corresponding figure for small and extra-small firms by 10 per cent. When the percentage of firms that introduce new products is considered, innovation differences based on size are even more marked. This is an expected result, as few small or extra-small Taiwanese firms have the capability to engage in product innovation.

4 ACTIVITIES THAT INFLUENCE INNOVATION

4.1 Knowledge Inputs to Innovation

4.1.1 R&D activities

In Taiwan, R&D intensity (gross expenditure on R&D (GERD) as a percentage of GDP) has grown significantly since the 1980s, rising from 0.85 per cent in 1981 to 2.3 per cent in 2002.[3] However, Taiwan is still far behind the international front-runners, Sweden, Finland and Korea, in R&D intensity. As Figure 2.1 shows, from the 1990s onwards, more than half of Taiwan's R&D investment was made by the private sector. Business expenditure on research and development (BERD) has shown the fastest rate of growth, whereas private research institutes (non-business) have shown the greatest decline.

By 2002, 62 per cent of R&D expenditure was undertaken by the private sector. Taiwan's BERD as a percentage of GDP reached 1.43 per cent in 2002, with an annual growth rate averaging 8.4 per cent between 1995 and 2002. Despite this fast growth rate, BERD intensity lags behind those of

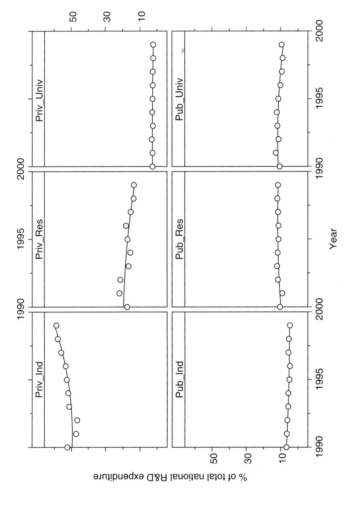

Note: Priv_Ind: private industry; Pub_Ind: public industry; Priv_Res: private (science and technology) research institutes; Priv_Univ: private universities; Pub_Res: public (science and technology) research institutes; Pub_Univ: public universities.

Source: Hwang (2001).

Figure 2.1 Changes in the R&D intensity of six sectors in Taiwan

Table 2.3 Comparison of BERD as a percentage of GDP

	1999	2000	2001	2002	1995–2002 annual growth rate (%)
FI	2.20	2.41	2.42	2.41	11.8
JPN	2.10	2.12	2.26	2.32	4.0
KR	1.76	1.96	2.23	2.18	7.3
USA	1.98	2.04	2.00	1.87	3.9
DK	1.42	n.a.	1.65	1.75	9.9
TW	1.31	1.30	1.37	1.43	8.4
SG	1.20	1.17	1.33	1.32	13.6
NL	1.14	1.11	1.10	1.03	2.8

Source: OECD (2004).

other OECD (Organisation for Economic Co-operation and Development) countries (Table 2.3).

The government's support for R&D accounted for 0.86 per cent of GDP in 2003, a figure similar to that of South Korea (0.90 per cent), but considerably less than in most of the OECD countries. The majority of the funds went to universities and research institutes. Public research institutes have shown a slight increase in the national share of GERD. However, university research has fallen to less than 10 per cent of total R&D expenditure. The share of basic research, an activity usually performed by universities, is one of the lowest in the industrialized world.

A rapid change in Taiwan's BERD composition by sector may reflect structural changes in the knowledge base underpinning industrial transformation in Taiwan. The increasing need to upgrade the high-tech manufacturing sectors (electronics, optoelectronics, computing, pharmaceuticals, aerospace and instruments) has increased the gap between R&D investments in these sectors and the rest of manufacturing. In 2001, these sectors totalled 66 per cent of the total manufacturing BERD, topping the list of the countries shown in Table 2.4.

In terms of output in peer-reviewed scientific and technical journals, Taiwan was ranked eighteenth in the world in 2001. It accounted for 1.1 per cent of worldwide production, which represented a significant improvement on its share of 0.9 per cent in 1996.[4] However, higher growth trends are observed in other newly industrialized and developed countries such as Korea, Singapore and China.

Taiwan's scientific publication profile is typical of a recently industrialized country, in that there is a strong focus on basic research in the natural and engineering sciences. However, Taiwan's publishing presence is

Table 2.4 Structure of sectoral BERD (1995 and 2001)

(%)	Total manufacturing		Hi-tech total		Hi-tech manufacturing									
					Pharma.		OA & comput.		Electronics		Aerospace		Instruments	
	1995	2001	1995	2001	1995	2001	1995	2001	1995	2001	1995	2001	1995	2001
TW	97.1	92.4	51.0	66.0	1.2	1.1	15.8	18.2	33.9	43.7	0.1	0.7	1.5	2.3
JPN	96.2	90.6	37.8	40.5	6.8	7.1	9.0	13.0	17.5	15.3	0.7	0.8	3.8	4.3
SE	87.5	87.4	47.5	55.1	14.3	17.9	1.4	0.8	19.9	28.9	5.1	2.7	6.9	4.8
FI	88.0	84.6	42.3	56.4	5.1	6.0	2.1	0.2	31.2	47.5	0.1	0.3	3.9	2.4
KR	83.3	82.8	37.0	51.4	1.4	2.2	1.8	7.8	31.6	36.2	1.5	3.8	0.7	1.4
NL	82.5	75.9	34.4	39.5	6.8	8.9	12.1	25.7	11.1	0.3	2.7	0.1	1.7	4.5
IR	89.8	75.0	50.9	51.5	16.2	10.5	5.3	5.1	23.9	30.6	0.4	0.4	5.1	5.0
USA	78.9	64.9	47.9	39.4	7.7	6.5	6.7	5.2	11.6	12.9	12.8	5.2	9.1	9.6
DK	67.9	64.1	33.2	34.0	20.0	23.0	0.9	0.8	6.3	4.0	0.0	0.0	6.0	6.1
NO	40.5	48.0	25.7	21.4	n.a.	6.3	1.5	1.0	15.8	13.5	0.4	0.4	1.8	2.1

Note: OA = optoelectronics.

Source: Wu and Lin (2004).

relatively weak in the life and medical sciences. Despite fast growth rates for such publications and an increasing share of total world scientific publications, the quality of publications measured by citation rate does not reach the world's average. Although specialization has decreased gradually (the shape of the figure has become slightly broader), revealed comparative advantage values (see Figure A3.1 in the Appendix) for the fields of computer science, materials and engineering are considerably higher than the world average (value 1). Taiwan's research system shows a greater specialization in highly competitive national industries, such as electronics equipment manufacturing.

Recently, patenting activity has expanded in Taiwan. Considering the total number of patents approved by the US Patent and Trademark Office, Taiwan's ranking rose from eleventh in 1989 to fourth in 2002. In terms of patent intensity (patents granted per million people), Taiwan ranked third, behind only the USA and Japan in 2002. The distribution of utility patents across technological fields shows Taiwan's unique performance in the fields of semiconductors, electronics, electrical appliances and components; 50 per cent of such patents relate to ICT, especially semiconductor technology.[5] Most of them are the outcome of process innovation by OEM firms.

Specialization tends to persist for long periods. However, persistent efforts to change the industrial structure and the technological base at the national level may change the patterns of technological specialization, as is shown by patenting activity. As Figure A3.2 in the Appendix shows, Taiwan has deepened its technological specialization in semiconductors and electronics. The revealed technological advantage value of 3.5 for the period 1994–2001 corresponds to a jump in the absolute number of patents from 172 to 5264, which began in the 1987–93 period and continued during 1994–2001.[6] By contrast, Taiwan's specialization in textiles, apparel and other manufacturing has declined over the 1990s.

The use of these patents may, however, tell a different story. Table 2.5 provides a comparison of the number of US patents received by country with the ratio of expenditure-to-revenue of technological trade. When the ratio is greater than one, it means that the country's technological revenue exceeds expenditure, or that the country is a net exporter of technology. As Table 2.5 suggests, among the 'big six' US patent-holding nations, only Taiwan has a disproportionately small expenditure-to-revenue ratio compared with the number of its US patents. The ratio is significantly lower than the other five countries and it showed no growth between 1997 and 1999. Although Taiwan has a large number of US patents, it has not managed to earn a commensurately large amount of royalty fees from them. Indeed, Taiwan remains a major net importer of technology.

Table 2.5 Number of US patents versus coverage ratio of technology trade, 1996–99

	1996		1997			1998			1999		
	Number of US patents[1]	Technology trade: receipt/ payment (%)	Number of US patents	Technology trade: receipt/ payment (%)		Number of US patents	Technology trade: receipt/ payment (%)		Number of US patents	Technology trade: receipt/ payment (%)	
USA	61 104	4.14	61 707	3.50		80 291	3.09		83 905	2.75	
JPN	23 053	1.56	23 179	1.90		30 840	2.13		31 104	2.34	
GER	6 818	0.76	7 008	0.84		9 095	0.85		9 337	0.77	
TW	1 897	–	2 057	0.05		3 100	0.02		3 693	0.03	
FR	2 788	0.75	2 958	0.71		3 674	0.83		3 820	–	
UK	2 453	1.60	2 678	1.72		3 464	1.80		3 572	–	

Note: [1] The number of patents excludes new design patents.

Source: NSC (2001).

It is likely that this low rate of return from technology trade is related to the nature of Taiwanese patents. Taiwan's patenting activity (and its technological emphasis) is related mainly to the production process, as discussed earlier. As a result, it is heavily biased towards development rather than new research. Lin et al. (2001) further analysed the patent stock of Taiwan and suggested that most of the patents granted to Taiwanese firms before 1970 were of the new-design type.[7] Only after 1990 did the number of utility patents begin to increase.

4.1.2 Competence building

Taiwan's industrialization has been supported by high-quality human capital with a strong work ethic. Traditionally, Chinese people are highly education-oriented. The ethics of 'respect for education', inherited from Confucian values, have facilitated family support. The government's education policy has adjusted over time in accordance with the demands of industrial development. During the import-substitution and land-reform periods, the focus of educational policy was on the establishment and support of industrial and agricultural vocational schools. The primary objectives of educational policy were to improve literacy and ensure that the supply of labour met the requirements of new industries. Nine-year compulsory education was enforced in 1968, and it resulted in the upgrading of overall national competence.

The rapid growth of export-oriented industries created a great demand for qualified technical personnel, which led to a swift expansion of the vocational schools. In 1970, the ratio of general high-school students to vocational-school students was 1:1, but it had risen to 1:2 by 1980. This strong demand created pressure for reform of the traditional education system, which was carried out in the 1990s. The number of students enrolled in tertiary education as a percentage of Taiwan's population changed dramatically between 1981 and 2002, rising from 1.97 per cent to 5.51 per cent.[8] However, the proportion of the workforce returning to higher education remains at a low level. This may suggest that lifetime learning takes place mainly through on-the-job training rather than formal education.

Table 2.6 shows that tertiary education in Taiwan has focused on engineering, which accounted for 23.9 per cent of total graduates in 2000. Similar trends exist for mathematics and computer science graduates. In these disciplines, the percentage of graduates was almost double the OECD average.

In the past, funding for public education was entirely subsidized by the government and private education organizations depended mainly on students' tuition fees. Recent changes in the public higher educational

Table 2.6 Tertiary graduates (%), by field of study and level of education (2000)

Field[1]	Edu.	Art	Social	Services	Engin.	Agr.	Health	Life S.	Science	Math.	Comp.	N
Column	(1)	(2)	(3)	(4)	(5)	(6)	(7)	(8)	(9)	(10)	(11)	(12)
TW	7.2	10.3	26.9	1.0	23.9	3.4	7.3	4.2	1.2	2.6	6.3	5.7
JPN	6.3	18.1	37.2	x(12)[2]	21.3	3.4	5.2	4.4	x(8)[2]	x(8)[2]	x(8)[2]	4.0
KR	5.6	20.9	22.8	2.5	27.4	3.2	6.6	2.1	4.4	2.1	2.4	a[3]
USA	13.1	14.2	42.2	2.4	6.5	2.3	9.8	4.1	1.5	0.9	2.8	0.3
DK	1.0	23.6	44.7	0.3	8.9	3.2	5.5	4.2	4.3	1.0	1.8	n[4]
FI	8.2	12.4	23.5	2.6	24.0	2.3	19.3	1.9	2.7	1.0	2.2	n[4]
IR	9.0	20.2	30.8	1.4	9.3	1.7	7.8	6.9	3.3	1.1	8.4	0.2
NL[5]	16.8	7.3	34.8	2.6	10.4	2.3	20.9	1.1	1.9	0.3	1.5	n[4]
NO	22.7	7.2	22.1	3.3	6.8	1.4	25.3	1.2	1.4	0.3	3.3	4.9
SE	18.8	5.7	21.3	1.0	20.5	1.0	22.8	2.3	2.4	0.6	3.1	n[4]
OECD mean	13.2	12.3	33.5	33.5	13.2	2.3	11.5	3.1	3.0	1.1	3.1	0.9

Notes:
1. Edu.: education; Art: humanities and arts; Social: social sciences, business and law; Engin.: engineering, manufacturing and construction; Agr.: agriculture; Health: health and welfare; Life S.: life sciences; Science: physical science; Math.: mathematics and statistics; Comp.: computing; N.: not known or unspecified.
2. The symbol 'x' indicates that data are included in another column. The column reference is shown in brackets after 'x'; for example, x(12) means that data are included in column 12.
3. The symbol 'a' means data not applicable because the category does not apply.
4. The symbol 'n' means magnitude is either negligible or zero.
5. The Netherlands' data exclude advanced research programmes.

Sources: Ministry of Education (2003b, p. 35); OECD (2002, p. 61).

organizations have forced students to take partial responsibility for their own financial support. Total educational expenditure in current values rose from NTD$56 907 million (US$1428 million) in 1985 to NTD$608 629 million (US$17 603 million) in 2002, an elevenfold increase. The ratio of public sector to private sector expenditure on education decreased from 4.3:1 in 1985 to 2.6:1 in 2002.

However, total educational expenditure as a percentage of GDP has decreased in recent years, falling from 5.5 per cent in 1995 (slightly above the OECD average) to 4.3 per cent in 2000 (below the OECD average of 5.2 per cent). A similar trend has been observed in tertiary sector expenditure, which was well below the OECD average in 2000 (see Table 2.7).

Foreign education systems and overseas work experience have been an important complement to the education system of Taiwan. More recently, local universities have begun to accept overseas students and provide various international exchange schemes. This has stimulated foreign scientists and postgraduate students to work in Taiwan and helped domestic graduates to gain a more international perspective. The number of Taiwanese students going overseas increased dramatically from 8178 in 1988 to 33 791 in 2002. However, it decreased in 2003. The Hart–Celler Act of 1965 in US-registered factories allows immigration based on possession of skills and family ties to citizens or permanent residents (Saxenian and Hsu, 2001), which has made the USA a favourable destination for Taiwanese students. Although the USA has continued to be the first choice for many Taiwanese students, recently growing numbers have gone to countries such as the UK, Australia and Canada.

Studying abroad has become a common pathway for those students and professionals who want to progress quickly in the local labour market. Returnees have had an enormous impact on the industries of Taiwan, particularly the electronics industry. The late 1980s witnessed a considerable return of former emigrants to Taiwan (Chang, 1992). It has been estimated that 33 per cent of students returned in 1988 after the successful completion of their studies abroad, a return rate that was three times higher than that of 1980 (Su, 1995). A survey based on Taiwan's 1990 population census suggests that around 50 000 emigrants returned during the period 1985–90. About 43 per cent of returnees had at least undergraduate education, and more than 30 per cent of returnees were employed as professionals and managers. The HSIP, which already had the largest concentration of Taiwan's high-tech industries, absorbed the best-educated returnees, in particular those with graduate education (Tsay and Lin, 2003).

The inflow of highly qualified manpower in the 1990s was clearly influenced by government policies. The loss of talent that Taiwan experienced

Table 2.7 Total public expenditure on education as a percentage of GDP (1995, 2000)

	Taiwan[1,2]	Japan	Korea	USA	Denmark	Finland	Ireland	NLD	Norway	Sweden	OECD mean
2000											
All levels	4.3	3.6	4.3	5.0	8.4	6.0	4.4	4.8	6.7	7.4	5.2
Tertiary education	0.7	0.5	0.7	1.1	2.5	2.0	1.3	1.3	1.7	2.0	1.2
1995											
All levels	5.5	3.6	n.a.	n.a.	7.4	7.0	5.1	5.0	9.0	7.2	5.4

Notes:
[1] Expenditure on educational administration and social education are not included after 2000 for Taiwan.
[2] Subsidies for private sector education are included in Taiwan.

Sources: OECD (2002, see Table B3.1; 2003, see Table B4.1); Ministry of Education (2003b, see Table 4-3).

during the 1960s, when highly educated emigrants remained overseas, called for nationwide attention as it had negative effects on economic development. The government reacted to this 'brain drain' by establishing a multinational community between Hsinchu and Silicon Valley in order to attract highly skilled workers (mainly of Chinese origin) to return to Taiwan from the USA. Scientists and engineers based in the USA and Taiwan were deliberately brought together through meetings and conferences sponsored by the Taiwanese government. These events helped to build up personal and professional relationships between engineers, entrepreneurs, executives and bureaucrats on both sides of the Pacific (Saxenian and Hsu, 2001). In addition, a number of new venture capitalists linked to this community invested in Silicon Valley start-ups, which later commercialized their technologies in Taiwan. Such overseas cooperation further strengthened the competences that had already been built up in Taiwan.

The high mobility rate of highly qualified human resources between Taiwan and the USA, particularly in the fields of science and technology (S&T), led to the subsequent transfer of information and know-how as well as the development of business relationships, all of which were crucial elements in the creation of high-tech companies.

4.2 Demand-side Factors

The formation of new markets has depended on the ability of leading actors to coordinate the supply of local firms with demand from local and overseas markets. During the first stage of development, Japanese trading houses and American mass merchandise buyers exercised this role. The division of labour between domestic and foreign actors created rigidities that prevented Taiwanese firms from entering new markets. This limited the participation of domestic actors as suppliers of capital goods, such as semiconductors, for the advanced segments of the electronics industry.

As the ability of Taiwanese firms to compete in markets based on cheap labour eroded, the government stepped in to coordinate the formation of new markets. Government-sponsored research institutes, such as ITRI, provided support for private sector technological upgrading and applications. ITRI was given responsibility for promoting technology diffusion and market bridging in recognition of the fact that the private sector was not capable of carrying out such tasks. The Institute for Information Industry (III) was created in 1979 to develop and promote the information industry. Although it was not as successful as ITRI in building the technology base, the III was noted for its involvement in aspects of the IT industry's demand articulation and quality requirements, including technology training, promotion and market intelligence (III, 2004; Breznitz, 2005).

Networking initiatives such as the Core–Satellite Development Centre System were important in ensuring that the production of the SMEs met the requirements of the large enterprises and multinational corporations. The Taiwan External Trade Development Council created the programme 'Innovalue', supported by the slogan 'very well made in Taiwan', which provided a certification to products of excellent quality and innovative design.

As the activities related to new market formation became more complex, supporting activities evolved over time into a sophisticated set of networks. Over the years, the government dedicated an increasing proportion of public expenditures to technology diffusion and adoption programmes. These included:

1. Subsidies for technology adoption (e.g. tax incentives)
2. Information (demonstration and extension) programmes
3. Public technology procurement (e.g. acquisition of new products/systems)
4. The development of technical standards (e.g. settings for products/processes)
5. Government-mandated technology transfer (e.g. adoption by off-sets and reciprocal obligations).

Private participation in technology diffusion in the ICT sector has increased over time, most significantly in recent years. The participation spread from technology adoption towards technology procurement and consortia for technical standard setting. There have been increasing efforts in the public and private sectors to form consortia to address issues such as architectural principles and standards to speed and expand market reach. The private sector participates in the following types of technology diffusion and adoption activities:

1. Technology adoption and transfer (e.g. in- or out-licensing of patents)
2. Cooperative research organizations (e.g. collaborative research between public and private organizations and between users and producers)
3. Private technology procurement (e.g. the award of supply contracts involving a significant R&D component)
4. Private consortia for technical standard setting (e.g. system frameworks and interfaces for products and processes).

The results of public and private sector efforts in the development of new markets and the diffusion of new technologies have been mixed, as exemplified by the development of the information industry in the 1990s. Although the cooperation between public and private actors had positive results in the case of personal computers and peripherals assembly, the development of hard disk drives in Taiwan points to a failure (Hung, 2000, 2002).

In response to demand from industrial customers, Taiwanese firms have been successful in coordinating production networks in China and South-East Asia. It is estimated that more than 20 000 production lines have been set up in these regions. The adoption of international standards, architecture and norms has become the basis for quality articulation. Examples include the ISO (International Organization for Standardization) series and the product and process technology, logistics and marketing reference models. Compliance with these standards is a fundamental requirement for competing in local and international markets, and Taiwanese firms have taken decisive steps towards their adoption.

4.3 Provision of Constituents

4.3.1 Provision of organizations

Taiwan's NSI is rich in the diversity of its organizations. The popular view of Taiwanese firms is of aggressive SMEs led by empowered family managers working long hours. This is no longer what characterizes Taiwan's most successful businesses, nor has it been the source of economic growth over the last decade. A different breed of large firm has emerged as the engine of innovation and growth. This type of firm specializes in production and process innovation as a way to cut costs and achieve efficiency. Table 2.8 shows the ranking of Taiwan's large firms by sales for 1980 and 2003. The data illustrate that Taiwan's top firms specialize in non-brand manufacturing, chemical materials and financial services. Eleven of the firms in the top 25 in 2003 did not exist in 1980.

In terms of the overall trend of firm births and deaths, the total number of registered factories grew from 78 293 in 1988 to 98 865 in 2003. However, newly opened factories showed a clear downward trend in the same period, falling sharply from 10 312 to 5103.[9] Factory closures followed a cyclical trend, with peaks in 1998 and 2003, which may have been related to the 1997 Asian financial crisis and the increasing exodus of Taiwan firms to China, fuelled by the fact that thousands of SMEs have lost their comparative advantages based on low wages.

Taiwan's environment gradually became hostile to the entry of new firms in recent decades because the ability to compete in high-tech industries required a critical mass of financial resources, managerial and R&D capabilities. However, the government reduced technological uncertainty by targeting products and components that were mature by world standards (Amsden and Chu, 2003, p. 114). In some cases, these resources were

Table 2.8 Top 25 firms by sales in Taiwan, 1980 and 2003

Rank 1980	Rank 2003	Name	Type of firm[1]	Sales 1980 (US$m)	Sales 2003 (US$m)	Industrial sector	Est. (year)
14	1	Cathay Life Insurance Co. Ltd	S,NF	304	13 800	Financial services	1962
1	2	Chinese Petroleum Corp.	P,G	4 192	13 303	Petroleum	1946
	3	Bureau of National Health Insurance.	G	n.a.	10 360	Financial services	1947
2	4	Taipower Co. Ltd	P,G	1 795	9 747	Power generation	1947
NR	5	Hon Hai Precision Industry Co.	S,M	n.a.	9 521	OEM electronics	1974
NR	6	Chung-Hwa Post Life Insurance Co. Ltd	B,G	n.a.	9 250	Financial services	2003
NE	7	Quanta Computer Inc.	S,M	n.a.	8 492	OEM computers	1988
239	8	Nan Shan Life Insurance	NF	27	7 786	Financial services	1963
NE	9	Formosa Petrochemical Corp.	S,M	n.a.	6 872	Petrochemicals	1992
	10	Bureau of Labour Insurance	G	n.a.	6 151	Financial services	1950s
28	11	Shink Kong Life Insurance	P,N	175	6 116	Financial services	1963
NE	12	Taiwan Semiconductor Manuf.	S,M	n.a.	5 866	OEM semiconductors	1987
NE	13	Chung Hwa Telecom Co. Ltd	S,G	n.a.	5 205	Telecom	1996
NE	14	Compal Electronics	S,M	n.a.	4 713	OEM electronics	1984
6	15	China Steel Corp.	S,M	495	3 768	Steel	1971
3	16	Nan Ya Plastics Corp.	S,M	576	3 709	Chemicals & plastics	1958
NE	17	ING Life Insurance (Taiwan Branch)	NF	n.a.	3 297	Financial services	1987
NE	18	BenQ Corporation	S,M		3 158	OBM computers	2001
12	19	Formosa Chemicals and Fibres Corp.	S,M	353	3 092	Chemicals & fibres	1965
NR	20	Lite-On Technology Corp.	S,M	n.a.	2 896	OEM electronics	1974
NE	21	AU Optoelectronics Corp.	S,M	n.a.	2 836	OEM optoelectronics	2001
NE	22	Samsung Electronic Taiwan Co. Ltd	NF	n.a.	2 721	Electronics	n.a.

NE	23	United Microelectronics Corp.	S,M	n.a.	2 466	OEM semiconductors	1980
8	24	Formosa Plastics Corp.	S,M	476	2 454	Chemicals, plastics	1954
10	25	Tatung Co. Ltd	S,M	397	2 430	Manufacturing	1939

Notes: NE: non-existent in 1980; NR: not ranked in 1980; S: listed on the stock market; M: manufacturing company; NF: non-manufacturing company; G: government enterprises; P: public issue company; F: foreign shares exceeding 50%; B: banking or finance.

Source: CCIS (1981, 2004)

mobilized by large business groups that had accumulated wealth during the early stages of Taiwan's development. However, in most cases, firms started out small but rapidly gained wealth as they concentrated on highly profitable markets such as selling liquid crystal displays (LCDs) (Mathews, 2005). Such firms did not act in isolation but operated within networks and cooperative consortia, which facilitated the capture of these markets.

4.3.2 Networking, interactive learning and knowledge integration

Historically, Taiwan's large number of SMEs (91 per cent of firms have fewer than 50 employees – see Shieh, 1990, p. 47) has been linked to the formation of networks. The characteristics of the Chinese family business have been particularly influential in this pattern. The Chinese entrepreneurial spirit combined with the highly centralized nature of family businesses impeded the career development of non-family skilled workers (Poon, 1994, p. 16). These characteristics of the Chinese family business have resulted in a strong 'centrifugal tendency', as thousands of middle managers have decided to start their own businesses (Whitley, 1991, p. 15; Shieh, 1992). Politically, SME formation was favoured because of divisions between the political mainland elite and Taiwanese businessmen, as well as the bias of the Nationalist government against capital concentration.

The issue of networking arises because many of these SMEs have operated in a cooperative fashion, rather than individually. The causation of, and decisions within, Taiwanese production networks were influenced by both short-term cost motivation and long-term cultural relationships, with diversity as the norm. The limited amounts of capital, cheap labour and simple technology in Taiwan's small firms in traditional industries motivated a very efficient allocation of resources through subcontracting networks. The characteristics of demand were also an important reason for subcontracting. Orders varied considerably in terms of season, size and complexity. Firms subcontracted some tasks because of the lack of production capacity, machinery, or labour (Shieh, 1990, pp. 104–13).

The role of government in creating networks through the design of import-substitution policies around critical components is illustrated by the creation of the IC industry in the mid-1980s as a spin-off from the government-owned Electronic Research Scientific Organisation,[10] as well as by the decisive role played by the Ministry of Economic Affairs (MOEA) in selecting CD-ROMs for promotion. Other examples include the formation of Centre–Satellite Industrial Systems and science parks and ITRI in promoting the development and transfer of technology to private industry. Amsden and Chu (2003) concluded that the government led the networking process through policies such as import-substitution for high-technology products and critical components and parts, the promotion of government-sponsored and -subsidized R&D and the creation of spin-offs from government-owned research institutes.

The OEM system was of great importance for firms whose growth was focused on manufacturing activities, thus avoiding both risks of the consumer market and allocation of resources for product sales and marketing. As local firms that were part of an OEM arrangement became more sophisticated and gained more freedom in taking decisions, relationships between firms purchasing and supplying materials became stronger and more interdependent. Some important manufacturing networks were created within plastics, petrochemical and textile groups, as these groups diversified into electronics.

4.3.3 Provision of institutions

By the early 1980s, many policy makers in Taiwan had become conscious that the market price mechanisms were too slow to propel the kind of development that Taiwan needed (San, 1995, p. 12). Consequently, a number of strategic policies were initiated to drive Taiwan to a new stage of development. These strategic policies changed the existing priorities for industrial development by establishing new institutional frameworks, which have been influential not only at the level of corporate governance decisions, but also for SMEs seeking opportunities for growth and diversification (Hung, 1999).

Historically, the key strategist of industrial development has been the MOEA, which has implemented its policies mainly through the Industrial Development Bureau (IDB). The IDB was established in 1970 and has been responsible for establishing the framework for the creation of new industries and the upgrading of existing ones.

During the 1980s, the IDB focused on the development of strategic industries under the principle of 'two high, two large, two low'. This slogan encapsulated the IDB's effort to identify and support industries with high technological intensiveness and high value-added, large market potential and large industrial linkages (forward and backward), and low energy consumption and low levels of pollution (San, 1995, p. 12). During the 1990s, the focus changed to industrial restructuring and upgrading, motivated by the appreciation of the new Taiwan dollar (NTD) and the loss of competitiveness of traditional industry. In the 2000s, the Council for Economic Planning and Development established the National Development Plan 'Challenge 2008', which aims to focus on key emerging industries in order to transform Taiwan into an R&D, manufacturing and operational world centre for industries such as semiconductors, visual displays, digital content and biotechnology. IDB was given the specific role of coordinating government and private sector policies to promote these key emerging industries.

Concerns about the 'hollowing out' of Taiwan's manufacturing industry because of the massive migration of firms to China have prompted a strategy to keep or attract firms and high-value activities to Taiwan. Examples of this strategy include the programme 'Investment in Taiwan First' and the 'Operation Headquarters Development Plan', which encourages enterprises to locate their operational and decision-making centres in Taiwan.

S&T policies are based on the consensus positions reached at the National Science and Technology Conference which are a major component of the four-year National Science and Technology Development Plan that established the guidelines for S&T development. A key player is the Science & Technology Advisory Group, which provides advice about policy directions, including more specific issues such as support for emerging high-tech industries and legal and regulatory frameworks. Traditionally, this high-profile group liaises with international advisers including directors and leading scientists and multinationals.

Institutional changes have occurred in industrial relations and intellectual property regimes. Industrial relations evolved as a consequence of a process of political democratization, which has led Taiwan from a Chinese-style system of state corporatism to a societal corporatism, similar to Japan or Germany (Chen et al., 2003). However, in contrast to Korea or Japan, where large industrial conglomerates employed a great proportion of the workforce, SMEs with fewer than 30 employees employed 65 per cent of the workforce in Taiwan (Chen et al., 2003). In these SMEs, family relationships tended to determine industrial relations, and so unionization did not play an important role. Moreover, with the rise of high-tech manufacturing, the competition for skilled engineers and researchers forced firms to develop individual strategies for attracting and maintaining human resources. These strategies were a necessary complement to the government effort to combat 'brain drain' discussed earlier (see Section 4.1.2).

The deepening of political democratization in Taiwan has begun to alter industrial relations. Recently, workers have demanded greater participation in management decision-making processes.[11] The effects of this democratization on the labour movement are uncertain, but it is clear that public policy decision making has become more complex and that it has affected the government's ability to deal with issues such as unemployment.

The intellectual property rights system in Taiwan involves significant contradictions. The main issues are resolving the tension between the competing interests of diffusion versus protection and local interest versus compliance with worldwide standards. Taiwan's membership in the World Trade Organization (WTO) has meant that its regulatory framework has moved closer to international standards. However, ensuring the enforcement of property rights has continued to be a major worry for European and American interests.

4.4 Support Services for Innovating Firms

4.4.1 Incubating activities

In 1996, Taiwan implemented an incubator programme under the guidance and support of the Small and Medium Enterprise Administration of the MOEA (Hsu et al., 2003). The programme has supported entrepreneurs at the earliest stage of the technological entrepreneurship process and helped them to implement their ideas commercially. The incubators provided entrepreneurs with physical premises, financial resources, tools, professional guidance and administrative assistance. To date, a total of 83 incubators have been established. Of these, 66 were set up inside university campuses, 15 within research institutes and local governments, and two in the private sector.[12] Incubators in Taiwan have been instrumental in getting access to pilot trials and prototyping – a critical factor for many entrepreneurs. Most incubators have provided all types of expertise in enterprise formation, development of business plans and market research, along with connections to the business community, specialized attorneys, accountants and venture capitalists. Successful products or inventions have been actively diffused in the business and academic communities, which has stimulated additional experimentation and research with a renewed focus on commercial application. While it may be still too early to judge if this government-sponsored incubator programme is successful, it seems to have reached a good take-off point.

4.4.2 Financing

As Taiwan developed, funding requirements changed and financial organizations became more specialized. Typically, the government provided funds in the early stage of business development, with commercial banks becoming involved only at a relatively late stage. Specialist banks might become involved slightly earlier than commercial banks. Although individual investors could provide funds at any stage, venture capital companies tended to become involved in the early to middle stages of development whereas the capital markets did not have a funding role until much later.

The government began to promote venture capital (VC) funds for high-tech industries in 1983, with policy coming into effect in 1985. Over the past 20 years, Taiwan's VC industry has grown and developed, raising about US$500 million in funds, which has been invested in over 8000 domestic and foreign high-tech companies in the semiconductor, information, telecommunications, electronics and optoelectronics industries. Today, around 200 VC funds operate in Taiwan and venture funds have aided over 300 companies to go public in both Taiwan and abroad. Nearly 50 per cent of all listed companies on the Taiwan Stock Exchange (TSE) and the over-the-counter market are venture-backed.

Data from the Taiwan Venture Capital Association indicate that 73.5 per cent of the VC funds in 2003 were invested in the stages of ramp-up and maturity, whereas seeding and start-up accounted for only 25.7 per cent of the VC funds (Saxenian and Li, 2003). Venture funds established by independent venture capitalists provided less than one-third of total new funds. Consequently, only a small percentage of venture funds (25 per cent of less than 33 per cent) in Taiwan are invested in the stages of 'seed' and 'start-up'.

Due to the shortage of venture capital, firms in Taiwan used other ways to expand and diversify their businesses, such as reinvestments or buying non-controlling shares in other companies. Reinvestments were a useful instrument for firms that missed out on the electronics boom and were not able to set up their own subsidiaries. This is illustrated by the fact that large business groups in the traditional manufacturing sector undertook many reinvestments in an attempt to profit from the electronics boom.

Finally, the TSE, the only centralized securities trading market in Taiwan, was opened for business in 1962. In 1986, the weighted average index reached 1000 points for the first time, with an average of 130 million shares traded every day, and a daily trading volume of US$60 million. In February 1990, the index reached an all-time high, soaring to 12495.34 points, with an average of 993 million shares traded daily, and US$3.3 billion in daily market turnover, making the TSE one of the busiest markets in the world. This rapidly growing stock market capitalization has played an important role in assisting public listed companies to raise cheap capital through public funding (Hung, 2003).

4.4.3 Provision of consultancy services

Consultancy services have a long history in Taiwan. The first consultancy services were associated with the mission to China by the US Agency for International Development. In the 1960s, the Stanford Research Institute and Arthur D. Little International consultants with specialists from the Chinese Petroleum Corporation and government agencies outlined the opportunities for petrochemical development (Balaguer, 2000). The China Productivity Centre, established in 1955 by the government, was created with the objective of disseminating modern management and technology for the enhancement of industrial productivity. It has been a pillar organization in productivity improvement and the incorporation of new techniques and standards, particularly in SMEs.

Consulting services have been effective in traditional industries. The textile industry emphasized the Taiwan Textile Federation as an organization that had a major impact in upgrading the industry (Balaguer, 2000, p. 199). The Taiwan Association of Machinery Industry, which was established for regulating exports to the USA, has become a leading organization in supporting machinery manufacturers to adopt new technologies. In technology-intensive industries, the government-sponsored institutes have been influential providers of consulting services. The role of ITRI is well known in the establishment of the IC industry (Hobday, 1995; Mathews, 1997). More recently, the III has actively promoted the development of software design through the establishment of a strategic alliance with IBM. The Development Centre for Biotechnology and Union Chemical Laboratories of ITRI has worked on developing generic drugs for transfer to the local pharmaceutical industry. Before the 1990s, most of the major knowledge-intensive business service (KIBS) providers in Taiwan were addressing MNCs. Not until recently did the government recognize the importance of KIBS and start to put efforts into cultivating knowledge-intensive service capabilities domestically. One of the most important mechanisms for providing relevant information to the industry has been the Industrial Technology Information Service that collects, researches and analyses worldwide market and technical information. This information is made available to local companies.

University research results have also become an important source of knowledge in the corporate innovation process. Consequently, Taiwan's NSC has allocated funding to help universities to organize technology transfer and licensing offices. These offices have served as intermediary units to link industrial companies to universities. They monitor the progress of university researchers in order to increase the possibility of subsequent patent filing and commercialization. A similar system of 'integrated' technology transfer offices has been organized with assistance from the Ministry of Education.

Recently, Taiwan's government has put tremendous effort into stimulating the R&D and intellectual property service system, which will assist firms in outsourcing particular areas of their R&D activities and industrial design, as well as collecting information on worldwide patents and trading intellectual properties. However, the government has yet to decide whether it should provide all of these services or leave some to the private sector.

5 CONSEQUENCES OF INNOVATION

Long-term labour productivity is one of the best indicators of how an economy introduces and diffuses innovation, as this process alters the output/worker ratio. Long-term labour productivity is related to the incorporation of new technology, the substitution of labour for capital and the improvement of the output/worker ratio via investment in human capital (for example, better schooled or trained workers). Data available from the Director-General of Budget, Accounting and Statistics on labour productivity between 1988 and 2003 show that Taiwan had a high rate of labour productivity, averaging 6.09 in this period. Dividing this period into two shows that there was little change in this value over the whole period: the values were 6.17 and 5.97 for 1988–95 and 1996–2003, respectively. Kaitila (2003) showed similar trends using data on average growth rates of GDP per hour for the periods 1980–90 and 1990–2001. Of the ten countries surveyed by Kaitila, Taiwan registered the fastest growth rate in the last period.

There are conceptual difficulties and data limitations involved in attempting to measure total factor productivity (TFP) growth, as pointed out by Mansfield (1990), among others. Nevertheless, TFP growth is interesting because it is not due to measured increases in inputs but instead measures productivity gains due to innovation, better technology, better organization, and specialization (World Bank, 1993). Another reason that TFP growth is of interest in the East Asian context is the debate on 'paper tigers' initiated by Krugman (1994). Krugman argued that 'The newly industrializing countries of East Asia, like the Soviet Union of the 1950s, have achieved rapid growth in large part through an astonishing mobilization of resources' (Krugman, 1994, p. 70). Krugman's point was that growth in inputs (capital and labour) was the main cause of productivity growth and that this type of growth could not be sustained for a long time. Krugman's argument has been hotly debated, and today there is little support for it. Taiwan's TFP growth indexes do not support Krugman's hypothesis. The World Bank Report, *The East Asian Miracle*, estimated that technological efficiency change in Taiwan was 0.8 per cent between 1960 and 1989, a value that indicates that Taiwan was catching up with world best practice (World Bank, 1993). The contribution of TFP to total growth was greater than 33 per cent, which is not far off the contribution in advanced economies (World Bank, 1993, pp. 57–8). Other studies have estimated even greater contributions from TFP. Hou and San (1993) pointed out that the so-called Solow residual (the rate of growth of TFP) accounted for 54 per cent of growth between 1952 and 1979. Singh and Trieu (1996) estimated that the TFP contribution to growth was 42 per cent between 1978 and 1990. Fu (2002) calculated that TFP contributed to 33 per cent of manufacturing growth, and Sun (2004), using a methodology that included market imperfections and net taxes, calculated that the TFP contribution to sectoral growth was more than 40 per cent during the period 1979–99 in a number of industries.

TFP includes productivity gains due to innovations, but there are no data available relating productivity growth to the type of innovation, depending on whether it is product or process innovation. In the Taiwanese case, it makes sense to speculate on the crucial role of process innovation in sectoral productivity growth. As discussed previously, there is no doubt that the strong growth and competitiveness of the OEM/ODM sector have resulted from the ability of Taiwanese firms to implement changes enabling them to cut costs and increase the scale of production very effectively. Many of these changes have been accompanied by the development and adoption of process innovations. As many of these were incremental in nature, it is unlikely that they were registered as R&D spending. As Wang and Tsai (2003) showed, there was no correlation (at the 5 per cent level of significance) between R&D spending and TFP growth in large Taiwanese manufacturing firms.

6 GLOBALIZATION

In contrast to Japan and Korea, Taiwan's industrial development has been characterized by the presence of substantial numbers of foreign MNCs, which have influenced market demand and technological learning. For example, in Taiwan's personal computer industry, there were two types of MNCs. The first type based their operational activities on the island; examples include AST in 1987 and DEC in 1993. The second type involved foreign buyers or vendors who relied on Taiwanese OEM or ODM firms for the supply of parts and integrated systems (Hung, 2003). Compared with the inward globalization activities, outward globalization has tended to be relatively low, particularly in R&D activity. International R&D activities pursued by nationally controlled organizations have been rare. This is partly because Taiwanese firms specialize in manufacturing and OEM/ODM production, and partly because diplomatic isolation and political tensions have discouraged international activity. Although offshore production has reached considerable levels, its primary goal has been to exploit resources rather than technological knowledge and skills.

However, in recent years, the emergence of the Chinese market has increasingly driven a considerable trend to outward globalization. This has been particularly significant for firms producing electronics and electrical appliances, which now account for more than 40 per cent of Taiwan's outward investment to China (Chen, 2004). In considering the impact of globalization as mainly derived from China, it is worth noting that Taiwan tends to be severely restricted by political factors. The rapid growth in

Chinese investment made by the Taiwanese since the late 1980s, coupled with the rise of Taiwanization in the island, seems to have strengthened the political tensions between the straits. This has led to greater state attention to, and control of, direct investment in China, despite the fact that pressure to join the WTO has resulted in China's policy shifting from being extremely rigid to being more flexible. This shift was evidenced in the 'no haste, be patient' announcement in 1996 to the 'aggressive opening, effective management' policy after 2001. Nevertheless, the entry of high-tech firms (e.g. the semiconductor wafer fabrication industry) into China has been strictly controlled, partly because of the partial dismantling of the Taiwanese state and partly because of increasing fear about a rising China. Thus Taiwan's policy attitude towards industrial problems related to the rise of China has remained conservative and defensive, rather than anticipatory and active.

In comparison with the state's reactive approach, the private sector has tended to be more entrepreneurially oriented, venturing rapidly into the Chinese market over the past 15 years or so. As a result, there has been a tug-of-war between the export-oriented sectors and the state regarding the initial entry into the Chinese market (Hung, 1999).

Now, we consider the effects of globalization on innovation. Globalization is considered to be a double-edged sword when analysing its effects on the development of Taiwanese industry. On the positive side, the arrival of the digital age, together with continued economic dependence on the US economy, has led Taiwan to develop into a global centre of IT systems design and manufacturing. The emergence of converging digital technology products, together with the destructive impact of the 1997 Asian economic crisis on Japan and Korea, has reinforced Taiwan's advantages in information technology. Further, the rise of China gives the Taiwanese a chance to continue competing on the cost and manufacturing sides.

However, during recent years, a growing China has become a source of competition that threatens to weaken Taiwan's ability to capture the returns on investment in innovation. For example, China's strong tendency towards piracy could endanger Taiwanese dominance in DVDs. Highly attentive to the successful model of the HSIP, the Chinese state has promoted many policies to actively attract high-tech businesses, particularly in semiconductors, thus creating concerns in Taiwan about the downfall of industry competitiveness.

This emphasis on increasing outward globalization does not deny that the Taiwanese state played a distinctive role as a regulator and leader of economic activities and systems of innovation (SI), as we have shown in this chapter. It follows that continuing diplomatic isolation, together with the limited significance of capital market internationalization (particularly in

China), has constrained Taiwanese firms' exploitation of international economic, social and financial resources. These firms have continued to rely heavily on the national frontier as a source of resources and legitimacy, particularly with the steady reinforcement of the Taiwanese system of production and innovation and the continued expansion of national financial systems. Consequently, the nation-state has remained the primary focus of resource dependence as well as the dominant regulating and policy-making agency. This, in turn, has continued to induce coherence in the functioning of the Taiwanese NSI as a whole in respect of innovation activity.

7 STRENGTHS AND WEAKNESSES OF THE SYSTEM AND INNOVATION POLICIES

7.1 Strengths and Weaknesses

Our analysis has shown that Taiwan's NSI has fostered a breed of large firms that have excelled in competing in mature high-tech markets. This type of firm has been the main engine of Taiwan's economic growth over the last decade. Such firms have gained a 'second-mover advantage' because of the capacity to increase the scale of production and to excel at production technology and management. These second movers have allowed Taiwan to specialize in OEM/ODM manufacturing and to create strong relationships with advanced and sophisticated industrial customers. This relationship has been a powerful channel for the introduction of new technology and skills as the multinational corporations develop new products and standards.

Even though scale has become a critical element for Taiwanese firms' competitiveness in international markets, flexibility and fast adjustment are two characteristics well embedded in Taiwan's NSI. Fast adjustment to a product life cycle has been a key characteristic of both large firms and SMEs in the electronics industry. This has allowed firms to handle abrupt crises flexibly.

In terms of competence building, Taiwan has developed an effective mechanism for technology diffusion and learning. The public sector has played a critical role in building up basic competences in strategic areas. This capability has been rapidly transferred to the private sector through spin-off ventures and by staff relocating to private firms. Large private firms also have a long tradition of building in-house capabilities, including training, attracting and retaining qualified staff, and in strengthening R&D. These mechanisms have facilitated the absorption of foreign technology, but also the development of endogenous technology.

Some mismatches and disadvantages remain in Taiwan's NSI. Although Taiwan is already an important producer of intellectual property in the world, its protection of such property (including enforcement when standards are infringed) does not match world standards. This will continue to have important implications for attracting FDI in high-tech industries, a policy that the government is actively promoting.

Taiwan lacks effective institutional forms of R&D collaboration that facilitate the appropriation of innovation. Private firms often complain about the difficulties involved in negotiating intellectual property rights, such as those related to patents or licensing agreements in partnerships. As government licensing of these patents was done almost entirely on a non-exclusive basis, many patents were not developed into commercial uses. Non-exclusive licensing does not give the industrial firms the required level of protection to justify the costs of development.

Moreover, Taiwan's NSI is generally weak when it comes to product innovation, marketing and distribution, all of which support the development and growth of successful brand names. Most Taiwanese companies are technology followers, undertaking only minor modifications to the latest product designs developed elsewhere. More than 60 per cent of Taiwan's exports are OEM/ODM products that are not associated with Taiwanese firms' own brand names. The value-added of an increasing number of businesses lies in the design, marketing and distribution stages of production, which are usually controlled by multinational firms.

Taiwan's OEM products are known for their low prices, but price competition in the industry and rising labour costs at home are squeezing profit margins for local manufacturers. Additionally, Taiwanese high-tech firms are encountering new constraints in accessing foreign technology as they approach the technological frontier in a number of sectors.

Although Taiwan's firms and individuals have shown a surprising capacity for industrial inventions (for example, there is a large number of Taiwanese patent holders in the USA), these inventions are rarely linked to basic knowledge or research. And Taiwan ranks poorly in relation to the OECD in terms of science linkage of the patent activity.[13] Finally, Taiwan's traditional industry is at a crossroads because the erosion of existing comparative advantages is irreversible. To overcome the problems of labour shortages and increasing wages, local firms are trying to increase their levels of factory automation and adopting advanced production techniques. However, most firms are attempting to move labour-intensive production offshore because the differential in labour costs between Taiwan and the mainland is significant. However, despite the large migration of SMEs to Mainland China, Taiwan's R&D intensity has not grown in its traditional industries. This means either that firms in traditional sectors are moving the little R&D they perform to China or that the research capability in those sectors in Taiwan is so limited that is not reflected in national statistics.

7.2 Summary and Evaluation of the Innovation Policy Pursued

In Taiwan, the governing principle of innovation policy has been upgrading. While in the past this was very much a matter of technological catching up, today innovation policy faces a more complex environment. The government's new vision of Taiwan as Green Silicon Island, which involves the concepts of a knowledge-based economy, a sustainable environment and a just society as guidelines of policy making, means in practice that innovation policy has become more difficult as it deals with many different issues and objectives.

Globalization is influencing Taiwanese innovation policy in two directions. The first direction is related to the rise of China as a manufacturing power and the increasing need for Taiwan to specialize in areas where complementarity rather than cost competition exists. The second direction aims to make Taiwan an 'international hub for R&D innovation'. In doing so, policies aim to attract international leading firms to set up long-term R&D regional centres, overseas highly qualified R&D personnel and establishing joint ventures with local firms.

The issues of balancing regional development and sustainability are also integral parts of new innovation policies, particularly those related to infrastructure. The massive investment in the extension of the (Hsinchu) Science Park model to Southern and Central Taiwan represents a serious attempt to create high-tech growth and employment in areas where traditional industries dominate. Furthermore, the 'Challenge 2008' National Development Plan, which aims to transform Taiwan into a biotech hub in the Asia-Pacific region, has put forward specific policies to upgrade agricultural counties by the use of biotechnologies and the establishment of agricultural biotechnology parks (NSC, 2005). The policies aiming to create a biotech hub in the region are not unique to Taiwan – Singapore has moved earlier than Taiwan in this kind of initiative (see Chapter 3 in this book). However, the agro-biotech focus rather than life sciences may offer grounds for regional specialization.

Taiwan innovation policy also includes targets. The ambitious 'Two Trillion, Twin Star' industry strategy aims by 2006 to generate a total output of NT$3.58 trillion (US$110 billion) in four high-tech sectors: semiconductors, display, digital content and biotechnology industries (IDIC, 2006).

Finally, the weak linkages between industry and academia and SMEs' suboptimal investment in R&D are continuous focuses of innovation policy. To the existing University & Industry Research Cooperation

Programme the government has added two new programmes, the Technology Development for Academia Programme and the Centre for Regional Industry & Academia Cooperation Programme, which attempt to increase cooperation and encourage technology transfer. In terms of R&D targets, the National Development Plan objective is to achieve the target of total R&D expenditure at about 3 per cent of GDP by 2008.

Coming back to the concept of SI and its application in public policy, can we legitimately talk about a Taiwan NSI as guiding principle for innovation policy action? Freeman's (1987) definition of NSI also relates to the national capacity to identify key technological areas and mobilize large resources in pursuing strategic priorities. In this (catching-up) sense, although not explicitly used in the Taiwan policy setting, the NSI was a *de facto* policy approach applied effectively by policy makers during the first three decades of Taiwan industrialization. Since the late 1990s the NSI approach has been applied by policy makers and academics more conscientiously in reference to issues of articulation and connectivity in SI such industry–academia links. However, in contrast to the Finnish and Swedish cases (see Chapters 10 and 7), the NSI approach has been mainly a tool for policy criticism rather than policy design or implementation.

7.3 Future Innovation Policy

Taiwan's track record of sound policy making for industrial innovation represents a solid foundation, but is by no means a guarantee of future success. Increasing competition and rapid technological change represent both threats and opportunities for the renewal of the Taiwanese NSI.

Considering the National Plan 'Challenge 2008' and the conclusions of the last National Conference on S&T as a main policy framework, six main themes are likely to dictate the future of innovation policy in Taiwan. The first theme is related to the quality of interactions within the NSI, particularly between industry and university. Policies encouraging industry–academia cooperation aim to promote industrial innovation. These policies are particularly concerned with overcoming the existing weak link between industrial inventions and basic research, as we discussed in Section 7.1. Recent policy changes also focus on loosening legal restrictions to allow universities more freedom to collaborate with industries.

Although policies seem to have acknowledged this situation with the development of new programmes that deal explicitly with industry–academia linkages, some of these programmes not only have very limited funding but they have also been decreasing in the last few years. The National Science and Technology Programmes, on the contrary, which attempt only indirectly to foster industry–academia links, are much better supported and will probably have a more direct impact on university–industry links at least in some selected fields.

A second theme is the increasing emphasis on upgrading academic excellence in Taiwanese universities. The great expansion of universities has raised doubts about their quality and led to calls for the government to direct future educational development. Recent reforms in the universities have centred on the integration of universities and the development of specialty areas. The government has initiated several projects in pursuit of world-class universities and high-calibre research centres as well as policies to cultivate and attract local and foreign talent.

However, rules for hiring foreign professionals are inflexible and not advantageous; for example tax treatment is not attractive for a period of less than six months, and salary packages are considerably inferior to those in advanced countries. Although much lip service has been paid to the need to attract world-class researchers and university professors, the existing legislation and the salary level do not represent an appropriate incentive to attract high-calibre professionals from all around the world. The successes of IC industry in attracting Taiwanese overseas engineers in the 1990s have not applied to universities and research institutes. Culture and family links are still the main reasons why many Taiwanese with overseas experience come back to Taiwan research and teaching organizations. Hong Kong and Singapore, for example, have been much more successful in creating more international academic and research communities (see Chapters 6 and 3 in this book).

The third theme is the design of policies to face the 'hollowing-out' effects of the emerging Brazil, Russia, India and China bloc. This theme concerns innovation policy because Taiwan is increasingly forced into competition based on innovation. The Taiwan government has developed a comprehensive investment strategy to attract international firms recognized for their innovation capabilities. The plan has shown some success, particularly in the display and optoelectronics sector, where some important international players have located facilities in Taiwan's new science parks.

The fourth theme is policies for developing a KIBS sector.[14] In an economy with a strong tradition of OEM manufacturing competitiveness, innovative services' firms are lacking, as we have shown in Section 3, particularly in comparison with Scandinavian countries. Policy emphasis has been on the extending of the use of e-business and upgrading network technologies in SMEs, but the government and large firms are still important providers of infrastructure-related KIBS. High growth (8.1 per cent) of KIBS was registered between 1988 and 2001;[15] however, KIBS are not ubiquitous activities in the Taiwan economy but a phenomenon related to

ICT goods manufacturing and infrastructure development. The promotion of KIBS should be a key policy for the creation of new business opportunities and skilled jobs.

The fifth policy theme is the priority of S&T and innovation policy in Taiwan. While, in the past, technology absorption, diffusion and the accumulation of capabilities in the industry were core elements of economic and industrial policy at the highest level, at present Taiwan lacks a well-articulated policy approach where innovation could be promoted in a highly strategic fashion from a top executive level. This contrasts, for example, with the Korean experiences of the creation of the Office for Science, Technology and Innovation at very high executive level and the appointment of the Minister of Science and Technology as vice prime minister.

The final theme is sustainability and environmental performance. After six years of the 'Green Silicon Island' vision, Taiwan is performing far better as a silicon rather than as a green island. This is partly because of the lack of appropriate policy. One example is energy policy, where Taiwan has not been able to develop a coherent policy framework for stimulating new sectors based on renewable energy, albeit, market conditions seem to be very favourable. Taiwan, although not a signatory to the Kyoto Protocol, is under strong pressure to reduce CO_2 emissions. This strong need to become a more sustainable economy plus its existing strength in innovative manufacturing may represent a golden opportunity for the renewal of the Taiwan NSI.

ACKNOWLEDGEMENTS

This research was funded by Science & Technology Policy Research and Information Centre, National Applied Research Laboratories, Taipei, Taiwan.

NOTES

1. The important technology-based industries designed by the government were: (1) communications, (2) computers, (3) consumer electronics, (4) integrated circuits, (5) precision machinery, (6) aircraft, (7) advanced materials, (8) special chemicals and pharmaceuticals, (9) healthcare, and (10) pollution control. For details on tax incentives, see Article 8 of the Statute for Upgrading Industries.
2. TTIS industry disaggregation is at the two-digit level, so comments refer to the most representative industries.
3. Defence R&D is included.
4. The worldwide rankings are derived from unpublished CWTS (the Centre for Science and Technology Studies) data.

5. This is a patent issued for inventions that perform useful functions. Most inventions fall into this category. A utility patent lasts for 20 years from the patent application's filing date.
6. This means 3.5 times more specialized than the world level.
7. This is a patent issued on a new design, used for purely aesthetic reasons, that does not affect the functioning of the underlying device. Design patents last for 14 years from the date the patent is issued.
8. Source: Ministry of Education (2003a), p. 39.
9. Source: http://210.69.121.6/gnweb/statistics/statistics01/reports/E04.xls (last visited on 12 December 2004).
10. Rather than imposing import restrictions and high tariffs on high-tech components, the government undertook and facilitated strategic R&D programmes into the substitution of critical imported products and components. Examples of these programmes were: (1) Development of New Industrial Products Programme (DNIP) 1984–91, (2) Development of Targeted Leading Products Programme (DTLP) 1991–present and (3) Development of Critical Components and Products (DCCP) programme 1992–present.
11. Between 1997 and 2002, the number of labour management committees in Taiwan almost tripled, rising from 1013 committees in 1997 to 2701 committees in 2002.
12. Source: http://www.moeasmea.gov.tw/ (last visited on 30 May 2005).
13. The science linkage (SL) is a measure of the number of citations to the scientific literature in a patent. This measure provides an indicator of the dependence of a patent on the scientific research base.
14. There are obvious difficulties in a precise definition of KIBS that apply to the Taiwan case.
15. Source: http://www.cepd.gov.tw/encontent/index.jsp (last visited on 30 May 2005).

REFERENCES

Amsden, A.H. and W.-W. Chu (2003), *Beyond Late Development: Taiwan's Upgrading Policies*, Cambridge, MA: The MIT Press.
Asian Development Bank (1998), *Asian Development Outlook 1998*, Oxford: Oxford University Press.
Balaguer, A. (2000), *Learning and Growth in Organized Markets: A Commodity Chain Perspective of Petrochemical Development in Taiwan*, PhD dissertation, Perth, Australia: Murdoch University.
Breznitz, D. (2005), 'Innovation and the limit of state power: integrated circuit design and software in Taiwan', in S. Berger and R.K. Lester (eds), *Global Taiwan: Building Competitive Strengths in a New International Economy*, Armonk, NY: M.E. Sharpe, pp. 194–227.
CCIS (China Credit Information Services) (1981), *Top 500: The Largest Corporations in Taiwan*, Taipei.
CCIS (China Credit Information Services) (2004), *Top 500: The Largest Corporations in Taiwan*, Taipei.
CEPD (Council for Economic Planning and Development) (1997), *Taiwan Statistical Data Book*, Taipei: CEPD–Executive Yuan.
Chang, S.-L. (1992), 'Causes of brain drain and solutions: the Taiwan experience', *Comparative International Development*, **27**(1), 27–43.
Chen, S.-H. (2004), 'Taiwanese IT firms' offshore R&D in China and the connection with the global innovative network', *Research Policy*, **33**, 337–49.
Chen, S.-J., J.-J. Ko and J. Lawler (2003), 'Changing patterns of industrial relations in Taiwan', *Industrial Relations*, **42**(3), 315–40.

Dedrick, J. and K.L. Kraemer (1998), *Asia's Computer Challenge: Threat or Opportunity for the United States and the World?*, New York: Oxford University Press.

Freeman, C. (1987), *Technology and Economic Progress: Lessons from Japan*, London: Pinter.

Fu, T.-T. (2002), *Total Factor Productivity Growth: Survey Report*, Tokyo, Japan: Asian Productivity Organization.

Gold, T.B. (1988), 'Colonial origins of Taiwan capitalism', in E.A. Winckler and S.M. Greenhalgh (eds), *Contending Approaches to the Political Economy of Taiwan*, Armonk, NY: M.E. Sharpe, pp. 101–17.

Hobday, M. (1995), *Innovation in East Asia: The Challenge to Japan*, Aldershot, UK and Brookfield, USA: Edward Elgar.

Hou, C.-M. and G. San (1993), 'National systems supporting technical advance in the industry: the case of Taiwan', in R.R. Nelson (ed.), *National Innovation Systems: A Comparative Analysis*, Oxford: Oxford University Press, pp. 384–413.

Hsu, P.-H., J.Z. Shyu, H.-C. Yu, C.-C. You and T.H. Lo (2003), 'Exploring the interaction between incubators and industrial clusters: the case of ITRI incubator center in Taiwan', *R&D Management*, **33**(1), 79–90.

Hung, S.-C. (1999), 'Policy system in Taiwan's industrial context', *Asia Pacific Journal of Management*, **16**(3), 411–28.

Hung, S.-C. (2000), 'Institutions and systems of innovation: an empirical analysis of Taiwan's personal computer competitiveness', *Technology in Society*, **22**, 175–87.

Hung, S.-C. (2002), 'On the co-evolution of technologies and institutions: a comparison of Taiwanese hard disk drive and liquid crystal display industries', *R&D Management*, **32**(3), 179–90.

Hung, S.-C. (2003), 'The Taiwanese system of innovation in the information industry', *International Journal of Technology Management*, **26**(7), 788–800.

Hung, S.-C. and R. Whittington (1997), 'Strategies and institutions: a pluralistic account of strategies in the Taiwanese computer industry', *Organization Studies*, **18**(4), 551–75.

Hwang, G.-Y. (2001), 'R&D trend in Taiwan during the 1990s', *Benchmarking Sci-Tech Development*, **1**(2), 9–15 (in Chinese).

III (Institute for Information Industry) (2004), *Yearbook of Information Industry*, Taipei: Market Intelligence Centre, Institute of Information Industry.

IDIC (Industrial Development and Industrial Centre) (2006), *Beyond all Investment Choices: The Investment Environment of the ROC on Taiwan*, Taipei, IDIC, MOEA.

Kaitila, V. (2003), 'GDP and productivity indicators for small European and Asian economies, and large reference countries', paper presented at the European Science Foundation/Eurocores Taipei Workshop (September), Taipei, Taiwan.

Krugman, P. (1994), 'The myth of Asia's miracle', *Foreign Affairs*, November/December, 62–78.

Lin, C.-Y. (1973), *Industrialization in Taiwan: 1946–72*, New York: Praeger.

Lin, L., M.-Y. Hsu, F.-S. Wu and M.-M. Huang (2001), *The Industrial Impacts of Patent Protection*, research project report of the effects of patent protection, Taipei: Taiwan Intellectual Property Office (in Chinese).

Mansfield, E. (1990), 'Comment on Griliches, Zvi, and Jacques Mairesse, "R&D and productivity growth: comparing Japanese and U.S. manufacturing firms"', in C.R. Hulten (ed.), *Productivity Growth in Japan and the United States*, Chicago, IL: University of Chicago Press.

Mathews, J.A. (2005), 'Strategy and the crystal cycle', *California Management Review*, **47**(2), 6–32.

Ministry of Education (2003a), *Education Statistics of the Republic of China*, Taipei.

Ministry of Education (2003b), *Educational Indicators: An International Comparison*, Taipei.

NSC (National Science Council) (2001), *Indicators of Science and Technology, Republic of China*, Taipei.

NSC (National Science Council) (2003), *Taiwan Technological Innovation Survey*, Taipei (in Chinese).

NSC (National Science Council) (2005), *Yearbook of Science and Technology, Taiwan, ROC*, Taipei.

OECD (2002), *Education at a Glance*, Paris: OECD.

OECD (2004), *Main Science and Technology Indicators*, Paris: OECD.

Poon, T.S.C. (1994), *Comparing the Subcontracting System: Towards a Synthesis in Explaining Development in Hong Kong and Taiwan*, mimeo, Perth: Asia Research Center, Murdoch University.

San, G. (1995), *Technology Support Institutions and Policy Priorities for Industrial Development in Taiwan, R.O.C.*, Taipei: Chung-Hua Institution for Economic Research (CIER) (in Chinese).

Saxenian, A. and Y.-J. Hsu (2001), 'The Silicon Valley–Hsinchu connexion: technical communities and industrial upgrading', *Industrial and Corporate Change*, **10**(4), 893–920.

Saxenian, A. and C.-Y. Li (2003), 'Bay-to-bay strategic alliances: the network linkage between Taiwan and the US venture capital industries', *International Journal of Technology Management*, **25**(1/2), 136–64.

Shieh, G.S. (1990), 'Manufacturing "bosses": Subcontracting Networks under Dependent Capitalism in Taiwan', unpublished PhD thesis, Berkeley, CA: University of California.

Shieh, C.S. (1992), *The Boss Island: The Subcontracting Network and Micro Entrepreneurship*, New York: Peter Lang.

Singh, N. and H. Trieu (1996), *The Role of R&D in Explaining Total Factor Productivity Growth, in Japan, South Korea, and Taiwan*, Working paper no. 361, Department of Economics, University of California, Santa Cruz.

Su, J.-C. (1995), 'The factor analysis of brain inflow and the effect of technology in Hsinchu science-based industrial park', unpublished thesis, National Central University, Taiwan (in Chinese).

Sun, C.-H. (2004), 'Market imperfection and productivity growth: alternative estimates for Taiwan', *Journal of Productivity Analysis*, **22**(1), 5–27.

Tsay, C.-L. and J.-P. Lin (2003), 'Return migration and reversal of brain drain to Taiwan: an analysis of the 1990 census data', in R. Iredale, C. Hawksley and S. Castles (eds), *Migration in the Asia Pacific*, Cheltenham, UK and Northampton, MA, USA: Edward Elgar, pp. 273–92.

Tung, A.-C. (2001), 'Economic development and trade – Taiwan's semiconductor industry: what the state did and did not', *Review of Development Economics*, **5**(2), 266–88.

Wang, J.-C. and K.-H. Tsai (2003), 'Productivity growth and R&D expenditure in Taiwan's manufacturing firms', NBER Working Paper No. 9724, Cambridge, MA: National Bureau of Economic Research.

Whitley, R.D. (1991), 'The social construction of business systems in East Asia', *Organization Studies*, **12**(1), 1–28.
World Bank (1993), *The East Asian Miracle: Economic Growth and Public Policy*, New York: Oxford University Press.
Wu, R.-I. and H.-Y. Lin (2004), 'Mapping Taiwan's competitiveness of R&D and innovation', International Conference on 2004 Industrial Technology Innovation: Growth Engines of the Innovation-driven Economy, 19–20 August, Taipei.

3. From technology adopter to innovator: Singapore
Poh Kam Wong and Annette Singh

1 INTRODUCTION

As a newly industrialized nation, Singapore has been very successful in developing its technological capability in the past 40 years since political independence. This success has been based on evolving a national system of innovation (NSI) that emphasized attracting and leveraging global multinational corporations (MNCs) to transfer increasingly advanced technological operations to Singapore, and developing infrastructure and human resources to absorb and exploit new technologies rapidly. In the last decade or so, however, the country has started to shift towards a more balanced approach, with increasing emphasis on developing indigenous research and development (R&D) and innovation capability. While the government has acted as a 'developmental state' in guiding science and technology (S&T) capability development as an integral part of Singapore's overall economic development strategy, the emergence of a more vibrant technology-entrepreneurial community is likely to be critical to Singapore's continuing transition from technology adopter to innovator.

2 MAIN HISTORICAL TRENDS

Among developing economies, Singapore has achieved one of the most impressive economic growth records in the last four decades since its political independence in 1965, averaging 7 per cent GDP growth per annum over the 1960–2005 period (Table 3.1). Despite an economic slowdown in 2001–3 (with a strong recovery in 2004), Singapore's per capita GDP of US$29 111 in 2005 (measured as purchasing power parities (PPP)) is still the third-highest in Asia, at about 70 per cent of the US level (IMD, 2005). Singapore's 2004 per capita GDP (on a PPP basis) was not only higher than that of Korea and Taiwan, but also than that of some small advanced European countries, such as Finland, the Netherlands and Sweden (see

Table 3.1 Aggregate economic growth performance, 1960–2005

	% real growth p.a.					
	1960–70	1970–80	1980–90	1990–2000	2000–2004	2000–2005
GDP	8.7	9.4	7.1	7.5	3.3	4.0
Labour productivity	n.a	4.3	4.8	3.4	2.1	2.6

	S$ at current prices					
GNI per capita[1]	1970	1980	1990	2000	2004	2005
	2820	9900	20100	39600	41500	42983

Note: [1] GNP per capita before 1997.

Sources: Calculated from Department of Statistics, *Yearbook of Statistics Singapore* (various years); Ministry of Trade & Industry, *Economic Survey of Singapore* (various years). Per capita GNI obtained from Singstat website http://www.singstat.gov.sg/keystats/hist/gnp.html. Mid-year population estimate for 2000 obtained from Singstat website, http://www.singstat.gov.sg/FACT/KEYIND/keyind.html

Appendix Table A2.3). In considering broader measures of development as captured by the Human Development Index, Singapore ranked a reasonably high 25th out of 177 countries, although it was second lowest of all the countries in this study (see Appendix Table A1.2).

The rapid economic growth of Singapore has been achieved through continuous industrial restructuring and technological upgrading. In the first decade following independence, growth was led largely by labour-intensive manufacturing. In the two subsequent decades, it was propelled by the growth of increasingly technology-intensive manufacturing activities by foreign MNCs, with high-technology products contributing an increasing share of total value added (see Appendix Table A2.1). The development of Singapore into an increasingly important business, financial, transport and communications services hub in the Asia-Pacific region has provided additional engines of growth since the 1980s (Table 3.2). Nevertheless, manufacturing has remained important to the economy, with its share of GDP remaining above 25 per cent for most years in the last two decades. Since the 1990s, knowledge-intensive business services (KIBS) and manufacturing have become the key drivers of Singapore's economic growth.

The development and growth of Singapore's NSI has been strongly influenced by its overall economic development strategy. As highlighted by Wong (2003), the evolution of Singapore's NSI over the last four decades can be analysed as proceeding through four phases (see Figure 3.1)

Table 3.2 Singapore's GDP distribution by sectors, 1960–2005 (%)

Industry	1960	1970	1980	1990	2000	2004	2005
Agriculture & mining	3.9	2.7	1.5	0.4	0.1	0.1	0.1
Manufacturing	11.7	20.2	28.1	28.0	25.9	28.6	27.8
Utilities	2.4	2.6	2.1	1.9	1.7	1.7	1.8
Construction	3.5	6.8	6.2	5.4	6.0	4.5	3.9
Commerce	33.0	27.4	20.9	16.3	19.1	16.8	17.5
Transport & communication	13.6	10.7	13.5	12.5	11.1	11.4	14.7
Financial & business services	14.4	16.7	18.9	25.5	25.3	24.8	23.4
Other services	17.6	12.9	8.7	9.9	10.9	12.1	10.9
Total	100.0	100.0	100.0	100.0	100.0	100.0	100.0

Note: Figures may not add up to 100 due to rounding.

Sources: Calculated from Department of Statistics, *Yearbook of Statistics Singapore* (various years); Ministry of Trade & Industry, *Economic Survey of Singapore* (various years).

1. The industrial take-off phase (from 1965 to the mid-1970s): this period was characterized by high dependence on technology transfer from foreign MNCs to establish Singapore as a labour-intensive offshore manufacturing base in South-East Asia.
2. Local technological deepening (from mid-1970s to late 1980s): this period was characterized by rapid growth of local process technological capabilities brought about by new and upgraded MNC operations in Singapore, and the concomitant emergence of a critical base of local supporting industries in precision engineering and components assembly.
3. Applied R&D expansion (from late 1980s to late 1990s): this period was characterized by the rapid expansion of applied R&D activities by global MNCs in Singapore, alongside the establishment and growth of new public R&D institutions geared primarily to support MNC product and process innovation activities.
4. Shift towards high-tech entrepreneurship and basic R&D (from the late 1990s onwards): this period is characterized by the emerging emphasis on indigenous technological innovation capabilities, the formation of local high-tech start-ups, and an increasing shift towards basic R&D and the development of new science-based industries, particularly those related to life sciences.

Like Korea and Taiwan, Singapore has achieved significant technological capability development over the last 40 years. However, unlike Korea and Taiwan, Singapore's technology development was, until recently,

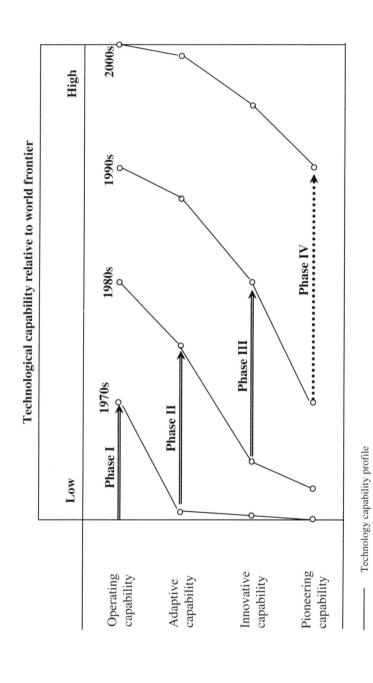

Figure 3.1 Stylized profile of technological capabilities of Singapore over the four phases of the NSI development

Table 3.3 Comparative R&D indicators, Singapore and selected OECD/Asian NIEs

Grouping	Country	Year	R&D/GDP (%)	Researchers per 10 000 labour force
G-5	Japan	2004	3.2	102
	Germany	2005	2.5	65
	USA	2004	2.7	91[1]
	UK	2004	1.7	56[2]
	France	2004	2.1	73
Industrialized small countries	Finland	2005	3.5	150
	Switzerland	2004	2.9	58
	Sweden	2005	3.9	117
	Ireland	2005	1.3	55
	Netherlands	2004	1.8	45[3]
	Denmark	2004	2.5	91
	Norway	2005	1.5	91
	Australia	2004	1.8	79
	New Zealand	2003	1.2	76
Asian NIEs	Korea	2005	3.0	76
	Taiwan	2004	2.4	71
	Hong Kong	2002	0.6	n.a.
	Singapore	1996	1.4	56
	Singapore	2000	1.9	66
	Singapore	2005	2.4	99

Notes:
[1] 2002 figure.
[2] 1998 figure.
[3] 2003 figure.

Sources: OECD (2006a; 2006b) and various national sources.

largely dependent on foreign MNCs rather than on indigenous companies like the *chaebols* in Korea and small and medium-sized firms in Taiwan (see Lim, Chapter 4 and Balaguer et al., Chapter 2, this volume). About three-quarters of Singapore's manufacturing output in recent years came from MNCs, and more than 60 per cent of equity in its manufacturing sector was foreign (Wong, 2003). Technology transfer from MNCs was therefore the major source of technological upgrading in Singapore for much of the industrial catch-up, not indigenous R&D. Publicly funded R&D was also on a smaller scale than in Taiwan and Korea throughout the 1980s and 1990s (Table 3.3).

Public policy has played a major role in influencing the development dynamics of Singapore's NSI. Before 1990, the Singapore government had put little emphasis on the development of indigenous R&D capabilities. This orientation changed significantly in the 1990s, with the establishment of a National Science and Technology Board (NSTB) and the launching of two five-year National Technology Plans (NTPs). Besides significantly increasing the scope and level of public R&D activities, these plans also channelled significant resources to building up R&D infrastructures and providing incentives to attract private sector R&D. In the early 2000s, the pace of S&T development has been further intensified, supported by a stronger commitment to R&D funding by the government through the Third National Science and Technology Plan (NSTP) 2001–5, and an increasing policy focus on high-technology entrepreneurship and basic research, particularly in life sciences. In 2006, the government further intensified the level of public funding of long-term strategic R&D with the launch of the National Research Foundation (NRF).

From a historical perspective, Singapore's NSI can thus be described in terms of a transformation from a primary emphasis on technology adoption – particularly the assimilation and diffusion of technology through leveraging inward MNC investments – to a more balanced approach that involves significant promotion of indigenous innovation capability, including the creation of local high-technology firms (Wong, 2003; 2006).

3   INNOVATION INTENSITY

Only one national innovation survey based on the Oslo Manual has been conducted in Singapore so far, covering manufacturing and selected KIBS branches in 1999 (Wong et al., 2003; Wong and Singh, 2004). The survey showed that only about one-third (32 per cent) of manufacturing companies in Singapore were innovating, having introduced new products/processes within the previous three years, whereas more than half of the companies (57 per cent) within the KIBS sector were innovating (Table 3.4). In the manufacturing sector, product innovations (introduced by 24.1 per cent of companies) were marginally more common than process innovations (22.4 per cent), whereas the reverse was true for the KIBS sector, where 44.4 per cent of firms had introduced product innovations versus 49.4 per cent for process innovations (Table 3.5). The important role of the electronics/information and communication technology (ICT) cluster in the Singapore NSI was evident in the sectoral breakdown of innovating companies. The electronics industry had the highest incidence of innovating companies within the manufacturing sector (68.8 per cent), while the

Table 3.4 Proportion of innovating companies in Singapore manufacturing and KIBS sectors, 1999

	Innovating companies (%)
Manufacturing	31.7
Electronics	68.8
Chemicals	38.0
Precision & process engineering	28.5
Transport engineering	18.2
KIBS	56.9
IT & related services	73.0
Market research, business & management consultancy	58.0
Architectural, engineering, land surveying, other technical	40.0
R&D, advertising, publishing, exhibitions & conferences	70.0

Source: Survey data from the Singapore National Innovation Survey: Manufacturing; and the Singapore National Innovation Survey: Knowledge-Intensive Business Services.

Table 3.5 Incidence of product and process innovation in the Singapore manufacturing and KIBS sectors, 1999 (% of companies)

	Product innovation	Process innovation
Manufacturing	24.1	22.4
KIBS	44.4	49.4

Source: As for Table 3.4.

information technology (IT) services industry had the highest incidence of innovating companies within the KIBS sector (73 per cent).

The levels of innovation intensity, measured in terms of expenditure on innovation activities as a percentage of total sales, were still relatively modest for most of the manufacturing firms, with over half of the firms that engaged in some form of innovation activities having innovation intensity below 5 per cent. There was, however, a small core (just over 9 per cent) of highly innovation-intensive firms that invested over 20 per cent of their revenues in innovation activities. Interestingly, KIBS firms appeared to exhibit higher average innovation intensities than manufacturing firms (Table 3.6).

The predominant innovation activity in Singapore in 1999 involved adopting innovations previously introduced to the market rather than

Table 3.6 Innovation and R&D intensities of innovating firms, 1999

	Total expenditure on innovation as a percentage of sales					
	<2%	2%–4.9%	5%–9.9%	10%–19.9%	20%–39.9%	≥40%
Manufacturing	28.2	26.4	20.9	15.5	7.3	1.8
KIBS	18.5	10.5	18.3	21.5	21.2	9.9
	Total expenditure on R&D as a percentage of sales					
	<1%	1%–2.9%	3%–4.9%	5%–9.9%	10%–19.9%	≥20%
Manufacturing	36.1	25.0	20.8	9.7	4.2	4.2
KIBS	8.7	19.4	15.9	12.3	25.3	18.4

Source: As for Table 3.4.

Table 3.7 Innovation activities engaged by companies, 1999 (% of innovating firms)

	Manufacturing	KIBS
R&D	66.4	43.3
Acquisition of R&D services	18.3	11.5
Acquisition of machinery & equipment	80.2[1]	31.1
Acquisition of software, external technology	22.4[2]	61.5
Industrial design, market research & marketing expenses for innovations	37.9	n.a.
Preparations to introduce new or significantly improved services or methods to deliver them	n.a.	75.7
Training	61.2	58.3
Market introduction of innovations	n.a.	50.0
Adoption of e-commerce applications	n.a.	46.7

Notes:
[1] Acquisition of machinery, equipment and software.
[2] Licensing of external technology.

Source: As for Table 3.4.

developing new ones. The most common innovation activity for manufacturing innovators was the acquisition of machinery, equipment and software (80 per cent of innovating manufacturers), while for KIBS firms it was acquisition of software or other external technology (61.5 per cent of innovating KIBS firms) (Table 3.7). This highlights one key characteristic of Singapore's NSI at the end of 1990s: adoption and assimilation of new

Table 3.8   New/improved products/services as percentage of total sales, 1999

	<10%	10%–24%	25%–49%	50%–74%	≥75%
Manufacturing	32.3	28.8	19.3	7.2	12.4
KIBS	21.7	25.3	20.5	14.5	18.1

Source: As for Table 3.4.

technology was relatively advanced, but the creation of technology was still somewhat lacking. Indeed, the level of R&D intensity was found to be relatively low even among companies engaging in innovation activities. In the manufacturing sector, more than one-third of innovating companies spent less than 1 per cent of their sales revenue on R&D activities, and fewer than one in five spent more than 5 per cent of their sales revenue on R&D (Table 3.6). The innovating KIBS firms exhibited higher R&D intensities in this regard, with almost 20 per cent spending at least 20 per cent of their turnover on R&D.

KIBS firms also seemed to exhibit a higher propensity in commercializing their innovations, with more than half deriving at least one-quarter of their turnover from new/improved services introduced within the previous three years. The comparable figure for manufacturing innovators was only 39 per cent (Table 3.8).

The prevalence of innovation among manufacturing firms in Singapore was found to lag behind the small advanced European economies, except Norway, as well as Taiwan, although it is comparable to that of Korea (see Appendix Table A4.4). Although the six European countries are shown as having lower shares of innovative KIBS firms (49–52 per cent) than in Singapore (Appendix Table A4.4), it should be noted that the figure for Singapore may be biased upward, since the coverage of KIBS firms in the Singapore survey appeared to be narrower.

4   ACTIVITIES THAT INFLUENCE INNOVATION

4.1   Knowledge Inputs to Innovation

4.1.1   R&D activities

As mentioned earlier, R&D in Singapore was minimal until the late 1980s, with a gross expenditure of R&D (GERD) to GDP ratio of only 0.86 per cent in 1987 (Table 3.9), significantly below the norm of advanced countries. Since then, however, R&D investment intensity in Singapore has increased significantly, with GERD experiencing a twelve-fold increase between 1987

Table 3.9 Growth of R&D in Singapore, 1978–2005

Year	GERD (S$ m)	GERD/GDP (%)	RSEs	RSE/10 000 labour force
1978	37.80	0.21	818	8.4
1981	81.00	0.26	1 193	10.6
1984	214.30	0.54	2 401	18.4
1987	374.70	0.86	3 361	25.3
1990	571.70	0.85	4 329	27.7
1991	756.80	1.01	5 218	33.6
1992	949.50	1.17	6 454	39.8
1993	998.20	1.06	6 629	40.5
1994	1 174.98	1.09	7 086	41.9
1995	1 366.55	1.15	8 340	47.7
1996	1 792.14	1.38	10 153	56.3
1997	2 104.56	1.49	11 302	60.2
1998	2 492.26	1.82	12 655	65.5
1999	2 656.30	1.90	13 817	69.9
2000	3 009.52	1.88	14 483	66.1
2001	3 232.68	2.10	15 366	72.5
2002	3 404.66	2.15	15 654	73.5
2003	3 424.47	2.13	17 074	79.4
2004	4 061.90	2.25	18 935	86.7
2005	4 582.21	2.36	21 338	90.1
Compound average growth rate per annum (%)				
1978–1990	25.4		14.9	
1990–1995	19.0		14.0	
1995–2000	17.1		11.7	
2000–2005	8.8		8.1	

Sources: National Survey of R&D Expenditure and Manpower (various years), Science Council of Singapore (before 1990); *National Survey of R&D in Singapore* (various years), National Science & Technology Board (for 1990–2000) and Agency for Science, Technology & Research (2001–5).

and 2005, and the GERD/GDP ratio more than doubling to reach 2.4 per cent in 2005. Although still behind many advanced OECD countries and Korea, Singapore's GERD/GDP ratio has overtaken more than a dozen OECD countries, and is now at parity with Germany (Table 3.3).

While both the public and private sectors contributed to this rapid increase in R&D intensity in Singapore, Table 3.10 shows that private sector R&D grew faster from the mid-1980s to the mid-1990s. Foreign companies in particular played a key role, as an increasing number of global MNCs,

Table 3.10   R&D expenditure by sectors, 1978–2005 (%)

Year	Private sector	Higher education sector	Government sector	Public research institutes	Total
1978	67.5	21.7	10.8	n.a.	100.0
1985	54.6	30.0	15.4	n.a.	100.0
1990	49.8	32.5	17.7	n.a.	100.0
1995	62.0	15.8	10.7	11.6	100.0
2000	61.6	12.3	12.0	14.1	100.0
2004	63.8	10.5	10.9	14.9	100.0
2005	66.2	9.7	10.4	13.8	100.0

Sources: As for Table 3.9.

many of which already had prior manufacturing operations in Singapore, started to establish R&D activities as well. Some of the more technology-intensive local firms also started to invest in applied R&D, particularly a number of government-linked companies (GLCs) established by the government in 'strategic' defence-related high-tech industries (Wong, 2003).

After reaching a peak of 64.5 per cent in 1997, the share of private sector R&D in total GERD had stabilized at around 61–3 per cent, although it rose again to 66 per cent in 2005. The balance of GERD is split fairly evenly between the three public R&D sectors: higher education, public research institutes/centres (PRICs) and various government sectors. Until 1991, Singapore's public R&D had been concentrated in the higher education sector. In 1991, with the formation of the NSTB and the launch of the nation's first NTP, a number of new PRICs were established, while some existing R&D centres in a number of ministries were reorganized as PRICs under the NSTB. R&D spending by these PRICs has grown rapidly since 1991 and now exceeds the R&D spending by the higher education sector (Tables 3.9–3.11).

Until the late 1990s, most R&D in both the public and private sector of Singapore focused on incremental, applied work. As can be seen from Table 3.11, basic research represented only 16 per cent of total R&D expenditure in Singapore in 1993 (the first year when data became available), with applied R&D and experimental development accounting for 39 per cent and 45 per cent respectively. The share of basic R&D actually declined a little in the period 1993–2000, due to the higher growth of private sector and PRIC R&D that had greater emphasis on applied work. With the possible exception of the Institute of Molecular and Cell Biology, the PRICs set up by NSTB in the 1990s all focused initially on conducting R&D to complement and support MNC operations in Singapore. It was

Table 3.11 Deepening of Singapore's R&D system, 1993–2005

	1993	1997	2000	2004	2005
Percentage of Masters and PhD holders among RSEs (FTE)	39.3	41.6	43.8	43.9	44.3
Percentage breakdown of R&D exp. (%)					
Basic research	16.1	12.8	11.8	18.8	20.6
Applied research	39.1	43.8	35.0	29.8	32.5
Experimental development	44.9	43.3	53.2	51.4	46.8

Sources: As for Table 3.9.

only from the late 1990s that their R&D missions were shifted towards more strategic, longer-term R&D, while several new life-science R&D institutes were established in the early 2000s. Thus basic R&D only began to receive an increasing share of total R&D spending in Singapore from 2001, rising to 21 per cent in 2005 (Table 3.11).

In terms of sectoral distribution, it is no surprise that the manufacturing sector accounted for the bulk of private sector R&D, given the significant role of manufacturing in Singapore's economy (Table 3.12). Manufacturing R&D was highly concentrated in a number of sectors, with almost two-thirds in the electronics sector alone in 2005, followed by engineering (16.3 per cent). This is consistent with the fact that electronics and IT have been the most important and dynamic sectors in the Singapore economy since the 1980s, which in turn stimulated a certain amount of R&D in the precision engineering industry (Wong, 2003). Private sector R&D in life sciences remains small, even though public R&D in life sciences has grown quite rapidly in recent years.

There has also been a noticeable increase in the share of private sector R&D going to the services sector in the early 2000s, reflecting the growing sophistication of Singapore's KIBS. In particular, ICT services have been major contributors to service R&D, although part of it was reclassified under 'other services' in the annual R&D survey from 2001.

Among private firms performing R&D, foreign MNCs continue to account for around two-thirds of R&D spending, after dropping below 60 per cent in the late 1990s and early 2000s (Table 3.13). Reflecting the concentration of MNCs in electronics and chemicals, a large share of their R&D (45.3 per cent) is in these two sectors. In contrast, local enterprises have more diversified R&D activities. The two largest industries, electronics and precision engineering, together accounted for only 43.9 per cent of local enterprise R&D in 2004, as there was also sizeable R&D in ICT and financial and business services.

Table 3.12 Distribution of private sector R&D expenditure by industry, 1993–2005 (%)

	1993	1998	2004	2005
Primary industries & construction	n.a.	n.a.	0.5	0.1
Manufacturing	81.1	86.9	63.7	65.2
Electronics	51.4	48.3	44.5	39.0
Chemicals	5.6	10.8	2.5	4.7
Engineering	16.8	22.7	13.4	10.6
Precision engineering	11.2	19.2	11.1	8.1
Process engineering	1.2	0.6	n.a.	n.a.
Transport engineering	4.3	2.9	2.4	2.5
Life sciences	4.0	4.2	2.1	3.0
Light industries/other manufacturing	3.4	1.0	1.2	7.9
Services	18.9	13.1	35.8	34.8
R&D	n.a.	n.a.	12.3	10.9
IT & communications	3.2	9.2	5.5	4.9
Finance & business	4.3	1.4	7.9	8.6
Other services	11.3	2.5	10.1	10.3
All industry groups	100.0	100.0	100.0	100.0

Sources: As for Table 3.9.

Table 3.13 Foreign companies' share of industry R&D expenditure, 1993–2005 (%)

Year	Share in total private R&D (%)
1993	67.6
1994	74.5
1995	64.3
1996	67.0
1997	61.2
1998	55.8
1999	55.8
2000	57.9
2001	57.6
2002	52.9
2003	59.8
2004	64.0
2005	66.8

Sources: As for Table 3.9.

84   *Fast growth countries*

The changing intensity and composition of R&D activities in Singapore, as discussed above, is reflected in Singapore's patenting output trend. The absolute number of Singapore-based patents is still low; until 2000, the total number of US patents granted to Singapore residents was among the lowest of the countries in this study (see Appendix Table A3.1). Nevertheless, Singapore's patenting performance has improved dramatically over the last few years.[1] As can be seen from Figure 3.2, the number of patents to Singapore-based inventors granted by the US Patent and Trademark Office (USPTO) averaged less than ten per year in the 1970s to mid-1980s, and was still less than 50 annually up to 1992, but had since jumped to over 500 in 2002, reaching an all-time high of 593 in 2004. Indeed, more than half of the total cumulative number of US patents granted by 2005 were recorded in the last four years from 2002 to 2005 alone.

Until 2000, foreign companies accounted for more than half of all US patents granted to Singapore-based inventions, reflecting the dependence of Singapore on R&D by foreign MNCs. However, from 2000 to 2003, patents assigned to Singapore companies outnumbered those assigned to foreign companies, reflecting the growth in indigenous innovation capabilities in both the public sector and the local private sector, including the emergence of local high-tech start-ups. Nevertheless, from 2004, patents assigned to foreign companies once again outnumbered those to local companies, suggesting the growing importance of Singapore as a regional R&D hub for global MNCs.

4.1.2 Competence building

A distinctive feature of Singapore's NSI development is the early and sustained emphasis on building human resource competences geared to absorbing and assimilating new technologies. While the expansion of education at all levels has been a priority public expenditure focus of the government throughout the years, the relative emphasis has changed over time. In addition, the government has played a critical role in promoting industrially relevant workforce development. This included the establishment of vocational and technical training institutes for upgrading industrial workers' operative skills in Singapore's first decade after independence in 1965, complementing the on-the-job training received by workers employed by MNCs (Wong, 2003). In the mid-1970s to mid-1980s this focus shifted to developing more advanced technicians and engineers, through rapid expansion of polytechnical education and through specialized technical training programmes, many of them collaborative ventures between the government and reputable overseas partners (MNCs and highly regarded foreign industrial training institutes) (Table 3.14). From the late 1980s, the emphasis shifted again to increasing enrolments in technology-based university

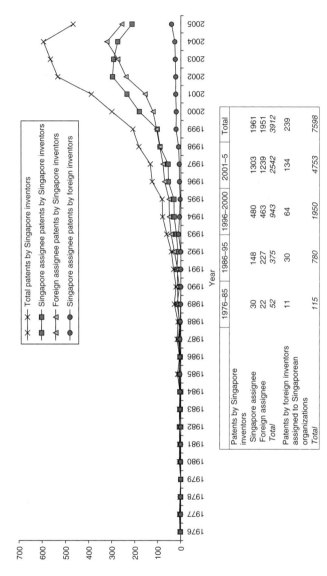

	1976–85	1986–95	1996–2000	2001–5	Total
Patents by Singapore inventors					
Singapore assignee	30	148	480	1303	1961
Foreign assignee	22	227	463	1239	1951
Total	*52*	*375*	*943*	*2542*	*3912*
Patents by foreign inventors assigned to Singaporean organizations	11	30	64	134	239
Total	*115*	*780*	*1950*	*4753*	*7598*

Notes: Where a patent is assigned to more than one country, it is allocated according to the country of the first-named company. Patents by Singapore inventors include all patents with at least one inventor who is a Singapore resident.

Source: Database of the US Patent and Trademark Office (USPTO) (various years).

Figure 3.2 Growth in Singapore patents, 1976–2005

Table 3.14 *Average output of technical graduates from tertiary education institutions in Singapore, 1970–2005 (number of graduates per year)*

	1970–79	1980–84	1985–89	1990–94	1995–99	2000–2005
University level[1]	680	1 040	2 162	3 215	5 027	8 463
Polytechnic level[2]	1 516	2 463	4 836	6 686	8 595	11 372
Total	2 197	3 504	6 998	9 901	13 622	19 835
	University graduates as percentage of total					
	31.0	29.7	30.9	32.5	36.9	42.7

Notes:
[1] Includes degree courses from ISS.
[2] Includes diploma courses from ISS.

Sources: Calculated from Ministry of Labour, *Singapore Yearbook of Labour Statistics* (various years); Ministry of Manpower, *Singapore Yearbook of Manpower Statistics* (various years).

degree courses, which again complemented the on-the-job learning obtained by engineers and technicians working in MNC manufacturing subsidiaries. Increasing outsourcing by many of these MNCs also gave local supporting industries greater opportunities to 'learn by transacting' (Wong, 2003). Finally, beginning in the mid-1990s and accelerating through the early 2000s, increasing engagement in R&D in Singapore by both global MNCs as well as some local high-tech firms, coupled with expanding R&D in the higher education sector and the increasing focus of PRICs on longer-term strategic research, has greatly increased the pace and scope of R&D personnel development in Singapore.

The workforce training role of PRICs had been given particular emphasis under the Third NSTP 2001–5, with the launching of various new programmes that emphasized human resource development, including new scholarship schemes for postgraduate education at leading universities overseas, a scheme for postgraduate research students to undergo internships at the various local research institutes, and the funding of PRIC researchers to be attached to local high-tech firms to develop their technology commercialization experience. Likewise, the various industry-promoting government agencies such as the Infocomm Development Authority (IDA) (previously known as the National Computer Board) continued to play a significant role in promoting technical workforce development even though their industrial development promotion role changed over the years (Wong, 2002).

Overall, Singapore appears to have done well in increasing the supply of technical graduates over the years (see Table 3.14). From an output of about 2200 per year in the 1970s and 5200 per year in the 1980s, the annual flow increased to 11 600 in the 1990s and nearly 20 000 by the 2000s. The proportion of university graduates (versus polytechnic graduates) also increased from about 30 per cent to 43 per cent in the same period.

In contrast to Korea, where the education system is still rather weak (see Lim, Chapter 4, this volume), Singapore has received high marks for the quality of its S&T education and the technical competence of its workforce (see, e.g., ranking by the annual *World Competitiveness Yearbook (WCY)*). Singapore has also consistently ranked among the top countries in terms of performance in international mathematics and science tests among primary and high-school pupils, such as the Trends in International Mathematics and Science Study. For university education, the National University of Singapore (NUS) has recently been ranked among the world's top 20 universities overall and the top ten engineering schools in particular by the *Times Higher Education Supplement*.

Nevertheless, in terms of the absolute number of research scientists and engineers produced cumulatively, Singapore's technical human resource development may still have some way to go to catch up with other small advanced countries with similarly small domestic population bases. Among RSEs (research scientists and engineers) engaged in R&D activities in Singapore, the proportion with Masters'/PhD degrees remained relatively low, at around 40–44 per cent, during the 1990s and 2000s (Table 3.11). The *WCY* has generally rated the availability of skilled technical labour in Singapore as behind most of the advanced OECD countries, although its ranking has improved somewhat as of 2004 (Wong, 2003; 2006).

One key aspect of Singapore's competence-building policy is the efforts made at attracting foreign talent. To supplement the local supply of skilled labour, the government has consistently adopted a liberal immigration policy to attract overseas skills. While precise statistics on the immigration of qualified technical labour are not available, the annual R&D surveys indicate that foreigners typically accounted for over one-quarter of the total pool of RSEs in Singapore in recent years (A*STAR, various years). Even this figure, however, is a gross underestimation, as it does not include the sizeable number that had been offered permanent residence. Similarly, over one-third of Singapore's IT workforce in the late 1990s was found to consist of foreigners (Wong, 2002). The proportion is even higher in the emerging life-science fields (Wong, 2006). While Malaysia was a major source of foreign talent in the early years, China and the Indian sub-continent have provided the bulk of the foreign technical professionals working in Singapore since the mid-1990s. Despite experiencing a number of economic slowdowns (in 1985, 1998 and 2001), the government has not backed down from its policy of importing foreign talent. Indeed, the Economic Review Committee (ERC), set up in December 2001 to chart the future direction of Singapore, outlined policy initiatives to accelerate the attraction of foreign talent to Singapore (Wong, 2003; ERC, 2002).

4.2 Demand-side Factors

4.2.1 Leveraging foreign MNCs to create new markets and articulate demands

Given its rather small domestic market, Singapore's economy has been highly dependent on the external regional and global markets for growth. Indeed, a cornerstone of Singapore's public policy has been to promote openness to external trade and investment, essentially relying on the external market forces to provide the signals for new market formation. This does not mean, however, that the state plays no role in shaping the formation of new markets; unlike the largely *laissez-faire*, hands-off role of the state in Hong Kong (see Sharif and Baark, Chapter 6, this volume), the Singaporean state has pursued an active, opportunistic role in identifying new market trends that have emerged. It has quickly devised policy incentives and invested in supporting infrastructure resources to attract global players that are well placed to capitalize on these new market development trends to locate part of their activities in Singapore, thereby allowing Singapore to reap an 'early entry' advantage.

This fast market trend follower strategy has enabled Singapore to gain a significant role in two major global manufacturing markets: electronics/ICT and pharmaceutical/life sciences. In the late 1960s and early 1970s Singapore was among the first of the developing countries to recognize the trend of outsourcing of labour-intensive electronics manufacturing by US and Japanese firms. By focusing its foreign direct investment (FDI) promotion programme on electronics firms, Singapore was able to ride the first wave of export-oriented, offshore electronics manufacturing growth. In the early 1980s, when the mass market, personal-computer-related manufacturing industries emerged, Singapore was again quick to seize the opportunity to attract the major PC-related manufacturing firms to Singapore, particularly the disk drive industry (Wong and McKendrick et al., 2000). In the 1990s Singapore further expanded into semiconductor wafer fabrication, ink-jet printer and portable information devices.

Despite rising labour costs throughout the 1980s and 1990s, Singapore was able to maintain its regional electronics manufacturing hub role by constantly attracting new waves of global electronics MNCs to establish operations in Singapore and encouraging existing MNCs to shift into higher-technology-intensity products. Besides offering investment incentives such as tax reduction and investment credits, Singapore's government proactively invested in logistics infrastructure such as airport and seaport facilities, as well as actively promoting various related supporting industries, component suppliers and contract manufacturers to increase the agglomeration economies of the key electronics clusters.

The large number of electronics MNCs operating in Singapore has facilitated the emergence of local electronics companies since the 1980s. As technology transfer from foreign companies largely took the form of process technology, indigenous companies tended to specialize in contract manufacturing and supporting industries, such as precision parts and components, printed circuit boards and industrial automation equipment, rather than in original design manufacturing (as in Taiwan (see Balaguer et al., Chapter 2, this volume)), or original brand manufacturing. Stable customer/buyer relationships with the technologically more sophisticated MNCs helped motivate these specialized suppliers to upgrade their own technological processes.

Besides local electronics supporting industries, a small number of indigenous electronics firms in semiconductor and computer-related sectors also emerged in the early and mid-1990s. With the exception of Creative Technology, however, these were mainly GLCs, such as Chartered Semiconductor and ST Electronics. In the late 1990s, the Internet boom and a government initiative promoting the creation of high-tech start-ups facilitated the emergence of ICT firms. Although the bursting of the Internet bubble and the consequent meltdown of Nasdaq dampened the rate of technology start-ups (as happened worldwide), the number of spin-offs from university and public R&D institutes continued to grow in the 2000s (Wong, 2003).

In the case of the pharmaceutical/life-science industry, Singapore initially focused on attracting large pharmaceutical firms to locate their manufacturing operations in Singapore in the 1990s, as it started doing with electronics MNCs earlier. By the early 2000s, however, with the anticipation of rapid growth of life-science industries beyond the advanced countries, and the recognition that commercialization of upstream research through dedicated biotech firms plays a much more important role in life-science business, the government decided to adopt a different approach to promoting life-science-based industrial clusters (Finegold et al., 2004). Under a US$1 billion new Life Science Initiative announced in 2001, the government signalled its intention to turn Singapore into an integrated regional hub for biomedical sciences, with a critical mass of capabilities across the entire value chain, from basic research to clinical trials, product/process development, full-scale manufacturing and healthcare delivery (Biomed-Singapore, 2003).

A three-pronged approach has been adopted: A*STAR (the Agency for Science, Technology and Research, revamped from the previous NSTB) is charged with establishing new life-science-related PRICs to increase biomedical research and human resource development, while the Economic Development Board (EDB) is to attract foreign investments in biomedical R&D by global pharmaceutical MNCs. Both EDB and A*STAR are also assigned to promote the various supporting industries for the life-science cluster, including attracting specialist life-science venture capital (VC) funds to invest in local life-science companies, as well as investing their own funds. Notwithstanding the broader approach to life-science industry development, leveraging of global pharmaceutical MNCs remains an integral part of the Singaporean government's strategy to create new growth markets.

Because of the high openness of the economy and the strong reliance on external market forces, local Singaporean firms are by and large highly exposed to global competitive market pressures and the demand for quality that they transmit. In particular, the high presence of many leading 'world-class' global MNCs in Singapore itself contributes significantly to the articulation of demands for quality and process improvement in manufacturing and logistics services (Wong, 2001; 2003). It is thus not surprising that Singapore has developed among the best air- and sea-transport infrastructures and logistic support industries in the world: they had to innovate or risk losing customers to other regional competition. For example, the Port of Singapore had been among the world's leading users of ICT to improve operational productivity and efficiency in the search to stay ahead of lower-cost competitor ports in the region.

Besides leveraging global MNCs, the government attempts to develop some independent capability in identifying emerging market opportunities around the world through the establishment of 'listening posts' in key 'lead-user' cities in the USA, Europe and Asia. In addition, the government (through an agency called International Enterprise (IE)), also subsidizes the cost of business development activities in overseas markets by local firms. However, compared to the scale of other countries, for example Taiwan and Switzerland, Singapore's investment in this regard is probably quite modest. Moreover, unlike countries such as Taiwan and Israel, which have very extensive overseas entrepreneurial and technical professional diasporas intimately linked to the leading high-tech regions in the world, including in particular Silicon Valley, Singapore's overseas talent diaspora is much smaller.

4.2.2 'Lead-user' role of government to promote domestic diffusion of new technologies

While the relatively small domestic market has necessitated the general emphasis on external market opportunities, the Singapore government has intervened significantly in helping to promote the formation and growth of a number of domestic technology markets, particularly those that are deemed to have strong multiplier impacts on the rest of the economy. Prime examples of such domestic-market-oriented diffusion policies are the promotion of adoption and usage of ICT and advanced manufacturing technologies. Indeed, among developing economies, Singapore scored particularly high in terms of indicators of adoption of ICT, licensing of foreign technologies, use of advanced technologies in production and process management capabilities according to the annual *WCY* (Wong, 2003).

In addition to devising programmes to accelerate technology diffusion in the private sector (see Section 4.3.3), the government itself has been a lead user in the adoption of new technology, especially ICT. Many of the major statutory bodies in Singapore have become lead users of technology not just locally but probably in the world. For example, before its privatization, Singapore Telecoms was a fast adopter of new telecommunications technologies, resulting in Singapore having one of the most advanced telecommunications infrastructures in the Asia-Pacific region. The Ministry of Environment was among the first in Asia to adopt incineration technology for waste disposal. The Ministry of Transport was also the first in Asia to deploy electronic road pricing. Similarly, Singapore was among the first in the world to automate trade document submission and approval using electronic data interchange.

4.3 Provision of Constituents

4.3.1 Provision of organizations

As highlighted earlier, Singapore's industrial development has until recently been largely based on attracting global MNCs to locate their operations in Singapore. Thus relatively little attention was given to developing local innovative firms until the mid-1990s. Most local manufacturing start-ups that were established before then were primarily suppliers and contract manufacturers to MNCs, while the PRICs produced few spin-offs up to the late 1990s, as R&D was aimed more at complementing and supporting MNC operations in Singapore.

The late 1990s, however, saw a drastic shift in public policy emphasis in this regard. The Asian financial crisis in mid-1997, which led to a severe regional economic downturn, raised concerns about the need to diversify markets and achieve greater penetration of European and North American markets. This clearly required Singapore to have a higher technological competitive edge. In addition, growing competition from China and India meant that Singapore would be subject to severe cost pressures. Finally, the leadership had become increasingly impressed by the Silicon Valley model of high-tech innovation (including the successful Israeli and Taiwanese variants) as key to success in the global knowledge-based economy. All these factors motivated the government to launch a new economic development programme called the Technopreneurship 21 (T21) initiative in 1999.

T21 covers four areas affecting high-tech entrepreneurship: education, facilities, regulations and financing. To encourage more Singaporeans to become technology entrepreneurs, the T21 initiative aimed to inject an entrepreneurial dimension into the educational system in schools and universities, so that they would serve as generators not only of trained employees but also of graduates who are entrepreneurial. The T21 initiative has also sought to create more conducive facilities that provide an intellectually stimulating and creative environment for 'technopreneurs'. A major step in this direction has been the development of One North, a major new R&D complex to attract international talent working in high-tech R&D (see Section 4.4.1 for more details). In addition, rules and regulations have been reviewed to remove obstacles to technopreneurship. Finally, a US$1 billion Technopreneurship Investment Fund has been launched to help jump-start the development of a VC industry in Singapore (see Section 4.4.2 for more details).

It is unclear how much impact the T21 initiative has had on spurring technopreneurial development in Singapore. The number of high-tech start-ups in Singapore more than doubled from 1998 to 2000 (Table 3.15), but this was largely in tandem with the global Internet boom and probably would have taken place anyway, without any public policy inducement. Moreover, shortly after the T21 initiative was announced, the Nasdaq meltdown in April 2000 began to have a severe knock-on chilling effect on the high-tech start-up phenomenon in 2001–3, not just in Silicon Valley, but also worldwide. This probably explained the drastic drop in the new high-tech start-up formation rate in 2001, notwithstanding the T21 initiative. The rate has since remained below the level of 1999. Nevertheless, the Singapore government appears to remain firmly committed to promoting the growth of local high-tech firms, although the T21 initiative was no longer mentioned by 2004.

Reflecting the more realistic mood after the worldwide dotcom crash, a more comprehensive government initiative in promoting innovation and entrepreneurship was proposed by an Entrepreneurship and Internationalization Sub-Committee (EISC) as part of an overall ERC set up by the Singapore government in 2002 to formulate a new economic

Table 3.15 High-tech enterprise formation rate, 1998–2005

	1998	1999	2000	2001	2002	2003	2004	2005
Total no. of high-tech start-ups	2064	3235	4547	2929	3006	2875	2770	2991
High-tech manufacturing	768	932	897	784	768	687	787	843
ICT services	1268	2271	3561	2051	2124	2060	1860	1997
R&D services	28	32	89	94	114	128	123	151
Total no. of start-ups	29870	34604	36457	33202	36675	39337	41164	42556
% of high-tech start-ups	6.9	9.3	12.5	8.8	8.2	7.3	6.8	7.0

Notes: Defined as start-ups in high-tech industries according to the classification of the US Bureau of Labor Statistics (BLS). ICT services includes postal services.

Source: Wong et al. (2007b).

development strategy for Singapore in the new millennium. Released in 2003, the EISC report identified six elements to strengthen the spirit of entrepreneurship and innovation in Singapore, and to foster the growth and internationalization of Singapore-based companies, including the GLCs (Wong, 2006). These elements included: culture (creating opportunities for young people to develop their entrepreneurial potential, particularly though educational channels); capabilities (developing enterprise capabilities at both individual and industry levels by reducing impediments to labour mobility, attracting foreign entrepreneurial talent, increasing collaboration between companies, encouraging and assisting companies to venture abroad); conditions (adopting a more enterprise-friendly approach to regulation, including the role and management of GLCs); connections (increasing internationalization of Singapore enterprises); capital (rectifying gaps in enterprise financing, especially for start-ups); and catalysts (providing incentives to channel more capital towards enterprise).

Besides continuing emphasis on developing indigenous R&D capabilities via A*STAR and attracting foreign MNCs to set up R&D operations in Singapore via EDB, two other statutory boards were assigned the primary responsibilities to carry out the EISC's recommendations: the Standards, Productivity and Innovation Board to promote innovation among the small and medium-sized enterprises (SMEs), and IE Singapore to help Singapore-based companies to internationalize.

The EISC also resulted in the establishment of a quasi-public organization called the Action Community for Entrepreneurship (ACE), a collaborative forum involving representatives from both the private and public sector involved in new venture development. Launched in May 2003, ACE was intended to be a pro-enterprise movement to create a more business-friendly environment in Singapore, by providing opportunities for networking between entrepreneurs, business angels, VC, bankers, lawyers and other professionals. It also engages in educational efforts to increase awareness of entrepreneurship and to encourage more entrepreneurial thinking among Singaporeans (Wong et al., 2004).

While it is too early to assess how much impact these new public policies might have on high-tech entrepreneurial developments in Singapore, given the long gestation period required for some of the policy tools (for example, changing the culture of people towards taking risk), there are hopeful signs that the composition of firms started up in recent years has been quite different from existing SMEs. Unlike their earlier counterparts, the new start-ups are based more on product innovation and increasingly focused on IT, software, Internet applications, biotechnology and life sciences (Wong, 2006). Spin-offs from universities have also begun to increase in frequency in the last few years (Wong et al., 2007a).

4.3.2 Networking, interactive learning and knowledge integration

An essential feature of an NSI is the pattern of linkages and knowledge flows among the different organizations and their innovation activities. In the case of Singapore, there has been a clear progression in the pattern of linkages and interactive learning among the constituent organizations in Singapore's NSI over time. In the initial period of the 1960s to the 1970s, technology transfer from foreign companies to their local subsidiary operation in Singapore facilitated 'learning by using' among local MNC employees. The government also offered incentives to MNCs to send Singaporean engineers to headquarters to acquire new technical skills. In contrast, there were few knowledge links between the MNCs and the rest of the economy in this period, as few local supporting industries then existed.

The mid-1970s, however, saw the growth of a base of local supporting industries, which began to invest in acquiring and exploiting imported technologies on their own, in addition to learning from their MNC customers through 'learning by transacting' (Wong, 2003). Such interfirm linkages between local suppliers and MNC buyers had been further stimulated by a deliberate government support programme called the Local Industry Upgrading Programme, targeted at encouraging foreign MNCs to help their local suppliers to upgrade technologically. As found by Wong (1992), such supplier–buyer relationships have contributed significantly to the technological development of local firms, less through the deliberate efforts of MNCs to transfer technology than through exposure to their procedures and technologies in the buyer–supplier relationship. Long-term supply relationships also helped reduce the risk of investing in new technologies for the suppliers, contributing to greater technological effort by local supporting industries.

The 1980s saw a further intensification of such links between manufacturing MNCs and local supporting firms, including the development of new services firms in logistics, IT and design. In addition, another form of flows between MNCs and local SMEs has also appeared since the mid-1980s: an increasing number of employees from the MNCs left to start their own manufacturing SMEs, often as suppliers to their former employers. Moreover, as an increasing number of MNCs located their R&D operations in Singapore, a new phenomenon of R&D scientists and engineers leaving to start their own high-tech firms to innovate their own products and services for the global market is beginning to emerge in the 2000s.

Linkages between industry and education and training institutions have also evolved over time. In the initial years, the linkages were quite strong at the polytechnic and industrial training level. Close consultation with industry and the anticipatory planning and rapid response of the government to meet industrial skill needs have been important contributing factors to the rapid industrialization of Singapore (Soon and Tan, 1993). The government did not hesitate to recruit expatriates with significant MNC experience to head new training institutes. Indeed, as mentioned earlier, many of the early industrial training institutes were run jointly by MNCs to establish a reputation for their programmes.

In contrast, linkages between the universities/PRICs and the enterprise sector were less well developed, at least until the late 1990s, due to the long gestation time needed for the PRICs to establish core capabilities relevant to industry, and the lack of focus on industrially relevant research at the universities until the late 1990s. Many of the MNCs in their turn looked to their headquarters and associate companies for technological needs rather than local PRICs/universities. They also preferred to tap public R&D subsidies offered for in-house R&D, so that they would own the intellectual property generated. Nevertheless, linkages between universities and enterprises appear to have strengthened considerably since the early 2000s, due to increasing emphasis by the universities on commercializing technologies (Wong et al., 2007a).

Last but not least, interfirm innovation linkages among local firms appear to have been much weaker. There are few reported cases of joint R&D among local firms, and the kind of industry-wide R&D consortia found in Taiwan and Japan have been largely absent in Singapore. There have also been few reported cases of industry-wide collaboration in technology deployment (Wong, 2003). Overall, there appears to have been inadequate policy attention given to promoting innovation collaboration among local enterprises in Singapore, compared to Taiwan and Finland, and this appears to be a major weakness in Singapore's NSI.

4.3.3 Provision of institutions

Before 1990, Singapore's main policy focus had been on promoting technology adoption, and public involvement in R&D activities was low and confined largely to scientific research in public universities and defence R&D, both of which had little commercial linkage to industry (Wong, 2003).

The first significant recognition of the economic importance of R&D came in 1989 when a Committee of Ministers of State was formed to outline the long-term strategy and direction of Singapore's development. The result was a 'vision' document called *The Next Lap*, which highlighted the need to focus on R&D and specialize in high-tech niches in order for Singapore to catch up with the advanced countries over the next 20 years (Government of Singapore, 1991). The importance of innovation gained more recognition in a subsequent Strategic Economic Plan (SEP) formulated in 1991 (MTI, 1991). As part of the recommendations of SEP, the first five-year NTP was released, with the simultaneous establishment of an entirely new statutory board, NSTB, in 1991. The key objectives of the NTP were to promote industrially relevant R&D, build up S&T human resources, and develop S&T support infrastructure. A S$2 billion allocation was given to NSTB to implement the NTP.

A key outcome of the NTP was the establishment of a series of PRICs, which were to be funded and managed by the newly established NSTB. This was done through a combination of creating *de novo* institutes, as well as reorganizing and transferring a number of existing research institutes from the higher education and government sectors. The NTP was followed by the formulation of a second five-year plan in 1996, the second NSTP, where the budget allocation was doubled to S$4 billion, and where the importance of investing in science was recognized in addition to technology. Despite this recognition of the importance of science, the NSTP was still heavily skewed towards applied R&D promotion rather than basic research. Indeed, the initial mission of most of the PRICs established under the NTP was to develop the applied technologies deemed critical for Singapore's industrial clusters (Wong, 2003), and this applied focus continued into the late 1990s.

The dotcom boom and the growing success of Silicon Valley in the late 1990s led the government to launch the T21 initiative in 1999 (see Section 4.3.1). However, with the bursting of the Internet bubble in 2000, policy makers realized the need for start-ups to have truly innovative technologies and defensible intellectual property, and hence the need to raise basic research and innovative capabilities. At the same time, the government recognized the need to make a big push into life sciences. Both these factors led to the decision to restructure NSTB in 2001. Under T21, NSTB had been made the lead agency to implement the government's 'technopreneurship' drive, but this had resulted in conflicts with its original focus on building the nation's research capabilities. With the restructuring of NSTB and its renaming as A*STAR, the responsibility for nurturing technopreneurship was transferred to EDB. A*STAR was refocused on developing Singapore's R&D capabilities, particularly the attraction and training of R&D workforce. To do this, it set up two councils: the Bio-Medical Research Council (BMRC), responsible for promoting R&D and developing human capital in the life sciences, and the Science and Engineering Research Council (SERC), which does the same in targeted science and engineering clusters such as ICT, chemicals and engineering clusters (Wong et al., 2004). Seven of the 12 PRICs under the auspices of A*STAR came under SERC, while the other five (including three brand new institutions) came under BMRC.

Planned public spending on S&T was also increased to S$7 billion in the third NSTP for 2001–5, with S$5 billion allocated to A*STAR to fund public research and to develop postgraduate research personnel, while the remaining S$2 billion was managed by EDB to support R&D in the private sector. The Plan also allocated a larger proportion of the public R&D budget to long-term strategic and basic research (Wong, 2003; A*STAR, 2001). This shift in emphasis towards building long-term basic research capabilities further intensified in 2006, when the government set up a new National Research Foundation (NRF) with a S$5 billion allocation to fund new research areas not covered by existing A*STAR PRICs, including water and environmental engineering and interactive digital media.

Figure 3.3 serves as a useful summary of the overall institutional framework that has evolved for managing Singapore's NSI in recent years. As can be seen from the earlier discussion, a characteristic feature of Singapore's approach to S&T policy implementation is the relatively top-down approach to technology policy formulation that is strategic in nature, yet flexible in terms of actual implementation. Indeed, Schein (1996) described this approach of Singapore policy making as 'strategic pragmatism'. Although his work was focused on the EDB alone, much of what he found appears to be applicable to the S&T policy arena in general. In essence, Singapore's political leaders at the Cabinet level formulate broad, long-term strategic economic development initiatives, but delegate much of the detailed implementation to the designated implementation agencies. Moreover, the government has been quite prepared to revise substantially

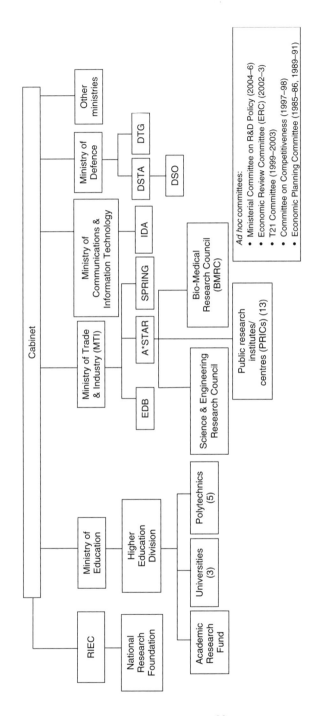

Figure 3.3 Emerging institutional framework for S&T policy in Singapore as of 2006

an earlier strategic plan and replace it with a newer one, if and when it perceives that the environmental opportunities or threats have changed materially. This change in strategic direction is typically not due to changes in specific political leadership, since the government has been under a one-party rule since political independence in 1965.

One consequence of this top-down approach is that S&T policy is typically not formulated in isolation, but as an integral part of a larger economic development strategy. This 'developmental state' approach has meant that S&T policies are strongly integrated within ministries with significant economic development roles, particularly the Ministry of Trade and Industry (MTI), and the Ministry of Information, Communications and the Arts (MICA). Indeed, it is telling that, until today, Singapore does not have a separate Ministry of Science and Technology. Instead, these policy-making and implementation functions have been subsumed by the economic-development-oriented ministries. Thus A*STAR, which oversees the PRICs, comes under MTI, while IDA, which promotes ICT deployment and administers the Innovative Development Scheme for ICT-related companies, comes under MICA.

A related characteristic of Singapore's institutional framework for S&T policy implementation is the involvement of multiple agencies. For example, the promotion of ICT development involves both EDB and IDA. Although they have generally different responsibilities (EDB for investment promotion, IDA for innovation and diffusion promotion and telecoms regulations), there are various innovation-related activities where the two overlap (for example EDB provides incentives to a foreign firm to establish R&D activities in Singapore, while IDA may provide innovation grants to the same company). As highlighted by Wong (2003), the government has so far been able to manage such multi-agency coordination relatively well, with relatively little of the 'turf fighting' seen in many developing countries. To further highlight the strategic importance of research and innovation and to strengthen coordination of different programmes and initiatives by different ministries and agencies, the government in 2006 set up the Research, Innovation and Enterprise Council (RIEC), chaired by the prime minister, with the new NRF reporting directly to this council and serving as its policy-making secretariat.

Other aspects of institutions within the Singapore NSI have been discussed elsewhere in this chapter and will not be repeated here. For example, trade and competition in Singapore's open economy – particularly the prevalence of foreign MNCs – and its effect on corporate governance is outlined in Sections 4.2, 4.3.1 and 7.1. Similarly, S&T employment relations, including the liberal use of foreign S&T talent, is discussed in Sections 4.1.2, 4.3.2 and 7.1.

4.4 Support Services for Innovating Firms

4.4.1 Incubating activities

Singapore's first science park was set up under a government initiative in 1980 to provide infrastructure for R&D in Singapore. With a total land area of 30 hectares, the park was fully occupied by the mid-1990s with a mix of tenants including government agencies and numerous private companies. Development of Science Park II, with a land area of 20 hectares, began in 1993, with tenants mainly comprising ICT companies and related PRICs.

Following the completion of these two science park programmes, the government embarked in the early 2000s on a much larger infrastructure development project called One North, which would house a new innovation-based city comprising R&D facilities, campuses for new higher educational institutions, living quarters for research scientists, hotel, convention and restaurant facilities, as well and offices for venture-related services such as IP (intellectual property) law firms and VCs. Occupying 190 hectares of land near the campus of NUS, One North represents by far the most ambitious R&D infrastructure support project attempted by the government to date.

Central to this new infrastructural development is the establishment of a biomedical hub called Biopolis, a 'city within a city' that specifically caters to the unique research needs of the biomedical sciences. Opening its first phase in June 2003, Biopolis not only houses A*STAR's five biomedical research institutes as the anchor tenants, but also aims to attract biomedical MNCs, start-ups, and support services such as biotech patent agents and law firms, so as to create a dense cluster of life-science-related activities that will, it is hoped, facilitate the formation of informal networks for knowledge sharing and accelerate the growth of a critical mass of biomedical expertise in Singapore (Finegold et al., 2004).

Besides physical facility development, the government also launched in 2002 a new initiative to facilitate interactions among existing incubators for high-tech start-ups in Singapore. Called HOTSpots (Hub Of Technopreneurs), the initiative comprises a network of seven technopreneur incubation centres across the city run by a mix of private and public sector operators, with the aim of providing common activities such as mentoring workshops, industry roundtables, brokerage events, fundraising platforms, networking sessions and social functions for the 400-plus technology-related companies operating in these incubators.

4.4.2 Financing

A key component of an NSI is institutions providing risk capital for technology commercialization activity. As MNCs provided risk capital in the early years of Singapore's high-tech development, there was little need to develop financial institutions to support high-tech investment. The financing needs of local SMEs were met mainly through the conventional financial institutions, although the government did provide a number of subsidized loan-financing schemes to encourage SME investment in new technology and innovation (Wong, 2003).

The government first began to promote the development of a VC industry in the 1980s, although the effort was very modest and consisted mainly of investment incentives to attract a number of foreign VCs to set up in Singapore, with very small-scale injection of public funds into the industry. This was quite unlike the case of Taiwan, where the government played a more prominent role, directly supporting VC through tax incentives and financial assistance programmes (see Balaguer et al., Chapter 2, this volume). In the early 1990s, the government began to play a more direct role in VC industry development by creating a number of new funds such as Vertex Management and EDB Ventures, but the real growth of the VC industry occurred only in 1999, with the establishment of a US$1 billion Technopreneurship Fund by the Singapore government under the T21 initiative. This 'fund of funds' aimed to induce leading VCs to use Singapore as their regional operation hub and train a core of experienced VC professionals. Although the fund was successful in attracting several leading US VC firms to Singapore (for example Draper Fisher Jurvetson, Crimson Ventures), others that received sizeable funding from the fund did not follow suit.

The above initiatives have undoubtedly enabled Singapore to establish itself as the preferred location for VC regional hub operations in South-East Asia, but in terms of the volume of venture deal flows Singapore still lags considerably behind Taiwan (Wong, 2006). Indeed, while the cumulative amount of funds managed in Singapore has grown impressively in recent years, reaching over S$16 billion in 2004 (EDB, 2005), the real impact this has had on high-tech start-up formation in Singapore has actually been quite modest, for two reasons. First, most of these funds were targeted at more mature rather than early-stage ventures. Second, most of the funds under management made relatively little investment in Singapore-based start-ups, preferring to invest in the larger regional market for high-tech ventures, particularly China. As can be seen from Table 3.16, the total amount of VC fund invested in Singapore-based ventures has been around 0.1 per cent of GDP in recent years, considerably lower than that in the USA and Israel (Wong et al., 2005).

Besides the relatively low level of entrepreneurial activities in Singapore in general compared to the USA, another contributing factor to the low level of VC-funded deals is the low level of business angel investing at the

Table 3.16 VC investment in Singapore-based ventures (10^6 S$), 2000–2005

	2000	2001	2002	2003	2004	2005
VC investments in Singapore companies	601.3	384.4	155.0	185.0	143.2	319.2
GDP*	157 700	152 100	155 700	159 100	180 600	194 360
VC Investment/GDP	0.38%	0.25%	0.10%	0.12%	0.08%	0.16%
Average VC per investment	3.3	5.3	2.4	2.3	2.7	6.8
No. of companies invested	182	73	65	82	53	47

Note: * GDP at current market price.

Source: Wong et al. (2007b).

seed stage, which is typically needed to fund early start-ups to grow to a later stage fundable by VC. The propensity for business angel investment in Singapore is relatively low compared to the USA (Wong, 2006). Recognizing this, EDB introduced in 2002 a public co-investment scheme (Startup EnterprisE Development Scheme or SEEDS) to stimulate early-stage business angel investment. The scheme has made a visible impact, with over 150 new start-ups having received SEEDS funding since its launch.

4.4.3 Provision of consultancy services

The availability of a wide range of consultancy services has been recognized as a key element in the development of a knowledge-based economy by the economic policy makers in Singapore. Indeed, Singapore's EDB has been among the first FDI promotion agency in the world to target services promotion, including consultancy services. The fact that Singapore has generally pursued a much more open competition policy towards the services sector than most countries in South-East Asia has also contributed towards a larger presence of foreign consultancy services firms, which typically use Singapore as a regional hub in South-East or East Asia. In this sense Singapore is similar to Hong Kong, where foreign consultancy firms also play an important role, whereas other regional countries place a greater emphasis on direct government involvement in consultancy, for example Taiwan's Industrial Technology Information

Service Office (see Sharif and Baark, Chapter 6; Balaguer et al., Chapter 2, this volume).

The presence of a wide range of consultancy services firms, many of which are foreign in origin, means that access to specialized services has generally not been a problem for most Singapore-based firms. It is interesting to note, however, that the development of specific IP-related advisory services has been particularly singled out by the ERC. As a result, a new unit was established in EDB in 2003 to spearhead the promotion of IP-related services industries, while a new IP Academy has been established to train specialized IP professionals such as patent engineers and licensing agents.

4.5 Summary of the Main Activities Influencing Innovation

In summary, Singapore's NSI has shifted over the last four decades from emphasizing technology usage to technology creation. As can be seen from Figure 3.1 earlier, this shift has occurred over four distinctive phases, with each successive phase built upon the resources accumulated earlier but involving new actors, new activities and new forms of linkages among existing actors. In particular, there has been a phased building up of MNCs, local manufacturing enterprises, PRICs and university R&D, and, in the last phase, local high-tech start-ups pioneering new products. In terms of technology capability development, there has been a sustained shift from learning to use (with high reliance on internal transfer by MNCs) to learning to adapt and improve (via 'learning by doing' within MNCs as well as 'learning by transacting' in local firms acquiring external technology), learning to innovate (mainly applied R&D in product or process) and, finally, learning to pioneer (creating indigenous intellectual property and commercializing it in the marketplace).

5 CONSEQUENCES OF INNOVATION

Despite the significant changes in the level and composition of innovation activities in Singapore over the years, there have been few systematic empirical studies that assess the impact of innovation activities on the country's overall economic development performance. Many factors influence overall economic growth besides innovation capabilities. Singapore achieved a high level of economic growth in the early years without significant investment in innovation activities. Indeed, an early study by Young (1992) came to the controversial conclusion that Singapore achieved almost zero total factor productivity (TFP) growth over the 1960–80 period, raising questions about the longer-term viability of Singapore's development model. However, a subsequent study by Hsieh (2002) has questioned the validity of Young's findings, highlighting problems with the capital data used by Young. Hsieh's own estimate of Singapore's TFP (using a different estimation model) was substantially higher. More recent estimates of Singapore's TFP growth in the 1980s and 1990s all yield substantially higher values (Collins and Bosworth, 1996; Rao and Lee, 1996; Department of Statistics, 1997; MTI, 2002). On balance, empirical evidence appears to support the view that TFP performance in Singapore was low but not zero in the 1980s, adding about 1.5 per cent to annual economic growth (about 20 per cent of the total economic growth of 7.1 per cent over the period). TFP performance improved substantially in the 1990s, adding about 2.5 per cent economic growth per annum, or about one-third of the total growth of 7.5 per cent for the period (Wong, 2003).

While other factors may contribute to TFP growth besides technological innovation, the fact that Singapore only began to register higher TFP in the 1990s is consistent with the empirical observation that Singapore did not begin to invest substantially in R&D until after the mid-1980s. More recent econometric estimates by Ho et al. (2007) show that R&D investment in Singapore had a significant impact on its TFP performance in the last 20 years. However, compared to the OECD nations, the impact of R&D investment on economic growth in Singapore is not as strong, as evidenced by lower estimated elasticity values. The long-run elasticity of output with respect to R&D was computed to be 8.1 per cent for Singapore versus over 10 per cent estimated by other researchers for OECD countries.

It should also be noted that, even if Singapore's growth in the 1960s and 1970s can be accounted for largely by growth in inputs, with little TFP contribution, this does not negate the importance of investment in technology absorption capability during that period. Indeed, the rapid growth achieved by Singapore in those two decades required a rate of absorption of new capital and labour inputs rarely witnessed elsewhere. Such high rates of sustained growth and development would not have been possible without absorption capability being developed correspondingly (Wong, 2003). Arguably, it was rational for Singapore to have emphasized investment in using technology (diffusion) in the earlier period when skills were low, and to have invested increasingly in creating technology (innovation) in later periods when the knowledge base of the population had deepened.

In summary, the available evidence indicates that Singapore's economic development in the last 20 years has been accompanied by a corresponding increase in TFP, much of which could be attributed to increasing innovation capability. Nevertheless, Singapore still has some way to go in catching up with the advanced nations in terms of R&D productivity.

6 GLOBALIZATION

As highlighted throughout the above discussion, Singapore's NSI development has been significantly linked to the globalization of trade, investment and talent flows. Throughout its development, Singapore's economy has been oriented towards the external export market. Inward FDI stocks amounted to 73.6 per cent of GDP during the 1980s, rising to 161.3 per cent during the early 2000s. Again, this was higher than any of the countries in this study, with the exception of Hong Kong (see Appendix Table A2.6). At the end of 2003, the stock of foreign direct equity investment in Singapore was S$223.1 billion. Most of this is concentrated in the manufacturing and financial services sectors.

Through its external export market orientation and openness to foreign capital and talents, Singapore's NSI has benefited substantially in terms of leveraging both demand (global markets) and supply (talents, capital and technology inflow) factors. It is possible, however, that Singapore's NSI may have over-relied on using technologies created elsewhere, particularly through attracting global MNCs to locate in Singapore, resulting in a lower level of indigenous technological development than may have been achieved by Taiwan and Korea. This is illustrated by its high level of foreign ownership of domestic patents (51 per cent in 2002, higher than any of the other countries in this study except Ireland) (see Appendix Table A2.7).

Singapore's outward investment has also increased rapidly, with outward FDI stocks rising from 24.8 per cent of GDP in the 1980s to 99.5 per cent in the 2000s (see Appendix Table A2.6). Its stock of foreign direct equity investment stood at S$130.4 billion in 2003 and was heavily slanted towards financial services. The recipients are mainly in Asia, showing a distinct trend toward regionalization in Singapore's outward FDI.

While much of Singapore's outward FDI has not been technology-related, it is interesting to note that there has been a slight increase in overseas R&D by some Singaporean high-tech companies. As could earlier be seen from Figure 3.2, while few US patents invented overseas were assigned to Singapore organizations up to the mid-1990s, there has been a noticeable increase in recent years. Although the cumulative number of patents from overseas R&D remains small (239 as of 2005) and represents only 6 per cent of patents from R&D in Singapore itself, it does suggest an emerging trend for increasing internationalization of R&D activities by Singaporean firms.

7 STRENGTHS AND WEAKNESSES OF THE SYSTEM AND INNOVATION POLICIES

7.1 Strengths and Weaknesses

One of Singapore NSI's greatest strengths is its focus on fast adoption of technologies. A policy of openness to external trade, investment and talent flows has contributed not only to Singapore's economic development in general, but also to its ability to access and use the latest technologies globally. In particular, public policies in Singapore have been well developed to leverage global MNCs to facilitate technology transfer and diffusion. A sound infrastructure for rapid exploitation of new technology has thus been established, and a culture has emerged where the use of advanced technology is the norm at all levels of society – individual, business and governmental.

The above asset is somewhat offset by a continued weakness in the indigenous R&D system, especially in basic research. This is a concern given Singapore's desire to operate in more knowledge-intensive industries, such as life sciences and advanced materials, where knowledge is often closely guarded and remains highly concentrated in selected regional innovation clusters in the advanced countries. Singapore's success in this arena is thus largely dependent on whether it can shift from excessive reliance on adopting existing technologies to a more balanced approach that combines the ability to use available technologies with the capability to create and commercialize its own innovations.

In overcoming its weakness in indigenous R&D capability, Singapore can draw on one of the strengths of its NSI, which is the government's high commitment to attracting and developing talent. Not only is there strong public investment in educational institutions and other workforce development institutions dedicated to developing local technical talents in Singapore; there is also an openness to, and social acceptance of, foreign talent.

Another weakness is in the area of technology entrepreneurship. Notwithstanding the recent efforts of the government, entrepreneurship in Singapore is hindered by a cultural norm that views failure as a stigma. Consequently, many people prefer to work in 'safer' environments, such as MNCs or the public service, rather than take the risk of starting their own ventures. This has resulted in a relatively low entrepreneurial propensity among Singaporeans compared to countries like the USA: while only about 4–7 per cent of Singaporeans were found to be engaging in new start-up activities in the 2000–2005 period, the proportion was 11–17 per cent in the case of the USA (Wong et al., 2005). To a certain extent, Singapore may suffer from a paradox of success, whereby past high economic growth that created full employment and ample opportunities for rapid career advancement within MNCs or the public sector has resulted in perceptions of high opportunity cost among the highly educated youths who are most likely to pursue alternative career paths such as starting their own business. The small domestic market has also meant that most entrepreneurs need to be able to think and execute beyond the domestic economy. Finally, there is currently a relative lack of successful role models that can inspire and mentor a new generation of high-tech entrepreneurs.

7.2 Summary and Evaluation of the Innovation Policy Pursued

Our discussion of Singapore's innovation policy shows that it has been tightly interwoven with the wider industrial, trade and competition policy framework. This has had both advantages and disadvantages for the development of the country's NSI. On the one hand, the leveraging of foreign MNCs to jump-start local economic and technological growth was remarkably successful; without such a policy, it seems unlikely that Singapore would have developed as rapidly as it has. Moreover, the strategy of initially building operational and adaptive capabilities, rather than devoting too many resources to R&D, allowed firms to develop the capabilities relevant to them at the time, and laid a foundation for them to pursue innovative activities of their own later. On the other hand, the relative success of the MNC-leveraging economic development strategy may have delayed the shift towards building up indigenous innovation capabilities compared to the Taiwanese and Koreans.

Notwithstanding the growing role of government in promoting R&D and technological innovation activities in recent years, there has been limited research on the impact of government policies on the level of innovation activities by firms. A study by Wong and He (2003), using data from a cross-sectional survey of over 100 manufacturing firms performing R&D, shows that public R&D support programmes did have a significant impact on innovation performance, but only for firms that had a pro-innovation corporate culture.

The short history of some of the innovation policies such as the technopreneurship policy also makes it difficult to assess their impact. While it is fairly clear that the policy shift towards investment in R&D has had a positive impact on the intensity of R&D and innovative performance of Singaporean firms, the outcomes of policies geared towards promoting high-tech entrepreneurship are less obvious and likely to require a longer gestation period.

Finally, despite the comprehensive scope of policy shifts, some notable gaps can be observed. For example, despite growing public investment in funding R&D, commercialization of technology from PRICs has not increased at the same rate, due possibly to the lack of mechanisms such as the Small Business Innovation Research scheme in the USA and the Industrial Technology Research Institute in Taiwan to help bridge the 'valley of death' between R&D and seed investment by VC or angel investors. Another area of concern is the small domestic market and the lack of a critical mass of aggressive lead users of emerging technology, which make it difficult for Singapore-based technology start-ups to validate their product innovations by having their first customers in Singapore; instead, they need to go international to seek markets even from day one. The inadequate development of global connectivity has also been pointed out earlier. Last, but not least, an effective implementation of the recommendation to change the cultural mindset of Singaporeans towards entrepreneurship may involve more fundamental long-term policy changes related to the educational system, the social security system, and the public sector talent recruitment system, all of which are politically sensitive and hence require significant policy coordination at the highest level (Wong, 2006).

7.3 Future Innovation Policy

Compared with the NSI of other advanced small economies, it is clear that Singapore's NSI needs to shift from its past emphasis on technology adoption towards a greater emphasis on technological innovation in the future. While much progress has already been made in recent years, the following areas deserve greater policy intention in the future:

1. *Intensification of R&D investment*   Despite the steady growth in R&D in recent years, with the GERD/GDP ratio rising from less than 1 per cent in 1990 to 2.4 per cent in 2005, Singapore's R&D intensity needs to increase further, given the need to achieve minimum critical mass in most areas of scientific and technological endeavour. In this regard, it is instructive to look at how the rapid growth in R&D intensity in small countries like Finland and Israel has contributed to their global competitiveness in selected high-tech industries.
2. *Shift towards more basic research in the 'Pasteur' quadrant*[2]   Besides raising overall R&D, there is an urgent need to shift more resources towards the development of basic research capabilities in the 'Pasteur' quadrant. In contrast to the past emphasis on applied R&D, Singapore needs to emphasize the development of basic research capabilities that can provide more radical or breakthrough solutions, or that can better anticipate future problems of industry. It is through the tapping of such basic yet economically relevant R&D capabilities that Singaporean

companies can hope to achieve more durable competitive advantages. The leading regions of high-tech entrepreneurial vitality in the world, especially Silicon Valley and Israel, invariably feed on wellsprings of leading-edge technologies which are generated by a strong focus on basic research capabilities.

As indigenous local firms remain weak in general, the development of basic research capabilities must remain a major responsibility of the public research institutes and universities, which need to shift their R&D portfolio towards programmes with longer gestation but greater potential for high payoffs. In particular, there is a need to boost the basic R&D budget in local universities. Despite significant progress, the level of university R&D funding in Singapore remains significantly below those of the major state universities in the USA (Wong et al., 2007a). An increase in the university R&D budget is also needed to attract top foreign talent, which can then draw good doctoral students and postdoctoral fellows to build critical mass.

3. *Improving mechanisms for funding technology commercialization from PRICs and universities* With the significant increase in public R&D funding going to the PRICs and universities, there is a need to establish new mechanisms to promote and facilitate the transfer and commercialization of public R&D outputs into existing industry and new ventures. In this regard, Singapore should learn to adapt successful schemes like the Small Business Innovation Research programme in the USA, the incubator system in Israel, the collaborative R&D programmes of Tekes in Finland, and the R&D consortia programmes of the Industrial Technology Research Institute in Taiwan (see Kaitila and Kotilainen, Chapter 10 and Balaguer et al., Chapter 2, this volume).

4. *Strengthening S&T policy evaluation capabilities* The combination of consultation with industry and international advisers, benchmarking against best practice, setting clear performance objectives and frequent self-monitoring of programme relevance has been effectively used in the past by the government to develop its S&T policy. However, as the sophistication of policy instruments increases and the relationship between policy instruments and industry becomes more complex, these need to be supplemented by more rigorous policy research and independent impact assessment – especially in view of the role of NRF as the 'secretariat' to RIEC for national innovation policy.

5. *Promoting international R&D cooperation and S&T networking* For Singapore to become a viable player in global R&D competition, R&D institutions in Singapore need to develop more collaborative partnerships with leading R&D institutions overseas. This is particularly so in view of the latecomer nature of Singapore in many advanced R&D

areas, and the limited pool of domestic R&D talents. The government can play a catalytic role in helping local R&D organizations to establish closer networking and collaboration with targeted partner organizations in various advanced countries in Europe, Japan and North America. The recent initiatives by NRF to attract leading universities like MIT and ETH Zürich to establish international collaborative R&D centres in Singapore are moves in the right direction, but need to involve the local enterprises as well.

NOTES

1. For the purpose of this report, Singapore patents are described as those that have at least one Singapore-based inventor.
2. According to Richard Stokes's typology of research modes, research in the Pasteur quadrant is that of use-inspired basic research, that is, basic research having practical applications.

REFERENCES

A*STAR (various years), *National Survey of R&D in Singapore*, Singapore: Agency for Science, Technology & Research (previously National Science & Technology Board).

A*STAR (2001), 'NSTB chairman unveils Board's new focus & orientation', press briefing by Mr Philip Yeo, Chairman, NSTB, 15 February.

Biomed-Singapore (2003), downloaded from http://www.biomed-singapore.com/bms/gi_mc.jsp.

Collins, S. and B. Bosworth (1996), 'Economic growth in East Asia: accumulation versus assimilation', *Brookings Papers in Economic Activity*, **2**, 135–204.

Department of Statistics, Singapore (1997), 'Multifactor productivity growth in Singapore: concept, methodology and trends', Occasional Paper on Economic Statistics, Singapore: DOS.

Department of Statistics, Singapore (DOS) (various years), *Yearbook of Statistics Singapore*, Singapore: DOS.

Economic Development Board (EDB) (2002), 'EDB year 2001-in-review', press release, 22 January, downloaded from http://www.sedb.com/edbcorp/sg/en_uk/index/in_the_news/press_releases/2002/edb_year_2001-in-review.html.

Economic Development Board (EDB) (2005), '2004 investment commitments surpassed expectations', downloaded from http://www.sedb.com/edbcorp/sg/en_uk/index/in_the_news/press_releases/2005/2004_investment_commitments.html.

Economic Review Committee (ERC) Sub-committee on Enhancing Human Capital (2002), 'Realizing our human potential', downloaded from http://www.mti.gov.sg/public/ERC/frm_ERC_Default.asp?sid=132.

Finegold, D., P.K. Wong and T.C. Cheah (2004), 'Adapting a foreign-direct investment strategy to the knowledge economy: the case of Singapore's emerging biotechnology cluster', *European Planning Studies*, **12**(7), 921–41.

Government of Singapore (1991), *The Next Lap*, Singapore: Government of Singapore.
Ho, Y.P., P.K. Wong and M.H. Toh (2007), 'The impact of R&D on the Singapore economy: an empirical evaluation', submitted to *Singapore Economic Review*.
Hsieh, C.T. (2002), 'What explains the industrial revolution in East Asia? Evidence from the factor markets', *American Economic Review*, **92**(3), 502–26.
IMD (2005), *World Competitiveness Yearbook 2005*, Lausanne: International Institute for Management Development.
Ministry of Labour (various years), *Singapore Yearbook of Labour Statistics*, Singapore: Research and Statistics Department, Ministry of Labour.
Ministry of Manpower (various years), *Singapore Yearbook of Manpower Statistics*, Singapore: Manpower Research and Statistics Department, Ministry of Manpower.
Ministry of Trade and Industry (MTI) (1991), *Strategic Economic Plan: Towards a Developed Nation. The Strategic Economic Plan*, Singapore: MTI.
Ministry of Trade and Industry (MTI) (various years), *Economic Survey of Singapore*, Singapore: MTI.
Ministry of Trade and Industry (MTI) (2002), 'Total factor productivity with Singaporean characteristics: adjusting for impact of housing investment and foreign workers', *Economic Survey of Singapore, Third Quarter 2002*, Singapore: MTI.
Organisation for Economic Co-operation and Development (OECD) (2006a), *Main Science and Technology Indicators: Volume 2006 Issue 2*, downloaded from http://new.sourceoecd.org.
Organisation for Economic Co-operation and Development (OECD) (2006b), *OECD Factbook 2006: Economic, Environmental and Social Statistics*, downloaded from http://new.sourceoecd.org.
Rao, B.V.V. and C. Lee (1996), 'Sources of growth in the Singapore economy and its manufacturing and services sectors', *The Singapore Economic Review*, **40**(1), 83–115.
Schein, E. (1996), *Strategic Pragmatism: The Culture of the Economic Development Board of Singapore*, Cambridge, MA: MIT Press.
Science Council of Singapore (various years), *National Survey of R&D Expenditure and Manpower*, Singapore: Science Council of Singapore.
Soon, T.W. and C.S. Tan (1993), *Singapore: Public Policy and Economic Development*, Washington, DC: The World Bank.
USPTO, database of the US Patent and Trademark Office, downloaded from www.uspto.gov.
Wong, P.K. (1992), 'Technological development through subcontracting linkages: evidence from Singapore', *Scandinavian International Business Review*, **1**(3), 28–40.
Wong, P.K. (2001), 'The role of the state in Singapore's industrial development', in P.K. Wong and C.Y. Ng (eds), *Industrial Policy, Innovation and Economic Growth: The Experience of Japan and the Asian NIEs*, Singapore: Singapore University Press, pp. 503–79.
Wong, P.K. (2002), 'Manpower development in the digital economy: the case of Singapore', in M. Makishima (ed.), *Human Resource Development in the Information Age: The case of Singapore and Malaysia*, Tokyo: IDE/JETRO, pp. 79–122.
Wong, P.K. (2003), 'From using to creating technology: the evolution of Singapore's national innovation system and the changing role of public policy', in S. Lall and

S. Urata (eds), *Competitiveness, FDI and Technological Activity in East Asia*, Cheltenham, UK and Northampton, MA, USA: Edward Elgar Publishing, pp. 191–238.

Wong, P.K. (2006), 'The re-making of Singapore's high tech enterprise ecosystem', in H. Rowen, W. Miller and M. Hancock (eds), in *Making IT: The Rise of Asia in High Tech*, Palo Alto, CA: Stanford University Press, pp. 123–74.

Wong, P.K. and Z.L. He (2003), 'The moderating effect of a firm's internal climate for innovation on the impact of public R&D support programmes', *International Journal of Entrepreneurship and Innovation Management*, 3(5/6), 525–45.

Wong, P.K. and A. Singh (2004), 'The pattern of innovation in the knowledge intensive business services sector of Singapore', *Singapore Management Review*, 26(1), 21–44.

Wong, P.K., Y.P. Ho and A. Singh (2007a), 'Towards an "entrepreneurial university" model to support knowledge-based economic development: the case of the National University of Singapore', *World Development*, 35(6), 941–58.

Wong, P.K., L. Lee and Y.P. Ho (2007b), *Global Entrepreneurship Monitor (GEM) 2006 Singapore Report*, NUS Entrepreneurship Centre, National University of Singapore.

Wong, P.K., M. Kiese, A. Singh and F. Wong (2003), 'Pattern of innovation in the manufacturing sector of Singapore', *Singapore Management Review*, 25(1), 1–34.

Wong, P.K., L. Lee, Y.P. Ho and F. Wong (2005), *Global Entrepreneurship Monitor (GEM) 2004 Singapore Report*, NUS Entrepreneurship Centre, National University of Singapore.

Wong, P.K., F. Wong, Y.P. Ho and L. Lee (2004), *Global Entrepreneurship Monitor (GEM) 2003 Singapore Report*, NUS Entrepreneurship Centre, National University of Singapore.

Wong, P.K. and D. McKendrick et al. (2000), 'Singapore', in D.G. McKendrick, R.F. Doner and S. Haggard (eds), *From Silicon Valley to Singapore: Location and Competitive Advantage in the Hard Disk Drive Industry*, Palo Alto, CA: Stanford University Press, pp. 155–83.

Young, A. (1992), 'A tale of two cities: factor accumulation and technical change in Hong Kong and Singapore', *NBER Macroeconomics Annual 1992*, 7, 13–54.

4. Towards knowledge generation with bipolarized NSI: Korea

Chaisung Lim

1 INTRODUCTION

Korea has undergone a successful industrialization process from its beginnings as a poor, agricultural country exploited by Japanese colonization and subsequently devastated by the Korean War. Korea began a period of dynamic economic growth in the early 1960s and gross national income (GNI) per capita rose from $87 in 1962 to $12 197 in 1996 (current prices). After a two-year decline (1997–98) due to the financial crisis, Korea's economy recovered to reach $12 646 in 2003 (Figure 4.1). Economic growth has been led by large firms belonging to conglomerate groups known as *chaebols*, and based on accessing and exploiting the international market and international sources of technology.

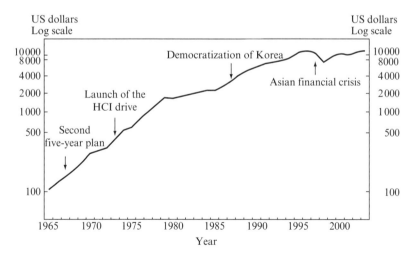

Source: OECD (2004b, p. 24).

Figure 4.1 Gross national income per capita in Korea

Korea's national system of innovation (NSI) is mid-sized, with a complexity and institutional range similar to that of other mid-sized countries such as Germany and the UK. Korea's NSI is typical of a catching-up country, in which it shifts from one of technology import to one of technology generation and effective utilization of emerging technological opportunities from abroad. Korea was in the low–middle-income group of the OECD countries in terms of GNI per capita in 2003 (OECD, 2005). The industrialization process in Korea was led by the government until the mid-1980s, when the private sector took over. In this process, firms accumulated the capabilities that allowed them not only to operate and improve production processes, but also to develop imitative products and even some original products based on these technologies. They relied on imported technology, components and raw materials. Korea thus displays a mix of advanced and developing country features in that its private sector comprises large advanced manufacturing firms, small manufacturing firms and service firms, which are still classed as laggard. In addition, the education and financial systems in Korea are also classed as laggard according to advanced country standards. The knowledge base of the knowledge-generating organizations, such as universities and government research institutes (GRIs) is weak, in spite of a recent upgrading in the R&D (research and development) competence of these organizations.

In the 1990s and early 2000s, Korea has been attempting to develop its NSI from one of a catching-up country, based on technology imports, to one of technology generation and effective utilization of emerging technology from abroad. The Korean NSI has undergone dramatic changes since the financial crisis of 1997. There has been a growing concern that although the intensity of R&D investment is one of the highest in the world, it is not being matched by performance at the national level. The aim of this chapter is to identify the features of the Korean NSI that are being challenged by globalization, and the changes that will be needed to allow Korea to become one of the group of advanced economies.

2 MAIN HISTORICAL TRENDS

Korea is one of those ex-colonized countries, such as Taiwan and Finland, that have demonstrated remarkable industrial growth since the Second World War. As in Taiwan, Japanese colonial rule provided a capitalist institutional base and introduced new infrastructures such as railway, ports, roads and irrigation systems. During its 36 years of colonization Korea underwent substantial industrial growth and structural change

(Kim and Kim, 1997, p. 6). However, this was not matched by a similar level of industrialization, which was below that of most manufacturing countries. This was mainly because the Japanese wanted Korea to remain an agricultural country. In addition, in spite of a 1.6 per cent annual growth in per capita GNP, the average living standards of Koreans declined. For example, per capita consumption of rice, and real wages per capita both declined (Kim and Roemer, 1979, p. 20).

The liberation in 1945, the period of US military rule from 1945 to 1948, and the Korean War of 1950–53 were all periods of political and economic chaos. As a result of its separation from North Korea, where most of the manufacturing facilities were concentrated, a war-devastated South Korea was left with almost no industrial infrastructure. The land reforms and the war that occurred between 1950 and 1953 led to the demolition of the landowner class, which had dominated agricultural society. The Korean people, most of whom were poverty-stricken, were forced to grasp the opportunities presented by industrialization.

Between 1953 and 1960, the country underwent a period of reconstruction with US aid (Kim, 1993). Political and economic institutions were slowly established and shaped. From the early 1960s onwards, under the Park government which lasted until 1979, there was a period of rapid economic development based on a series of five-year economic plans. The government's export policy in the 1970s stimulated firms to export production from labour-intensive industries, such as footwear, plywood and textiles, and in the 1980s encouraged exports of semiconductors and automobiles. This export policy was aimed at utilizing Korea's abundant workforce and exploiting technology, components, materials and capital goods from foreign sources. Kim (1993; 1997) maintains that the dynamic growth in Korea in the 1970s and 1980s was facilitated by the aggressive creation and accumulation of technological capabilities by domestic firms. The contribution of foreign direct investment (FDI) firms was not as important as in Singapore and Taiwan. Korean workers improved their skills through diligence and investment in learning. In this respect, they are not very different from workers in the other Asian countries of 'Confucian' culture, such as Taiwan and Singapore, which lay great emphasis on learning. Human resources adapted to imitating foreign products and improving production processes in firms were being supplied by a rapidly expanding education system.

Chaebol group firms, which diversified into related and unrelated business areas, were the organizations that created and accumulated technological capability through capital investment and investment in R&D and organizational learning. With the introduction in the mid-1970s of a policy aimed at increasing the importance of the heavy and chemical industries (HCI), the large firms belonging to *chaebol* groups grew substantially and were able to take advantage of economies of scale in large-scale investment and operation (Ahn and Kim, 1997, p. 372). Some economists interpret this promotion of the HCI as government intervention aimed at industrial upgrading for dynamic growth (Amsden, 1989), but it was also directed towards military objectives. It was recognized that there was a need to build self-defence capabilities in case of problems following the withdrawal of the US army (Kim, 1997). This policy meant that the HCI *chaebol* groups were cooperating with the political regime in the process of industrialization. The large *chaebol* firms were 'close' both economically and politically to the ruling regime. They therefore had access to both financial and human resources under advantageous conditions not enjoyed by small firms. This brought about a dual system of strong large firms and weak small firms, which became a chronic problem for the economy.

Efforts were made in the 1970s and 1980s to improve the public R&D infrastructure through the establishment of GRIs and the Daeduk science town. The financial sector favoured firms with large assets when conferring loans and, along with the tight government control of labour and the stable political and economic system in the 1970s, thus contributed to *chaebol*-driven economic growth until the mid-1980s.

In the late 1980s, Korea faced challenges from changes in both the local and the global environments (Ahn and Kim, 1997). With the wealth accumulated from successful economic growth, the *chaebol* groups, which had once been the economy's drivers, became non-market rent seekers speculating in land and other assets. This produced a nationwide phenomenon of asset speculation. The labour movement, in a bid for democratization in Korea, made demands for higher wages, the levels of which had been kept down during the process of industrialization. The political system became less stable as the authoritarian government began to be challenged by the democratization movement organized by students, trades unions and the opposition parties.[1] The economic system became progressively more open and liberalized as a result of pressure from the USA and international organizations such as the World Trade Organization (WTO).

In the 1990s, liberalization rendered the economic system unstable, and mismanagement of the process provoked the financial crisis of 1997 (OECD, 2000, p. 28). In this open economy, Korean firms were exposed to competition from both the less developed countries and the advanced countries. The 1990s, therefore, was a period when Korean economic growth lost momentum and the country underwent major reforms after the financial crisis. However, in spite of the unstable environment, some

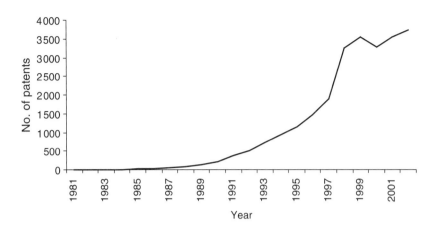

Note: Number of patents as shown by patents registered with the USPTO.

Source: USPTO website, 9 March 2003.

Figure 4.2 Korea's patenting trends

Korean large firms emerged as world-class manufacturers whose success in the world market was based on their technological capabilities, which had been continuously upgraded (see Figure 4.2). Polarization between a few large successful firms and a group of persistently laggard small firms made it difficult to chart the direction of Korea's NSI at that time. The GRIs and educational organizations were under pressure to change to meet the demands of firms for user-oriented research and for differently skilled labour. The government was increasingly confronted by 'difficult to manage' problems relating to the NSI, which was becoming more sophisticated as a result of the increased size of the economy and the dramatic changes that were occurring.

3   INNOVATION INTENSITY

The first innovation survey in Korea, modelled on the Community Innovation Survey (CIS), was conducted in 1996.[2] Propensity to innovate is illustrated here by various indicators based on the innovation survey. Share of innovative firms can be used as an indicator of Korea's innovation propensity. According to the 2002 survey, the share of innovative manufacturing firms with at least one successful innovation over the previous two years (2000, 2001) was 33 per cent, while for European countries this share

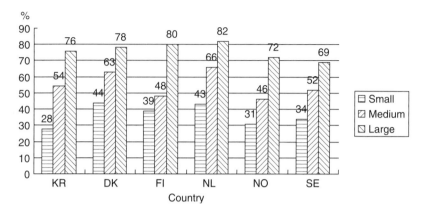

Note: 'Small' firms: 10–49 employees in Europe, 10–49 in Korea; 'medium' firms: 50–249 employees in Europe, 100–299 in Korea; 'large' firms: more than 250 employees in Europe, more than 300 in Korea.

Sources: Um and Choi (2004); European Commission (2004).

Figure 4.3 Share of innovative firms in the manufacturing sector

over the previous three years (1998, 1999 and 2000) was 44 per cent.[3] The more than 10 per cent gap between Korea and Europe clearly shows that Korean firms were less likely to innovate than European firms. Patents per million population as a measure of innovation propensity is a better indicator for international comparison and reveals that in 1998, Korean firms showed a lower propensity to innovate than firms in the advanced countries: 7.7 in Korea against 36.2 in the OECD countries, 52.2 in the USA, 80.9 in Japan, 69.9 in Germany, 41.4 in Denmark, 74.9 in Finland, 11.7 in Ireland, 49.8 in the Netherlands, 107.4 in Sweden and 26.4 in Norway (OECD, 2003, p. 178).[4] Large firms in the manufacturing sector were the most innovative – 75.5 per cent of large firms were innovators, compared with 53.7 of medium-sized firms and 28.8 per cent of small firms. According to the innovation survey, the gap between the share of 'large' and 'small' innovative firms depicted in Figure 4.3 is larger than for the European countries: the gap in Korea is 48 percentage points, compared with scores for Denmark of 34, Finland of 41, the Netherlands of 39, Norway of 41, and Sweden of 35.

The innovativeness of Korea's service sector is below (11 per cent lower) its manufacturing sector. It is also much lower than for the service sector in Europe (see Figure 4.4).

The prevailing pattern is one of imitative innovation relying on foreign technology. Among the product-innovating firms in the manufacturing

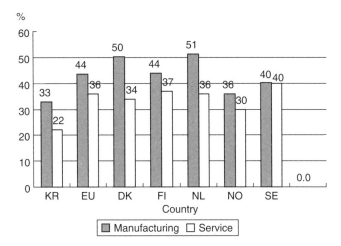

Note: The share of innovative firms in the EU is the share in manufacturing, mining and quarrying, electricity, gas and water supply. The share in Korea is only in manufacturing. Korea: innovation activities between 2001 and 2002 for the service industry; activities for 2000–2001 for the manufacturing industry. EU: innovation activities between 1998 and 2000.

Sources: European Commission (2004); Um and Choi (2004).

Figure 4.4 Share of innovative firms in Korea

sector in Korea, only 6.8 per cent produced product innovations 'new to the world'; most were either 'new to the firm' or 'new to Korea'. Access to foreign firms is the most useful way of acquiring technology. 'Purchasing the right to use invention or technological licensing from foreign firms' was rated as the most useful method of technology acquisition out of the nine methods listed, scoring 70.1 out of 100 points. The next most useful method used is 'Cooperation with customer firms' (64.1), followed by 'Purchasing the right to use invention or technological licensing from domestic firms' (61.6).[5]

4 ACTIVITIES THAT INFLUENCE INNOVATION

4.1 Knowledge Inputs to Innovation

4.1.1 R&D activities

R&D intensity tends to be high in Korea. According to the innovation survey, Korean firms spend 2.1 per cent of sales revenue on R&D. The proportion of gross domestic R&D expenditure in GDP was among the highest in the OECD countries (2.6 per cent) in 2003. In 2003 Korea

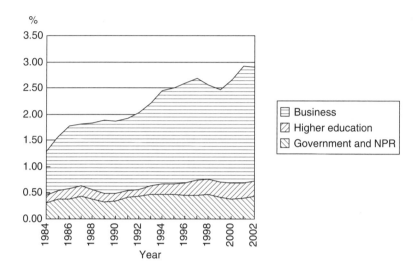

Note: NPR = non-profit research organization. In GRI and NPR expenditure, the majority is GRI.

Sources: KOITA (2003a); MOST (1991, 2002, 2003).

Figure 4.5 R&D as a fraction of GDP

ranked fifth for R&D intensity among the OECD countries (OECD, 2004a).

Korea's high R&D intensity is the result of 20 years of continuous increasing investment in innovation activities (see Figure 4.5). This intensified R&D activity has been led by large firms. Korea has a large business sector occupying 76.1 per cent of the total R&D expenditure in 2003, and with a share in total R&D larger than most of the OECD countries. In these respects, Korea is similar to Sweden. Domestic large firms in Korea invest in R&D for assimilation of foreign technology and to improve the ability to interpret and commercialize emerging technology from abroad. Firm R&D expenditure in total R&D expenditure dominated throughout the 1980s. The private sector accounted for 48 per cent of the total in 1980, 56 per cent in 1981, 73 per cent in 1983, and 79 per cent in 1984.[6] In the late 1990s, expansion of R&D investment made by small firms was stimulated by the expansion of venture businesses. The share of small firms in the business sector R&D doubled in the period 1995–2001, from 11.4 per cent of total business R&D expenditure in 1995 to 23.6 per cent in 2001 (Suh, 2004). See Box 4.1 for firm size and R&D activities in 2003.

> ### BOX 4.1 FIRM SIZE AND R&D ACTIVITIES
>
> In 2003, of all firms in all industries, 359 had more than 1000 employees, 2110 firms had 300–999 employees, and 20 317 firms had 10–299 employees. Of these, the top 20 firms in terms of R&D expenditure accounted for 51.7 per cent of total business R&D; the top ten firms in terms of R&D expenditure accounted for 43.7 per cent of total business R&D expenditure, and the top five firms accounted for 37.0 per cent. The top 20 firms in terms of number of researchers employed accounted for 35.9 per cent of total researchers in firms; the top ten firms for 30.9 per cent; and the top five firms for 27.5 per cent.
>
> *Source:* Korea National Statistical Office (2005); KITA (2004, p. 6).

The government R&D sector in Korea is relatively large, compared to that of Germany or the Netherlands (see Figure 4.6). Because university R&D is poorly developed, government R&D dominates the public sector, accounting for 13.8 per cent of total R&D expenditure in 2003. The sector consists of GRIs and national testing laboratories. National testing laboratories play a minor part in terms of their contribution to industry innovation activity. GRIs receive the major proportion of government R&D funding and also most of the new technology funding, and undertake mission-oriented research, R&D for strategic industries, and R&D for technology diffusion – most of which is applied research. The GRIs' role in technology diffusion is weak (Kim, 1997) and, compared with Taiwan, the contribution of GRIs to small firms is not well established. For instance, there are no examples in Korea of successful strategic national research projects carried out by GRIs to give momentum to technological capability accumulation by small firms. Government funding consists of government contributions and research contracts with government (Min et al., 2004). Contracts with firms provide additional sources of funding for GRIs. In 2001, 45.6 per cent of the salary costs in the GRIs were covered by research contracts with government, 11.3 per cent by contracts with firms and 33.5 per cent by government contributions (Cho et al., 2003).

As the R&D capabilities of universities and firms increase, the role of GRIs is being challenged. In the 1990s GRIs underwent considerable reorganization. Regular evaluation of performance (1991) and project-based management systems (1996) were introduced. However, establishing

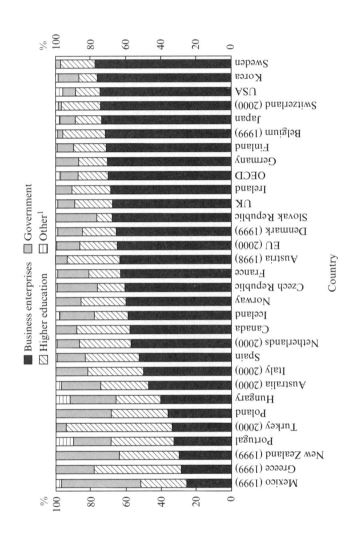

Note: [1] Private non-profit or not classifiable.

Source: OECD (2004b, p. 29).

Figure 4.6 R&D expenditures by performing sector (2001) (% of national total)

a new role for GRIs in the face of current requirements is an issue that is still being debated in science and technology (S&T) policy.

University R&D is weak. Although universities employ three-quarters of the nation's doctorates, the share of university R&D in total Korean R&D is small compared to the other OECD economies (Figure 4.6). Most universities have less well-equipped R&D facilities and less well-organized research teams than GRIs because, for several decades, the majority of R&D projects were concentrated in the GRIs. The quality of university research is well below the advanced countries' standards, and the impact of papers published between 1981 and 2001 was low: Korea ranked lowest among the countries examined in this book (Wang et al., 2003).

However, in the 1990s university R&D received increased funding, raising the share of university research in total R&D expenditure from 6.1 per cent in 1992 to 10.1 per cent in 2003. Universities have been under pressure to upgrade their R&D capabilities through a number of S&T policy programmes and education reforms. For example a programme for the establishment of centres of excellence (Science Research Centre, Engineering Research Centre), which started to be implemented in 1990–91, has not only made universities increasingly aggressive in competing for rewards for excellence in R&D, but has also intensified their R&D activities by encouraging a culture of more organized R&D teams and activities (Yoon and Hwang, 2000). Under the education reforms that were introduced, university evaluations since the mid-1990s have put greater weight on university R&D performance.

For a catching-up country where effective use of knowledge is more important than its generation, foreign firms are a significant source of knowledge. Foreign firms provide technology through technical licensing and capital goods. Imported technology continuously expanded over the 1980s and 1990s in Korea as technology imports rose from $107.2 million in 1980 to $2686 million in 1999.[7] However, R&D investments rose even more rapidly, driven by heavy investments by firms that, being exposed to increased global competition from both advanced countries and less developed countries from the late 1980s, wanted to upgrade technological capability. This increased R&D expenditure in the late 1980s was also influenced by government policy to reduce the tax on R&D expenditures (OECD, 1996, p. 29). Figure 4.7 shows that the share of royalty payments to business expenditure on R&D declined over the period 1976–2002, while the proportion of business R&D expenditure to sales increased over the same period (Suh, 2004, p. 24).

Korea's intensified R&D resulted partly from increased specialization in ICT (information and communications technology). The revealed

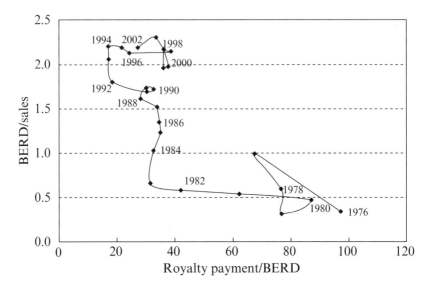

Note: BERD – business expenditure on R&D.

Source: Suh (2004, p. 24).

Figure 4.7 Changing relationships between royalty payments and R&D

technology advantage index, based on US patent data, shows that Korean patenting activities are concentrated in the ICT sector – semiconductors and electronics, telecommunications, computers, and 'heating, ventilation and refrigerators' (see Figure A3.2 in the Appendix). Korea ranked third among the OECD countries in terms of specialization in ICT in 1999 (OECD, 2003, p. 26). Therefore it is not surprising that the ICT sector accounts for the largest proportion of R&D expenditure. The 'electronic equipment' sector, according to the MOST (Ministry of Science and Technology) classification, including electronic components, telecommunication equipment and video equipment, accounted in 2002 for 53 per cent of total R&D in the manufacturing sector, with the automobile sector ranking second. The electronic equipment and automobiles sectors, which are dominated by large firms, together accounted for 68 per cent of business R&D expenditure (MOST, 2003). This reflects a shift in the industry structure since the 1980s from labour-intensive industries, such as textiles, to capital-intensive industries, such as ICT. According to data on value-added, in 1980 the major manufacturing sector industries were textiles and leather (5.5 per cent), food, beverages and tobacco (5.4 per cent), chemicals (3.4 per cent), and electricity and electronics

(3.0 per cent). In 2000, they were electricity and electronics (5.7 per cent), transport equipment (3.8 per cent), petroleum (3.5 per cent), chemicals (3.4 per cent), and food, beverages and tobacco (3.4 per cent).[8]

In terms of innovation activities, the Korean NSI is less open globally than the NSI of small advanced countries in the EU (see Table A2.6 in the Appendix). R&D and innovation activities of foreign affiliated firms remain peripheral in Korea. The share of R&D expenditure by foreign actors in Korea was negligible at 0.1 per cent in 2000. Among the top ten patent applicants (2001) in Korea, only one firm – LG Philips LCD (originally LG, which was sold to Philips during the financial crisis) – was a foreign-affiliated firm (KOITA, 2003b, p. 228). However, with the inrush of FDI following the financial crisis, the number of foreign firms' R&D institutes has increased dramatically: 62.8 per cent of foreign research institutes were established after 1995 (KOITA, 2002). The patenting activities of foreign-affiliated firms have increased rapidly since the end of the 1990s. For example, in electronics and electricity, the number of patents applied for by foreign affiliates in Korea (with more than 50 per cent owned by foreigners) increased from 4 in 1990 to 122 in 1995 to 1250 (1.1 per cent of all patents in electronics and electricity) in 2000.[9]

4.1.2 Competence building

The Korean education system provides for six years of primary schooling, three years in middle school, three years in high school and four years at university. Vocational training is provided through three-year courses at vocational high school and two-year technical college courses.

This system is a reflection of the nation's commitment to education. Korea's total spending on education at 7.1 per cent of GDP in 2000 is the highest in the OECD area (OECD, 2004b, p. 27). During the process of industrialization, the education system expanded based on private spending. Private spending on education as a share of GDP is the highest in the OECD countries (ibid.) (see also Table 4.1). Korean children received an education on a par with, or even better than, those in the advanced countries: the share of young adults aged between 25 and 34 with an upper secondary qualification is the highest among the OECD countries (see Figure 4.8). The Korean education system up to secondary level is regarded as very successful. In international tests, Korea ranked among the top three countries (PISA, 2000) for students of age 15, for science, reading and mathematics (OECD, 2004b).

The education policy introduced in 1981 encouraged expansion of the education system and increased student entry rates, resulting in entry into higher education doubling over five years: from 16 per cent in 1980 to 35.6 per cent in 1985.[10] However, despite this, the level of training received in

Table 4.1 Expenditure on educational organizations as a percentage of GDP (1999)

Country	Public[1,2]	Private[3]	Total
Korea	4.1	2.7	6.8
Norway	6.5	0.1	6.6
Japan	3.5	1.1	4.7
Sweden	6.5	0.2	6.7
USA	4.9	1.6	6.5
Finland	5.7	0.1	5.8
Denmark	6.4	0.3	6.7
Germany	4.3	1.2	5.6
France	5.8	0.4	6.2
Ireland	4.1	0.4	4.6
UK	4.4	0.7	5.2
Italy	4.4	0.4	4.8

Notes:
[1] Public, private: public and private sources of funds for educational organizations after transfers from public sources, by year.
[2] Includes public subsidies to households attributed to educational organizations.
Includes direct expenditure on educational organizations from international sources.
[3] Net of public subsidies for education organizations.

Source: OECD (2002, p. 171), as cited in Edquist and Goktepe (2003, p. 40).

Korean universities and technical colleges is not sufficiently high to satisfy the needs of large firms. Korean universities' investments in facilities, including equipment for scientific experimentation, are still not sufficient to allow them to use the most up-to-date equipment. In addition, the university curricula are outdated. Poor R&D capabilities in universities lead to poorly qualified new researchers. Technician training in accredited colleges is mostly based on textbook learning and does not include practical experimentation (Goh, 1998, p. 225). As a result, the tertiary education system has not been successful in providing well-qualified technicians. According to Woo (2002), based on ILO (International Labour Organization) statistics, the share of professional technicians in the manufacturing sector, 3.3 per cent in 1993, is much smaller than in Germany at 11.4 per cent in 1991, in Japan at 6.2 per cent in 1996, and in the USA at 12.0 per cent in 1994.[11]

Korea could obtain a supply of qualified researchers through having graduates study abroad. With the relaxation in the 1980s of the restrictions on this, graduate numbers have increased and the number of overseas trained doctoral graduates who return to Korea annually, which was 238 in

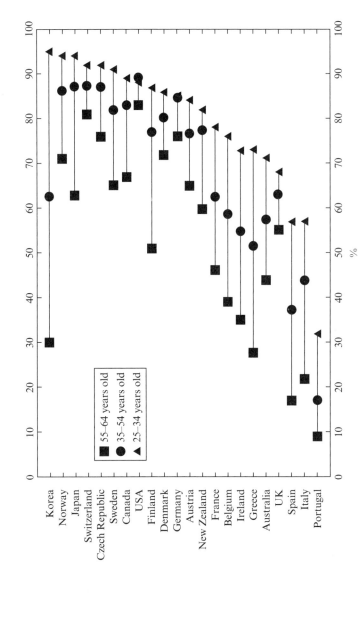

Source: OECD (2004b, p. 28).

Figure 4.8 Share of the population with at least an upper-secondary qualification (%, 2001)

1980, increased to 1510 in 2002.[12] Thirty-six per cent of these were doctorates in the natural sciences or engineering.

Alongside the formal education system, in 1993 there were 87 vocational training institutes and 232 enterprise-based training centres in Korea (OECD, 1996, p. 54). These institutes were established to try to satisfy industry's needs for a skilled technical workforce (Goh, 1998, p. 226).

In Korea, the drivers of competence building are the large firms. Therefore the emphasis on training has come from these large firms. In the 1990s, the gap between the skills required by large firms and what was being supplied by Korean education providers increased further. The training programmes and experimental equipment and testing equipment in large, advanced Korean firms are almost on a par with those in firms in the advanced countries. However, the training provided by Korean education institutes is far behind the standard of international training programmes.

Korean firms invest in training in part because of their need to upgrade the quality of graduates from the Korean education system, but also because they are obliged by law to make such investments. Firms with more than 1000 employees are obliged to provide workers with vocational training or to pay a training levy (OECD, 1996, p. 54). The ratio of training costs to total labour costs is high in large firms: in 2002 it was 0.4 per cent in small firms (10–299 employees) and 1.8 per cent in large firms (more than 300 employees). Over the 1990s, firms' expenditure on training increased roughly fourfold, and the share of training in total labour costs rose from 1.2 per cent in 1990 to 1.5 per cent in 2001, led by aggressive investment in training by large firms (Korea Research Institute for Vocational Education and Training, 2003, pp. 55–6).[13]

The Korean education system has been undergoing reform aimed at improving the quality of its provision. In the mid-1990s, the laws and regulations on higher education were revised, giving more responsibility to the higher education organizations. For example, a reform introduced in 1995 gave private universities autonomy over their admissions processes, and relaxed the regulations on the establishment of universities and colleges. It also required an evaluation of universities' performance in a bid to stimulate competition among universities (The Committee of Education Reform for the President, 1995). The results of this evaluation are made public.

In summary, the Korean education system has expanded greatly in terms of numbers in response to the demand from industry for higher levels of education while maintaining the quality of education up-to-the secondary education. Tertiary-level education in Korea is not good enough to satisfy the needs of advanced firms. The education system is under a reform process to improve its performance.

4.2 Demand-side Factors

There are various ways that governments can create markets. Large-volume markets can be created through investment in infrastructure, including telecommunications, high-speed trains and nuclear power generation plants. These markets provide opportunities for domestic firms and other relevant organizations in Korea to invest in R&D for adoption, application and development of technology (Song, 1999; Lee, 2004). Korea's ICT policies are an example of efforts directed to market creation.

Government policies in the 1990s focusing on rapid diffusion of ADSL (asymmetrical digital subscriber line) broadband Internet were responsible for the commercial success of broadband Internet, which at the time was not regarded as being commercially feasible by many advanced countries. In Korea, the number of subscribers expanded beyond break-even point. Government policies included competition policy, which meant that entry and pricing in the market was not regulated and resulted in facility-based service providers (FSPs) of broadband Internet facing strong competition; and policies of early commitment for the promotion of the Internet. In 1999 and 2000, the government provided public loans to FSPs to facilitate the roll-out of the market. This reduced the financial burden of the investments required to access networks. A third government policy was the cyber building certificate system, introduced in 1999, which facilitated diffusion of the broadband network to apartments by providing apartment construction companies with certificates that could be used for marketing purposes. In Korea, 48 per cent of households live in apartment complexes, which provided economies of scale for FSP's market operations (Hong and Ko, 2005). This led to an expansion in the market for broadband technology and broadband connection services (Kim and Jeon, 2003). As a result, Korea has one of the highest levels of Internet service subscribers (*The Economist*, 2003a).

The Government can also create a market by selecting a particular industry standard. For example, when the Korean government declared CDMA (code division multiple access) as the national standard for mobile telecommunications in the early 1990s, Korea was the first country to adopt CDMA as the national standard for mobile communication, and a huge market for CDMA telecommunications equipment was created. Domestic firms invested in CDMA technology to gain market share (Lee et al., 2005; Song, 1999).

Demand from export-oriented large firms for local supplies of imported components has created a market opportunity for local firms, which have been obliged to upgrade their technological capabilities in the process of developing and producing the products ordered. Korea has large firms that have to satisfy the specifications of advanced country customers in order to export. In 2003, 13 Korean firms were members of the *Fortune* 'Global 500'. Some of the largest firms, such as Samsung Electronics, have world-class 'brand power' (see Box 4.2). Due to intensified competition in the global market in reducing price and enhancing quality, Korean firms had continuously to seek out local and international suppliers for outsourcing components and services.

BOX 4.2 SAMSUNG ELECTRONICS – WORLD-CLASS BRAND POWER

The value of Samsung's brand is close to that of Sony, according to the consultancy firm Interbrand. The management of Samsung Electronics needed to improve brand power; they found that their products, although less expensive, were of the same quality as those of the top brand producers. Samsung invested heavily in advertising. The firm recruited Eric Kim, an expatriate in the USA, to run a global marketing office. In dealing with advertising channels, he unified fragmented channels of advertising that involved more than 50 different agencies. He launched a daring campaign, which resulted in the Samsung product appearing in *Matrix Reloaded*, and in Samsung being a principal sponsor of the 2004 Olympic games. The success of the Athens Olympics brought about a big increase in consumer awareness of the Samsung brand.

Source: The Economist, 15 January 2005.

FDI firms have also played a part in influencing quality upgrades. For example, Volvo Korea, formerly Samsung Heavy, sent a specialist team from its headquarters in Sweden to inspect the quality of the work of all the local subcontractors and select those producing the best-quality products. This put pressure on all domestic subcontracting firms to try to meet Volvo's specifications (Lim, 2003).

Regulation can also produce better-articulated demand. Regulations in Korea relating to the environment and health and safety were tightened in the late 1990s (Regulatory Reform Committee, 2003, p. 522). Tighter environmental regulation was motivated by pressure from international organizations such as the OECD and the WTO. The Korean government set environmental targets for 2002 in line with those in the USA and other advanced countries. All of this increased the pressure on industry to produce more environmentally friendly products.

4.3 Provision of Constituents

4.3.1 Provision of organizations

Korea has been aggressive in generating organizations in order to capture opportunities from new technologies. The US venture business boom in the mid-1990s stimulated Korean policy makers to find ways to incubate such new businesses. As a result of the financial crisis in 1997, the Korean government attempted to revitalize the economy through the encouragement of venture businesses. To provide an attractive environment for these, heavy investment was made in the IT infrastructure. Korean information infrastructure investment between 1996 and 1999 was 1.9 per cent of GDP, higher than in the USA (0.5 per cent), Japan (0.3 per cent) and Singapore (0.6 per cent) (OECD, 2000, p. 257). Korean venture promotion policy included a set of criteria identifying those venture businesses that should be offered incentives.[14] In 2000, the law on establishing firms was amended to allow investors in firms that fitted the criteria to receive tax incentives.[15] Physical facilities were provided for venture businesses. All of these policy changes contributed to an increase in the number of newly established ventures from 422 in 1996 to 878 in 1999 (OECD, 2000, p. 150). Some of the new ventures were launched by large firms that were finding it difficult to react to the burgeoning commercial opportunities in the venture business area: 161 spin-off companies were set up by ex-Samsung employees, and 98 by former Hyundai staff (see Table 4.2) (Suh, 2002). The large firms also set up 'corporate ventures', which were managed by one of their employees, in order to capture the commercial opportunities offered by new technologies.

With the influx of venture businesses, many new firms were established. According to the statistics on new organizations, mostly firms, the number of new organizations established in 1997 was 21 057, reaching a peak in 2000 of 41 460. As the venture business boom subsided, the number of new organizations decreased to 36 157 in 2002. However, the ratio of births to deaths, the number of new firms divided by the number of bankrupt ones, increased from 3.4 in 1997 to 20.2 in 2002 (Small and Medium Business Administration, 2003, p. 18). This shows that there was a continuous increase in the number of new firms over this period. Bearing in mind that the period between 1997 and 2002 included the financial crisis and subsequent slow economic growth, there has nevertheless been a positive change in the business environment and in the attitude towards launching new firms. According to the General Entrepreneurship Monitor report, Korea

Table 4.2 Spin-offs from the chaebols

	No. of mother companies	No. of spin-off companies						No. of employees
		1997	1998	1999	2000	2001	Total	
Samsung	16	0	115	29	5	12	161	17 235
Hyundai	12	36	27	18	8	9	98	16 937
LG	15	5	18	51	14	6	94	21 443
SK	11	3	11	11	13	7	45	3 650
Hanjin	5	0	0	4	1	0	5	2 866
POSCO	1	0	0	0	1	0	1	40
Hanwha	2	0	0	1	0	0	4	2 636
Doosan	1	0	0	0	3	1	4	103
Ssangyong	1	0	0	0	2	0	2	880
Dongbu	1	2	5	1	1	0	9	144
Dongyang	2	0	0	2	1	0	3	227
Hyosung	1	0	2	0	0	0	2	52
CJ	3	1	0	0	1	4	6	643
Kolon	3	0	0	0	3	0	3	289
Hyundai Dept	1	0	0	0	0	1	1	658
Daewoo E.	1	0	0	4	0	0	4	60
Total	76	47 (10.6)	178 (40.3)	124 (28.1)	53 (12.0)	40 (9.0)	442 (100.0)	67 863

Note: Spin-offs are confined to cases of management buy-out and employee buy-out.

Source: Data from Federation of Korean Industry cited in Suh (2002).

ranked high (fourth in 2001) in the total entrepreneurial activity index (Small and Medium Business Administration, 2003, p. 19; Harding, 2002).

The venture business firms, i.e. excluding venture capital firms, invested heavily in R&D for innovation, 8.5 per cent of R&D expenditures in 2001 (Small and Medium Business Administration, 2003). These venture firms contributed greatly to the dynamic growth of Korean industry. For example, the share of exports by venture businesses rose from 2.4 per cent in 1999 to 4.3 per cent in 2002 (Small and Medium Business Administration, 2003, p. 212).

As Korea was not at the world frontier in science and engineering in new industries, it was not able to fully exploit the technological opportunities from new discoveries in the area. The venture company policy was aimed at stimulating the growth of technology-based small firms in both new and traditional industries.[16] Of the venture businesses created between 1995 and 1999, 40 per cent were in the computer and telecommunication equipment sectors (OECD, 2000, p. 152), the remaining 60 per cent including traditional industries such as machinery.

There were some problems related to funding of venture businesses. The government's financial incentives related to firms classified as venture businesses by recognized public or private organizations. Therefore funding was not based on performance in the market. This led to overheated venture financing, which channelled financial resources to firms even when they were not successful in the market, enabling non-competitive firms to survive. After the decline in 2000–2001 of the venture business bubble in the USA and then in Korea, venture businesses that were not competitive were not able to survive in the even tougher market that had developed. The performance of some of these venture businesses was not in line with the government's strategic goals to use them as engines of economic growth.

However, venture businesses are undoubtedly influential actors in new industries such as ICTs and biotechnology. In spite of their sometimes poor levels of performance they are recognized by policy makers as being one of the three major firm groups (the other two being large firms and FDI firms), since the growth of venture businesses is still seen as a solution to the problem of weak competitiveness in the small-firm sector (Knowledge Management Team, 2003).

4.3.2 Networking, interactive learning and knowledge integration

Interactive learning among firms is one of the micro-foundations of a system of innovation (SI) (Lundvall, 1988; 1992). However, in Korea, interaction among domestic firms was less important for learning than interaction with firms in the advanced countries, with which Korean firms needed to liaise closely to achieve successful adoption and assimilation of foreign technology. As a consequence, networking among domestic firms in Korea was not well developed. As the technological capability of domestic actors has risen alongside increasingly successful economic growth, domestic sources of knowledge are becoming more important. Citation analysis of US patent data shows that the share of Korean patents cited in Korean patents registered in the USA showed an increase in the 1990s (Park, 2004).

One key characteristic of the Korean NSI is the existence of strong *chaebol*-led networks in the major industry sectors. Vertical linkages among firms are common in these industries. Large firms belonging to *chaebol* groups have subcontracting relationships both with their affiliated firms and with other small firms that are not affiliated. According to a Korean innovation survey of the manufacturing sector conducted in 2002 (Shin et al., 2002, p. 127), the second most important partner is the 'mother firm/*chaebol* affiliated firms/their affiliated firms', while the greatest contributors as partners for collaboration in innovation processes are

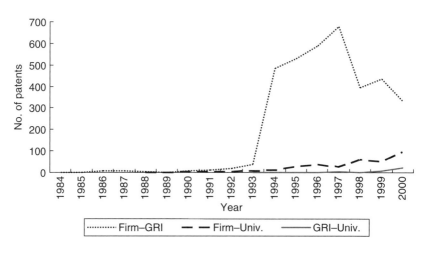

Source: Korea Institute of Patent Information.

Figure 4.9 Trends of co-invented patents

customers/user firms. With the rapid expansion of venture businesses, venture business firms are becoming established as a group that collaborates with large firms in ICT and biotechnology. These venture business firms differ from the traditional small firms in that they have advanced technological capabilities in specialty areas, and can therefore establish horizontal relationships, including strategic alliances, with large firms to help reduce the cost and risks involved in the process of innovation (Suh, 2004).

Cooperative R&D activities are becoming more common. Figure 4.9, depicting trends in cooperative patents, shows that cooperative R&D really took off in the 1990s. Cooperation between GRIs and firms is the main trend in cooperative R&D. A study of published collaborative papers in the ISI database shows that the number of papers with authors from different organizations in Korea increased from 41.5 per cent in 1991–95 to 55.8 per cent in 1997–2001 (Ahn and So, 2003). This can be seen as reflecting the 1990s policy drive towards cooperative R&D. In national R&D projects, proposals for collaborative research projects between GRIs, firms and universities receive priority. Most of the specially designated national R&D programmes and industrial generic technology R&D programmes in the 1990s were implemented as collaborative projects (Kim et al., 2000). As a result, universities have had more opportunity to become involved in cooperative R&D.

Although funding is seen as a way of encouraging collaborative R&D, collaboration in its turn is regarded as a relatively easy way to obtain research funding from government. Frequently such research involves little real collaborative work. Of those firms that were nominally part of a government collaborative research programme, 48.8 per cent carried out the research independently and only 26.7 per cent were involved in a regular exchange of knowledge and information as part of their research (Kim, 2003). Therefore the amount of interactive learning that results from collaborative research projects is small.

There are some barriers to interactive learning by firms in Korea. For instance, firms that belong to a *chaebol* group may be reluctant to interact with firms from a different *chaebol* group. Interactive learning between large firms and small firms is impeded by the low level of competitiveness of small firms in Korea which is the result of the dual structure in Korea of strong large firms and weak small firms, a problem that has not been resolved during the catching-up process.[17] In order to overcome the weak competitiveness of small firms, in 1990 the government introduced efforts to encourage innovation by small firms through collaborative R&D with GRIs and universities.[18] However, there are no examples to date of successful small firm/GRI collaborative R&D in relation to strategically important technology – the successes of CDMA and digital TV achieved by large firms in Korea and the examples of collaboration between small firms and GRIs found in Taiwan have not been repeated (see Chapter 2 in this volume). In any discussion of networking and interactive learning, networking with international actors should not be overlooked, and is discussed further in Section 6.

4.3.3 Provision of institutions

Historically, the Korean government was led by bureaucrats and the role of parliament was small. The Korean government's S&T policy has been driven by central government. In Korea, where local autonomy was only introduced after 1995, the role of local governments is small, although it is beginning to increase as a result of recent industry policy emphasis on regional development. The Korean government has traditionally been the regulator and driver of the industrialization process. In the 1960s and 1970s, government intervention in the early stages of industrialization was direct. From 1980 to 2000 this direct intervention was superseded by market mechanisms. Over the ten years from 1992 to 2002, successive governments engaged in deregulation, which left little room for direct policy actions. In the five years from 1998 to 2002, 5888 of 11 125 legal regulations were abolished (Regulatory Reform Committee, 2003, p. 44).

After the financial crisis, initially as a result of pressure from the International Monetary Fund, the government became the driver of reforms in corporate governance, labour relationships, the financial system and the government sector. These reforms introduced major institutional changes in these sectors. The corporate governance reforms forced large firms to focus on profitability and shareholders' interests, rather than increased size (OECD, 2000, p. 29). The legal framework for corporate governance has improved significantly since the financial crisis, and the ceiling on foreign ownership has been abolished.

As a result of the financial crisis, a significant proportion of the labour force was laid off, and lifelong employment in Korea is no longer a reality. In 1998, a policy for enhancing the flexibility of labour markets was introduced. Layoffs for 'urgent managerial needs' including mergers and acquisitions, and the creation of temporary work agencies to transfer workers to other companies, were allowed (OECD, 2000, p. 191). Payment practices changed so that performance was reflected in wages. The percentage of companies adopting systems of contract-based employment and performance-based payment increased significantly from 1.6 per cent and 5.7 per cent respectively in 1996 to 27.1 per cent and 21.8 per cent in January 2001.[19]

After the financial crisis, several policies were introduced, designed to enhance market competition, or to reinforce policies that were already being implemented. First, the trade liberalization that had begun in the 1980s was continued. Trade barriers were reduced, and non-tariff barriers were dismantled; the import diversification programme, which restricted the import of certain items, was abolished in 1999. The 220 quotas in place in 1994 had been reduced to nine by 2000 (OECD, 2000, p. 175). Second, FDI policy was reformed in a bid to attract foreign investors. Investment of foreign capital was regarded as not only good for stimulating competition in Korea, where an oligopoly of three or four large firms was common in the main industry markets, but also desirable for stabilizing the newly liberalized Korean economy. Third, to encourage fair competition, there was a strengthening of fair trade policy. In 1996, the Korea Fair Trade Commission (KFTC) was given ministerial status. After the financial crisis, its role was further reinforced by its being awarded more investigative power (OECD, 2000, p. 21). The KFTC has tried to control the structure of *chaebol* investment in other firms through direct intervention (ibid., p. 173).

Protection of intellectual property rights (IPR) was strengthened from the late 1980s. The share of Korean-owned patents registered in Korea increased from 11.4 per cent in 1980 to 32.9 per cent in 1990. As Korean patenting activity was intensifying, the Korean government realized that safeguarding IPR was important not only to protect foreign inventors, but also to protect domestic inventors whose numbers were increasing rapidly. Also, the USA was pressing for IPR protection. As the result of a series of negotiations between Korea and the USA, the patent law was amended in 1987 to protect IPR in computer software and materials. To enhance the protection of inventions, and expand the scope of protected inventions, a product patent system was introduced in 1987. The term of patent right protection was extended from 12 to 15 years. The penalties for infringing patent rights were increased. IPR protection became an important issue under TRIPS (the agreement on trade-related aspects of intellectual property rights) in 1995, and this became increasingly important for encouraging innovation among domestic inventors, who were responsible for the more recent patenting activity in Korea. The share of Korean patents registered in Korea in 2003 was 69.1 per cent (Korea Intellectual Property Office, 2004).

Certain Korean institutions, habits and norms are problematic for the development of its NSI. Networks among Koreans tend to be 'clannish', constituting what might be called a society of low 'social capital'. *Chaebol*-group-affiliated firms tend to interact only with firms within their own *chaebol* group. Government officials tend to work only to better their own departments or ministries.[20]

Korean government policies in the past were based upon a bureaucratic, top-down approach. Over the 1990s there was a trend towards a more interactive, bottom-up approach. For example, several national R&D programmes, including the '21st frontier R&D programme' were announced, which enabled non-government official groups to lead the planning and implementation processes (Kim, 2002). However, Korean government policy still has a long way to go in terms of the interactive approaches necessary if solutions to the increasingly sophisticated problems of the NSI are to be found.

The late 1990s and early 2000s have been a period of major institutional change as the beliefs, values, norms and rules of the 1970s and 1980s are being subjected to very radical changes. In spite of the changes that have already taken place, a positive outcome at the national economy level has yet to be achieved.

4.4 Support Services for Innovating Firms

4.4.1 Incubating activities

In order to encourage establishment of venture firms, various policy drives aimed at stimulating new firms have been launched. The main focus of these policy initiatives has been on providing physical facilities and facilitating the flow of investment towards new small firms (Kim and Lee, 2005). This has included tax incentives, provision of services, relaxation of the requisites for establishing a corporation, and granting temporary leave to university and other R&D institute professors and researchers to establish venture businesses (Jeong, 2005).

Table 4.3 Business incubators in Korea

Programme	No.	Ministry
Small firm business incubators	293	Ministry of Industry and Energy (Small and Medium Business Administration)
Software support centre	52	Ministry of Information and Telecommunication
Technology incubator	1	Ministry of Science and Technology
Centre for Supporting the Culture Industry	2	Ministry of Culture and Tourism
Total	348	

Source: Small and Medium Business Administration (2003).

Business incubators to nurture these new firms were set up under the initiatives of different ministries, including the Small and Medium Business Administration of the Ministry of Commerce, Industry and Energy (see Table 4.3). The first incubator was established in 1993, and 281 were established between 1998 and 2002. The eventual 348 incubators resulted in 4989 firms at the end of 2002.

The majority of business incubators are located in universities. Of the 291 incubators approved by the Small and Medium Business Administration in 2005, 83.2 per cent (242) are located in universities, 5.5 per cent (16) are in public research institutes, 3.1 per cent (9) are at the Small Business Corporation[21] and 2.7 per cent (8) are local government owned.[22]

Although Korea has a similar number of incubators to other advanced countries (Kim and Lee, 2005), they do not perform well in terms of providing services that satisfy the demands of firms, lacking experts to give advice (Kim and Lee, 2005; Yang, 2004). The incubator policy in Korea is not regarded as a success although the policy had a positive impact in providing an infrastructure for new entrepreneurial businesses (The Research Group on Innovative Small Firms, 2005, p. 57).

4.4.2 Financing

The financial system in Korea is based mainly on the banking system. Finance from banks is seen by Korean firms as more important than that derived from the securities market. Financial organizations were once controlled by government. Credit analysis and internal risk control mechanisms were poor. The lending system, which gave credit on the basis of mortgages, advantaged large firms because of their greater assets. However, after the financial crisis, government policy was to reduce the debt ratio of large firms, thereby liberating some finance. The financial organizations tightened their conditions for lending money to firms. A new supervisory framework for the financial sector, which was devised in imitation of the US system, was established to scrutinize the behaviour of financial organizations in a bid to avoid a future financial crisis (OECD, 2000, p. 29). Financial organizations have been learning to build their own lending systems on the basis of credit information and estimated risks. However, the new framework has not had a positive outcome, and financial organizations have been reluctant to assume the risks involved in lending money. This, it is argued, has reduced monetary flows and is one of the major causes of low investment in the economy (Cho, 2004; Jeong, 2004).

All these changes are the result of globalization and the reforms that followed the financial crisis. From the latter half of the 1980s, regulation of financial organizations was eased and the market was opened to foreign firms. However, the poorly coordinated liberalization process was unable to prevent excess inflows of short-term overseas borrowing led by large firms, which increased at an accelerating rate and produced the financial crisis (OECD, 2000, p. 28).

The financial reforms put in place after the crisis included a loosening of financial sector regulation on foreign investments. The share of foreign ownership in the banking sector increased from 7 per cent in 1997 to 27 per cent in 2002. The largest shareholders in the three commercial banks in Korea, and in four other banks, were foreign investors (OECD, 2004b, p. 113). As a result of the opening up of the securities market, in 2003 foreign investors controlled 36 per cent of the stock market, compared to 15 per cent before the crisis (*The Economist*, 2003b).

With the venture business boom in the late 1990s, the role of financial organizations in funding new technology-based firms (NTBFs) became increasingly important. The policies for supporting the financing activities for NTBFs are the following: (1) a policy of supporting funds for start-up companies with less than a three-year history; (2) a policy to provide guarantees of credibility for competent venture firms; (3) a policy to favour venture business firms in the allocation of funds to small firms; and (4) efforts to channel financial resources to NTBF through stimulating growth of venture capitals (Jeong, 2005).

The number of venture capital firms increased dramatically after 1997, with 111 firms being established between 1998 and 2002. In 2002 the total number of venture capital firms stood at 128. The capital of these firms totalled over 2000 billion won (US$1.6 billion) in 2000 (Small and Medium Business Administration, 2003, p. 43). Government contributed to the establishment of investment funds. For example, in 2002, government funding represented 240.4 billion won (US$204 million) while private firms' annual investment for venture businesses reached 568 billion won. Annual investment, excluding financial loans, by venture capital firms in venture businesses increased from 642.6 billion won (US$383 million) in 1997 to 3055.9 billion won (US$2576 million) in 2002 (ibid., p. 47).[23] As venture businesses have begun to decline, the government has increased its investment in venture capital funds from 15.3 per cent in 1999 to 39 per cent in 2004. It has been stated that, even so, resources tend to go to businesses that have been established for two years or more. The share of new businesses (established for less than two years) that received funding declined from 57 per cent in 2000 to 26 per cent in 2003 (The Research Group on Innovative Small Firms, 2005, p. 49).

4.4.3 Provision of consultancy services

Knowledge-intensive business services (KIBS) are not well developed in Korea. According to Lee et al. (2002), knowledge-intensive services in 2002 accounted for 1.8 per cent of the sales of the whole industry, a slight increase over 1.6 per cent in 1996.[24] Among KIBS firms, computer-related services have expanded rapidly, with an annual rate of increase in sales of 35.5 per cent in the 1996–2000 period, while the rates of increase in R&D services and business services (including consultancy) were 8.2 per cent and 5.7 per cent respectively. The rate of increase in technical services, which includes construction technology and 'engineering, S&T services', declined (the annual increase rate was minus 1.5 per cent). R&D services, business services and technical services are traditional KIBS sectors that closely interact with manufacturing firms in process or product innovation. It can be seen, therefore, that traditional KIBS activities directly relevant to innovation have been growing relatively slowly, in contrast to the fast-growing computer-related services.

These trends indicate that consulting services, which are regarded as important to the innovation process, are poorly developed in Korea. According to the 2002 Korean innovation survey, in terms of importance (on a five-point scale) of sources of innovation information, consulting firms ranked seventh among the external sources of organizations. Consulting firms are regarded not only as less important than those firms that companies routinely interact with,[25] but also less important than other public R&D organizations.

4.5 Summary of the Main Activities Influencing Innovation

R&D inputs have been discussed in relation to both R&D activities and technology imports. R&D activities in Korea are led by the private sector and dominated by large firms, while small firms are weak despite the recent increase in small firms' innovation activities due to the emergence of venture businesses. Imported foreign technology, which was once the major technology input, is still important, but now plays a minor role compared to R&D activities. Korea has a relatively large GRI sector, which is responsible for most public R&D activity and expenditure. The role of GRIs is being challenged by the increasingly intensive R&D activities undertaken by universities and firms. Korea's university sector is among the weakest in the OECD economies, although R&D investment has recently increased. The education system in Korea produces a large pool of human resources but is under increasing pressure to upgrade the quality of education being offered.

On the demand side, the government's investment in infrastructure has provided market opportunities, and government ICT policy has created a market for ICT technology and services. Export-oriented large firms and FDI firms help to articulate demand for local supplier firms.

There was a shift in the policy focus in the late 1990s from large firms to venture businesses. As a result, venture businesses and incubators have become influential actors in new industry sectors such as ICT and biotechnology. Networking is not well developed in Korea. However, as the technological capabilities of these actors have risen over the last decade, cooperation has increased. The dominant type of interfirm networks in Korea's major industries are the *chaebol*-led networks based on subcontracting relationships between large and small firms. There has been some progress in universities–industry–GRI networks, but more is needed. Following the financial crisis in 1997, major institutional changes involving investment of foreign capital, labour relations and the financing and governance of large firms were introduced, which has led to radical changes in the beliefs, norms and rules in Korea. The market environment has become more open, in line with government policy to promote competition.

The features of the Korean NSI show that there are disparities in the development of innovative capabilities, and mismatches between the actors and the institutions, while networking among the domestic actors is slowly developing. These features reflect characteristics of a catching-up country moving from an NSI of knowledge import to one of knowledge generation and effective utilization of foreign technology from abroad. However, they also are indicative of the number of policy issues that remain to be tackled.

5 CONSEQUENCES OF INNOVATION

Krugman (1994) argues that the fast economic growth of the Asian 'tigers' was the result of mobilizing resources such as labour and capital. Although opinions may vary depending on different methods and approaches used in analysing economic growth, it is clear that Korea over the last two decades has been progressively moving towards an economy where innovation activities are increasingly important for economic growth.

According to firm-level data, a substantial proportion of sales is driven by innovation activities. According to the 2002 innovation survey, among successfully innovating firms in 2001, 18.2 per cent of the sales were generated by new products developed in 2000–2001 and 34.2 per cent of the sales were generated by improved products.

The effect of innovation can be discussed in relation to productivity growth. A study was conducted on the relationship between R&D and labour productivity (Lee and Kim, 2003). According to this study, a 1 per cent increase in R&D investment (stock of R&D investment) is estimated to have increased labour productivity (per labour hour) by an average of 0.13 per cent over the 1980s–1990s, 0.12 per cent in the 1980s and 0.16 per cent in the 1990s. Patenting activity, as an approximate measure of innovation activity, has an impact on total factor productivity (TFP). A Korean study on patenting and TFP using data from 1964–2000 estimated that a 1 per cent increase in patenting application in any one year produces an increase in TFP of 0.11 per cent in total productivity in the subsequent five years (Youn et al., 2003).

The contribution of capital investment (capital stock net of stock of R&D investment) was higher than the R&D investment in the 1980s and 1990s, although R&D activities had became increasingly important for enhancing productivity over the 1980s and 1990s. Suh (2002) estimated that the contribution of R&D intensity (number of researchers in total labour) to labour productivity in the period 1999–2000 was higher than in 1995–96. All of this implies that Korean productivity growth has increasingly relied on innovation activities.

Korea's relatively high R&D intensity is not matched by economic performance. R&D intensity (R&D expenditure/GDP) as an input measure, as discussed in Section 3, is high in Korea. However, this is not matched by innovative outputs. In addition, Korean labour productivity per hour worked in total manufacturing is the lowest among the ten countries studied in the ESF (European Science Foundation) project (Kaitila, 2003). In terms of GDP per hour worked, as an indicator of productivity, in 2002 Korea was at 37 per cent of the level of the USA, while Germany and the Netherlands were at 101 per cent and 106 per cent respectively (OECD, 2003a). GDP per capita is low in comparison with the OECD countries, and is the lowest among the four Asian 'tigers' (see Table A2.3 in the Appendix). All of this indicates possible problems in the Korean NSI. Recent policy initiatives to address these problems have been introduced and are discussed in Sections 7.2 and 7.3.

6 GLOBALIZATION

In Korea, where the level of FDI is lower than in other East Asian 'tigers' such as Singapore or Taiwan (see Chapter 2 on Taiwan, this volume), access to global markets and knowledge has been mainly driven by large Korean firms. Korean firms have enhanced their technological capabilities through accessing foreign firms via technical licensing arrangements, import of capital goods and components and exporting under original equipment manufacturing and, more recently, original design manufacture and own brand-name arrangements. Korean firms are increasingly involved in cross-licensing and the sharing of intellectual property, acquiring firms, strategic alliances in R&D and marketing (Kim, 1997; KOITA, 2004). Korea's large firms, which have become multinational companies, have foreign subsidiaries and branch offices for production, R&D and marketing. Korea can be said to be well connected within the global production network (Hobday, 1995; Ernst and Ravenhill, 1999; Ernst and Guerrieri, 1998).

In addition, Korean large firms are increasingly searching for competitively priced supplies to substitute for domestic suppliers (Lee et al., 2005). For example, even firms affiliated to the Samsung *chaebol* group find it increasingly difficult to sell their products to Samsung Electronics. FDI

Table 4.4 The Korean NSI with high R&D intensity and low performance

I–O	Criteria	Compared to OECD	Compared to EU	Compared to Asian tigers
Input	R&D intensities: R&D per GDP (see Section 3)	Highest level	Highest level	Highest
Output	Innovativeness: innovation survey (see Section 3)	n.a.	Lower	n.a.
	Innovativeness: patenting per population (see Section 3)	Low	Low	n.a.
	Productivity	n.a.	Lowest	Lowest
	GDP per capita1	n.a. Low–middle in terms of GNI per capita (OECD, 2005)	Low	Lowest

firms are also leading the trend in global outsourcing. This is a big challenge for Korean small firms with weak competences.

In comparison with its global production network, Korea's R&D network is less well connected. Korea has a poor record in co-authorship of scientific articles and co-invention of patents compared to the advanced countries. Korea is ranked very low in terms of percentage of patents with foreign co-inventors (OECD, 2003a, p. 127). The extent of international cooperation in scientific research co-authorship is low. The share of co-authored articles in international journals in the ISI database for Korea in 1999 was 26 per cent, which is higher than Japan (18 per cent) but lower than advanced Western countries such as France, Germany, the UK and Canada, and also lower than Malaysia and Thailand, with 56 per cent and 61 per cent respectively (Lee et al., 2003). In addition, Korea's patenting activity abroad is weaker than that of the other Asian 'tigers'. Korea's domestic ownership of inventions made abroad in the 1980–2002 period was only 3 per cent, compared with Taiwan at 9 per cent and Singapore at 8 per cent (see Table A2.7 in the Appendix).

The government policy swing to attract foreign capital and the positive change in attitude towards foreign capital increased the influx of FDI into Korea in the late 1990s to its highest level. The financial crisis provided the momentum for opening up the Korean NSI to foreign actors. The government's policy to attract foreign capital is exemplified in the Foreign Investment Promotion Act of 1998, which allowed for a reduction in the regulation on FDI. It provided a basis for tax exemptions and incentives for foreign investment. In addition, in order to improve the environment for foreign firms at the regional level, in 2003 Incheon, Busan and Gwangyang were designated free economic zones. As a result of these policy changes, the influx of FDI expanded dramatically. Four years (1998–2002) after the financial crisis there was a period of remarkable influx of foreign investment – $35 billion in actual inflows was more than double the amount received during the previous 35 years (OECD, 2004b, p. 140). In terms of investment, this reached a peak in 1998–99 and was succeeded by a period of adjustment during 2000–4. Figure 4.10 clearly shows that there was a persistent trend of increased investment after 1997.

With the increased FDI, R&D activities by foreign firms also increased, as discussed in Section 4.1.1, although the share of R&D expenditure by foreign actors is still minimal. The perception on FDI firms has changed as a consequence of the government's new vision of making Korea into one of the major hubs of manufacturing and R&D activities, and logistics services for the Asian region – a vision that emphasizes the importance of attracting foreign capital (Knowledge Management Team, 2003).

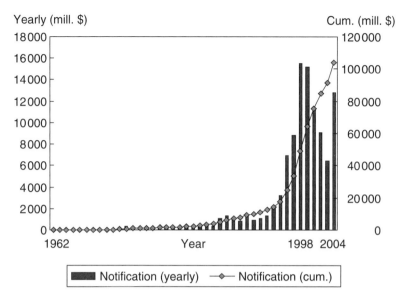

Source: MOCIE (2005).

Figure 4.10 Foreign direct investment in Korea, 1962–2004

Korea's pattern of catching up can be described as 'catching up by specializing in new industries' – that is, the ICT industries. Freeman (1989; 1995) discussed the pattern of a catching-up country utilizing the technological and market opportunities of new industries belonging to the new technological paradigm of electronics and information industries. Globalized production activities, R&D and marketing of firms are well developed in these industries (Ernst and Ravenhill, 1999; Ernst and Guerrieri, 1998). Recently, Korea and Taiwan have demonstrated a new pattern of innovation by gaining market through rapid commercialization of emerging technology from abroad (Albert, 1998). The recent examples of CDMA mobile phones and digital TV show that large firms in these fields are increasingly moving towards developing original products by rapidly commercializing emerging technology from abroad (Lee et al., 2005).

The Korean economy is becoming increasingly integrated into the Asian economy. China, whose capital is only a two-hour flight from the capital of Korea, is growing rapidly and has been a major partner for trade and FDI. In addition, Korea is negotiating free trade agreements with Singapore, Japan and the USA. This will give Korea wider access to foreign markets but will expose weak small firms to international competition. Korea is also increasingly reliant on European FDI – the amount invested between 1962 and 2004 was higher than the amount received from Asian investors (MOCIE, 2005). The USA and Japan were the traditional partner countries for Korea in terms of trade, FDI and sources of knowledge, but these recent changes have broadened Korea's interaction to include other countries.

7 STRENGTHS AND WEAKNESSES OF THE SYSTEM AND INNOVATION POLICIES

7.1 Strengths and Weaknesses

The main strength of the Korean NSI is its large firms. Korea has a strong manufacturing base in scale-intensive industries, such as electronics and automobiles, all of which are led by large firms. Large firms are capable of rapidly developing and producing commercial products from emerging technology through access to local and international sources of knowledge and markets. These large firms have been able to react to the challenges of a global environment. Another strength of the Korean NSI is its heavy investment in IT infrastructure and high diffusion rate of IT devices, both of which have provided a positive environment for Internet businesses and innovative transformation of the service sector. In addition, Korea's capability for rapid commercialization in fast-growing new industries such as electronics and IT is a great advantage. Successful experience of managing R&D consortia in emerging technologies has been accumulated in firms, GRIs and ministries.

The GRIs with their accumulated R&D experience, based on their past role as major performers of public R&D, could be a source of strength. If their operations were to be organized effectively, the GRIs could provide a strong knowledge infrastructure for the industry. The Korean labour force is well known for its level of education and hard work. A highly educated labour force will be a source of strength in the age of the knowledge-based economy.

The weaknesses of the Korean NSI can be summarized as follows. The corporate sector has not been sufficiently transformed. The *chaebols* have adopted the reforms rather reluctantly. If the reform plan were to fail, the large firms affiliated to *chaebols* would continue to be a major burden on the Korean economy but would survive in spite of poor competitiveness due to support from other firms in the *chaebol* groups.

A chronic weakness of the Korean NSI is the prevalence of non-competitive small firms. The weakness of small firms is a barrier for large firms that are searching for competitive local suppliers. GRIs and universities are supporting small firms through technology-diffusion-oriented research but despite recent improvements these programmes are weak. The problem of the bipolarized economy due to the dual structure of 'strong large firms and weak small firms' has yet to be resolved and is likely to be aggravated by the opening of the economic environment.

The science base in Korea is weak and therefore the competitiveness of the components and devices arising from science based R&D is weak. This low quality undermines the competitiveness of final-goods producers.

The service sector is more important than manufacturing in Korea in terms of value-added. However, the competitiveness and innovativeness of the service sector is poor. Poor performance of the service industry is likely to be a barrier to further economic growth. In addition, due to the small pool of KIBS, the innovation activities in the manufacturing sector are not as well supported as those in advanced countries with a strong KIBS sector.

Although Korea has a large pool of highly educated human resources, the quality of education does not satisfy the needs of the large firms. This shortcoming will undermine the competitiveness of the nation.

Koreans tend to work in closed networks, including informal regional networks, networks of alumni, of families etc. This reduces the information flow and flexible work organization necessary for innovation processes.

The financial system is new and it is not sufficiently well developed to channel financial resources to those firms that display good performance because there is a limited pool of knowledge on the credibility and performance of firms. Banks and other organizations are reluctant to take risks in making loans, which is reducing firms' investments.

In spite of the introduction of various measures designed to attract foreign capital, the conditions in North Korea produce uncertainties about the stability of Korea's business environment that will reduce Korea's attractiveness.

Korean government practices are bureaucratic and follow a top-down approach to policy. The move to more interactive policy making is slow, which makes it difficult to resolve the increasingly sophisticated problems inherent in an evolving SI.

7.2 Summary and Evaluation of the Innovation Policy Pursued

This section focuses on the most recent policy change because many aspects of earlier Korean innovation policies have been discussed in the previous sections. In Korea, science and technology policies have been distinct from economic and industrial policies. The S&T policy has been driven mainly by the perspectives of the science and engineering fields, and has ignored the economic and social aspects involved in the process of innovation.

Therefore the policies for R&D input, competence building (the education and training system), the financial system, labour relations and incubation activities have not worked together to encourage innovation activities.

As more and more ministries are becoming interested in encouraging innovation activities, the conflicts and overlaps in the policies of these different ministries are becoming more apparent. There has been pressure to resolve this problem.

In 2004, there was a remarkable change in S&T policy. The government announced 'A Plan to Construct a National Innovation System' (OSTI, 2005). This plan reflected the government's commitment as economic policy emphasized the country's shift to an innovation-based economy which made S&T policy a major national priority. The plan clearly differs from previous policies in the following ways: (1) the policies relevant to S&T across ministries are to be coordinated and monitored by MOST, which was given power to coordinate budgets relevant to S&T across ministries in accordance with medium- and long-term R&D investment strategy and planning (OSTI, 2005); (2) MOST was given a newly created special administrative body to implement coordination and monitoring work in close interaction with the National Science and Technology Council.

Although the plan includes the phrase 'national innovation system' in its title and tries to tackle policy issues relevant to the economic and social aspects of innovation, the new ideas and approaches are in the area of S&T in its narrowest sense. Even so, it has something in common with a systems of innovation approach in that it tries to enhance the effectiveness of an SI at the national level.

7.3 Future Innovation Policy

The goals of the plan reflect the direction of future Korean innovation policy in (1) enhancing the innovation capability of three core innovation actors (universities, firms and GRIs); (2) increasing R&D expenditure and enhancing its effectiveness; (3) searching for and developing future strategic technologies and industries that could drive industrial growth; (4) reinforcing linkages and cooperation among both domestic and foreign innovation actors (universities–industry–GRIs); and (5) organizing the system for coordination and planning of technological innovation policy, and information networks in S&T.

The challenge to the implementation of this plan is the existence of barriers presented by existing policy programmes and laws that were designed to achieve goals set within the boundaries of individual ministries. Although the declared plan emphasizes coordination of policies and collaborative work among ministries for future innovation policies, ministries are still likely to work towards achieving the goals of only their particular ministries.

In order to change the behaviour of government organizations, mechanisms to reinforce continuous feedback of the results of policy implementation will be necessary. As a result of the ever bigger and increasingly sophisticated NSI, capturing the scale and scope of the system will become more difficult and the effects of policies will be more difficult to measure. The system for evaluating the effectiveness of government policy programmes at the NSI level needs to be continuously revised, experimented with and improved upon, using sophisticated evaluation and statistical systems. The results of evaluations of policy programmes need to be reflected in improved programmes through ministerial or cross-ministerial efforts.

Korean S&T policies are slowly shifting towards an interactive model. However, the policy body needs to be more targeted to take on the role of manipulating the conditions of systems of innovation on the basis of interactive policy approaches. One example of the complex measures necessary to facilitate innovation activities is the task of removing barriers to networking among actors and simultaneously removing barriers to investment by nurturing technological competences.

Finally, in the ever more globalized world, strengthening the capabilities of small firms, which are increasingly vulnerable to international competition, will be crucially important for future policy. The characteristics of the Korean NSI can be summarized in six words: 'strong large firms, weak small firms'. In order to resolve the chronic problems of small firms, future innovation policy needs to find strategic ways of enhancing the technological capabilities of small firms and the networking of small firms with domestic and international actors for knowledge and market access. The importance of strategic policies to ensure that the poor competences of small firms do not become a barrier to upgrading the competitiveness of the nation cannot be over-emphasized.

ACKNOWLEDGEMENTS

The research on which this chapter is based was financed by the Korea Research Foundation during 2003–4 (code number H00004) and sponsored by the Science and Technology Policy Institute (STEPI) during 2002–3 and 2004–5. The author gratefully acknowledges the funding and support provided by these organizations. The chapter was written with the assistance of an advisory board, whose members are: Joon-Kyung Park (Korea

Development Institute), Keun Lee (Seoul National University), Kongrae Lee (Science Technology Policy Institute), Wichin Song (Science Technology Policy Institute), Joonghae Suh (Korea Development Institute (KDI)) and Cheonsik Woo (KDI). The author thanks the board for their insightful and valuable comments. The author is also indebted to Yongrak Choi, President of STEPI, who was not only supportive of the project, but also provided continuous encouragement through stimulating discussions at the international workshops of the project. The author is also grateful to Dr Mijong Um at STEPI for her advice on the Korean innovation survey and indicators work, and to Dr Karpsoo Kim at KOTEF and Dr Hyeran Hwang at Daejeon Development Institute for their helpful comments. Thanks go to all those involved in providing data without which this research could not have been conducted. Any remaining errors, and the views expressed, are solely the author's responsibility.

NOTES

1. The Park regime ended in 1979. Power was handed over to a party that inherited the tradition of the Park regime until 1998, when political power went to the opposition party.
2. Korean innovation surveys were carried out in 1996 (CIS1, manufacturing), 1998 (CIS2, manufacturing and service), 2002 (CIS3, manufacturing) and 2003 (CIS3, service).
3. CIS3 reports on innovation activities between 1999 and 2000. Data from European Commission (2004).
4. These figures are based on patents filed with the European Patent Office, the US Patent and Trademark Office and the Japanese Patent Office.
5. Shin et al. (2002, p. 132).
6. MOST (several years).
7. Source: http://www.koita.or.kr/, website of Korea Industry and Technology Association, last accessed 20 May 2006.
8. Major manufacturing sector industries are defined as those accounting for more than 3 per cent of the amount in value-added for all industry, including service and primary industries. Data from the KDI database of input–output tables by Bank of Korea.
9. Data from Korea Institute of Patent Information.
10. KDI (1995), as cited in Goh (1998).
11. Professional technicians are defined as those technicians and associate professionals belonging to professions, according to the International Standard Classification of Occupations.
12. Approximate data for overseas trained doctorates returned to Korea: number of doctorates registered in the Korea Research Foundation website (http://www.krf.or.kr/). Students who have obtained doctorates abroad, and who are looking for a job in Korea, are normally requested to register the details of their degree and institution on this website.
13. However, recently there has been increasing scepticism from firms about training investment: they have found that the incentive for investing in employee training has reduced because employee mobility has gradually risen, due to the breakdown of the lifetime employment tradition following the financial crisis.
14. The criteria for being a venture business are being a venture capital firm, an R&D-intensive firm (R&D investment more than 5 per cent of sales), or a firm utilizing patents or new technology (more than 50 per cent of sales). These firms enjoyed financial benefits

(tax incentives, bank loans) and other benefits (allowing stock options, enabling employment of faculty from national or public universities). At the end of 2002, the designation criteria were changed: evaluation became a two-step process that included assessment by an independent organization of the firm's innovation capability. The current designation system was discontinued in 2005 (OECD, 2004b, p. 184).

15. For example, applications for establishing firms that were not processed within 45 days were automatically approved (Small and Medium Business Administration, 2003).
16. See footnote 14 for the criteria for venture businesses.
17. This is in contrast with Japan, which recognized the problem of a dual structure of strong large firms, weak small firms in the 1960–70s and solved it through policies providing an environment encouraging close linkages between large and small firms, and building prefecture (provincial) laboratories to support small firms' testing and development activities. Because of poor linkages between large and small firms, Korea is now experiencing a strange phenomenon: rapid growth of large firms in strategic industries and recession among most of its small firms. Koreans refer to this as a 'bipolarization phenomenon' in the economy. In 2003, the effect of 15.7 per cent export growth led by large firms failed to overhaul domestic demand. Domestic demand (investment in production equipment and private consumption) decreased to minus 1.5 per cent (Institute for Monetary and Economic Research, 2004). Small firms, relying mostly on domestic demand, experienced a recession.
18. For instance, a programme of R&D consortium projects was initiated in 1993 to enhance R&D networking. In 1993, 328 firms participated in 19 consortia with 2 billion won government funding. In 2002, 2787 firms participated in 197 consortia with 38.1 billion won government funding (Small and Medium Business Administration, 2003, p. 73).
19. Ministry of Labour as cited in KPMG Consulting (2001).
20. As a result, policy measures that conflict or overlap often emerge. As more ministries have become interested in innovation policy, the number of conflicting and overlapping policies has increased. For recent policy changes for enhancing coordination of policies, see Section 7.2.
21. A non-profit Korean government agency to implement government policies for the promotion of small and medium-sized businesses.
22. From http://www.bi.go.kr/, as cited in Kim and Lee (2005).
23. US$ in terms of average exchange rate of the year.
24. KIBS include computer-related services, R&D services, business services and technical services.
25. Customers/user firms, competitors, suppliers of raw materials or components, suppliers of machinery or equipment, firms in the same corporate group.

REFERENCES

Ahn, C. and J. Kim (1997), 'The outward-looking trade policy and the industrial development of South Korea', in D. Cha, K. Kim and D.H. Perkins (eds), *The Korean Economy 1945–1995: Performance and Vision for the 21st Century*, Seoul: Korea Development Institute, pp. 339–82.

Ahn, K. and M. So (2003), 'The current status of collaborative research in science and technology', mimeo.

Albert, M. (1998), *The New Innovators: Global Patenting Trends in Five Sectors*, Office of Technology Policy, US Department of Commerce.

Amsden, A. (1989), *Asia's Next Giant: South Korea and Late Industrialisation*, Oxford: Oxford University Press.

Cho, B. (2004), 'The economic effects and transformation of banking management', in C. Lee (ed.), *Korean Economy Disappears*, Seoul: 21 Century Books, pp. 117–42.

Cho, H., E. Lee, J. Han, B. Son and J. Lee (2003), *Upgrading Quality of Science and Technology and Alleviating the Imbalance*, Seoul: Science and Technology Policy Institute (STEPI).
The Committee of Education Reform for the President (1995), *The Suggestion for Education Reform to Establish New Education System Suitable for the Age of Globalization and Informatization*, Seoul.
The Economist (2003a), 'Seriously wired', *The Economist*, 17 April.
The Economist (2003b), 'Survey: South Korea', *The Economist*, 17 April.
The Economist (2005), 'Special report: Samsung Electronics', *The Economist*, 15 January, pp. 60–62.
Edquist, E. and D. Goktepe (2003), *Indicators on Education & Competence Building, ESF/EUROCORES Project: National Systems of Innovation in a Globalizing, Knowledge-based Economy: A Comparative Study of Small Countries in Europe and Asia*, Lund: University of Lund.
Ernst, D. and P. Guerrieri (1998), 'International production networks and changing trade patterns in East Asia: the case of the electronics industry', *Oxford Development Studies*, **26**(2), 191–212.
Ernst, D. and J. Ravenhill (1999), 'Globalization and convergence, and the transformation of international trade networks in electronics in East Asia', *Business and Politics*, **1**(1), 35–62.
European Commission (2004), *Innovation in Europe*, Brussels: EC.
Freeman, C. (1989), 'New technology and catching up', in R. Kaplinsky and C. Cooper (eds), *Technology and Development in the Third Industrial Revolution*, London: Frank Cass & Co., pp. 85–99.
Freeman, C. (1995), 'The information economy: ICT and the future of the world economy', in S.A. Roseu (ed.), *Changing Maps: Governing in a World of Rapid Change*, Ottawa: Carleton University Press, pp. 163–86.
Goh, S. (1998), 'Nurturing human resources and education system', in K. Lee and W. Song (eds), *National Innovation System of Korea*, Seoul: Science and Technology Policy Institute (STEPI), pp. 215–35.
Harding, R. (2002), *Global Entrepreneurship Monitor United Kingdom 2002*, London: London Business School.
Hobday, M. (1995), *Innovation in East Asia: The Challenge to Japan*, Aldershot, UK and Brookfield, USA: Edward Elgar.
Hong, D. and S. Ko (2005), 'Building an information infrastructure for knowledge economy', in J. Suh and J.-E. Aubert (eds), *Korea as a Knowledge Economy*, Seoul: Korea Development Institute, pp. 68–90.
Institute for Monetary and Economic Research (2004), *The Cause of Bipolarization of the Economy and Policy Issues*, Institute for Monetary and Economic Research, Seoul: Bank of Korea.
Jeong, J. (2005), 'The status of venture businesses and the suggestions for improvement of the policies issues', in J. Kim (ed.), *The Role of Small Firms in the Shift Towards Innovation Driven Economy*, Seoul: Korea Development Institute, pp. 389–461.
Jeong, S. (2004), 'The implication of profit maximization of share holders', in C. Lee (ed.), *Korean Economy Disappears*, Seoul: 21 Century Books, pp. 350–74.
Kaitila, V. (2003), 'GDP and productivity indicators for small European and Asian economies and large reference countries', paper presented at the Taipei Workshop in the ESF Project, November.

Kim, J. and S. Lee (2005), 'The status of operation of business incubators and the future policy issues', in J. Kim (ed.), *The Role of Small Firms in the Shift Towards Innovation Driven Economy*, Seoul: Korea Development Institute.

Kim, K. (2002), *A Study on Best Practice Model of R&D Planning System: Comparison between Korea and Japan*, Seoul: Science and Technology Policy Institute.

Kim, K. (2003), *Firms' Participation in the Government R&D Program: Current Status and Future Direction*, a report for Ministry of Science and Technology, Seoul: Science and Technology Policy Institute (STEPI).

Kim, K., J. Suh and S. Han (2000), *The Current Status and Issue of Policies for Cooperative R&D*, Seoul: Science and Technology Policy Institute (STEPI).

Kim, K.S. and J. Kim (1997), 'Korean economic development: an overview', in D. Cha, K.S. Kim and D.H. Perkins (eds), *The Korean Economy 1945–1995: Performance and Vision for the 21st Century*, Seoul: Korea Development Institute.

Kim, K.S. and M. Roemer (1979), *Growth and Structural Transformation*, Cambridge, MA: Harvard University Council on East Asian Studies.

Kim, L. (1993), 'National systems of innovation: dynamics of capability building in Korea', in R. Nelson (ed.), *National Systems of Innovation*, New York and Oxford: Oxford University Press, pp. 357–83.

Kim, L. (1997), *From Imitation to Innovation*, Cambridge, MA: Harvard Business School Press.

Kim, Y. and H. Jeon (2003), 'Fit between market uncertainty and patterns of innovation system; a case study on the broadband Internet access industry', *Journal of Technology Innovation*, **11**(2), 239–82.

Knowledge Management Team (2003), *Research on the Competitiveness of Industries in Korea*, Seoul: Korea Development Institute.

Korea Industrial Technology Association (KOITA) (2002), *Survey on the Current Status of Foreign Research Institutes in Korea*, Seoul: Korea Industrial Technology Association (KOITA).

Korea Industrial Technology Association (KOITA) (2003a), accessible at http://www.koita.or.kr/archive/r&d/r&d_2003.htm, last accessed 5 February 2005.

Korea Industrial Technology Association (KOITA) (2003b), *White Paper on Industrial Technology*, Seoul: KOITA.

Korea Industrial Technology Association (KOITA) (2004), *White Paper on Industrial Technology*, Seoul: KOITA.

Korea Intellectual Property Office (2003, 2004), *Annual Report of Statistics on Knowledge Property*, Daejon: KIPO.

Korea International Trade Association (KITA) (2004), *Main Indicators of Trade*, Seoul: KITA.

Korea National Statistical Office (2005), 'Number of firms in different sizes', available at http://kosis.go.kr, last accessed 5 February 2005.

Korea Research Institute for Vocational Education and Training (2003), *Human Resources Development Indicators in Korea*, Seoul: Korea Research Institute for Vocational Education and Training.

KPMG Consulting (2001), *Foreign Direct Investment in Korea*, Seoul: KPMG Consulting.

Krugman, P. (1994), 'Myths of Asia's miracle', *Foreign Affairs*, **73**(6), 62–78.

Lee, K. and C. Lim (2001), 'Technological regimes, catch-up and leapfrogging: findings from the Korean industries', *Research Policy*, **30**, 459–83.

Lee, K., C. Lim and W. Song (2005), 'Emerging digital technology as a window of opportunity and technological leapfrogging: catch-up in digital TV by the Korean firms', *International Journal of Technology Management*, **29**(1/2), 40–63.

Lee, K., J. Park, J. Hwang and E. Kim (2002), *Innovation and Strategy of the Knowledge-intensive Service Sector in Korea*, Seoul: Science and Technology Policy Institute (STEPI).

Lee, S., H. Lee, J. Yun and Y. Lee (2003), *Report for Building Information Infrastructure of Science and Technology for Science and Technology Hub in East Asia*, Daejon, Korea Institute of Science and Technology Information.

Lee, T. (2004), 'CANDU-type nuclear fuel', *Technovation*, **24**(4), 287–97.

Lee, W. and B. Kim (2003), 'Analysis of impact of R&D investment on productivity', *Monthly Bulletin*, May, 24–51.

Lim, C. (2003), 'Changwon machinery industry cluster', paper for the workshop on innovative clusters and regional economic development, Korea Research Institute for Human Settlements, Korea, 14 April.

Lundvall, B.-Å. (1988), 'Innovation as an interactive process: from user–producer interaction to the national system of innovation', in Giovanni Dosi, Christopher Freeman, Richard Nelson, Gerald Silverberg and Luc Soete (eds), *Technical Change and Economic Theory*, London: Pinter, pp. 349–69.

Lundvall, B.-Å. (1992), *National Systems of Innovation: Towards a Theory of Innovation and Interactive Learning*, London and New York: Pinter.

Min, C., W. Song and B. Hyun (2004), 'Strategic cooperative system between regional university and GRI', STEPI report 2004-16.

Ministry of Commerce, Industry and Energy (MOCIE) (2005), *Foreign Direct Investment in 2004 and Prospects for 2005*, Gwachon, Korea: MOCIE.

Ministry of Science and Technology (MOST) (2002), *Analysis on the Research Achievements in Science & Technology Fields Using 2001 SCI Database*, Seoul: MOST.

Ministry of Science and Technology (MOST) (1991, 2002, 2003), *Directory of Science, Technology and R&D Activities*, Seoul: MOST.

OECD (1996), *Reviews of National Science and Technology Policy – Republic of Korea*, Paris: OECD.

OECD (2000), *Economic Survey – Korea*, Paris: OECD.

OECD (2002), *Education at a Glance*, Paris: OECD.

OECD (2003), *Science, Technology and Industry Scoreboard*, Paris: OECD.

OECD (2004a), *Main Science and Technology Indicators*, Paris: OECD.

OECD (2004b), *OECD Economic Surveys: Korea*, Paris: OECD.

OECD (2005), *New GDP Comparisons Based on Purchasing Power Parities for the Year 2002*, Paris: OECD.

Office of Science and Technology Innovation (OSTI) (2005), *Strategy and Vision*, Seoul: Office of Science and Technology Innovation.

Park, K. (2004), *Knowledge Technological Regime and Technological Catch Up: Comparing Korea and Taiwan using US patent data*, PhD thesis, Seoul: Seoul National University.

PISA (The OECD Programme for International Student Assessment) (2000), *The Results from PISA 2000*, OECD.

Regulatory Reform Committee (2003), *White Paper on Regulatory Reform*, Seoul: Regulatory Reform Committee.

The Research Group on Innovative Small Firms (2005), *The Final Report*, Seoul: Korea Industrial Technology Foundation.

Shin, T., W. Song, M. Um and J. Lee (2002), *Korean Innovation Survey 2002: Manufacturing sector*, Seoul: Science and Technology Policy Institute (STEPI).

Small and Medium Business Administration (2003), *Performance of the Policy for Nurturing Venture Businesses over the Last Five Years*, Daejon: Small and Medium Business Administration.

Song, W. (1999), 'The politics of national technology development program: a case study on the CDMA Technology Development Program', *Korean Academy of Studies on Administration*, **33**(1), 311–29.

Suh, J. (2002), 'The emerging pattern of SMEs' innovation networks in Korea and its policy implications', mimeo, Korea Development Institute.

Suh, J. (2004), 'Enhancing productivity through innovation: Korea's response to competitiveness challenges', paper prepared for the KDI 33rd anniversary conference on industrial dynamism, 22–23 April, Seoul: Korea Development Institute.

Um, M. and J. Choi (2004), *The Analysis of Innovation Activities in the Service Sector*, Seoul: Science and Technology Policy Institute (STEPI).

Wang, K., M. Tsai, C. Liu, I. Luo, S.A. Balaguer, S. Hung, F. Wu, M. Hsu and Y. Chu (2003), *Intensities of Scientific Performance: Publication and Citation at a Macro and Sectoral Level of All the Nine Countries*, Taipei: STIC, pp. 11–20.

Woo, C. (2002), 'Upgrading Korea's education and HRD in the age of the knowledge economy', paper for the workshop 'Upgrading Korean education in the age of knowledge economy: context and issues', sponsored by Korea Development Institute and the World Bank, 14–15 October, Seoul, Korea.

Yang, H. (2004), 'Evaluation of business incubator policies and suggestion for improvement of the policy', *Venture Business Management Research*, **7**(3), 135–59.

Yoon, J. and H. Hwang (2000), 'The evolution of policies for supporting R&D activities of universities for expanding knowledge base', *Science and Technology Policy*, **10**(4), 104–17.

Youn, T., C. Park and Y. Choi (2003), *A Study on the Impact of Intellectual Property Rights on Economic Growth*, Daejon: Korea Intellectual Property Office.

5. High growth and innovation with low R&D: Ireland

Eoin O'Malley, Nola Hewitt-Dundas and Stephen Roper

1 INTRODUCTION

For a long time Ireland was a relatively poor country by Western European standards, but since the late 1980s its average productivity and income levels have caught up quickly with the rest of Europe (see Appendix Tables A2.3 and A2.4). The most rapidly growing sectors in Ireland during most of this period were those generally identified as being R&D-intensive or 'high-tech'. Today, there are concentrations of these high-tech industries in Ireland that are greatly disproportionate to the country's small size (see Appendix Table A2.6).

Given the importance of high-tech industry in Ireland's economic renaissance, it may seem somewhat paradoxical that Ireland has relatively low levels of domestic R&D expenditure. Gross expenditure on R&D as a percentage of gross national product (GNP) and business expenditure on R&D as a percentage of GNP are both a good deal lower in Ireland than in the EU. Thus Ireland has a record of fast productivity and output growth, and a relatively large presence in normally R&D-intensive industries, while its levels of expenditure on R&D remain relatively low. Furthermore, the available evidence suggests that Ireland also has quite high rates of innovation by international standards, despite low levels of R&D expenditure (see, e.g., Appendix Table A4.2).

This unusual combination of characteristics raises questions about how this situation has come about. A significant point in our analysis is that inward technology transfer has been important, particularly in association with the large amounts of inward foreign direct investment (FDI) that Ireland has received. By this means, many innovations that are derived from R&D carried out elsewhere are introduced into Ireland. This in turn raises further interesting questions. For example, is it sustainable for Ireland to rely so much on this type of process that is driven by FDI? And, are there indications of progress in the development of Irish indigenous enterprises and in the upgrading of indigenous competencies and innovation capabilities?

2   MAIN HISTORICAL TRENDS

2.1   The Evolution of Industry

In most of the island of Ireland, apart from the north-east (which is now Northern Ireland), there was no sustained process of industrialization in the nineteenth century.[1] When Ireland became independent from the UK in the early 1920s, less than 5 per cent of its labour force was employed in manufacturing. The country then adopted a strong protectionist stance in order to develop new industries, and industrial employment grew quite rapidly.[2] However, by the 1950s most industries were still selling only to the protected domestic market and there had been little development of competitive or specialized industries.

The failure of the period of protectionism to produce internationally competitive industry led to a fundamental reorientation of Irish industrial policy in the late 1950s. This change involved (1) promoting the development of exports as a priority; (2) attracting internationally competitive FDI, by means of tax and grant incentives; and (3) embracing free trade with the UK from 1966 and with the EU from 1973 onwards. Under this new policy, imported goods won a continuously increasing share of Ireland's domestic market while many existing Irish-owned or indigenous industries struggled to develop competitive exports. Thus, much of Irish indigenous industry did not fare very well under these outward-looking policies, until the late 1980s (O'Malley, 1989, ch. 6).

Ireland did, however, have substantial success in attracting FDI by multinational enterprises (MNEs) that located in Ireland to produce for export markets (e.g. Roper and Frenkel, 2000). At first the MNEs starting up in Ireland were largely involved in mature and often labour-intensive industries. In the 1970s and 1980s, however, they were increasingly involved in newer, more technologically advanced products such as electrical and electronic products, machinery, pharmaceuticals and medical instruments and equipment. Initially, and for some time, FDI was concentrated in the production phase of the value chain, placing little demand on local technological inputs, skills and suppliers, but there were signs that this was changing gradually over time.

The motivation for foreign MNEs to invest in Ireland also shifted over time. In the 1960s, tax and grant incentives, and low labour costs by Western European standards, were key attractions. Access to major markets became a significant additional attraction after Ireland joined the EU in 1973. Access to skills became increasingly important from the 1970s since Ireland specifically aimed to provide a skilled labour force for industries such as computers and other electronic products, pharmaceuticals, medical and scientific instruments and software. Furthermore, from the 1970s onwards, Ireland became proactive and selective in targeting and seeking out the industries and companies that it wanted to attract, aiming to create industrial clusters with vertical and horizontal linkages (Ruane, 2003; Gorg and Ruane, 2000a).

2.2 Recent Trends and Industrial Structures

Beginning in the early 1990s, the Irish economy grew exceptionally fast. Real GNP grew by 6.3 per cent per year in the 1990s while manufacturing production grew by 13.8 per cent per annum and manufacturing employment increased by 2.5 per cent per annum (see Appendix Table A2.3). There was particularly rapid growth in three broad sectors of manufacturing – electrical and optical equipment, chemicals and chemical products, and paper products, publishing and printing – which accounted for 39 per cent of manufacturing gross output in 1991, rising to 68 per cent in 1999.[3]

These three sectors are mostly foreign-owned, but since the late 1980s Irish-owned industry has also had a considerable revival. It has grown relatively quickly by international standards, it has experienced significant upgrading in terms of skills and R&D performance, and it has generally had higher growth rates in high-tech sectors than in more traditional industries (O'Malley, 1998 and 2004; O'Riain, 2004). It is also worth noting that industry in Ireland is highly export-oriented, with about three-quarters of output being exported. Foreign-owned or high-tech industries are especially highly export-oriented, while Irish-owned or traditional industries are less so (see also Appendix Table A2.6).

3 INNOVATION INTENSITY

Comparing levels of innovation activity in Ireland and elsewhere is made difficult for recent years because of the lack of Community Innovation Survey 3 (CIS3) data for the 2000–2002 period. Earlier information, from CIS1 and CIS2, is available for Ireland relating to the period up to 2000. In addition, longitudinal data on innovation in Irish manufacturing are available from the Irish Innovation Panel covering the period 1991–2002.

CIS1 data were collected in 1993 and relate to innovation activity in 1991–93. They provide two useful innovation activity measures: the proportion of establishments introducing either a technologically changed product or process during the previous three years and the proportion of innovating plants' sales derived from innovative products (Table 5.1). For these indicators, Ireland had the highest proportion of innovating plants in manufacturing and, with the Netherlands, was in the middle of the league table for shares of innovative sales. The CIS2 was undertaken in 1997 in EU member countries and Norway. Here, we focus on three indicators derived from CIS2 relating to: the proportion of firms introducing innovative

Table 5.1 International innovation comparisons from CIS1 and CIS2

	Innovators (% of plants)				Innovative sales	
CIS1	Ireland	72			Spain	52
	Germany	67			Germany	50
	Belgium	61			Denmark	44
	Netherlands	57			Belgium	40
	Denmark	56			Ireland	36
	Norway	53			Netherlands	36
	France	39			Luxembourg	32
	Spain	37			Norway	32
	Luxembourg	37			Italy	29
	Italy	34			France	27
	Product innovators (%)		Process innovators (%)		Innovative sales	
CIS2	Ireland	66	Ireland	54	Germany	50
	Germany	65	Germany	53	Spain	44
	Austria	60	Denmark	51	Italy	43
	Denmark	57	Austria	49	Ireland	41
	Netherlands	56	Netherlands	46	Austria	40
	UK	52	Italy	41	Portugal	40
	Sweden	48	Norway	40	Sweden	37
	France	38	Sweden	38	Netherlands	33
	Italy	37	UK	37	Finland	33
	Norway	35	France	31	Norway	33
	Luxembourg	32	Luxembourg	29	UK	31
	Belgium	31	Spain	25	Denmark	29
	Finland	29	Finland	25	France	29
	Spain	24	Portugal	23	Belgium	28
	Portugal	15	Belgium	22		

Notes and sources: CIS1 figures are extracted from Eurostat CD-ROM. Indicators are: percentage of innovative manufacturing plants and percentage of sales of changed products of innovative plants. CIS2 figures are derived from tables provided by Eurostat. Indicators are: percentage of plants with product innovation; percentage of plants with process innovation; and percentage of sales derived from changed products of innovative plants.

products; the proportion of firms introducing innovative processes; and the proportion of innovative plants' sales derived from innovative products. As in CIS1, Ireland had a larger proportion of innovating plants than any other country included in the survey, and in this period was higher up the league table in terms of innovative sales – in fourth place and eight percentage points above the Netherlands (Table 5.1). Thus the evidence from CIS1 and CIS2 suggests that exceptionally high proportions of Irish plants were innovating, while the proportion of sales derived from innovative products was quite high, although closer to the survey average.

One difficulty with the data in Table 5.1 is the changes in innovation definitions between CIS1 and CIS2. This makes it hard to assess the overall trend in innovation activity in Ireland from CIS data, a better source here being the Irish Innovation Panel. Using these data, it is important to take into account the changing economic climate in Ireland. For example, rapid manufacturing growth in Ireland in the 1991–2000 period largely reflected the rapid expansion of output in the high-tech sectors, driven mainly by inward FDI. In 2000–2002, however, the rate of growth of output in Ireland slowed considerably. This reflected the global downturn in high-tech industry and its impact on other sectors. Thus, for Irish firms investing in innovation, the period 2000–2002 was markedly less enticing than previous periods.

In general, data from the Irish Innovation Panel suggest that the proportion of manufacturing plants undertaking R&D, product innovation and process innovation remained relatively stable through the 1990s, but declined in 2000–2002 (Table 5.2). This is what was anticipated, given the economic

Table 5.2   Innovation indicators for Ireland, 1991–2002

	1991–93	1994–96	1997–99	2000–2002	1991–2002
R&D in plant (% plants)	50.8	48.3	52.6	44.6	49.4
Product innovation (% plants)	62.8	65.9	65.3	56.7	62.8
Process innovation (% plants)	n.a.	57.7	65.8	53.9	59.8
New products in sales (mean %)	18.1	13.8	17.1	13.5	15.7
New/improved products in sales (mean %)	32.1	30.5	29.2	27.3	29.7

Notes and sources: Figures relate to manufacturing plants with ten or more employees. Survey observations are weighted to give representative results. Product innovators are those plants introducing new or improved products during the last three years. Process innovators are plants introducing new or improved processes over the same period. n.a. = not available. Data from the Irish Innovation Panel (see www.innlab.org).

conditions in this latter period. The surprising thing is that each of these measures, as well as the sales from new and new and improved products, dropped to the early 1990s levels. This shows how important the influence of the economic environment can be for plants' innovation investments.

Within Irish manufacturing, larger plants and foreign-owned plants are significantly more likely to undertake both product and process innovation than small plants and indigenously owned plants (Table 5.3). However, note that the figures for Irish indigenous plants would still be high by international standards, being just three or four percentage points below the national figures for Ireland. Levels of innovation activity are also notably higher in the high-tech sectors than in more traditional industries.

4 ACTIVITIES THAT INFLUENCE INNOVATION

4.1 Knowledge Inputs to Innovation

4.1.1 R&D activities
Policy in Ireland, in recent years, has emphasized public support for R&D in the higher education and private sectors. Despite this, levels of R&D investment remain low by international standards.

Gross expenditure on R&D (GERD) Gross expenditure on R&D (GERD) has remained relatively constant in Ireland since the early 1990s, from 1.3 per cent of GNP in 1993 to a high of 1.45 per cent in 1997 and 1.39 per cent in 2001 (Forfás, 2004). Ireland has consistently lagged behind both EU and OECD averages for GERD, with this gap becoming more pronounced between 1997 and 2001, when GERD continued to increase in the EU and OECD countries while declining in Ireland. Ireland now lags behind countries such as Norway, the UK, the Netherlands, Denmark, Finland, Sweden, the USA and Japan, as well as the EU and OECD averages.

Public R&D investment Public investment in R&D can take a variety of forms – direct investment by government, support for R&D in higher education, and support for business R&D. The last of these is discussed in the next section; the focus here is on intramural and higher education institution (HEI) R&D investment by government.

In 2002 the government allocation to R&D in Ireland was 57.5 per cent higher than in 1999. Government intramural expenditure on R&D rose from 0.08 per cent of GNP in 1999 to 0.13 per cent in 2001.[4] This dramatic increase was largely due to increased allocations to research in third-level

Table 5.3 Innovation activity in Ireland, 2000–2002, by sector, size and ownership

	Innovators (%)	Product innovators (%)	Process innovators (%)	Product innovation intensity (%) median value	Plants with new to market products (% of all plants)
Ireland	71.8	56.7	53.9	25.0	47.2
Industrial sector					
Food, drink & tobacco	70.2	65.6	51.5	20.0	51.7
Textiles & clothing	66.8	57.9	48.8	34.5	56.0
Wood & wood products	76.3	68.4	57.9	15.0	57.9
Paper & printing	65.7	28.3	51.5	40.0	16.2
Chemicals	82.3	69.9	61.3	16.6	56.7
Metals & metal fabrication	64.4	43.0	49.2	32.6	32.0
Mechanical engineering	71.8	64.4	53.5	40.0	61.9
Electrical & optical equ.	81.7	64.1	64.5	20.8	54.6
Transport equipment	82.3	73.5	73.5	69.2	55.6
Other manufacturing	77.9	62.6	53.7	25.0	56.7
Sizeband					
10–19	53.5	40.8	37.4	25.0	31.8
20–99	76.4	58.5	56.2	25.0	50.2
100+	91.9	79.9	75.9	23.1	66.7
Ownership					
Locally owned	68.2	53.5	51.4	25.0	46.5
Foreign-owned	83.4	66.5	61.1	25.0	49.2

Notes and source: Figures relate to manufacturing plants with ten or more employees. Survey observations are weighted to give representative results. Product innovation intensity is calculated as the number of new products introduced in the previous three years as a proportion of the total number of products sold by the plant. Data are derived from the IIP (see www.innlab.org).

organizations (via the Higher Education Authority), to the Department of Enterprise, Trade and Employment (DETE) for the establishment of a Technology Foresight Fund aiming to establish Ireland as leading in research in biotechnology and information and communication technologies (ICT), and additional funding through the DETE via Enterprise Ireland for R&D grants to industry.

Strategic allocation of public sector support for R&D has become a key element in the recent history of R&D funding by government. In 2000, Science Foundation Ireland was formed with a budget allocation of over €634.9 million to fund research in niche areas of the biotech and ICT sectors. The 'strategic rationale behind this initiative [was] the need to stimulate a greater level of top class research in the economy in support of high-technology sectors and to ensure that a sufficient supply of good researchers become available to drive a more sophisticated research performance in the business sector' (Forfás, 2000, p. 7).

The higher education sector is perceived as being important both through the production and application of new knowledge through the R&D it performs and through the skilled people it produces. One example of the interrelationship between education and industry is the Programmes for Advanced Technology (PATs), which were first established in the 1980s as working partnerships between the universities, the government/ development agencies and industry. PATs are subsidized centres to apply new technologies in niche areas of relevance to industry, and in many cases were established to meet the needs of traditional industry and enhance its competitiveness internationally (Industrial Policy Review Group, 1992). PATs have been important because they represented a turning point in the type of research undertaken in the third-level sector (Lenihan et al., 2004, p. 21), which is becoming increasingly focused on industry. As PATs were funded on a competitive basis, this led to an increase both in the quality of research proposals and research undertaken.

Although expenditure on R&D in Higher Education (HERD) increased throughout the 1990s from €133 million in 1993 to €294 million in 2001 (Forfás, 2003a), as a share of GNP it has decreased recently, from a high of 0.33 per cent in 1999 to 0.31 per cent in 2001. This remains below average levels of HERD in the EU (0.4 per cent) and the OECD (0.4 per cent) in 2001.

Business expenditure on R&D (BERD) As in most other countries, GERD in Ireland is dominated by industry (68 per cent of total GERD in 2001). BERD investment has mirrored that of GERD, increasing from 0.89 per cent of GNP in 1993 to a high of 1.04 per cent in 1997 before declining to 0.95 per cent of GNP in 2001.

Again, this lags considerably behind the EU and OECD, being 78.5 per cent of the EU average and 60.9 per cent of the OECD average in 2001. Ireland's BERD lags even further behind the figures for other economies with a substantial high-technology sector, such as 2.9 per cent in Sweden, 2.1 per cent in Japan and Korea, and 2.0 per cent in the USA and Finland (Forfás, 2003b).

Two sectors dominate BERD in Ireland, namely electrical and electronic equipment (37 per cent of BERD) and software and computer-related services (28 per cent of BERD) (Forfás, 2000). These two sectors, together with pharmaceuticals, instruments, and food, drink and tobacco, accounted for 85 per cent of BERD in 2001 (Forfás, 2003b). Furthermore, BERD is dominated by foreign-owned firms (65 per cent of BERD in 2001). Thus BERD, measured in absolute terms, is dominated by foreign-owned MNEs mainly in high-tech sectors. However, R&D intensity – defined as R&D as a percentage of sales – is higher for the indigenous sector (0.8 per cent) compared to the foreign-owned sector (0.6 per cent) (Table 5.4).[5] Even in the high-tech sectors where BERD is dominated in absolute terms by foreign MNEs, R&D intensity is higher among the indigenous plants.

At the same time, for the majority of sectors, whether Irish-owned or foreign-owned, relative R&D intensities in Ireland were below OECD levels in 2001 (Table 5.4). On the other hand, in 1997 and 1999 Irish indigenous plants had R&D intensities that were higher than OECD levels in half or more of the sectors. Even in those years, total Irish indigenous industry was much less R&D-intensive than total OECD industry, but that was because it was relatively concentrated in low R&D intensity sectors.

Patent activity　In Ireland, patenting activity remains low by international standards (see Appendix Table A3.1). In 2000, triadic patent families per million population amounted to 11.9 compared to 94.53 in Finland, 91.40 in Sweden and 92.63 in Japan. Where patent applications are made, they are distributed across the world, with patent applications in 2002 being made in Ireland (32 per cent), the USA (32 per cent), European countries (24 per cent) and elsewhere in the world (12 per cent).

Foreign-owned enterprises are much more likely to apply for patents in the USA than Irish-owned enterprises. For example, in 2001, 53 per cent of foreign-owned businesses in Ireland that applied for a patent did so in the USA as compared to only 15 per cent of Irish-owned businesses. In contrast, of all Irish-owned businesses applying for a patent in 2001, almost half (47 per cent) were made in Ireland while only 14 per cent of the foreign-owned patent applications were made in Ireland. The importance of foreign-owned multinationals in total patent applications is reflected in comparisons of cross-border ownership of inventions (see Appendix

Table 5.4 *BERD intensity (as % of industry output) by sector and nationality of ownership compared to the OECD sectoral average, 2001*

	Irish-owned			Foreign-owned		
	R&D intensity	OECD average R&D intensity	Irish intensity as % of OECD intensity	R&D intensity	OECD average R&D intensity	Foreign owned intensity as % of OECD intensity
Chemicals	0.4	3.2	12.5	0.1	3.2	3.1
Transport equipment	0.7	4.5	15.6	0.4	4.5	8.9
Pharmaceuticals	2.3	11.5	20.0	1.2	11.5	10.4
Instruments	1.8	7.0	25.7	1.2	7.0	17.1
Other manufacturing	0.3	1.0	30.0	0.2	1.0	20.0
Paper, print & publishing	0.2	0.4	50.0	0.0	0.4	0.0
Non-metallic mineral products	0.6	0.8	75.0	1.4	0.8	175.0
Electrical & electronic equipment	4.2	5.6	75.0	1.2	5.6	21.4
Machinery & equipment	1.8	2.1	85.7	1.0	2.1	47.6
Basic & fabricated metal	0.6	0.7	85.7	0.3	0.7	42.9
Food, drink & tobacco	0.3	0.3	100.0	0.2	0.3	66.7
Rubber & plastics	1.2	1.2	100.0	0.5	1.2	41.7
Textiles/clothing	0.7	0.3	233.3	0.7	0.3	233.3
Wood & wood products	1.0	0.2	500.0	1.5	0.2	750.0
Total	0.8	2.4	33.3	0.6	2.4	25.0

Source: Forfás (2003b, Tables 3.1 and 2.3).

Table A2.7). In 2002, 71 per cent of domestic inventions had foreign ownership as compared to only 15 per cent in Finland, and 22 per cent in Sweden. Therefore, while the overall level of patenting activity is comparatively low in Ireland, the patenting activity that does occur is dominated by foreign individuals and companies.

4.1.2 Competence building

The global context for educational development has been major changes in the nature of technology and a rising demand for skilled labour (Nickell and Bell, 1995). In Ireland, inward FDI has added to this demand (Fitzgerald, 2000), while the growth of private services (which have always demanded more skilled employees) has also led to much greater demand for skilled and educated employees.

At the same time, the view exists that linkages between education and its partners remain relatively disjointed and weak. For example, Smyth and Hannan (2000) state that – relative to other systems such as the Netherlands and Germany – linkages between education, training and the labour market in Ireland are weak. The general nature of education in Ireland leads to education acting as a 'signal' in securing employment with little correlation between educational attainment and type of occupation. However, the evidence suggests that investments in education, increased participation rates and the subsequent rise in educational attainment have had a positive impact on productivity (Durkan et al., 1999).

This positive effect from domestic investment in education has been reinforced recently by a net inflow of migrants into Ireland. Increasing international flows of students have also increased the dynamism of the Irish labour market, with increasing numbers of Irish students studying abroad, and Irish universities attracting increasing numbers of foreign students.[6]

On average in 1998, OECD countries allocated 5.8 per cent of their gross domestic product to educational organizations. Educational investments in Ireland as a proportion of GNP – 5.4 per cent in 1998 – were below this level and below the level of most Scandinavian countries, although above the level prevailing in the Netherlands (4.8 per cent of GDP) (OECD, 2003, pp. 163–9). Nevertheless, levels of investment per student in Ireland were actually lower than those in the Netherlands – as well as being significantly lower than those in Scandinavia in each part of the education system.

In terms of participation in education (both full- and part-time), Ireland performs fairly well in comparative terms; however, Scandinavian countries have higher proportions of older age cohorts in education than Ireland (Figure 5.1).

Analysis of graduation rates as a proportion of each age cohort provides an indication of the performance of the educational system. Ireland has a

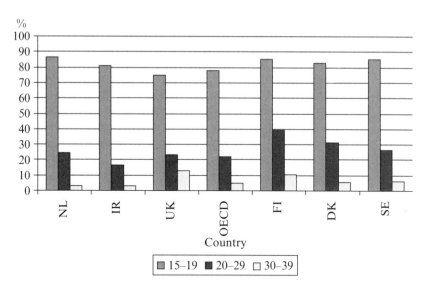

Source: OECD (2003, Table C1.2 p. 258).

Figure 5.1 Percentage of the population enrolled in full-time and part-time education by age group (2001)

Table 5.5 Tertiary graduation rates as percentage of population at each age, 2001

	Tertiary type B (1st time graduation)	Tertiary type A (1st time graduation)	Advanced research programmes
IR	19.0	29.3	0.9
UK	11.5	37.4	1.6
OECD	11.0	30.3	1.1
DK	8.0	38.8	1.0
FI	7.3	40.7	1.8
SE	4.0	29.6	2.7

Source: OECD (2003, Table A2.1).

significantly higher proportion of graduates with Tertiary type B qualifications than other North European countries. At the same time, however, Ireland has a lower proportion of its population obtaining Tertiary type A or advanced research awards (Table 5.5). This suggests that while participation and educational attainment have increased in recent years, the comparative advantage has mainly come in Tertiary type B

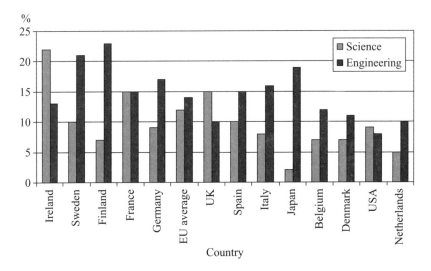

Source: OECD (2003).

Figure 5.2 Science and engineering graduates as a proportion of all disciplines, 2000

qualifications. There is still scope to 'catch up' with other OECD countries, and in particular with the Scandinavian countries in terms of graduates with Tertiary type A and advanced research awards.

In 2000, approximately 22 per cent of graduates in Ireland were in the sciences, which was an exceptionally high figure among OECD countries, and 13 per cent were in engineering, which was about the EU average (Figure 5.2). However, Ireland looks weaker on some related indicators, since Finland and Sweden have far more S&T PhDs per head of population and far more researchers per head of workforce.

4.2 Demand-side Factors

4.2.1 International demand and foreign direct investment

The rapid growth of FDI in Ireland was partly attributable to international demand. Since Ireland joined the EU in the early 1970s, many foreign MNEs have chosen Ireland as a location where they can produce for and sell into EU markets (O'Malley, 1989, ch. 7). The role of EU demand as an incentive to choose such a location was further enhanced by the introduction of the Single European Market in the early 1990s (Enterprise Strategy Group, 2004). Thus EU demand helped to draw FDI into Ireland, increasing and sustaining innovation.[7]

However, EU demand cannot be the only explanation for FDI in Ireland, since FDI has been proportionately much greater in Ireland than in Spain, Portugal or Greece. Other explanations for the scale of FDI in Ireland include Ireland's supportive and proactive industrial policies and the development of agglomerations or clusters of similar industries (Barry et al., 1999; Gorg and Ruane, 2000a; Ruane, 2003).

4.2.2 Domestic demand in general

Studies that have considered the effects of domestic demand on enterprises in Ireland are of two main types. First, a small number of studies have considered the influence of domestic demand in general, among other factors, in fostering competitiveness or growth. Second, there are some studies that have examined the impact of demand emanating specifically from foreign-owned MNEs on other enterprises in the host economy.

In the first of these categories, studies of three relatively successful indigenous sectors in Ireland – software, dairy processing and the popular music industry – are summarized in Clancy et al. (2001). They examined the role of domestic demand as one of four potential influences that might have fostered competitive success in these industries – the other three potential influences being factor conditions, related and supporting industries, and firms' strategy, structure and rivalry. They concluded that in various ways domestic demand conditions had a positive influence in developing the competitive advantage of these industries, including by means of stimulating innovation. For example, companies in the Irish indigenous software industry often said that they had been helped to develop new or improved software products or services by their interaction with sophisticated local customers – including foreign-owned MNEs. As regards the dairy processing industry, sophisticated and demanding local customers, such as large Irish retail chains and foreign-owned MNEs in the food industry, often pushed the dairy processors to upgrade their standards across various functions and to improve their cost efficiency. At the same time, however, with many Irish firms being highly export-oriented and in the context of a small domestic market, it was quite often the case that foreign demand had greater influences than domestic demand on Irish firms in these industries.

4.2.3 Domestic demand emanating from foreign-owned MNEs

It has long been recognized that foreign MNEs in Ireland could be a significant and expanding market for existing or potential Irish suppliers of inputs to these firms. In 1985 a policy measure called the National Linkage Programme was introduced to strengthen linkages between foreign MNEs and suppliers in Ireland by enhancing the technical and business competence of suppliers in cooperation with MNE customers (Crowley, 1996). This policy was quite successful and the foreign MNE sector has proved to be a source of particularly rapid growth in demand for its suppliers. There is evidence that foreign firms in Ireland have tended to purchase an increasing proportion of their materials locally over time (Gorg and Ruane, 2000b). Foreign non-food manufacturing MNEs purchased 15 per cent of their material inputs in Ireland in 1988 and this rose to 21 per cent in 1998 (Ruane, 2003; O'Riain, 2004).[8]

As regards the effect of demand from MNEs on the suppliers' propensity to innovate, Jacobson and Mottiar (1999) present a case study of the software manual printing industry in Dublin which shows that, at least in some cases, there can be a profound effect of this type. They found that a specialized software manual printing sector 'came into existence entirely because of the establishment in Ireland of the software MNEs. The production processes, quality control and delivery times have all been determined by the buyer firms. To be a supplier in this industry, high quality product on the basis of just-in-time delivery was a prerequisite' (Jacobson and Mottiar, 1999, p. 436). In this case, however, the manual printing firms were eventually left vulnerable to a decline in demand for their specialist product.

In a study covering all manufacturing, Hewitt-Dundas et al. (2002) found that foreign MNEs in Ireland represent a potentially important channel through which world-class knowledge can be transferred to supplier businesses. This is seen in the fact that supplier companies tend to lag some time behind their MNE customers in the adoption and use of a range of best-practice management and control systems. On average, Irish suppliers lagged 3.6 months behind their local MNE customers. Hewitt-Dundas et al. (2002) also examined the nature and intensity of interactions between MNE customers and their local suppliers that might provide the basis for knowledge transfer. They found that developmental interactions between MNE plants and their suppliers were common. For example, 79 per cent of MNE plants had collaborated with local suppliers on product development. Also, 58 per cent of MNE plants had assisted local suppliers with quality assurance systems.

The same study asked MNE establishments whether they had significantly enhanced the performance and competitiveness of their local suppliers in various respects. Most MNE plants said that they had done so in all the respects that were specified. For example, 73 per cent of MNEs said that they had significant impacts on their suppliers' sales. The percentages of MNEs reporting other significant impacts on their suppliers included: on suppliers' productivity – 73 per cent; on suppliers' product quality – 77 per cent; on suppliers' service quality – 86 per cent.

To conclude on the overall influence of demand, domestic demand has had little effect on foreign-owned MNEs in Ireland, whereas international demand, especially from the EU, has been a significant factor in drawing those particularly innovative companies into Ireland. Domestic demand, including demand emanating from foreign-owned MNEs in Ireland, can have some quite significant influences on the propensity to innovate among Irish-owned firms. There is little information available about the role of public procurement or standards setting in influencing demand and hence innovation in Ireland. Our impression is that, while there may be specific examples where public procurement policy has such effects, this does not appear to be of great significance.

4.3 Provision of Constituents

Previous analyses of the Irish NSI in comparison with 'best practice' elsewhere have tended to emphasize the weaknesses of the Irish system. For example, in a report produced for the Science, Technology and Innovation Advisory Council (STIAC) in 1995, the Circa group observed:

> Overall . . . the national science and technology system on the product side is now largely based on an under-funded university sector and a much rationalised state sector; it is dispersed, lacks critical mass in many areas and is weakened by historical neglect and drift. On the 'user side' indigenous industry is highly fragmented, of small scale and has low innovative capability in general. (STIAC, 1995, p. 44)

Much has been done in recent years, however, to address these issues through both organizational and policy innovation. Policy development is discussed in detail in Section 7 below, and we focus here on highlighting the key organizational and systemic developments.

The lead body in the governance of the Irish NSI is the Office of Science and Technology (OST) which develops, promotes and coordinates science, technology and innovation policy. This policy covers all aspects of innovation, including basic research, applied research, industry R&D, technology transfer, funding for innovation and public awareness of science and technology (S&T).[9] The OST is advised by the Irish Council for Science, Technology and Innovation (ICSTI), which was established in 1997. This is an independent body – although funded by government – and comprises 25 experts from industry, academia and public sector organizations. Forfás is the agency charged with the promotion of S&T in Ireland along with the promotion and development of enterprises. Forfás also provides an input to the OST and through its operational agencies it implements Irish science, technology and innovation policy. Three of these agencies are particularly important: IDA (Industrial Development Agency) Ireland and Enterprise Ireland, the agencies for foreign-owned and Irish indigenous industry respectively, and Science Foundation Ireland (SFI). SFI was launched in 2001 with a budget of €635 million for the period to 2006 to support basic research in strategic fields relevant to Ireland's industrial development in biotechnology, ICT and other areas (STI, 2002). Support for basic research in higher education in Ireland is handled by a separate organization – the Higher Education Authority (HEA). This operates the Programme for Research in Third Level Institutions (PRTLI) with a broadly similar budget to that of SFI. The aim of the PRTLI is to enhance the quality and relevance of graduate training in Ireland and to support outstanding researchers.

University expansion has been accompanied by the expansion and development of the geographically dispersed regional technology colleges/ institutes of technology. These developments have created a local capacity for human resource development, technology transfer and business incubation in previously underdeveloped regions of the economy. In the Border, Midlands and West region, for example, New University of Ireland (NUI) Galway is the only third-level organization (i.e. university), and the region's institutes of technology actually provide the majority of higher education places (Roper et al., 2003). The development of the institutes of technology has helped with population retention, created a local supply of graduates and provided a focus for business and technology development in those areas lacking a university campus.

4.3.1 Provision of organizations

Entrepreneurship and business start-ups are the main means through which new indigenously owned private sector organisations enter the NSI, although, as indicated earlier, for Ireland inward investment has also been a key driver of growth. International comparisons of entrepreneurial activity are provided by the Global Entrepreneurship Monitor or GEM project. The rate of total entrepreneurial activity (TEA) in Ireland, as measured by GEM, is the highest of the European countries in this study (Table 5.6).[10] Opportunity entrepreneurship was also relatively high in Ireland, as was high potential TEA. It has been suggested that the buoyant economic environment in Ireland has been conducive to entrepreneurial activity, and that 'the personal attributes and cultural supports present within Ireland are factors which contribute to its relatively high rate of entrepreneurial activity. The one area of weakness would appear to be in the area of appropriate skills and experience' (Fitzsimmons et al., 2004, p. 27).[11]

Evidence from the GEM study suggests that nearly two-thirds of the entrepreneurial activity in Ireland is concentrated in 'business services', 'consumer services', retailing and mining and construction (Fitzsimmons

Table 5.6   International context – 2002

	TEA	Opportunity TEA	Necessity TEA	High potential TEA
IR	9.1	7.8	1.4	1.52
NL	4.6	4.0	0.5	0.73
DK	6.5	5.9	0.4	0.96
SE	4.0	3.3	0.7	0.63
NO	8.7	7.4	0.4	1.25
FI	4.6	4.1	0.3	0.72
SG	5.9	4.9	0.9	1.33
KR	14.5	8.6	4.1	2.11
TW	4.3	3.8	0.7	0.89

Source: Wong (2003).

et al., 2004). Perhaps more important, however, is that 'the majority of new ventures being started by entrepreneurs on the island of Ireland are not innovative, a relatively small but significant proportion of entrepreneurs throughout the island (13 per cent) believe that they are innovative' (ibid., p. 19). This links into other evidence which suggests that the innovative dynamic in Ireland is only weakly related to entrepreneurial activity, and more strongly linked to inward investment, technology transfer, and its downstream effects (e.g. Roper and Frenkel, 2000; Roper and Hewitt-Dundas, 2004).

4.3.2   Networking, interactive learning and knowledge integration

Since the Culliton Report of 1992, the notion of innovation networks has been part of the rhetoric of innovation policy in Ireland. Following the recommendations of STIAC (1995), a pilot programme on Inter Firm Cooperation Networks was established in 1997 for an initial six-month period. This was judged successful but was never developed into a fully fledged programme. Subsequent network programmes have been largely skills-based (e.g. Plato, Skillnets) (Forfás, 2004). Despite the lack of specific initiatives designed to develop innovation networks, levels of innovation cooperation in Ireland are towards the middle of the range within Europe, above the Netherlands and Sweden but below Norway and Finland (Table 5.7). Unlike the other European countries, levels of innovation cooperation in Ireland are higher in services than manufacturing.

In Ireland, innovation cooperation is more common among larger and externally owned plants (Table 5.8). In addition, marked differences are evident between different sectors. Sectors with high levels of innovation cooperation are: food, drink and tobacco, chemicals, mechanical engineering,

Table 5.7 *Proportion of enterprises with cooperation arrangements on innovation*

	Manufacturing (%)	Services (%)
IR	38	53
NL	24	24
FI	52	48
SE	34	29
NO	49	37

Source: CIS3, IIP4.

Table 5.8 *Percentage of plants in Ireland with innovation linkages by plant size, sector and ownership*

A. By plant size (% of plants)	
10–19	23.4
20–99	39.5
100+	59.4
B. By industrial sector (% of plants)	
Food, drink & tobacco	45.7
Textiles & clothing	26.8
Wood & wood products	18.4
Paper & printing	29.4
Chemicals	52.9
Metals & metal fabrication	31.0
Mechanical engineering	37.3
Electrical & optical equipment	50.1
Transport equipment	43.6
Other manufacturing	39.5
C. By ownership (% of plants)	
Indigenously owned	34.9
Externally owned	49.3

Notes: Figures relate to all manufacturing plants with ten or more employees. Survey responses are weighted to give representative results.

Table 5.9   Percentage of manufacturing plants with innovation links to different firms and organizations

	% of plants
Links to:	
Other group companies	16.1
Clients or customers	20.8
Suppliers	24.7
Competitors	8.4
Joint ventures	5.5
Consultants	15.1
Government	5.3
Higher education organizations	11.1
Industry labs	5.6
Private research institutes	5.0

Notes: Figures relate to all manufacturing plants with ten or more employees. Survey responses are weighted to give representative results.

electrical and optical equipment, and transport equipment. Recent policy reports have, however, emphasized the need for measures to promote innovation networks (e.g. Enterprise Strategy Group, 2004).

Analysis by type of innovation partner suggests that producer–user relationships are the most common type of linkage, with 20.8 per cent of plants collaborating on innovation with clients or customers and 24.7 per cent with suppliers (Table 5.9). While it might be expected that horizontal innovation linkages would be less common than vertical innovation linkages, the former are particularly low in Ireland, largely due to industry labs and private research institutes not being particularly well developed. Industry-operated labs and private research institutes were among the least common type of external partners.

Innovation networks have developed organically in Ireland with little targeted policy support. Some successful examples have been identified (see the case studies in Forfás, 2004), but more could probably be done to develop the degree of association. This is important because, as Roper and Love (2004) have shown, Irish firms' participation in innovation networks has positive and statistically demonstrable benefits in terms of boosting levels of successful innovation.

4.3.3   Provision of institutions

Institutions, as understood here, relate to the context of laws, rules, practices and conventions within which firms in the NSI are operating. In

Table 5.10 Summary indicators of regulatory regimes

	Burden of regulation	Labour market regulation	Competition legislation	Levels of bureaucracy
IR	3.7 (2)	5.962 (2)	6.296 (6)	3.96 (4)
DK	3.1 (4)	6.831 (1)	7.485 (2)	5.46 (2)
FI	4.4 (1)	5.205 (4)	8.595 (1)	6.89 (1)
KR	3.0 (5)	2.612 (7)	5.184 (7)	3.00 (7)
NL	2.8 (7)	3.871 (5)	6.839 (3)	3.74 (5)
SE	2.9 (6)	3.379 (6)	6.500 (4)	4.97 (3)

Notes and original sources: Competition legislation, *source: IMD World Competitiveness Yearbook* (2003); Burden of regulation, 1 is burdensome, 7 not burdensome, *source:* World Economic Forum, *Global Competitiveness Report 2002–03*; Level of bureaucracy, 1 hinders activity, 7 does not hinder activity, *source: IMD World Competitiveness Yearbook* (2003); Labour market regulations, *source: IMD World Competitiveness Yearbook* (2003).

Source: National Competitiveness Council (2003).

general terms, Ireland is characterized by relatively light regulatory frameworks by international standards. Bureaucratic burdens on firms are, however, heavier than those in some other countries. Table 5.10 summarizes a number of international indicators of this type of measure, with rankings given in parentheses. In terms of the regulatory burden on firms – both in general and in labour market terms – Ireland comes out relatively well compared to the other countries considered here. However, in terms of the legislative and bureaucratic burden on firms the comparison is less flattering to Ireland. Competition legislation seems to impose a significantly more important barrier to growth in Ireland than in some other countries.

Corporate governance in Ireland largely follows the US and UK model, although board sizes of quoted companies tend to be smaller (around nine members) than those in the UK (12–13 members), and interlocking directorships are not as common in Ireland as in some other countries (Brennan and McDermott, 2002). Despite some tightening up of financial regulation, auditing and public reporting in recent years (e.g. the establishment of the Office of the Director of Corporate Enforcement in 2001), doubts remain over whether Ireland's 'light touch' regulatory approach to corporate governance continues to be appropriate. Brennan (2006), for example, cites a *New York Times* report that refers to Ireland becoming 'the wild west of European finance', and argues that further developments in corporate and financial legislation are required in Ireland to preserve its position as a desirable location for inward investment and international business operations. This view on corporate governance contrasts with Ireland's more rapid progress in terms of developing its intellectual property rights (IPR) legislation. Since 2001 Irish IPR law has been consistent with the requirements of the agreement on trade-related aspects of intellectual property rights (TRIPS) and Ireland is a member of the World Intellectual Property Organization and party to the International Convention for the Protection of Intellectual Property. The US State Department argues that this means that Ireland has 'one of the most comprehensive systems of IPR protection in Europe'.

4.4 Support Services for Innovating Firms

4.4.1 Incubating activities

The European Commission is concerned to increase the level of entrepreneurial activity in the EU, and this objective has also been part of the policy rhetoric in Ireland (see Forfás, 2000), with a particular emphasis on the need to increase entrepreneurial activity in knowledge-based industries (Fitzsimmons et al., 2004, pp. 4–8). This reflects both the general perception that such firms are the most likely to be fast-growing 'gazelles' but also the weakness of Irish support for such firms until very recently. In particular, although there have been some notable successes, incubation and commercialization services for high-tech start-ups and spin-outs have been relatively late to develop in Ireland compared to other countries.

Both innovation and enterprise are now seen as key policy objectives in Ireland, and this is reflected in a plethora of public initiatives. The 'enterprise' and 'innovation' agendas are strongly government- or agency-driven, with less private sector involvement. For example, levels of public intervention in industrial development are relatively high, with Fitzsimmons et al. (2004) identifying 44 individual support measures in Ireland for innovation and enterprise and a further nine measures available on a cross-border basis (with Northern Ireland). Conversely, GEM reports suggest relatively low levels of informal investment in entrepreneurial ventures as well as relatively low levels of more formal venture capital investment.

Both the level of venture capital activity and more formal small business incubation capacity have developed relatively rapidly in recent years, however. Across the country many of the institutes of technology provide foci for business and technology development, often operating incubator facilities and sometimes providing consultancy and research support to local firms as well as participating in joint ventures. Most of the universities also offer a range of technology and business support services to local firms, which include providing business incubation facilities, as well as participating in collaborative research projects with local firms, encouraging spin-out businesses, and providing access to the resources, expertise and amenities of the college.

Although recent developments in incubator activity in third-level colleges have been very positive, more mainstream commercialization activities which might lead to spin-out or start-up businesses remain limited. A recent study examined the commercialization activities of universities, institutes of technology and research institutes, and it found that a total of 16 people (on a full-time equivalent basis) are involved in commercialization activities in Ireland (InterTradeIreland, 2002). Moreover, there was a very strong bias towards the South and East region which accounted for 15.5 full-time equivalents compared to 0.7 in the Border, Midlands and Western region.

4.4.2 Financing

As indicated earlier, business R&D spending in Ireland is strongly concentrated in larger multinational plants. The largest funding source for this investment was the firms' internally generated funds. In 2001, this accounted for €560.7 million of the €598.1 million invested in R&D by externally owned firms, with a further €6.0 million coming from government funds. For Irish-owned businesses, a larger proportion of R&D funding came from government, €18.6 million out of €318.7 million, with €274.8 million coming from firms' own resources.

The traditional means by which the government has provided support for R&D and innovation has been selective grants.[12] The specific schemes through which support has been provided have changed over the years, but there has been a recent trend away from grant support towards equity in the form of ordinary shareholdings and preference shares (Forfás, 2000). Currently R&D is assisted through the Research, Technological Development and Innovation scheme and the R&D Capability scheme, while the Innovation Management scheme supports company training in R&D management.

In addition to nationally funded innovation support measures, most of Ireland, excluding Dublin, has also been eligible for EU Structural Funds support. This has been used to co-fund regional programmes for R&D and innovation support programmes (see, e.g., Roper et al., 2003).[13] Cogan and McDevitt (2000, p. 11) summarize the key impacts of the Structural Funds in the area of R&D, and argue that there have been positive impacts in terms of

- significantly expanding the support available for applied R&D projects in less well-developed areas;
- stimulating a process of organizational and institutional learning, enhancing the capability of Ireland to support future R&D and innovation activity;

- leading to a disciplined evaluation of policy, something that – they argue – was missing from policy before this.

Not everyone, however, has been so positive about the impact of EU support on R&D funding in Ireland. Yearley (1995), for example, suggests that the scale and pervasiveness of EU support and influence in Irish S&T policy has a less welcome aspect: 'Something like the nineteenth-century pattern may be reasserting itself. Research performed in Ireland may allegedly be for the good of the country, yet the selection of research is made outside the country, this time at the European level' (p. 191).

In addition to the Structural Funds, firms and other organizations in Ireland have also benefited from participation in the EU Framework programmes and other collaborative R&D projects such as Esprit (the EU IT programme). Grimes and Collins (2003), for example, note the impact of participation in the Esprit programme in terms of formalizing previously informal technology networks and strengthening the technology linkages between Ireland and other EU countries.

Evaluations of technology programmes in Ireland have generally highlighted relatively high levels of additionality and leverage of additional private investments in R&D and innovation (e.g. Roper, 1998). Evidence from a recent paper by Roper and Love (2004) also suggests that innovation success in manufacturing firms is strongly related to R&D activity in the company, networks both within and outside the supply chain, and intermediate skills. Also statistically significant and positive is grant support for product innovation and the level of capital investment in the firm. Both factors provide support for a significant policy effect working through specific grants for R&D and innovation and through stimulating investment.[14]

4.4.3 Provision of consultancy services

Consultancy and advisory support for innovation in Ireland is widely available from public, higher education and private organizations. In part, this reflects participation in pan-European initiatives such as the Innovation Relay Centres, but it also reflects attempts to provide support for innovation, particularly among indigenously owned and smaller firms. As part of this policy, Enterprise Ireland provides a range of advisory and consultancy services to industry on a contract – and often subsidized – basis to help firms identify, implement and optimize new technologies. A particular focus in recent years has been assistance with environmental impact assessments and the development of environmentally friendly products. Enterprise Ireland provides a range of information, advisory and environmental audit services for small and medium-sized enterprises as well as maintaining a specialist environment awareness website (www.EnviroCentre.ie). In addition, financial support is available for developing environmental management strategy to the ISO14001 standard, as is support for the development of environmentally superior products.[15]

Universities also offer consultancy and technology transfer services to local firms. NUI Galway, for example, has an innovation centre that provides eight nursery units as well as access to the resources, expertise and amenities of the college. The Galway campus also hosts the International Services Park, established in partnership with IDA Ireland. Also potentially important in terms of technology diffusion are other organizations such as the six innovation centres – part of the EC-BIC network – that operate throughout Ireland. These centres provide a range of training and business services with the aim of contributing to the development of innovative capacity among indigenous enterprises. In the Border, Midlands and Western region, WESTBIC, for example, operates through a series of seven local offices providing advisory, training and informational services (Roper et al., 2003). In the South and East region, innovation centres are based in Dublin, Cork, Waterford and Limerick. Other, more recent, initiatives related to knowledge diffusion include the Technology Transfer Initiative, a collaborative scheme being operated by University College Cork (UCC), NUI Galway and University Limerick (UL) with the aim of encouraging smaller firms to include an element of R&D in their operations (see, e.g., www.technologytransfer.ie).

4.5 Summary of the Main Activities Influencing Innovation

Overall, expenditure on R&D in Ireland continues to fall below that of other countries with a similar industrial (high-tech) structure. This reflects a genuine underinvestment in R&D, although in the case of Irish indigenous industry it also reflects the relative concentration of indigenous industry in low R&D-intensity sectors. In absolute terms, business R&D spending is dominated by externally owned companies – mainly in the high-tech sectors. However, when measured in terms of R&D intensity (R&D as a percentage of sales), foreign-owned companies in Ireland actually tend to be less R&D-intensive than Irish-owned companies in the same sectors, and far less R&D-intensive than OECD companies. Irish-owned companies, and some of the more 'traditional' sectors, sometimes have quite high levels of R&D intensity compared to OECD levels in the same sectors.

Levels of education investment in Ireland remain below those of the Scandinavian countries expressed both relative to GDP and on a per capita basis. Participation in education, particularly among older age cohorts, is lower than in Scandinavia, although broadly in line with the OECD average. Graduation rates for higher-level tertiary qualifications also remain relatively low by international standards, although there is a strong bias within the higher education system towards science graduates.

Domestic demand has had very little influence on the foreign-owned MNEs, which are very important in Ireland. However, domestic demand can have some quite significant influences on the propensity to innovate among Irish-owned firms. In particular, there is evidence that domestic demand emanating from foreign-owned MNEs can enhance innovation among suppliers in Ireland.

Previous analyses of the Irish NSI in comparison to 'best practice' elsewhere have tended to emphasize the weaknesses of the Irish system, at least up to the mid-1990s. However, innovation has been moving closer to the top of the industrial development agenda in Ireland, and this has been reflected in increasing maturity of the innovation support regime.

5 CONSEQUENCES OF INNOVATION

Evidence on the consequences of innovation in Ireland is available at both the enterprise and macroeconomic levels. Both are reviewed here, for as Cassidy (2003) remarks: 'Generally speaking, improvements in productivity are realized at the level of the firm; economy wide productivity levels (growth rates) are, therefore, primarily an aggregation of the productivity levels (growth rates) of individual firms' (p. 83).[16]

5.1 Plant-level Evidence

Roper and Love (2004) report results derived from the Irish Innovation Panel and found that, on average, plants in Ireland derive about one-sixth of their sales from products that were newly introduced. Using this measure – the proportion of sales derived from innovative products – as an indicator of a plant's innovation success, they found a strong positive link between product innovation success and business growth. Immediate productivity impacts were negative, however, due to disruption effects, although longer-term productivity effects were positive. Process innovation had a more immediate and enduring productivity benefit. These results imply that the relatively high incidence of innovation success in Irish firms noted earlier is likely to be making a positive contribution to enhanced productivity and sales growth.

Kearns and Ruane (1998 and 1999) used plant-level data provided by Forfás to examine some effects of R&D performance and innovation on plants.[17] Kearns and Ruane (1998) found that technological activity within plants is an important determinant of their probability of survival. They focused on indigenous manufacturing plants that existed in Ireland in 1986 and found that technologically active plants had higher probabilities of surviving until 1996 than comparable plants that were less technologically active. This higher survival performance held true with respect to several different variables that were used to measure 'technological activity', namely scale of R&D, R&D intensity and sales of innovative products developed within the plant. Kearns and Ruane (1999) found that foreign-owned MNE plants that undertook R&D in Ireland were likely to survive in Ireland for longer than MNE plants that did not undertake R&D. They also found that job losses were lower and job persistence was greater among R&D-active MNE plants than among those that did not undertake R&D.

In general, therefore, the plant-level evidence for Ireland suggests that innovative plants are growing faster, have higher productivity and have greater longevity than non-innovating plants. As the evidence reported earlier suggests a relatively high incidence of innovation activity in Ireland by international standards, this is likely to be a positive contributor to national economic performance. It is difficult, however, from the micro-economic evidence to make any assessment of the macro-impact of these effects. Here, evidence from macroeconomic studies is more appropriate.

5.2 Aggregate Economic Performance

Consistent with the idea that innovation has a significant influence on economic performance, Ireland has had relatively fast industrial growth compared to other European countries. Thus Ireland's share of EU industrial production rose from 0.9 per cent in 1991 to 2.1 per cent in 2001. This was not simply an effect of foreign-owned MNEs in a limited range of high-tech sectors. Ireland's share of EU industrial production increased in every individual sector except textiles, clothing and footwear. In addition, Irish indigenous industry, without the foreign MNEs, increased its share of EU production from 0.41 per cent to 0.44 per cent in 1991–2001, and it increased its share in a large majority of individual sectors (O'Malley, 2004).

Research on the sources of growth in Ireland – and the importance of innovation – in this process is relatively underdeveloped. Growth accounting studies such as Felisberto (2003), however, have concluded that Irish growth over the 1990s has been dominated by total factor productivity (TFP) growth: 'Factors such as increased labour productivity and efficiency, use of labour in an intensive way and foreign investment (whose repercussions are spread to the whole economy and through time) have largely been contributing to Irish productivity growth' (p. 1). More specifically, Felisberto (2003, p. 25) suggests that half of Irish productivity growth over the 1960–2002 period is accounted for by increases in capital and labour inputs, with half being due to TFP growth.[18] Recent EU studies also place a similar weight on the growth in manufacturing productivity in Ireland. Of labour productivity growth of 8.4 per cent from 1996 to 2000, for example, 7.3 percentage points were attributed to growth in manufacturing productivity in Ireland compared to 0.7 percentage points of total growth of 1.7 per cent in the EU (EU Commission, 2003, p. 86). Hence Cassidy (2003), in his recent review of Irish productivity growth, concludes that 'Ireland's superior productivity performance in manufacturing has been largely a consequence of two factors, namely higher productivity growth in the high-technology sectors than the European average and also a greater degree of specialisation in these sectors' (p. 93).[19]

ICT expenditure has also contributed more positively to growth in productivity in Ireland than in the majority of other EU economies. Daveri (2002), for example, examines the impact of ICT investment on productivity. While ICT spending in Ireland has been broadly in line with that in the EU as a whole, Daveri (2002) concludes that for Ireland, ICT capital added 0.35 percentage points to the growth in GDP per man-hour through the later 1990s in addition to adding 0.59 percentage points to TFP growth. Only three EU countries (Ireland, Portugal and Greece) experienced positive growth effects from ICT through both effects.

6 GLOBALIZATION

It is clear that aspects of globalization have played a major part in shaping Ireland's economy and its NSI. This was seen in the 1960s and 1970s, when the move from a protectionist policy to free trade opened the economy to an increase in imports and presented new opportunities to export. Since 1973 this more open international trade environment has been governed by Ireland's membership of the EU, which has also had other impacts on Ireland's laws, policies and attitudes. From being largely internally oriented, and to some extent UK-oriented, Ireland has become much more aware of, and influenced by, the wider world.

The aspect of globalization that we have focused on most in reviewing the period since the early 1990s is FDI, specifically FDI coming into Ireland, since this has been a major influence on the economy and the NSI. The reasons why so many foreign MNEs have invested in Ireland have shifted over time. Relatively low labour costs (by Western European standards) are no longer an incentive for FDI, although they used to be. More enduring reasons for FDI have been Ireland's ready access to EU markets, low tax on profits (probably the single most important factor), the availability of skilled labour required by many of the most rapidly growing industries, an English-speaking population (attractive to US companies), proactive industrial policies that sought out desirable industries and companies, and the consequent development of clusters of similar and related industries.

Note that in saying that the availability of skilled labour was an important attraction for FDI, this is not a matter of claiming that the Irish system of education and training is much superior in quality to that in other European countries. Rather, the point is that the system usually – and quite deliberately – managed to produce enough people with the skills required by the most rapidly growing industries, while shortages of the skills concerned occurred more commonly elsewhere. In addition, a contribution to the quality of the labour force in Ireland since the early 1990s was the return of many former emigrants who had acquired experience and skills abroad.

Inward FDI in Ireland contributed directly to the growth of employment, production and exports, and it had quite a significant influence in terms of developing purchasing linkages. FDI also had significant effects on the innovation performance of Ireland – partly because the MNEs concerned have often been particularly innovative companies themselves, but also because they have given rise to knowledge spillovers affecting innovation in the host economy, as discussed in Section 4.2, for example.

7 STRENGTHS AND WEAKNESSES OF THE SYSTEM AND INNOVATION POLICIES

7.1 Strengths and Weaknesses

Our analysis of the Irish case suggests a paradox. On the one hand, the analysis of its internal capabilities suggests a system that is not particularly strong in some respects, and that is clearly well behind international best practice in other respects. Levels of R&D investment, in particular, are far below those in the Scandinavian countries, while investments in skills are somewhat lower than international best practice. On the other hand, evidence from CIS1 and particularly CIS2 (e.g. Roper and Hewitt-Dundas, 2003), and material reviewed earlier, emphasize the strengths of the innovation and economic outputs from Ireland. In particular, Ireland tends to lead most of the benchmark countries both in the extent and quality of innovation achieved and in terms of productivity and economic growth.

As indicated earlier, a major factor in explaining this apparent contradiction is inward technology transfer, which has accompanied inward investment to Ireland over the last three decades. It is this – largely undocumented – inward technology transfer, from mainly US technology companies, that provides much of the current strength of the Irish NSI. In effect, a good deal of innovation takes place in Ireland that owes relatively little to local technological capability.

The creation and development of this dynamic of innovation based on inward technology transfer has not, however, been a random event. Instead, it has been the result of a systematic and effective policy of 'innovation by invitation' adopted by the development agencies (Roper and Frenkel, 2000). A key element of this has, of course, been the attraction of inward investment, but also important have been other measures designed to embed inward investors within the Irish economy, strengthen local linkages and build clusters of interrelated companies.

This 'strength' of the Irish NSI has also brought with it, however, an important dependence on the global economy and the willingness of external companies to continue to transfer technology to Ireland (see Yearley, 1995 for a general discussion). This is neatly illustrated by the time profile of innovation in Ireland over the 2000–2002 period, when innovation levels fell sharply as a result of the global downturn.

Apart from the major impact of FDI and inward technology transfer, it was also indicated (in Section 3) that firms in Irish indigenous industry tend to be quite active innovators compared to firms in the other European countries, while Irish indigenous industry has also grown faster than EU industry in most sectors (Section 5). By way of explanation for this, it is worth noting, first, that the level of R&D expenditure in indigenous industry is not really as inadequate as it may appear. Rather, the low level of R&D expenditure in indigenous industry at the aggregate level is quite largely due to the fact that indigenous industry is relatively concentrated in naturally low R&D-intensity sectors. Taken sector by sector, Ireland's R&D intensity has tended to be broadly comparable to, or at least not very far behind, OECD levels. Second, there are indications that the innovation and growth performance of Irish indigenous industry has benefited from interactions with advanced foreign-owned MNEs (e.g. Section 4.2). Third, compared to the other European countries, it seems that Ireland has a relatively strong regime of financial assistance and other state services for enterprises that have been an important strength of the system for indigenous industry (see also O'Riain, 2004).

Despite these remarks about indigenous industry, it should be recognized that a continuing weakness of the Irish NSI is that a significant amount of manufacturing capacity remains in smaller firms in the traditional sectors. These firms – in common with similar companies elsewhere – face considerable challenges in absorbing and implementing new technologies and innovation. Improving the innovation capability of this group of firms remains a challenge.

7.2 Summary and Evaluation of the Innovation Policy Pursued

Little attention was paid to the role of S&T in the Irish economy until the 1960s. Then the report 'Science and Irish Economic Development' (Howie et al., 1966) said that 'industrial research is relatively non-existent' and that much of it was 'plant and process adaptation development and barely merited the title research and development'. Following that report, the government established the National Science Council to advise on policies for research, development and technology. Grants to assist R&D in industry were introduced in 1969.

In the early 1970s, Cooper and Whelan (1973) pointed out that business expenditure on R&D in Ireland was very low compared to other OECD countries, and argued that the growth of industrial output and exports in Ireland had not depended to any significant extent on Irish R&D (Cooper and Whelan, 1973; Science Policy Research Centre, 1973; Yearley, 1995). During the 1980s, S&T issues were drawn more into the mainstream of industrial policy following a series of official reports.[20] By the end of the 1980s, in addition to the previously existing R&D grants, new policy measures had been introduced, including technology acquisition grants, subsidized 'technology audits', subsidized placement of S&T personnel in companies and a variety of measures to strengthen S&T in third-level colleges and their links with industry.

A number of further policy steps in the 1990s reflected greater recognition of the economic importance of S&T. In 1994, STIAC was established to assess policies, objectives, structures and components of the national S&T system (see Quinlan, 1995). The Council's report (STIAC, 1995) stressed the importance of the concept of the NSI. That report led to the publication of Ireland's first government White Paper on Science, Technology and Innovation in 1996, which led, in turn, to the establishment of a permanent organization, ICSTI. ICSTI's mandate is to advise on the direction of S&T policy, including higher education, technology and R&D in industry, financing of innovation and public awareness. More recent policy developments have sought to strengthen the indigenous innovative capability of Ireland through an upgrading of HEI investments in R&D, and measures designed to leverage higher levels of private R&D spending and to develop new R&D capabilities. This should lead – in the longer term – to greater indigenous innovation capability.

7.3 Future Innovation Policy

Four key themes seem set to dominate future innovation policy in Ireland. First, continuing investments in increasing Ireland's knowledge-generating capability seem essential, bringing Ireland's R&D investments more in line with those of its leading international competitors and at the same time reducing its dependence on inward technology transfer. This strand of innovation policy has two main elements – investment to increase the knowledge creation capability of Ireland's higher education sector and public initiatives to develop knowledge generation capability within private sector organizations. In terms of universities, a key development has been the establishment of SFI, which is investing €646 million between 2000 and 2006 in research teams, university and collaborative research centres, primarily in ICT and biotechnology. The model for SFI is the US National Science Foundation (NSF), and the SFI Director General, Bill Harris, was formerly Director of the Mathematical and Physical Sciences Division of the NSF. SFI has worked by providing large-scale grant funding for existing areas of Irish research excellence, establishing research partnerships with major international companies (including Intel, Procter and Gamble, Wyeth, Bell Labs, HP, and Medtronic) and attracting leading researchers to Ireland to head new research institutes (SFI, 2005). This reflects academic research such as that by Zucker et al. (1998a and 1998b), which emphasizes the role of such academic 'stars' as the key conduit of knowledge from laboratory to marketplace.

It is too early at this point to evaluate even partially the impact of the SFI initiative, but it is clear that through SFI and other initiatives levels of both public and private R&D investment in Ireland are increasing. Questions remain, however, about the extent to which these developments will feed through into increased innovation activity, particularly in Irish SMEs. Some authors have, for example, doubted the relevance of some SFI-supported research to such firms, while others have focused on Irish universities' limited historical commitment to commercialization (e.g. InterTradeIreland, 2002), and limitations in the absorptive capacity of Irish SMEs, in particular (e.g. Forfás, 2005). Alongside the developments in Ireland's university system, therefore, capacity upgrading in Irish SMEs is also a clear priority not only to boost their in-house knowledge-generating capacity but also to increase absorptive capacity (e.g. Veugelers and Cassiman, 1999). In this respect it is hard to disagree with the diagnosis of Forfás (2005) that stresses the need to strengthen Irish SMEs' awareness, intelligence-gathering and network capabilities as well as their internal human resources and learning capabilities.

The second major theme in future innovation policy in Ireland is likely to be the closer fusion of Ireland's search for inward investment and its development of an innovation agenda. In particular, in alliance with SFI, this suggests an emphasis on the attraction of the R&D and innovation functions of firms with an existing manufacturing or service facility in Ireland as well as new R&D activities of new inward investors. This is, of course, not a new agenda. Indeed, as IDA Ireland points out, over 300 externally owned firms already have some R&D based in Ireland, and this activity accounts for two-thirds of all business R&D spend in the country. The scope for Ireland to increase this level of R&D activity by multinational firms is likely to increase in future years as R&D internationalization itself increases. The international 'marketplace' for such developments, however, is also likely to become increasingly competitive, with Ireland facing strong competition for mobile R&D and innovation projects from Eastern Europe and the Asian economies. Initiatives such as SFI, and recently introduced tax credits for R&D activity, are both likely to increase Ireland's attractiveness as a potential location for internationally mobile R&D.

A third theme in future innovation policy in Ireland is likely to be an increasing emphasis on developing innovation capability in Irish service sector businesses. This reflects services' role as the dominant growth sector in terms of both wealth creation and export earnings. Recent work undertaken for Forfás, for example, has emphasized the challenges and potential rewards of policy devoted to boosting service innovation in Ireland. In particular, the study emphasized the relatively low proportion of innovative firms in the tradable services sector in Ireland – at least compared to manufacturing – and the potential export and export-enabling role of knowledge-based services. The study also emphasized, however, the weakness of the existing evidence base on the Irish service sectors as well as the diversity of innovation processes in different service sectors (see also Gallouj, 2002). A key issue, therefore, in this area of Irish innovation policy is strengthening the knowledge base to enable the design and implementation of effective evidence-based policy measures. Data from the fourth EU CIS will contribute to this process but more specific assessments of innovation processes in the service sector are also likely to be necessary.

Finally, as the NSI in Ireland becomes more knowledge-rich, measures designed to promote knowledge diffusion and adoption will increase in importance, with the aim of maximizing the local economic benefits of technology investments. One element of this, discussed above, would be efforts to develop the absorptive capacity of individual enterprises. Other measures are likely to focus on network development, encouraging stronger association particularly between knowledge-generating and knowledge-applying organizations. This is one area where policy in Ireland has lagged significantly behind international best practice in Sweden and Denmark, for example, where network development and a systemic approach to the development of the NSI have been adopted (e.g. Forfás, 2004). Key policy initiatives in Ireland are likely to include prioritizing network development as a focus of innovation policy, with related skill development, developing financial incentives to encourage collaborative R&D and innovation projects, and encouraging enhanced cooperation between higher education organizations.

ACKNOWLEDGEMENTS

Thanks are due to the editors and other contributors to this volume for data and comments on earlier drafts. Thanks are also due for the financial support provided for the Irish team by Enterprise Ireland, InterTradeIreland and Invest Northern Ireland. The views expressed in this chapter are those of the study team alone and are not necessarily shared by the sponsoring organizations. Errors remain our own.

NOTES

1. See O'Malley (1981) or O'Malley (1989, ch.3) for a discussion of explanations of the experience of Irish industry in the nineteenth century.
2. For further details on policies and economic developments at this time see Kennedy, et al. (1988), Girvin (1989) and O'Malley (1999).
3. These rapidly growing industries mainly produce 'high-technology' products (even in the paper products, publishing and printing sector, the major growth area has been production of software included in that sector). See O'Malley and Roper (2003) for further details on trends discussed in this section.
4. Despite the significant increase in intramural expenditure by government, this still lagged behind the EU (0.25 per cent of GDP) and the OECD (0.24 per cent of GDP).
5. The only exception to this is the non-metallic mineral products and wood and wood products sectors, where R&D intensity is higher for the foreign-owned sector.
6. In 2001, 4.93 per cent of third-level students in Ireland's universities were foreign. This percentage is higher than in Finland, Netherlands or Norway, yet lower than in Denmark, Sweden or the UK. Source: OECD (2003, Table C3.2, pp. 282–3).
7. Since foreign-owned industry in Ireland exports over 90 per cent of its output, Irish domestic demand could have had little or no influence on that part of industry. This is an important point since foreign-owned firms in Ireland accounted for 78 per cent of manufacturing gross output and 48 per cent of manufacturing employment by 2000 (Census of Industrial Production).
8. However, van Egeraat (2002) finds that, at least in the case of the microcomputer industry, this type of data can overstate the level of purchasing from producers located in Ireland since part of it can be purchasing of imported products from specialist distribution companies that are located in Ireland.
9. Department of Enterprise, Trade and Employment website, accessed 15 December 2003.
10. The TEA rate is the percentage of the adult population who are involved in the start-up process or engaged as owner-managers of firms less than 42 months old.
11. Other aspects of the environment for entrepreneurship in Ireland largely reflect those existing elsewhere, with experts stressing a lack of start-up finance, deficiencies in government programmes, poor entrepreneurship education and in Ireland, at least, poor physical infrastructure (Fitzsimmons et al., 2004, pp. 24–34).
12. In addition, firms in Ireland also benefit from fiscal incentives, which have been an important element of the incentive package for inward investment. The corporation tax rate is currently just 12.5 per cent. R&D tax credits were also introduced in Ireland in 2003.

13. A range of case studies illustrating the co-funding principles adopted by the Structural Funds in Ireland is available at: http://www.csfinfo.com/htm/case_studies/case_study_00_06.htm.
14. The positive effect of the policy variable is reassuring but requires some care in interpretation. In particular, the estimate will give an unbiased indication of the effect of grant support only if support is randomly distributed across the population of plants. For example, if product development assistance were targeted at firms more likely to be successful innovators even without assistance, the variable would overestimate the true assistance effect.
15. A useful gateway to information about the Irish development agencies including Enterprise Ireland is provided by www.forfas.ie.
16. Other studies, however, have considered the wider implications for Irish society of the information economy (e.g. Grimes, 2003; Grimes and Collins, 2003).
17. Of course, R&D is not the same thing as innovation, but they are related. Roper and Love (2004) found that plants' R&D intensity has a strong positive effect on their innovation success in the sense outlined above.
18. By contrast, in Portugal she attributes almost all of the growth in aggregate productivity to an increase in factor inputs. Notably, however, Felisberto (2003) shows no awareness of the potential importance of the transfer pricing issue in the case of Ireland.
19. Indeed, whether one looks at GDP per capita, GDP per person employed or GDP per hour worked, since at least the late 1980s Ireland has generally had faster productivity growth than the Nordic countries or the Netherlands (see Appendix Table A2.9).
20. That is, the review of industrial policy by Telesis (1982) and the National Economic and Social Council (1982) and the subsequent White Paper on Industrial Policy (1984).

REFERENCES

Barry, F., J. Bradley and E. O'Malley (1999), 'Indigenous and foreign industry: characteristics and performance', in F. Barry (ed.), *Understanding Ireland's Economic Growth*, Basingstoke: Macmillan; New York: St Martin's Press, pp. 45–74.

Brennan, N. (2006), 'Time to raise the bar on corporate governance', *Sunday Business Post*, Dublin, 26 March.

Brennan, N. and M. McDermott (2002), *Are Non-executive Directors of Irish plcs Independent?*, Institute of Directors/Centre for Corporate Governance, University College Dublin.

Cassidy, M. (2003), 'Productivity in Ireland: trends and issues', *Quarterly Bulletin*, Irish Central Bank, Spring, 83–106.

Clancy, P., E. O'Malley, L. O'Connell and C. van Egeraat (2001), 'Industry clusters in Ireland: an application of Porter's model of national competitive advantage to three sectors', *European Planning Studies*, **9**(1), 7–28.

Cogan, D. and J. McDevitt (2000), 'Science, technology and innovation policy & science and technology policy evaluation: the Irish experience', paper presented at the Converge Workshop Madrid, 5 October.

Cooper, C. and N. Whelan (1973), *Science, Technology and Industry in Ireland*, Dublin: Stationery Office.

Crowley, M. (1996), *National Linkage Programme: Final Evaluation Report*, Dublin: Industry Evaluation Unit.

Daveri, F. (2002), 'The new economy in Europe, 1992–2001', *Oxford Review of Economic Policy*, **18**(3), 345–62.

Durkan, J., D. Fitzgerald and C. Harmon (1999), 'Education and growth in the Irish economy', in F. Barry (ed.), *Understanding Ireland's Economic Growth*, Basingstoke: Macmillan; New York: St Martin's Press, pp. 119–35.
Enterprise Strategy Group (2004), *Ahead of the Curve: Ireland's Place in the Global Economy*, Dublin: Forfás.
EU Commission (2003), 'Drivers of productivity growth: an economy wide and industry level perspective', in *EU Economy 2003 Review*, Brussels, pp. 63–118.
Felisberto, C. (2003), 'On the importance of labour productivity growth', mimeo, University of Lausanne, Ecole des Hautes Etudes Commerciales, July.
Fitzgerald, J. (2000), 'The story of Ireland's failure – and belated success', in B. Nolan, P.J. O'Connell and C.T. Whelan (eds), *Bust to Boom? The Irish Experience of Growth and Inequality*, Dublin: IPA, pp. 27–57.
Fitzsimmons, P., C. O'Gorman, M. Hart and E. McGloin (2004), *Entrepreneurship on the Island of Ireland*, Newry: InterTradeIreland.
Forfás (2000), *Enterprise 2010: A New Strategy for the Promotion of Enterprise in Ireland in the 21st Century*, Dublin: Forfás.
Forfás (2003a), *Research and Development in the Public Sector, 2000 – Volume Two – The Research and Development Element of the Science and Technology Budget*, Dublin: Forfás.
Forfás (2003b), *Business Expenditure on Research and Development, 2001*, Dublin: Forfás.
Forfás (2004), *Innovation Networks*, Dublin: Forfás.
Forfás (2005), *Making Technological Knowledge Work: A Study of the Absorptive Capacity of Irish SMEs*, Dublin: Forfás.
Gallouj, F. (2002), *Innovation in the Service Economy: The New Wealth of Nations*, Cheltenham, UK and Northampton, MA, USA: Edward Elgar.
Girvin, B. (1989), *Between Two Worlds: Politics and Economy in Independent Ireland*, Dublin: Gill & Macmillan.
Gorg, H. and F. Ruane (2000a), 'European integration and peripherality: lessons from the Irish experience', *The World Economy*, **23**(3), 405–21.
Gorg, H. and F. Ruane (2000b), 'An analysis of backward linkages in the Irish electronics sector', *Economic and Social Review*, **31**(3), 215–35.
Grimes, S. (2003), 'Ireland's emerging information economy: recent trends and future prospects', *Regional Studies*, **37**(1), 3–14.
Grimes, S. and P. Collins (2003), 'Building a knowledge economy in Ireland through European research networks', *European Planning Studies*, **11**(4), 395–413.
Hayward, S. (1997), *Developing Ireland: Institutional Innovation and Technology Policy since the Mid-1980s*, Dublin: Forbairt.
Hewitt-Dundas, N., B. Andreosso-O'Callaghan, M. Crone, J. Murray and S. Roper (2002), *Learning from the Best: Knowledge Transfers from Multinational Plants in Ireland: A North–South Comparison*, Belfast: NIERC.
Howie, D.I.D., T.E. Nevin and A.V. Vincent (1966), 'Symposium on Science and Irish Economic Development', *Journal of the Statistical and Social Inquiry Society of Ireland*, **XXI**(V), 35–47.
Industrial Policy Review Group (1992), *A Time for Change: Industrial Policy for the 1990s*, Dublin: Stationery Office.
InterTradeIreland (2002), *Baseline Survey of Commercialisation Staff and Skills in Major Irish R&D Performing Institutions*, Newry: InterTradeIreland.
Jacobson, D. and Z. Mottiar (1999), 'Globalization and modes of interaction in two sub-sectors in Ireland', *European Planning Studies*, **7**(4), 429–44.

Kearns, A. and F. Ruane (1998), 'The post-entry performance of Irish plants: does a plant's technological activity matter?', Trinity Economic Papers Series, Technical Paper No. 98/20, Dublin.

Kearns, A. and F. Ruane (1999), 'The tangible contribution of R&D spending foreign-owned plants to a host region: a plant level study of the Irish manufacturing sector (1980–1996)', Trinity Economic Papers Series, Technical Paper No. 99/7, Dublin.

Kennedy, K.A., T. Giblin and D. McHugh (1988), *The Economic Development of Ireland in the Twentieth Century*, London and New York: Routledge.

Lenihan, H., B. Andreosso-O'Callaghan and G. Dooley (2004), 'Business R&D and innovation in the Republic of Ireland: strategy and public support', Department of Economics, University of Limerick.

National Competitiveness Council (2003), *Annual Competitiveness Report, 2003*, Dublin: National Competitiveness Council.

National Economic and Social Council (1982), *Policies for Industrial Development: Conclusions and Recommendations*, National Economic and Social Council Report No. 66, Dublin: NESC.

Nickell, S. and B. Bell (1995), 'The collapse in demand for the unskilled and unemployment across the OECD', *Oxford Review of Economic Policy*, **11**(1), 40–62.

O'Malley, E. (1981), 'The decline of Irish industry in the nineteenth century', *The Economic and Social Review*, **13**(1), 21–42.

O'Malley, E. (1989), *Industry and Economic Development: The Challenge for the Latecomer*, Dublin: Gill & Macmillan.

O'Malley, E. (1998), 'The revival of Irish indigenous industry 1987–1997', in T.J. Baker, David Duffy and Fergal Shortall (eds), *Quarterly Economic Commentary*, April, Dublin: Economic and Social Research Institute, pp. 35–62.

O'Malley, E. (2004), 'Competitive performance in Irish industry', in Daniel McCoy, David Duffy, Adele Bergin, Shane Garrett and Yvonne McCarthy (eds), *Quarterly Economic Commentary*, Winter, Dublin: Economic and Social Research Institute, pp. 66–87.

O'Malley, E. and S. Roper (2003), *A North/South Analysis of Manufacturing Growth and Productivity*, Newry: InterTradeIreland.

O'Riain, S. (2004), 'State, Competition and Industrial Change in Ireland 1991–1999', *The Economic and Social Review*, **35**(1), 27–53.

OECD (2003), *Education at a Glance*, Paris: OECD.

Quinlan, K. (1995), *Research and Development Activity in Ireland: A Spatial Analysis*, Centre for Local and Regional Development, St Patrick's College, Maynooth.

Roper, S. (1998), 'The principles of the new competition: an empirical assessment of Ireland's position', *Environment and Planning C*, **16**(3), 363–72.

Roper, S. and A. Frenkel (2000), 'Different paths to success – the growth of the Electronics Sector in Ireland and Israel', *Environment and Planning C*, **18**(6), 651–66.

Roper, S. and N. Hewitt-Dundas (2003), 'International innovation comparisons: insight or illusion?', mimeo, Aston Business School, Birmingham, UK.

Roper, S. and N. Hewitt-Dundas (2004), 'Innovation persistence: survey and case-study evidence', Working Paper, Aston Business School, Birmingham, UK.

Roper, S., N. Hewitt-Dundas and M. Savage (2003), 'Benchmarking innovation performance in Ireland's three NUTS 2 regions', Working Paper, NIERC, Belfast.

Roper, S. and J.H. Love (2004), 'Innovation success and business performance – an all-island analysis', draft report to InterTradeIreland.
Ruane, F. (2003), 'Foreign direct investment in Ireland', Lancaster University Management School, Working Paper 2003/005.
Science Foundation Ireland (2005), *Points of Excellence*, Dublin: SFI.
Science Policy Research Centre (1973), *Studies in Irish Science Policy*, Dublin: Stationery Office.
Smyth, E. and D. Hannan (2000), 'Education and inequality', in Brian Nolan, Philip J. O'Connell and Christopher T. Whelan (eds), *Bust to Boom? The Irish Experience of Growth and Inequality*, Dublin: IPA, pp. 109–26.
STI Outlook (2002), *Country Response to Policy Questionnaire*, Dublin: STI.
STIAC (1995), *Making Knowledge Work for Us – A Strategic View of Science, Technology and Innovation in Ireland*, Report of the Science Technology and Innovation Advisory Council, Dublin: Stationery Office.
Telesis Consultancy Group (1982), *A Review of Industrial Policy*, National Economic and Social Council Report No. 64, Dublin: NESC.
Van Egeraat, C. (2002), *New High Volume Production, Production Linkages and Regional Development: The Case of the Microcomputer Hardware Industry in Ireland and Scotland*, PhD thesis, Dublin City University Business School.
Veugelers, R. and B. Cassiman (1999), 'Make and buy in innovation strategies: evidence from Belgian manufacturing firms', *Research Policy*, **28**, 63–80.
Wong, P.-K. (2003), 'Entrepreneurship indicators', presentation at Taiwan workshop.
Yearley, S. (1995), 'From one dependency to another: the political economy of science policy in the Irish Republic in the second half of the twentieth century', *Science, Technology and Human Values*, **20**(2), 171–96.
Zucker, L.G., M.R. Darby and J. Armstrong (1998a), 'Geographically localised knowledge: spillovers or markets?', *Economic Inquiry*, **36**, 65–86.
Zucker, L.G., M.R. Darby and M.B. Bewer (1998b), 'Intellectual human capital and the birth of US biotechnology enterprises', *American Economic Review*, **88**, 290–306.

6. From trade hub to innovation hub: Hong Kong

Naubahar Sharif and Erik Baark

1 INTRODUCTION

In recent years Hong Kong has gone some way towards regaining its traditional position as the key transit point for exchange of both goods and services between China and the international economy. Sophisticated and reliable intermediary services occupy a key role in maintaining this status, and Hong Kong's future apparently turns on the capacity of its intermediaries to maintain a considerable share of business within Asia and between it and the global economy (Meyer, 2000, p. 247). As a trade hub linking China with global markets, Hong Kong's position in Asia has been unrivalled.

Hitherto, technological innovation in Hong Kong has however not been regarded as an important element of Hong Kong's developmental experience, and the few studies that have addressed the issue have emphasized the *laissez-faire* policies that have characterized the industrialization process in Hong Kong (e.g. Hobday, 1995). Hong Kong's entrepreneurs have been adept at exploiting available technology from the international market, but they have not generally carried out R&D for the purposes of creating proprietary technology on their own (Davies, 1999). For this reason, technological innovation has only recently begun to attract serious attention in Hong Kong, where the government in 1998 launched a new strategy in pursuit of knowledge-intensive economic growth.

Our point of departure for this report is that the transition to a new position as an innovation hub for China presents new challenges for its national system of innovation (NSI). In building its capabilities in innovation and technology development (including organizational and service innovations), Hong Kong can take advantage of the skills it acquired as a trading hub, combine them with the strong basic research capabilities contained in its universities, and apply them in order to become an innovation hub between the Pearl River Delta (PRD) and the rest of the world.

Following the return of Hong Kong to Chinese sovereignty and the Asian financial crisis of the late 1990s, the territory must further leverage its unique position as a gateway that provides high value-added services to global production chains linking China and the world. This task requires improvements in the R&D intensity of many sectors in Hong Kong's economy and the strengthening of innovative activities in the private sector. Accordingly, the Hong Kong government has adopted a more proactive approach to maintaining and further developing its competitiveness. The momentum of ongoing integration with the Chinese economy means that organizations in Hong Kong will increasingly locate many innovative activities on the mainland. For this reason, linkages with regional and global systems of innovation (SI) remain a key priority for Hong Kong.

2 MAIN HISTORICAL TRENDS

The Hong Kong story makes a fascinating tale of how what was a barren rock 150 years ago has emerged as a dynamic and vibrant world city.[1] In reality, Hong Kong's phenomenal economic growth has transpired over a shorter period covering the last four or five decades. Nevertheless, the foundation was laid over a longer period.

2.1 Early Twentieth Century

Studies of Hong Kong's economic development in the early part of the twentieth century have shown that a combination of informal institutions and state initiatives supported industrialization, relying primarily on small-scale manufacturers linked in familial or ethnic networks and connected with expanding markets for relatively low-technology products in China, South-East Asia and Europe/the USA (Clayton, 2000). Official British colonial history, reflecting primarily the perspective of the major British 'Hongs' or trading houses, which had little commercial interest in manufacturing and instead emphasized the promotion of the entrepôt trade, has largely neglected the growth of such industries in Hong Kong (Loh, 2002).[2]

2.2 The Cold War Period – 1950s to 1970s

The overthrow of the Kuomintang (KMT) regime of General Chiang Kai Shek in 1949 by the founders of the current government of the People's Republic of China (PRC) led to an exodus of about one million Mainland Chinese to Hong Kong. The people of Hong Kong, including its migrants, thus grew up and developed in a community that had Chinese roots but a

British administration. These migrants, in turn, accelerated the establishment and growth of manufacturing industries that further expanded Hong Kong's traditional role as an entrepôt. In the face of the declining power of the KMT in China, Shanghai textile barons transferred enormous amounts of capital and managerial expertise in textile manufacturing to the colony (Wong, 1988). Today it is estimated that over half of Hong Kong's more than 7 million citizens are descendants of post-1949 migrants.

2.3 The Opening of China – 1980s and 1990s

Given Hong Kong's singular position as a British Crown colony on the doorstep of the most populous country in the world, politics naturally shaped its NSI significantly. In this respect the two most significant events around 1980 were the modernization programme that the late Chinese leader Deng Xiaoping promulgated in 1978 and discussions between the Chinese and British governments that opened in 1982 over the future of Hong Kong. The latter negotiations ended in 1984 with the signing and ratification of the Sino-British Joint Declaration stating that Hong Kong would become a Special Administrative Region (SAR) of the PRC and that Hong Kong's capitalist system and 'way of life' would be preserved for 50 years. The 'one country–two systems' framework under which Hong Kong is presently governed was subsequently enshrined in the 'Basic Law', the present constitution of the HKSAR.

2.4 From Crown Colony to SAR

As Hong Kong approached its return to China in 1997, it was proudly boasting that no other society had more experience in investing and producing in China. Indeed, since the mid-1980s Hong Kong has been the largest source of foreign direct investment in China, and although the exact figures are impossible to determine, various statistical sources estimate that Hong Kong's contribution to realized foreign investment in China comprised by 1994 about two-thirds of the total (Berger and Lester, 1997, p. 5). It is on this basis that Enright et al. accurately describe how Hong Kong's historical role as a city of departure from China has laid the foundation for a reverse flow of business investments during the 1990s not only back to Hong Kong, but also to Mainland China through Hong Kong. They claim that this has 'helped Hong Kong become the de facto capital of the 50 million or more overseas Chinese who today play such an important role in the economic modernization of the Asian region and in the reconstruction of China's market economy' (Enright et al., 1997, p. 7).

The economic impact is considerable, since overseas Chinese investors – often Hong Kong companies or investors operating out of Hong Kong – now employ at least 14–15 million people in China. It is equally important that the migration of production facilities to the PRD in many ways represented growth, rather than decline, in Hong Kong's engagement in manufacturing. For political reasons such growth was however categorized as outside the territory even if it was, from a historical perspective, a reintegration into Chinese markets. This has also benefited the service industries in that most of the migration spurred further growth and increased the sophistication of producer business services (Tao and Wong, 2002). In establishing and upgrading these networks, Hong Kong firms have exploited their traditional strategies of imitation and followership, while emphasizing the development of organizational know-how rather than formal R&D for new products.

In summary, since the handover, Hong Kong and China – the PRD in particular – have entered a phase during which economic and political ties between the two have strengthened and the scope for collaborative innovation has widened.

3 INNOVATION INTENSITY

The data provided to indicate the propensity to innovate in Hong Kong are derived from an innovation survey administered in 2002 and 2003 according to criteria and specifications formulated by the Census and Statistics Department. This is a secondary source that does not conform to the precise requirements of the present comparative study of national SI, and such discrepancies should be kept in mind when interpreting the data.

Knowledge-intensive business services (KIBS) generally shows the greatest commitment to innovation and R&D activities, with manufacturing second, finance third and trade the least committed (see Table 6.1). Since the initiation of the open-door policy of China, Hong Kong has become an international financial centre as well as the service and information hub of Asia. The category of trade includes many firms that have manufacturing facilities in the PRD, while the Hong Kong office is responsible for planning, marketing and development. This relationship between Hong Kong and southern China is often described as *qian dian hou chang* (Hong Kong as the shop front and China as the factory at the back).

It is interesting that the figures for turnover due to products new to the firm demonstrate that the trade sector takes the largest advantage of new products. KIBS and manufacturing are in the second and third positions, respectively, although they commit more resources to innovation and R&D

Table 6.1 Indicators for innovation levels in Hong Kong SAR

	All firms	Sectors					Size	
		Manufacturing	Finance	KIBS	Trade	Others	Small	Large
ININT2K	0.0024	0.0064	0.0039	0.0246	0.0013	0.0011	0.0019	0.0032
RDINT2K	0.0011	0.0025	0.0001	0.0196	0.0006	0.0004	0.0010	0.0012
TURNIN	0.1170	0.0730	0.0260	0.1400	0.2260	0.0230	0.1480	0.1060
TURNINMAR	0.0004	0.0002	0.0001	0.0008	0.0007	0.0000	0.0004	0.0040
NTTM	0.0461	0.0184	0.3412	0.0699	0.0504	0.0282	0.0412	0.2096
INPCS	0.6892	0.5903	0.4118	0.6434	0.7047	0.7841	0.6908	0.6331
INNO	0.0622	0.1391	0.0087	0.0863	0.0567	0.0479	0.0610	0.1823

Variables: ININT2K: Innovation intensity in year 2000; RDINT2K: R&D intensity in year 2000; TURNIN: Turnover due to 'new-to-the-firm' products; TURNMAR: Turnover due to 'new-to-the-market' products; INPDT: Introduction of 'new-to-the-firm' products; NTTM: Introduction of 'new-to-the-market' products; INPCS: Introduction of new processes; INNO: Innovating firms.

Source: Own calculations based on data supplied by the Hong Kong SAR Census and Statistics Department figures.

than trade does. Most probably the reasons for the high proportion of new products in the trade sector are related to extensive subcontracting of new products to manufacturing in the mainland, with or without local R&D. In other words, many of the Hong Kong firms categorized under the trade sector are in fact undertaking the manufacture of new products in the mainland on behalf of clients overseas – where most of the R&D related to these new products were carried out by the overseas clients, not by the Hong Kong firm.

For the turnover due to 'new-to-the-market' products, KIBS is slightly ahead of trade while manufacturing remains third. This means that firms in the KIBS sector are likely to be launching more original new service products than those offered by trading companies. These figures also remind us that Hong Kong, with its information-intensive service industry, makes its living by providing services to an international clientele.

Compared with other small economies, the figure for the share of Hong Kong firms introducing new-to-the-market products – 0.0461 compared with an average of 0.3426 for small European countries – indicates that Hong Kong has performed poorly in product innovation. Only the finance sector has a proportion of firms introducing new-to-the-market products that is comparable to those of European countries. The relative weakness of product innovation recalls the popular wisdom about Hong Kong's economy that the territory is good only at reproducing others' innovations.

The figures for the introduction of new processes in various sectors of Hong Kong's economy are better than those for new products. The average figure of 0.6892 for Hong Kong is comparable to that obtained by other small economies, which scored an average 0.6619 in 1994–96 (see Appendix Table A4.7). Perhaps this is because Hong Kong, as a service economy, is more prone to introduce new processes than new products. New processes may simply be improvements to existing processes, and the indicator is a reflection of the fact that organizational change and innovative management remain important elements of competitiveness in Hong Kong. For example, efforts to improve quality (see Section 4.2.2) and raise total factor productivity (TFP) in both manufacturing and tradable services (see Section 5) have been a key concern of organizations such as the Hong Kong Productivity Council. This bias towards non-technological innovations is also reflected in Figure 6.1, where non-technological innovations far outnumber innovations focused on new technology.

Although, as a sector, KIBS has demonstrated a stronger potential to innovate, manufacturing nevertheless is slightly ahead of KIBS in exhibiting the largest proportion of innovative firms. However, the difference between the two sectors is very small (0.4 per cent). Instead, it is noteworthy that manufacturing and KIBS have more than twice the proportion of innovative firms as trade, and even more with respect to finance and other industries. These figures must be understood in context: trading firms focus on the marketing of new products, but do not carry out significant innovative activities or R&D on their own.

Finally we note that large enterprises, even if they constitute only a small proportion of all companies in Hong Kong, are far ahead of the small and medium-sized companies in terms of propensity to innovate. The indicators above, except turnover due to new products, illustrate that there is an immense need to revitalize the economy by assisting the small and medium-sized enterprises (SMEs) to become more innovative in a knowledge-based economy. In September 2004, there were about 282 000 SMEs in Hong Kong. They accounted for over 98 per cent of the total number of firms in Hong Kong. The majority of Hong Kong's SMEs are family-run enterprises, with much overlap between ownership and management, centralized decision making, high levels of family orientation (nepotism), the widespread use of personal networks, great flexibility and adaptability to changing market conditions, an emphasis on pragmatism over legalism and dependence on internal sources (as opposed to organizational external sources) for raising finance. In contrast, large firms, which constitute 2 per cent of all establishments in Hong Kong, undertake the largest amount of innovation expenditures (in dollar terms). Key statistics on the innovation activities of small, medium-sized and large firms for 2001 and 2002 are provided in Table 6.2.

As Table 6.2 shows, innovative activities are undertaken the least by small enterprises. Coinciding, however, with the government's effort to raise the awareness and importance of innovation among all Hong Kong firms, we see an increase in the percentage of SMEs undertaking innovative activities in 2002 as compared with 2001. While the number (and percentage) of large firms undertaking innovative activities dropped in 2002, their innovation expenditure continued to constitute over half of all innovation expenditure among businesses in Hong Kong.

The propensity to innovate among organizations in Hong Kong is heavily biased towards non-technological innovations. Figure 6.1 shows that although innovative activities have increased among Hong Kong firms surveyed in 2002 and 2003, efforts devoted entirely to technology innovation actually declined. Most innovation involved both technological innovation and non-technological innovation, and a substantial proportion was concerned only with non-technological innovation such as organizational change, marketing, etc.

Table 6.2 Key statistics on innovation activities in the business sector, 2001 and 2002

Size of establishment	Year	Total number of establishments	Number of establishments having undertaken innovation activities[1]	Innovation expenditure (HK$ million)[2]
Large	2001	5 781	771 (13.3%)	3 602.8 (53.4%)
	2002	5 083	662 (13.0%)	4 858.1 (55.0%)
Medium	2001	32 591	2 647 (8.1%)	1 987.2 (29.5%)
	2002	28 040	3 974 (14.2%)	2 562.1 (29.0%)
Small	2001	234 315	7 448 (3.2%)	1 156.4 (17.1%)
	2002	232 325	11 877 (5.1%)	1 415.1 (16.0%)
Total	2001	272 688	10 866 (4.0%)	6 746.4 (100.0%)
	2002	265 449	16 513 (6.2%)	8 853.3 (100.0%)

Notes:
[1] Innovation activities include product innovation, process innovation, ongoing innovation activities and abandoned activities. Figures in parentheses represent the percentages to total no. of establishments.
[2] Figures in parentheses represent the percentages to total innovation expenditure.

Source: Adapted from HKSAR Census and Statistics Department (2002, p. 38).

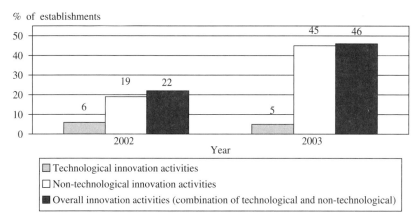

Note: Figures refer to the percentage of establishments that had undertaken, in the respective years, technological innovation activities or non-technological innovation activities or both.

Source: HKSAR Census and Statistics Department (2004), May.

Figure 6.1 Types of innovation activities undertaken in Hong Kong, 2002–3

4 ACTIVITIES THAT INFLUENCE INNOVATION

4.1 Knowledge Inputs to Innovation

4.1.1 R&D activities

R&D intensity (R&D as a percentage of gross domestic product (GDP)) in Hong Kong has been growing but, at a rate of 0.69 per cent in 2003, it remains very small in comparison with that of other countries with similar GDP per capita. Statistical information on R&D expenditures was not systematically collected in Hong Kong until the mid-1990s (see Table 6.3). In 2001, the Census and Statistics Department initiated annual surveys collecting more detailed data concerning R&D activities and innovation.

Over the period 1995–2001, higher education R&D (HERD) has constituted well over half (60 per cent or more) of total R&D expenditures. At this time, business expenditure on R&D (BERD) made up around one-quarter of the total, while the government sector expenditure on R&D (GOVERD) was responsible for only a tiny fraction. During 2002–3, the contribution of business R&D expenditure grew rapidly, and with a total BERD of HK$4.5 billion (approx. €400 million) in 2003 constituted almost the same amount as HERD (0.36 per cent compared with 0.39 per cent).[3] There are no figures available for the breakdown of R&D expenditures

Table 6.3 R&D expenditure by sector, as percentage of GDP, 1995–2003

Sector/year	1995	1996	1997	1998	1999	2000	2001	2002	2003
BERD	0.12[1]	0.14[1]	0.11[1]	0.12[2]	0.11[2]	0.09[2]	0.16[2]	0.20[2]	0.36[5]
HERD	0.25[4]	0.27[4]	0.29[4]	0.31[2]	0.35[2]	0.38[2]	0.38[2]	0.38[2]	0.39[5]
GOVERD	0.01[3]	0.01[3]	0.01[3]	0.01[2]	0.01[2]	0.01[2]	0.01[2]	0.02[2]	0.02[5]
Total (GERD)	0.38	0.42	0.41	0.44[2]	0.47[2]	0.48[2]	0.56[2]	0.60[2]	0.69[2]

Variables: BERD: business expenditure on R&D; HERD: higher education expenditure on R&D; GOVERD: government expenditure on R&D; GERD: gross domestic expenditure on R&D.

Notes:
[1] Figures from 'Feature article', Hong Kong *Monthly Digest of Statistics*, July 2001, p. FC5.
[2] Figures from 'Research and development statistics of Hong Kong, 1998–2002', Hong Kong *Monthly Digest of Statistics*, May 2004, p. FD4, Table 2.
[3] Estimates based on 1998 and onward figures, Census and Statistics Department.
[4] Percentage figures calculated from 'Government funding of R&D, innovation and technology upgrading, 1995/96–2001/02', Hong Kong *Monthly Digest of Statistics*, July 2003, p. FD9.
[5] Figures from HKSAR Census and Statistics Department (2005).

between domestic and foreign actors in the business sector, although some foreign firms such as 3M and Motorola are known to have conducted R&D activities in Hong Kong.

Higher education R&D activities are governed by the University Grants Committee (UGC), a government-appointed body that funds the eight organizations for higher education in Hong Kong. In 2000 and 2001 a negligible amount of HERD (less than 1 per cent of the total amount) (HKSAR Census and Statistics Department, 2004) came from parties outside Hong Kong. It can therefore be said that effectively all HERD activity is funded and conducted by domestic actors.

The insignificance of GOVERD reflects a strong belief in the virtue of maintaining small government agencies in Hong Kong, and although much of the territory's R&D funding comes from government sources, the actual expenditure and performance of R&D is generally done by organizations such as the universities or semi-public corporations.

Along with other science and technology (S&T) statistical indicators, data on patents can be regarded as performance indicators of R&D output. The number of patent applications increased from 1092 in 1991 to 9226 in 2001 (see Table 6.4), representing an annual growth rate of 24 per cent (HKSAR Census and Statistics Department, 2003a).

There are two types of patents in Hong Kong, namely the standard patent and the short-term patent. Subject to payment of a renewal fee, a standard patent in Hong Kong has a term of protection of up to 20 years, whereas a short-term patent has a maximum term of eight years. The vast majority of standard patent applications were filed by overseas firms, while 55 per cent of short-term patent applications were filed by Hong Kong residents or firms. At the same time, the number of patents granted by the US Patent and Trademark Office (USPTO) to Hong Kong residents almost doubled, from 279 in the five-year period 1990–94 to 570 in 1995–99 (Mahmood and Singh, 2003). The latest figures from 2004 indicate that 672 patents were granted to Hong Kong residents by USPTO, equivalent to 97.6 USPTO patents per million population (HKSAR Census and Statistics Department, 2005, Table 3.9). Hong Kong also saw 123.9 patents granted by the USPTO per 1000 full-time equivalent researchers, which was a higher ratio than in Taiwan, with 104.7, and Singapore, with 14.5 (Wong and Siu, 2004). These figures indicate that there exists a steady stream of high-quality innovative work carried out through advanced research in Hong Kong. This is a potential source of new inventions and economic growth, even if the results may be commercialized outside the territory.

Another R&D production measure is research output, not directly related to teaching, from the eight higher education organizations in Hong Kong. Research output includes scholarly books, journal articles, book

Table 6.4 Number of patent applications in Hong Kong by type, 1991–2001

Type of patent application/year	1991	1992	1993	1994	1995	1996	1997	1998	1999	2000	2001	2002
Standard							1 179 (49%)	19 139[1] (99%)	6 040 (97%)	8 295 (97%)	8 914 (97%)	9 130 (97%)
Short-term							30 (1%)	113 (1%)	175 (3%)	274 (3%)	312 (3%)	333 (3%)
Repealed[2]	1 092 (100%)	1 259 (100%)	1 195 (100%)	1 640 (100%)	1 961 (100%)	2 100 (100%)	1 215 (50%)					
Total	1 092	1 259	1 195	1 640	1 961	2 100	2 424	19 252	6 215	8 569	9 226	9 463

Notes:
[1] The surge of standard patent applications in 1998 was mainly due to the large number of applications filed under the Patent (Transitional Arrangements) Rules.
[2] Patent applications filed under repealed Registration of Patents Ordinance.
Figures in parentheses represent the percentages to total.

Source: HKSAR Census and Statistics Department (2003a, p. FC6).

Table 6.5 Research output at eight higher education institutions by broad subject area

Broad subject area/year	1993/ 1994	1996/ 1997	1997/ 1998	1998/ 1999	1999/ 2000	2000/ 2001	2001/ 2002
Biology and medicine	3 070	3 959	4 722	4 900	5 336	6 149	6 529
Physical sciences	1 092	1 749	1 894	1 910	1 951	2 649	2 764
Engineering	2 495	4 056	4 608	4 829	5 234	6 644	6 309
Humanities, social sciences and business studies	6 484	8 811	10 366	9 247	10 570	11 238	11 494
All subject areas	13 141	18 575	21 589	20 886	23 091	26 680	26 996

Source: Hong Kong *Annual Digest of Statistics*, 2003, p. 128.

chapters and other published papers. Since 1997, research output has numbered above 20 000 items annually (see Table 6.5).

We can discern the pattern of scientific and technological specialization for the higher education sector only. Yet HERD constitutes such a large proportion of overall GERD – well over 60 per cent – that it provides a fairly accurate picture. Most of the expenditure for the higher education sector from 1998 to 2002 was in the physical sciences (between 23 and 25 per cent), engineering and technology (between 22 and 23 per cent), and medicine, dentistry and health (between 18 and 19 per cent). The arts and humanities as well as social sciences each show a 10 per cent share (HKSAR Census and Statistics Department, 2004).

Since the year 2000, the government in Hong Kong has implemented measures to increase the amount spent on R&D because countries with a comparable level of per capita GDP commit 1.5–3 per cent of their GDP on R&D. The point is to leverage Hong Kong's position as a gateway linking China and the world, which the government has recognized as its greatest historical and present-day competitive advantage. Most notable among these measures has been the establishment of the Innovation and Technology Fund (ITF), set up with HK$5 billion (approximately €500 million) earmarked to provide funding for projects that contribute to innovation and technology upgrading in both new and established industries. The Innovation and Technology Commission (ITC) has also been set up to spearhead Hong Kong's drive to become a world-class knowledge-based society. The ITC manages the ITF and the Applied Research Fund (ARF), and supports such infrastructure projects as the Hong Kong Science Park.

4.1.2 Competence building

Hong Kong has been expanding its post-secondary education system since 1980. As part of its industrial policy the government has sought to facilitate the growth of industrial manufacturing by investing in infrastructure and human capital. To meet the increasing demand for skilled labour, the government has focused policy initiatives on vocational training. During the 1990s, the higher education sector was also expanded significantly. Recently, there have been increasing calls from lawmakers and academics to increase the proportion of students from Mainland China permitted to undertake tertiary education in Hong Kong.

In the higher education sector Hong Kong now provides 14 500 first-year first-degree places to about 18 per cent of the population, who range from 17 to 20 years of age. There are 11 degree-awarding organizations, eight of which are funded by the UGC and offer a total of over 45 000 degree places (in full-time-equivalent terms). Tertiary education constitutes approximately one-third of the government expenditure on education, which in turn is around 4–5 per cent of GDP (Education and Manpower Bureau, 2006).

The Vocational Training Council (VTC) was set up in 1982 to provide and promote a cost-effective and comprehensive system of vocational education and training to meet the needs of the economy. It operates the Hong Kong Institute of Vocational Education (IVE), including the VTC School of Business and Information Systems (SBI), industrial training and development and skills centres. It also administers the Apprenticeship Ordinance. IVE offers higher diploma, diploma, higher certificate, certificate and craft-level courses, which are designed to enable young people to build successful careers in industry and services. In November 2003, the VTC had enrolled 27 700 full-time and 26 300 part-time students (Vocational Training Council, 2004, p. 24).

Reflecting the territory's colonial past, many students from Hong Kong have pursued university education in the UK. It is also very popular to supplement basic educational degrees gained in Hong Kong with postgraduate degrees at universities abroad. In particular, there is a considerable market for MBA degrees, and several programmes are offered in a combination of local and overseas studies. Universities from the UK to USA to Australia have long operated a sophisticated higher education marketplace in Hong Kong, which has been a highly rewarding location in which to recruit students and deliver extension programmes. Nearly 30 000 students from Hong Kong study abroad each year. This market-based, transnational flow of university students represents perhaps the highest proportion within any post-secondary system in the world.

Public expenditure on education in Hong Kong has been approximately 23 per cent of the government budget in recent years. An expenditure of 4.7 per cent of GDP for education puts Hong Kong at a level of low spenders among OECD countries, similar to Spain and Japan (OECD, 2005, p. 176). About one-third of government expenditure towards education goes to tertiary education.

Largely because of the expansion of Hong Kong's university system, the past decade has seen a significant increase in the number of people who have obtained degrees from tertiary education organizations, as indicated in Table 6.6.

The introduction in 2001 of associate's degrees in post-secondary education programmes has also provided additional opportunities for people to develop their qualifications in Hong Kong and abroad. These new programmes are financed by private sources, in contrast to the bachelor's

Table 6.6   *Population aged 15 and over by educational attainment (highest level attended), 1991, 1996 and 2001*

Educational attainment	1991		1996		2001	
	Number	% of total	Number	% of total	Number	% of total
No schooling/kindergarten	557 297	12.8	480 852	9.5	469 939	8.4
Primary	1 100 599	25.2	1 146 882	22.6	1 148 273	20.5
Lower secondary	837 730	19.1	958 245	18.9	1 060 489	18.9
Upper secondary	1 169 271	26.7	1 403 211	27.7	1 473 681	26.3
Matriculation[1]	214 577	4.9	308 808	6.1	528 090	9.4
Tertiary						
Non-degree course	234 912	5.4	243 004	4.8	209 878	3.7
Degree course	255 979	5.9	525 516	10.4	708 622	12.7
Total	4 370 365	100.0	5 066 518	100.0	5 598 972	100.0

Note: [1] Figures include the equivalent educational attainment (highest level attended) of 'technician level (other further non-advanced education)' in the 1996 Population By-census and 'Diploma/certificate courses in institute of vocational education/former polytechnics' in the 2001 Population Census. The similar group 'Diploma/certificate courses in technical institutes/polytechnics' was included under 'Tertiary: non-degree course' in the 1991 Population Census.

Source: HKSAR Census and Statistics Department.

degree programmes already offered by the universities, which draw most of their funding from government sources. During the 2004–5 academic year, 90 associate's degree programmes were offered, together with 80 higher diploma programmes and 19 bachelor's degree programmes, under self-financing arrangements.

Until 20 years ago the colonial government deliberately maintained highly elitist access to the university system, funding fewer than 2000 new student enrolments annually in the UGC-run universities. A major change in policy was introduced by the then British governor David Wilson in 1989, with the expansion of tertiary education through the creation of new universities during the 1990s. The primary and secondary schools still tend to fall into several categories – from elite to basic – but Hong Kong students in general have been highly ranked with respect to educational performance in mathematics, in which they ranked first out of 41 nations in a test emphasizing mathematical comprehension (Grimm, 2004).

The most important recent policy shift was signalled by Chief Executive Tung Chee-Hwa in 2000, when he proposed a reconfiguration of the post-secondary system for training and general education. Under this policy 60 per cent of young people would continue to study beyond the secondary level by 2010, representing more than double the current participation rate and ten times the rate in 1989. Moreover, the government declared that this expansion of the educational system should be driven by self-financing schemes, not by direct government funding. This policy has had important implications for the educational services market in Hong Kong but it also opens up a range of issues relating to inequality of access to high-quality education and lifelong learning.[4]

4.2 Demand-side Factors

4.2.1 Formation of new markets

Although the formation of new markets can fuel the diffusion of innovation through the stimulation of demand for new products and processes, it has been largely neglected in policy measures throughout Hong Kong's history. A *laissez-faire* attitude, developed during the colonial regime, continues to characterize the community. This attitude in theory opposes all government intervention in a free market economy and therefore also any attempt on the part of the state to engage in the formation of new markets. This contrasts with conditions in Singapore, where the state has pursued an active, opportunistic role in identifying newly emerging market trends. Singapore's strategy is to reap a 'fast-follower' advantage by quickly funnelling resources to capitalize on these new market developments (Chapter 3 Singapore, this volume). The Hong Kong government departed occasionally from its free market principles to regulate or promote markets that were regarded as essential to society. The cases of property, infrastructure and information technology may illustrate how the state and the private sector have become involved in such initiatives. More significantly, the Hong Kong government has also participated unintentionally in the formation of new markets, apparently unaware of the far-reaching impacts of its policies.

The most notable avenue for market formation has been the government's land supply policies. Hong Kong's population is distributed almost evenly between public and private housing. The supply of land on which residential housing and commercial units can be built is strictly controlled by the government – a legacy of the British colonialists (and a major source of revenue for the government). Hong Kong covers a small geographical area, so land supply and plot ratios (the number of units that can be built on any given area of land) are important factors in determining not only housing supply but also, by extension, property prices. And in an economy where investment options are severely limited, property (both residential and commercial) serves as one of the most important investment and speculation vehicles.[5] As a result, large property developers are among the largest and most influential firms in the territory.

Because of the central nature of property in the economic livelihood of the territory, many economic sectors depend on the demand articulated in the property sector. These include most notably local and foreign construction and engineering companies required to design and construct buildings, banks and other financial service institutions that offer monetary lending services for buyers of property, property estate agents and brokers who serve as intermediaries in property transactions, law firms required to ensure the smooth handover of transactions between parties, architects and decorators involved with fitting out and decoration of units, and property management companies charged with the responsibility of ensuring the good upkeep, maintenance and security of buildings or estates. For all these business services, the rapidly expanding markets with increasing quality requirements provided opportunities for innovations in service products and organizational efficiency. For example, a number of consulting engineering firms became engaged in highly innovative activities (Baark, 2005).

Apart from its high level of regulation in the property sector, the government has also been a major actor in the formation of new markets through the provision of infrastructure for transportation facilities and the information technology sector. In terms of the transportation infrastructure, the government has continually tried to ensure that the public transportation network is up to date, complete and efficient – thus imposing significant requirements for quality and innovation on suppliers. As a result, the government annually spends large sums on the development and expansion of Hong Kong's road networks. In 1998, it completed the construction of the new airport at Chek Lap Kok, which today has become one of the busiest airports (in terms of both passenger and cargo traffic) in Asia. The government has also actively backed, and in some cases facilitated, the construction of container ports, railways and ferry services. These efforts have combined to aid and bolster Hong Kong's logistics and transportation sector, which has grown in importance as Hong Kong's role as a trade hub has grown and its role as a manufacturing centre has concomitantly declined since the 1980s.

In the information and communications technology sector, we note the launch of the 'Digital 21 Strategy' as a development blueprint to leverage Hong Kong's information technology (IT), Internet and telecommunications infrastructure as platforms from which to reposition and transform a heavily service-oriented economy into an innovation-led and technology-intensive economy. The aims of the strategy are to 'enhance and promote Hong Kong's information infrastructure and services' and to create an environment in which e-business can flourish (HKSAR Information Technology Strategy, 2004).

The government emphasized the development of mechanisms for electronic service delivery – popularly known as the ESD scheme – launched in 2000. In addition, this e-government programme would expand its focus on improving customer interfaces and promoting customer relationship management. Actively pursuing the IT upgrades, the government was also committed to outsourcing the proliferating IT projects starting in 2001. It also strongly emphasized the protection of intellectual property rights (IPR) to ensure promotion and development of IT.

4.2.2 Demand articulation of quality requirements

Within the general framework of an open, market-based economy, the Hong Kong government has traditionally believed that market forces will provide the necessary quality requirements. Both public and private organizations have nevertheless been active in attempts to improve quality requirements and ensure the safety of products, structures and services.

Several government organizations are involved in regulation and the setting of standards as the establishment of a firm institutional framework for maintainance of quality has been an important priority of the government, as will be discussed in Section 4.3.3. For example, the implementation of environmental regulations has developed gradually over some three decades. After years of studies and consultations, the Environmental Protection Agency (EPA) was formed in 1977 to formulate policies and coordinate other departmental activities to protect Hong Kong's environment.

In 1986, the government saw the need to replace the EPA with a separate and more powerful organization with executive powers. As a result, the Environmental Protection Department (EPD) was established.

The Consumer Council was established in April 1974 at a time of inflationary prices and widespread public concern about profiteering. Although the council receives a government subsidy, it enjoys total independence in formulating and implementing its own policy. The Consumer Council Ordinance came into force in July 1977 to provide for the formal incorporation of the Consumer Council. The Consumer Council acts as a watchdog in maintaining the quality of goods and services, and has also produced a number of studies investigating competition in selected sectors.

In response to a worldwide trend and the government's policy of promoting product safety in the territory, the Hong Kong Safety Institute Limited (HKSI) was established to meet perceived requirements in the field of product safety certification. Incorporated in Hong Kong on 7 August 1998, HKSI was set up to develop, implement and administer the unique third-party product safety certification programme in Hong Kong – the Hong Kong Safety Mark Scheme.

So far, public bodies have not engaged much in R&D related to the development of standards or regulations. They have instead emphasized keeping up with the most advanced testing and inspection methods, and this has required the adoption of advanced equipment and the recruitment of highly trained specialists rather than independent R&D. There has been no attempt to link the implementation of quality standards or regulations with technological development in local industry or services by means of public procurement. In this sense, the output of the services is 'adopted' from abroad rather than indigenously developed.

4.3 Provision of Constituents

4.3.1 Provision of organizations

Hong Kong's business landscape is dominated by SMEs, as opposed to that of Singapore, where industrial development had until the mid-1990s been powered largely by global multinational corporations that had located their operations there. In Hong Kong, SMEs are defined as non-manufacturing enterprises with fewer than 50 employees and manufacturing enterprises with fewer than 100 employees. Although the vast majority of SMEs operate with little attention paid to innovation, a number of small R&D-intensive firms have been formed during the last decade. Many of these are spin-off firms from university research via incubators or independent entrepreneurship. A well-known example is VTech, which has grown from its Hong Kong base to form a global network in R&D and manufacturing (see Box 6.1).

> **BOX 6.1   VTECH – ENTREPRENEURIAL INNOVATING HONG KONG FIRM**
>
> Founded in Hong Kong in October 1976 by two engineers, VTech began with only 2000 sq. ft of office space and 40 staff. Sales in the Group's first year were under US$1 million. Today, VTech has operations around the globe and approximately 20 000 employees worldwide. The Group's FY2004 annual results recorded revenue of US$915.2 million – and VTech is still growing rapidly. The VTech Group's three core businesses, including telecommunication products, electronic learning products and contract manufacturing services, incorporate state-of-the-art technology, unique features and value-for-money services.
>
> As a technology-driven company, VTech has placed much emphasis on research and development to maintain its leadership position in the market. To do this, the Group adopted a global R&D strategy. In 1987, it established an R&D centre in North America. In 1988, VTech became one of the first companies to establish R&D facilities in the PRC – tapping the vast resources of Chinese engineering talent. Today, VTech employs approximately 730 R&D professionals in R&D centres in Canada, Hong Kong and China.
>
> *Source:* Based on information from VTech company website, available at http://www.vtech.com.

A study of entrepreneurship in Hong Kong for 2004 indicated that only 3 per cent of the adult population have recently started businesses, which is a lower total entrepreneurship activity than is observed for countries in the same income bracket (Global Entrepreneurship Monitor, 2004). Furthermore, the new firms surveyed in the study were primarily exploiting existing technology instead of taking the risk of technological innovation. Most Hong Kong firms thus emphasize entrepreneurial learning and imitative strategies, seeking to exploit new market opportunities through flexible and fast reengineering of production networks rather than R&D-intensive product innovation (Yu, 2004).

However, large business groups in Hong Kong also provide a basis for new, technology-intensive ventures. Property developers who accumulated vast capital resources during the speculative real-estate boom of the 1990s have lately entered advanced technology sectors such as telecommunications and biotechnology. For example, the Cheung Kong Group, led by one

> **BOX 6.2 CK LIFE SCIENCES – A BIOTECHNOLOGY VENTURE**
>
> CK Life Sciences, listed on the Growth Enterprise Market of the Stock Exchange of Hong Kong in July 2002, is engaged in identifying needs and developing revolutionary biotechnology solutions for the improvement of human health and environmental sustainability. The group offers a range of environmentally friendly fertilizers that successfully improve crop yields comparably with chemical fertilizers, whilst also minimizing pollution to rivers, lakes and coastal reefs. A range of bioremediation products have also been developed to tackle pollutants. In the pipeline are a series of animal feed additives that address global concerns about the heavy use of growth hormones and antibiotics in intensive animal-rearing. In addition, the group is developing pharmaceutical applications for treatment of cancer and AIDS.
>
> Proprietary protection for CK Life Sciences' products, processes and know-how is key to the business and more than 100 product applications have been developed. CK Life Sciences' Intellectual Property portfolio consists of some already approved patents with many others being at different stages in the rigorous patent process in the United States or in other countries through the Patent Cooperation Treaty process.
>
> *Source:* Based on information from CK Life Sciences company website, available at http://www.ck-lifesciences.com.

of Hong Kong's most famous tycoons, Li Ka Shing, has diversified from property development into mobile communications (Hutchison) and Internet and media (Tom.com). It is also a key shareholder in CK Life Sciences, a biotechnology company led by Li Ka Shing's son Victor Li Tzar Kuoi (see Box 6.2).

The most important public organization for policy making related to innovation is no doubt ITC, established in 1999. Due to the history of various government initiatives, however, there are still a number of other policy-making bureaux connected with innovation, including the Commerce, Industry and Technology Bureau (CITB), the Financial Services Branch (for financial innovations), and to a lesser extent regulatory bodies such as the Telecommunications Authority or the Television and Entertainment Licensing Authority. These organizations are not formally connected in terms of their functions and missions, and since implementation of innovation policy takes place primarily through semi-public organizations such as the Hong Kong Science and Technology Parks Corporation (HKSTPC) or the Hong Kong Cyberport Management Company Limited, there is considerable scope for overlap and fragmentation, as we shall discuss in Section 7.2.

4.3.2 Networking, interactive learning and knowledge integration

The most prominent aspect of networking among actors in Hong Kong has been the successful development of international subcontracting for industrial firms. During the decades of industrialization from the 1960s through the 1980s, Hong Kong firms were able to learn and upgrade technology through their linkages with customers in the USA, Europe and Japan. Since the 1980s, these networks have been significantly extended to the PRD and other areas in the Chinese mainland.

With the expansion of local tertiary education in the early 1990s, universities sought to extend their cooperation with business firms in Hong Kong. This trend emerged with the establishment of the Hong Kong University of Science and Technology (HKUST) in 1991, but more recently all Hong Kong universities have started expanding their networks both inside their own universities and to other organizations overseas. Lately, this source of R&D cooperation partnership has been complemented by the creation of public research organizations such as the Applied Science and Technology Research Institute (ASTRI), established in 2003. For example, a photonics packaging technology from ASTRI has been commercialized in collaboration with a company called SAE Magnetics. ASTRI is expected to pursue further collaboration with other private firms.

Collaboration and interactive learning occurred among only 14.2 per cent of the private Hong Kong firms engaged in innovation surveyed by the Census and Statistics Department in 2003.[6] As shown in Figure 6.2, most of the R&D collaboration activities were directed at other business firms either within or outside enterprise groups. The higher education organizations were also popular partners in R&D cooperation, while other government organizations accounted for only a small proportion.

The same survey also revealed that firms considered cooperation with suppliers of equipment, customers and competitors as the most important sources of knowledge for technological innovation (see Table 6.7).

Through policies adopted by the Hong Kong government after 1998, attempts have been made to enhance cooperation and networking for technological innovation. For example, the University–Industry Collaboration Programme (UICP), created as part of the ITF, aims to expand network creation between universities and industries. Currently 49 projects in the Teaching Company Scheme, three projects in the Industrial Research Chair

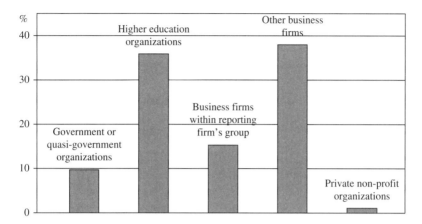

Source: HKSAR Census and Statistics Department (2003b, Table 2.9).

Figure 6.2 Cooperation arrangements related to R&D, 2002

Table 6.7 Top five sources of knowledge or information on technological innovation

Sources of knowledge	% of samples surveyed
Suppliers of equipment, materials, components or software	36
Clients or customers	26
Competitors and other firms of the same industry	18
Within the firm	17
Computer-based information networks (e.g. Internet)	14

Source: HKSAR Census and Statistics Department (2003b).

Scheme and 80 projects in the Matching Grant for Joint Research Scheme have been approved under the UICP.[7]

In addition, private firms not involved in UICP will collaborate with foreign organizations to enhance their competitiveness. The well-known local biotechnology firm CK Life Sciences (see Box 6.2 above) will conduct field trials of their eco-fertilizer in various countries through subcontracting networks with public organizations in various countries such as Australia and the USA. This firm is also known to actively seek assistance from universities to develop and test their herbal products. Yet, as we have noted, collaboration between universities and industry within Hong Kong is still very weak.

4.3.3 Provision of institutions

Hong Kong is a small and externally oriented economy that is already open to market competition. The government therefore sees no need to enact an all-embracing competition law. Instead, it has opted to issue a comprehensive competition policy framework through a policy statement and to reinforce this with sector-specific measures.[8]

Most policies and activities concerned with such institutional issues as competition and intellectual property rights are handled by the CITB of the HKSAR government. The CITB includes the Intellectual Property Department (IPD), founded in 1990. IPD aims to maintain and promote creativity and talent in the region, to ensure local awareness of the importance of intellectual property rights and respecting the rights of others and to accommodate the latest developments in technology.[9]

The government's support for patent applications is, on the other hand, administered and assisted by the ITC, also under the CITB. The enforcement of IPR is left to the Customs and Excise Department.

A variety of private services operates under either government-granted franchises that restrict entry or government-imposed schemes of control to regulate profits and prices. These services include telephone and telecommunications, broadcasting, television (terrestrial, satellite and cable), aircraft maintenance, air cargo terminals, air terminals, container terminals, buses and minibuses. Given the existence of franchises in several sectors of the economy, many have called for new legislation related to competition or anti-trust measures. The government has however remained content with regulating competition through sector-specific rules of the game and detailed management of pricing or mergers and acquisitions in each sector, in spite of apparent problems caused by resource allocation inefficiencies due to differences in institutional settings and the contradictory roles performed by regulatory agencies such as the Telecommunications Authority (Lin, 2002).

Before the handover of sovereignty, IPR were protected in Hong Kong mainly following the British model. Only trademarks were subject to local legislation in Hong Kong, whilst the other three branches, namely patent rights, designs and copyrights, depended greatly on acts of Parliament in the UK. Copyright legislation was enacted solely in accordance with the UK Copyright Act of 1956.[10]

The patent system in Hong Kong had previously been based on the Registration of Patents Ordinance. This was essentially a re-registration system involving a first registration in the UK. With the transfer of Hong Kong's sovereignty to the PRC on 1 July 1997, Hong Kong has localized its patent law and has introduced its own patent system, a move consistent with its efforts to develop into an innovation hub. The new Patents

Ordinance came into force on 27 June 1997. The ordinance provides for the establishment of an independent patents regime and the granting of both standard and short-term patents.

While the legal provisions for protection of IPR are clear and transparent, their implementation has affected primarily large businesses and public organizations. In contrast, the availability of counterfeit goods and pirated software or films has been difficult to restrict in practice, in part because of the constant flow of counterfeit goods from the Chinese mainland. Police campaigns to eradicate the trade in counterfeit goods tend to be sporadic and ineffective.

4.4 Support Services for Innovating Firms

4.4.1 Incubating activities

The Industry Department Technology Division launched various funding schemes in 1993 to support the development of new industries. These new funding sources were accompanied by the creation of facilities for incubation services for new high-tech firms. After a decade, the various programmes were eventually brought together in an enhanced incubation programme, called the Incu-Tech Programme, in April 2002.

Hong Kong has witnessed significant growth in public incubator activities for more than a decade, including the emergence of independent private incubator activities during the high-tech bubble of 2000–2001. The Hong Kong Institute of Biotechnology Ltd (HKIB) was founded in 1988 with a donation from the Hong Kong Jockey Club Charities Trust as a non-profit but self-financing downstream development centre for biotechnology products. The HKIB was formed to foster a successful biotechnology industry in the Hong Kong SAR through downstream R&D support and provision of an incubator facility for local entrepreneurs, but it has enjoyed only limited success.

Another early initiative was the establishment of the Hong Kong Industrial Technology Centre Corporation (HKITCC) in 1992, which constituted a publicly supported business innovation centre aimed at promoting technology development through three primary functions: technology-based business incubation and accommodation; the provision of technology transfer services; and the provision of product design and development and support services. In a similar vein, the HKSTP was inaugurated on 7 May 2001 to create a comprehensive organization for high-technology incubation in Hong Kong.

The Hong Kong government also launched the Cyberport project in 1999, with the intention to quickly create a strategic cluster of leading IT service companies in Hong Kong. Construction of the Cyberport was completed in 2004 and provides advanced facilities and office space for firms engaged in telecommunications, multimedia and Internet applications. The project has been controversial, however, and its role in actually incubating new high-technology IT firms has not yet emerged clearly (Baark and So, 2006).

University business incubators also expanded their activities during the 1990s. Currently, five out of eight universities are providing technology business promotion activity. In many respects, these incubating units make it possible to commercialize research results and technologies developed by faculty and graduate students. It is possible to interpret these schemes for support to university-based entrepreneurship as the fundamental thrust of innovation in Hong Kong (Mok, 2005). However, such conjectures remain theoretical rather than actual reality.

The statistical data related to incubator activities in Hong Kong are limited, providing few indicators of qualitative aspects.[11] Table 6.8 illustrates that most of the incubated firms belong to the information technology, telecommunications and electronics sectors.

Note that the data in Table 6.8 refer only to firms that have left the premises of incubator facilities, and do not indicate the actual survival rate for such 'graduated' firms. A spot check of 87 firms listed as graduated from incubator facilities indicated that at least 17 were not listed in current telephone directories, suggesting that they have stopped commercial business. In addition, only three out of 201 GEM-listed (growth enterprise market) companies are graduates of the incubation programme of the former HKITCC or the Incu-Tech Programme. Most high-technology companies launched on the GEM in Hong Kong have originated in Mainland China; a few well-known GEM-listed companies are technology-related spin-off companies of large corporations, such as the above-mentioned CK Life Sciences.

Table 6.8 *Cumulative number of graduated companies since 1992 and number of incubatees in incubation programme, 2004*

	No. of incubatees 2004	No. of incubated companies
Biotechnology	4	0
IT & telecommunications	29	74
Electronics	41	12
Precision engineering	2	0
Others	2	1

Source: Compiled from Incu-Tech Programme website.

4.4.2 Financing

The lack of appropriate financing sources was identified as a major bottleneck for Hong Kong firms and the most serious obstacle to innovation, with 47.4 per cent of respondents in the 2002 innovation survey citing this as a major hindrance.[12] Recent attempts by the government to improve the situation include co-financing R&D activities through the HK$5 billion ITF and financing investment and venture business through the ARF. By the end of 2003, ITF had supported more than 500 projects at a total funding of HK$1.53 billion. Most of the funding had gone to support technological development in the information technology and electronics industries (see Figure 6.3).

Among the parameters for assessing the impact of the ITF is the level of private sector contributions to R&D activities. Before 1999, the average annual level of industrial support funding and services support funding provided was HK$295 million, with private sector contributions amounting to about HK$24 million per annum. Following the establishment of the ITF in 1999, the average annual amount of ITF funding provided grew to HK$375 million. Since the launch of the ITF, the total amount of private sector contribution has increased to an average of HK$177 million per year (HKSAR, 2004). By and large, the ITF has not yet been instrumental in promoting diversification among property developers to invest in high technology (as discussed in Section 4.3.1). Rather, ITF funding has in large part been allocated to individual university researchers and specialized industry organizations (personal interview with Science Adviser to the Innovation and Technology Commission conducted on 23 August 2004).

The ARF is a government-owned venture capital fund formed to provide funding support to technology ventures and R&D projects that have commercial potential. The ARF has a capital endowment of HK$750 million. It is administered by the Applied Research Council, which appointed two private sector venture capital firms to manage the investment of funds from the ARF in November 1998. As of the end of March 2004, the ARF had supported 23 projects with approved funding of HK$387 million. Most of these projects had been in information technology (60 per cent) and telecommunications (28 per cent), while electronics (9 per cent) and biotechnology received much smaller shares. However, the poor performance of investments and lack of significant impact on new high-technology entrepreneurship in Hong Kong caused a decision by the governing board of ARF to cease investment in new projects in 2005 (Legco, 2005).

According to the findings of a recent survey commissioned by the ITC, each ITF project has on average generated 1.3 technologies or products, 0.55 patents have been filed per project and 0.15 patents per project have already been granted (HKSAR, 2004). However, as the ITF has been

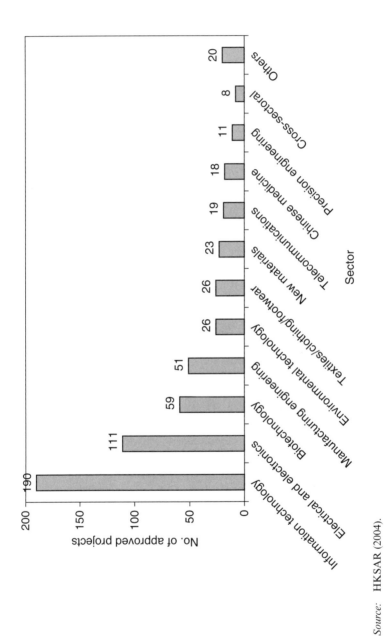

Source: HKSAR (2004).

Figure 6.3 Sectoral breakdown of the 562 approved ITF projects (involving a total funding of HK$1.53 billion as at 31 March 2004)

based largely on a bottom-up approach (initiated mainly by individual researchers and research groups), it has not proved conducive to building significant focus and clusters. After reviewing the existing innovation and technology programme, the government proposed in June 2004 to adopt a new strategy of innovation and technology development (see Section 7.3).

4.4.3 Provision of consultancy services

Given the importance of service industries in Hong Kong's economy, the availability of consultancy services represents a significant input into the NSI. Both public and private consultancy services have flourished during recent decades, and the role of foreign consultancy services is also important. Many services are related to trade with the Chinese market, but there are also a number of actors involved in providing consulting services related to technical innovation or management; the information in this section concerns such services.

The most important public organization providing consultancy is the Hong Kong Productivity Council (HKPC), which was established in 1967 to promote increased productivity and the use of more efficient methods across Hong Kong's business sectors. HKPC and its subsidiary companies employ about 600 highly skilled consultants and staff, providing a multitude of services to over 4000 companies each year. The HKPC has been expanding its service income gradually to an annual rate of HK$555 million in 2002–3, of which approximately HK$155 million came from government subsidy. The income from fees charged for services was HK$396 million in 2002–3, up from HK$287 million in 1998–99 (Hong Kong Productivity Council, 2003).

Research centres and individual faculty members at universities in Hong Kong also provide consultancy services for technological development and innovation. An important initiative for increasing the utilization of technical consulting services from universities in Hong Kong has been the UICP operating under the ITF, which distributed HK$7.3 million to 113 new and ongoing projects.

The private sector includes consulting engineering firms that primarily offer technical design and supervision services for construction in Hong Kong and on the Chinese mainland. Many of the large engineering service firms are local offices of major international consulting firms (Baark, 2004). The residential building sector was the largest end-use group of engineering services related to construction activities in 2002, followed by transport projects and service/commercial building projects. Virtually all of the world's leading management consulting firms are represented in Hong Kong, attracted by easy access to international skills and technology, a pool of experienced consultants, a rich client base and proximity to the mainland market. Together with other professional services such as accounting, insurance and financial services, management consultants form a vital component of KIBS in Hong Kong. As we have noted in Section 3, this sector has the highest propensity to innovate and it is also providing crucial inputs to other less innovative sectors of the economy.

The reliability and quality of such services have been generally recognized in the region by local and multinational firms operating in China, often using consulting services from Hong Kong firms, despite the higher fees charged by Hong Kong consultants.

4.5 Summary of the Main Activities Influencing Innovation

One may say that Hong Kong SAR's NSI is weak and emergent, with a low level of investments in R&D (0.69 per cent of GDP) and innovation. Given the very recent and incremental nature of the Hong Kong government's promotion of innovation, the influence of institutions as well as public organizations in Hong Kong's NSI has not yet reached significant levels. This is without doubt a legacy of Hong Kong's historical role as a trade hub coupled with low levels of government intervention.

Among the unique assets of Hong Kong's NSI is its innovation-oriented higher education sector (see Sections 4.1.1 and 4.1.2). The quality of university research in Hong Kong compares very favourably with respect to its neighbours in the region (as indicated by research output) (see Table 6.4), although recently proposed budget cuts have yet to take their effect on this part of the system. Increasing interaction with industry (in terms of technology and research transfer) has marked an important policy-initiated trend in this area over the past few years, as has increased emphasis on formal and informal integration measures with businesses and institutions in the PRD region.

In firms, R&D and innovation have been increasing over the past several years (see Table 6.3). This is due in large part to recent government initiatives. As Hong Kong firms face limits to profit maximization through cost reduction, they are increasingly looking towards innovation and R&D as a driver of future profits. Combined with the incentives the government is providing, it is expected that business R&D expenditures will continue to rise over the coming decade.

5 CONSEQUENCES OF INNOVATION

Science and technology did not play a large or formal role in Hong Kong's NSI until well into the mid-1990s. The colonial government, mostly concerned with Hong Kong's status as a trading hub, for the most part neglected formal innovative activity, research, development and investment directed towards the generation and commercialization of new knowledge (Parayil and Sreekumar, 2004). Hong Kong's economy nevertheless grew dramatically by cutting factor costs and developing organizational innovations (in particular non-R&D-based organizational innovations in firms).

The main thrust towards technological change in Hong Kong before the late 1990s was concentrated on improving productivity and the quality of products. Industry had remained competitive primarily by lowering costs through moving production to the PRD region in southern China and undertaking organizational innovations. As Hong Kong's economy moved away from assembly operations to higher value-added production, a constant flow of creatively applied technology was essential to stay ahead in competitive global markets. In other words, the key strategy for the last three decades has been to exploit technological knowledge and advanced equipment from overseas sources and to utilize such inputs together with organizational innovation to create flexible, low-cost production systems. The expansion of the higher education sector and the development of vocational training in Hong Kong have aimed to further improve the absorptive capacity in industry and services.

These efforts raised TFP in the Hong Kong economy by an average annual growth rate of 0.86 per cent during the two decades from 1981 to 2000. Yet while Hong Kong experienced high productivity growth in the second half of the 1980s, the rate of growth declined considerably in the 1990s. This is also the picture shown by figures reported in Appendix Table A2.2, which indicate that labour productivity growth in Hong Kong during 1995–2002 was 1.4 per cent. In the period 1996–2000, when the Asian financial crisis set in, Hong Kong actually achieved a negative contribution of TFP to output growth of -0.89, as shown in Table 6.9. The table also indicates that the development of TFP varied not only in terms of time periods but also between various sectors of the economy.

Li (2002) shows in particular that the tradable goods industries located in Hong Kong contributed directly to the GDP at an annual growth rate of 5 per cent in TFP during the period 1983–2000. Tradable services (including import/export trade, transport, storage, financial and business services) contributed at a lower TFP rate but remained positive at an average of 0.7 per cent during 1983–2000 and improved their TFP growth rate during the 1990s. In contrast, the non-tradable services (including construction, electricity, gas and water, communication, real estate and social services) experienced a negative TFP growth of -0.8 per cent during 1983–2000. The differences separating these three sectors of the economy are related primarily to the influence of labour productivity, which experienced stable

Table 6.9 Hong Kong's productivity development: 1981–2000
(% of annual growth rates)

Period	Output growth	Capital growth	Labour growth	Contribution to output growth		
				Total factor productivity	Labour	Capital
1981–85	5.25	8.71	2.61	−0.07	1.45	3.88
1986–90	8.04	7.50	1.11	3.98	0.60	3.46
1991–95	4.97	8.45	1.17	0.38	0.62	3.97
1996–2000	3.15	6.18	2.40	−0.89	1.36	2.68
1981–2000	5.34	7.71	1.82	0.86	1.00	3.49

Source: Based on Li (2002, Table 3, p. 7).

improvement in the tradable goods sector during the last two decades and rose in tradable services during the late 1990s, while labour productivity in non-tradable services actually declined slightly in the late 1980s and remained stable in the 1990s.

While innovation and new technology are likely to have contributed to the growth in TFP in tradable goods and services, productivity growth rates in these sectors have stagnated lately, and this problem indicates the need for further development and diffusion of technology. Such low productivity in non-tradable services is related to the lack of competition in several public utility sectors and a huge rise in speculative activities in real-estate services during recent decades. This latter phenomenon has created the illusion of economic growth without much value-added or technological input. Further exacerbating the problem, the availability of low-cost labour to manufacturers that moved to the Chinese mainland has discouraged investment in technological change and productivity improvement (Kwong, 1997). These factors continue to pose a threat to the competitiveness of Hong Kong's industries and services, and they constitute important reasons for the government's recent concern with improvement of innovative capabilities.

6 GLOBALIZATION

There is no doubt that Hong Kong is closely integrated in the international economic system and that globalization therefore has had a significant impact on the NSI in Hong Kong. Because of its pre-1997 colonial status, Hong Kong's development was linked to the policies of the UK, and with an open economy actors in Hong Kong sought opportunities in the international market. Since the return to Chinese sovereignty, the Hong Kong government has also consistently supported foreign direct investment in the territory and successfully made it a key priority to create a business environment that would encourage transnational corporations to set up regional headquarters in Hong Kong. Consequently, the number of overseas firms that have established their regional headquarters in Hong Kong grew from 602 in 1991 to 1167 in 2003, while overseas firms with regional offices in Hong Kong grew from 278 to 3798 during the same period (Hong Kong Trade Development Council, 2006).

Most of these firms hope to exploit Hong Kong's position in the growing Chinese market, and the major part of their activity is concerned with managing global production or supply chains. The rapidly expanding services located in Hong Kong are also serving global networks of production or trade. However, few transnational corporations have located significant R&D functions in Hong Kong, preferring instead to focus their overseas expansion of R&D on locations in the Chinese mainland. Meanwhile, foreign companies are setting up more R&D centres and service departments to serve the Chinese market, which has become a major market as well as a manufacturing base. Foreign investors had set up over 600 R&D centres in China as of June 2004, with a total investment of US$4 billion.[13]

The integration of Hong Kong into the economic system of China has further enhanced the trend in its economy towards globalization. At the same time, Hong Kong firms are actively seeking to extend their innovative networks in both the Chinese mainland and more advanced centres in industrialized countries. A recent survey of R&D in Hong Kong and the mainland indicates that many firms in Hong Kong were carrying out R&D in both Hong Kong and the PRD. Based on the information supplied by 229 firms (49 per cent of the sample of firms operating in both Hong Kong and the mainland), it was clear that the outsourcing of R&D and investments in R&D beyond the borders of Hong Kong was very significant.

As of 2003, only 17 per cent of the total R&D staff in Hong Kong firms were working in R&D units located in Hong Kong, while 53 per cent were employed in Guangdong Province, 3 per cent in the Yangtze River Delta, 19 per cent in other mainland provinces and 8 per cent overseas (Federation of Hong Kong Industries, 2003). The primary reasons for locating R&D on the mainland is the supply of talent and research facilities, while research costs rank only third. The majority of firms with mainland operations surveyed (78 per cent) indicated that they planned to continue or expand their R&D efforts, and almost half (46 per cent) planned to recruit more R&D staff in Guangdong. Only 13 per cent had plans to recruit more R&D staff in Hong Kong. Given the substantial R&D activity undertaken in Guangdong by Hong Kong firms, figures for the Hong Kong R&D expenditure probably understate the total R&D effort made by these firms.

Hong Kong government policies are oriented towards enhancing the global linkages of firms and organizations located in Hong Kong, while at the same time encouraging innovative overseas firms to move advanced R&D functions to the territory. Such a strategy is one that the Irish NSI has pursued with success, whereby technology transfer from abroad – particularly from US companies – has contributed to the technological levels of the Irish system more than its indigenous local capabilities (see Chapter 5, this volume). Both the CyberPort and the Hong Kong Science Park initiatives clearly aim to attract high-technology firms. Such a policy presents no contradictions to the government since it sees the location of advanced high-technology activities in Hong Kong – regardless of the origin of ownership of the organization – as a substantial benefit.

7 STRENGTHS AND WEAKNESSES OF THE SYSTEM AND INNOVATION POLICIES

7.1 Strengths and Weaknesses

Our analysis so far suggests that Hong Kong has a weak NSI, particularly if innovation is defined rather narrowly in terms of knowledge creation through R&D inputs and patentable technology as output. This weakness is chiefly a result of Hong Kong's historical place as a trade hub *vis-à-vis* China. The Hong Kong experience also demonstrates, however, how important it is to call attention to the exploitation of existing knowledge, emphasizing the use of existing technology transferred from abroad supplemented with many non-technological innovations. Hong Kong's NSI is able to absorb proven technology and carry out incremental improvements in products and especially production or organization processes that can provide competitive assets in terms of cost and flexibility of supply to global markets, as we have documented in Section 3. These assets in turn depend on the capacity of Hong Kong firms to organize or service global production chains deeply integrated in the Greater China region.

In this regard, Hong Kong has been most successful, and it is to this experience that Hong Kong can attribute much of its historical prosperity. By occupying the role of a trade hub, Hong Kong has amassed considerable experience in terms of exploiting existing knowledge for its own benefit. This experience has been successfully utilized to situate Hong Kong as a nexus of trade between Mainland China and the rest of the world. Hong Kong can continue to prosper by making use of the knowledge embodied in the capabilities it has as a trading centre, by transferring that knowledge and those capabilities to the innovation domain – that is, transforming itself into an innovation hub.

These strengths of Hong Kong's innovative efforts are most clearly observed in a few specialized low-tech sectors such as textile or garment products, watches and telephone handsets. They also characterize service sectors that have become prominent during the last couple of decades, such as financial services, logistics and management consulting and accounting services. A particular illustration of the ability to use innovative organization and networks to orchestrate global production chains is provided by the Hong Kong-based firm Li & Fung, described in Box 6.3. Such skills can be transferred effectively to the area of high-tech or innovation-intensive products. Institutions resting on fundamental market-oriented development in such areas as improved technical safety requirements or intellectual property protection, created during the last two decades, are acting to support these strengths in Hong Kong's NSI.

BOX 6.3   LI & FUNG – INNOVATIVE ORCHESTRATION SERVICES

One firm that has gained considerable fame on the basis of its high level of competitiveness in innovative services is Li & Fung, a Hong Kong trading company established in Canton in 1906 with sales amounting to US$4.2 billion in 2001. Li & Fung has developed a specialized role as the orchestrator of loosely coupled supply chain processes for a range of consumer products requiring labour-intensive manufacturing. Supplying well-known clients, like Levi Strauss, Reebok and Disney, the firm uses a wide network of more than 7500 suppliers in Asia and other continents to meet specific product needs, providing service along the entire chain of production through delivery of products to end customers – often packaged and marked with a price to be put directly on the shelf. This is achieved with the assistance of a hybrid organization that includes a highly advanced and sophisticated electronic trading system linking 5000 people supervising the manufacturing process and various clients globally (Brown et al., 2002).

At the same time Li & Fung utilizes more traditional networks of personal contacts and supervision to ensure quality assessment and on-time deliveries. This extensive network of human resources coexists with the information technology infrastructure to handle detailed design, production scheduling, logistics, final assembly and customer relations. A dedicated team is engaged in extremely knowledge-intensive 'disintegration' and optimization of supply chains, carrying out design and planning of distributed manufacturing and coordination of the vast network. But few of these activities require formal R&D and innovation is integrated into the development of new business processes and products. It is its specialized expertise in supply chain management that provides Li & Fung with its unique competitiveness in global markets.

Source: Based on information from Li & Fung company website, available at http://www.lifung.com/eng/global/home.php and Brown et al. (2002).

It is, however, also evident that weaknesses in the systems and organizations formed for the creation of new knowledge – requiring R&D-intensive development of products and services – pose a critical challenge to future economic development in the territory. After a decade of investment in higher education, universities in Hong Kong have upgraded research facilities and capabilities and are also trying to extend their networks to private industry and service sectors in order to commercialize potential new technologies. The university sector is likely to be the main public actor in the NSI even if current initiatives increase the number of government-sponsored research institutes serving specific sectors. We also observe that an increasing source of input for innovation in Hong Kong could be R&D carried out in organizations on the Chinese mainland. Maintaining crucial linkages to global networks, Hong Kong firms will be increasingly able to leverage their access to China's growing resources for innovation and thereby compensate for the relative weakness of local R&D organizations.

7.2 Summary and Evaluation of the Innovation Policy Pursued

In evaluating past innovation policies in Hong Kong, the most conspicuous point to be made is how 'late' the policies have been in their introduction. Not until 1999 did Hong Kong develop any kind of formal, co-ordinated innovation policy. Against almost any comparative benchmark – that of OECD countries, Asian tiger economies or countries of a similar size – that is far too late. Compare this with Singapore where, before 1990, the policy focus had been on promoting technology deployment while after 1990 the government shifted its attention to raising the indigenous R&D profile of local Singaporean companies through various long-term strategies and plans. In Hong Kong's case this delay may have been especially detrimental to the overall NSI because of the speed and intensity with which the PRC – the one economy on which Hong Kong uniquely depends for its economic livelihood – has been opening since 1979.

Lateness in tackling and introducing innovation policy and subsequently weak implementation have left many initiatives fragmented and ineffectual. Hong Kong could therefore benefit from an approach to its NSI whereby policy initiatives are better coordinated and understood in terms of a larger conceptualizing framework – the NSI approach. Indeed, in the latest government publication (see, e.g., HKSAR Innovation and Technology Commission, 2004) policy makers have explicitly integrated the SI approach as an aid to overall policy discussion and implementation. The moves towards adopting a system of innovation approach by the Hong Kong government must, however, be tempered by the observation that steps towards greater coordination and integration among the constituent elements of Hong Kong's NSI are in their infancy. There is a long way to go. The fear in Hong Kong remains that such bureaucratic changes may, ultimately, come too late to make a positive difference. The consensus among key policy advisers in Hong Kong is that a major weakness in innovation policy making has been a reluctance to address systemic relationships among the different policy areas. This hesitation has been associated with a lack of effective policy coordination. For example, despite the establishment of the ITC in 1999, there are still a number of other policy-making bureaux connected with innovation, including the Commerce, Industry and Trade Bureau, the Financial Services Branch (for financial innovations), and to a lesser extent regulatory bodies such as the Telecommunications Authority and the Television and Entertainment Licensing Authority that are not formally connected in terms of their functions and missions.

7.3 Future Innovation Policy

Hong Kong has made progress in transforming its role from that of an unrivalled trade hub into that of an innovation hub. This change, we argue, is the background for its newly initiated innovation policies and will influence their impact in the future. In mid-2004, the ITC proposed a new strategic framework for innovation and technology development underpinned by five core elements (see Box 6.4).

To implement the new strategy, priority has been given to the production of new knowledge. At the same time, however, the government recognizes the importance of 'leveraging the mainland' in order to become a facilitator or an innovation hub of technology inflows and outflows, particularly to the PRD. These two roles (producer and facilitator) are interconnected and co-evolving, especially as Hong Kong's overall integration (at cultural, economic and technological levels) into China gathers momentum.

> **BOX 6.4 KEY ELEMENTS UNDERPINNING THE HKSAR GOVERNMENT'S NEW INNOVATION AND TECHNOLOGY FRAMEWORK**
>
> (i) Focus: to identify key technology focus areas where HK is deemed to have an advantage for optimal use of resources to create greater impact
> (ii) Market relevance: to adopt a demand-led, market-driven approach in driving the innovation and technology programme to ensure that investments are relevant to industry and market needs
> (iii) Industry participation: to closely involve the industry in defining the focus areas and in other stages of innovation and technology development
> (iv) Leverage the mainland: to utilize the production base in the GPR Delta region as the platform for developing applied R&D and commercialization of applied R&D deliverables
> (v) Better coordination: to strengthen coordination among various technology-related institutions and the industry for enhanced synergy and impact
>
> *Source:* HKSAR (2004).

Still, Hong Kong's role as a trade hub has been in decline and will continue to weaken because of the continuing momentum of China's opening up, which began in 1979, and has resulted in its recent accession to the WTO. The rapid development of Chinese cities such as Shenzhen, Guangdong and Shanghai whose effectiveness in performing the trade-hub roles Hong Kong has historically monopolized is coupled with the increasing number of companies able and willing to set up shop directly in China. Only by becoming a facilitator for technology development can Hong Kong take advantage of the skills it acquired as a trading hub, combine them with the basic research capabilities of universities, and apply them in order to become an innovation hub between the PRD and the rest of the world to serve the interests of China's immense and rapidly developing economy.

Through its latest innovation and technology measures, Hong Kong is attempting to upgrade its existing capabilities in nine chosen focus areas. In a number of these areas – such as logistics, textiles and consumer electronics – Hong Kong can boast a wealth of expertise. In other areas, however, such as automotive parts, Chinese medicine and integrated circuit design, Hong Kong is looking to exploit its 'traditional' role as a facilitator between China and global networks.

Although the proposed set of policy initiatives looks elaborate, the available budget allocation for these measures from the remaining ITF funds (about HK$3.5 billion) is probably insufficient to produce a major turnaround. It is also difficult to predict the actual effectiveness of the R&D centres, which may be physical or 'virtual' centres. A final but significant problem revolves around the issue of developments across the border in southern China. While Hong Kong has made a point of 'leveraging the mainland', there are concerns that it is quickly falling behind. This means that Hong Kong may become isolated unless it acts promptly on its advantages as a hub. As integration with the mainland quickens and deepens, Hong Kong must pay close attention to the qualitative nature of the changes this integration is bringing about. In particular, changes that support the development of innovation-related links with China must be nurtured because of the decreasing importance of its trade-related links with China.

ACKNOWLEDGEMENTS

Research for this chapter was supported by a grant for High Impact Research on Technology Policy (HIA98/99.HSS05). We also wish to thank Charles Man Chi Wai and Raymond Ma Yin Nin for research assistance, and discussants at workshops in Lund, Sweden and Seoul, Korea for their valuable feedback. Any remaining errors are our own responsibility.

NOTES

1. When Hong Kong Island was ceded to the British in perpetuity, it was only a fishing community, inhabited by about 150 000 people and dismissed by the then British Foreign Secretary, Lord Palmerston, as 'a barren rock'.
2. According to Loh (2002), the British seized Hong Kong to serve as a base from which to penetrate China and other Asian nations for trade purposes rather than as new territory for its own sake.
3. Since 2001, the Census and Statistics Department has collected more detailed and systematic information pertaining to R&D, and therefore coverage of business R&D data improved dramatically. This may have contributed to the growth of the BERD figure, although it is likely that the figures reflect a genuine improvement of business investments in R&D following the recent policies promoting innovation.
4. See Post (2003).

5. The speculative nature of property was acutely illustrated in the late 1990s when prices rose dramatically to create a 'property bubble' in which asset prices were artificially high, only for the bubble to burst and result in the slashing of property prices.
6. See HKSAR Census and Statistics Department (2003b). Statistics quoted in this section are derived from this publication.
7. See data available from the Innovation and Technology Commission (http://www.info.gov.hk/itc/eng/funding/arf.shtml), accessed 31 January 2005.
8. See the 'Statement on Competition Policy' (http://www.compag.gov.hk/about/), accessed 24 January 2005.
9. See information on IPR in 'Protecting Intellectual Property Rights in Hong Kong' (http://www.hongkong.org/ehongkong 22/property.htm), accessed 30 August 2004.
10. Tackaberry (1997).
11. The data assembled for this section deal exclusively with activities of incubators, and do not include the general rates of birth and death of firms in the territory.
12. See HKSAR Census and Statistics Department (2003b, Table 3.14).
13. See, e.g., the announcement by the PRC Ministry of Commerce (http://english.mofcom.gov.cn/article/200408/20040800266847_1.xml), accessed 18 August 2004.

REFERENCES

Baark, E. (2004), 'Knowledge management, institutions and professional cultures in engineering consulting services: the case of Hong Kong', in Terrence E. Brown and Jan Ulijn (eds), *Innovation, Entrepreneurship and Culture: The Interaction between Technology, Progress and Economic Growth*, Cheltenham, UK and Northampton, MA, USA: Edward Elgar, pp. 65–86.

Baark, E. (2005), 'New modes of learning in services: a study of Hong Kong's consulting engineers', *Industry and Innovation*, **12**(2), 283–301.

Baark, E. and Y.S. Alvin (2006), 'The political economy of Hong Kong's quest for high technology innovation', *Journal of Contemporary Asia*, **36**(1), 102–20.

Berger, S. and R.K. Lester (1997), *Made by Hong Kong*, Hong Kong: Oxford University Press.

Brown, J.S., S. Durchslag and J. Hagel III (2002), 'Loosening up: how process networks unlock the power of specialization', *The McKinsey Quarterly* (Special Edition: Risk and Resilience), pp. 59–69.

Clayton, D.V. (2000), 'Industrialization and institutional change in Hong Kong', in A.J.H. Latham and Heita Kawakatsu (eds), *Asia Pacific Dynamism 1550–2000*, London and New York: Routledge.

Davies, Howard (1999), 'The future shape of Hong Kong's economy: why high-technology manufacturing will prove to be a myth', in P. Fosh et al. (eds), *Hong Kong Management and Labour: Continuity and Change*, New York: Routledge, pp. 43–58.

Education and Manpower Bureau (2006), 'Government expenditure on education', available from http://www.emb.gov.hk/index.aspx?langno=1&nodeid=1032 (accessed 5 March 2006).

Enright, Michael J., Edith Elizabeth Scott and David Dodwell (1997), *The Hong Kong Advantage*, Hong Kong and New York: Oxford University Press.

Federation of Hong Kong Industries (2003), 'Made in PRD: the changing face of HK manufacturers' (Part II and Full Report), Hong Kong.

Global Entrepreneurship Monitor (GEM) (2004), *Hong Kong & Shenzhen Report 2004*, Chinese University of Hong Kong.

Grimm, David (2004), 'Hong Kong, Finland students top high school test of applied skills', *Science*, **306**(5703), 1877, 10 December.
HKSAR (1998), *Commission on Innovation and Technology First Report*, Hong Kong.
HKSAR (1999), *Commission on Innovation and Technology Second Report*, Hong Kong.
HKSAR (2002), Innovation and Technology Commission, Innovation Expo, Hong Kong.
HKSAR (2003), *The Facts: Innovation and Technology*, Information Services Department, Hong Kong SAR.
HKSAR (2004), *New Strategy of Innovation and Technology Development*, Hong Kong.
HKSAR Census and Statistics Department (2001), 'Science and technology statistics section', *Hong Kong Monthly Digest of Statistics – July 2001*, Hong Kong SAR.
HKSAR Census and Statistics Department (2003a), 'Patent statistics for Hong Kong, 1991–2002', in *Hong Kong Monthly Digest of Statistics – January*, Hong Kong SAR.
HKSAR Census and Statistics Department (2003b), *Report on 2002 Annual Survey of Innovation Activities in the Business Sector*, Hong Kong SAR.
HKSAR Census and Statistics Department (2004), 'Research and development statistics of Hong Kong, 1998–2002', in *Hong Kong Monthly Digest of Statistics – May*, Hong Kong SAR.
HKSAR Census and Statistics Department (2005), *Hong Kong as a Knowledge-based Economy: A Statistical Perspective*, Hong Kong SAR.
Hobday, Michael (1995), *Innovation in East Asia: The Challenge to Japan*, Aldershot, UK and Brookfield, USA: Edward Elgar.
Hong Kong Productivity Council (2003), *Annual Report*, Hong Kong SAR.
Hong Kong Trade Development Council (2006), 'Economic & trade information on Hong Kong', available from http://hongkong.tdctrade.com/content.aspx?data=HK_content_en&contentid=173873&SRC=HK_MaPrFaSh#8 (accesssed 5 March 2006).
Kwong, K.S. (1997), *Technology and Industry*, Hong Kong: City University of Hong Kong Press.
Legco (2005), 'Legislative Council Panel on Commerce and Industry: The Applied Research Fund', LC Paper No. CB(1)2303/04-05(01), Hong Kong, available from http://www.legco.gov.hk/yr04-05/english/panels/ci/papers/cicb1-2303-1e.pdf
Li, Kui-wai (2002), 'Hong Kong's productivity and competitiveness in the two decades of 1980–2000', mimeo, City University of Hong Kong, September.
Lin, Ping (2002), 'Competition policy in East Asia: the cases of Japan, People's Republic of China, and Hong Kong', Hong Kong Lingnan University Centre for Asian Pacific Studies Working Paper No. 133 (17/02).
Loh, Christine (2002), 'Hong Kong SMEs: nimble and nifty', Hong Kong: CLSA, April.
Mahmood, I.P. and J. Singh (2003), 'Technological dynamism in Asia', *Research Policy*, **23**, 1031–54.
Meyer, David R. (2000), *Hong Kong as a Global Metropolis*, Cambridge, UK and New York: Cambridge University Press.
Mok, Ka Ho (2005), 'Fostering entrepreneurship: changing role of government and higher education governance in Hong Kong', *Research Policy*, **34**, 537–54.
OECD (2005), *Education at a Glance*, Paris: OECD.

Parayil, Govindan and T.T. Sreekumar (2004), 'Industrial development and the dynamics of innovation in Hong Kong', *International Journal of Technology Management*, **27**(4), 369–92.

Tao, Z. and Y.C.R. Wong (2002), 'Hong Kong: from an industrialized city to a center of manufacturing-related services', *Urban Studies*, **39**, 2345–58.

Vocational Training Council (2004), *Annual Report 2003/04*, available from http://www.vtc.edu.hk/ero/AR/index.html (accessed 5 March 2006).

Wong, Siu-lun (1988), *Emigrant Entrepreneurs: Shanghai Industrialists in Hong Kong*, Hong Kong and New York: Oxford University Press.

Wong, Y.C. Richard and Alan Siu (2004), 'Reviving Hong Kong's competitiveness', paper presented to Asia Economic Panel Meeting, 12–13 April, available from http://www.hiebs.hku.hk/aep/WongSiu.pdf (accessed 20 January 2005).

Yu, Tony Fu-lai (2004), 'From a "barren rock" to the financial hub of East Asia: Hong Kong's economic transformation in the coordinating perspective', *Asia Pacific Business Review*, **10**(3/4), 360–81.

PART II

Slow Growth Countries

7. Reconsidering the paradox of high R&D input and low innovation: Sweden

Pierre Bitard, Charles Edquist, Leif Hommen and Annika Rickne

1 INTRODUCTION

The notion of a 'Swedish paradox' has been central to recent innovation policy discussions in Sweden. When first formulated, it was as a reflection of a high research and development (R&D) intensity in Sweden coupled with a low share of high-tech (R&D-intensive) products in manufacturing as compared to the OECD (Organisation for Economic Co-operation and Development) countries. It was seen as a paradox between a high input and a low output measured by these specific indicators (Edquist and McKelvey, 1998).[1] In other words, it pointed to a low productivity of the Swedish national system of innovation (NSI) in this specific sense. Subsequently, the expression has been used widely, but often formulated as a general relation between inputs and outputs – e.g. that the investments in R&D in Sweden are very large, but that the 'pay-off' (in terms, e.g., of growth and competitiveness) is not particularly impressive (e.g. Andersson et al., 2002, ch. 2). Due to varying uses of the concept, and since many formulations have been based on rather partial data, it is not yet clear to what extent there exists a paradox or where the gap between input and output resides. In this chapter, we shall discuss the Swedish paradox in terms of a relation between inputs of R&D and innovation efforts and outputs of innovations of different kinds.

Those studies proposing that there is a paradox have also formulated a number of different hypotheses to explain it. One proposition is that the knowledge resulting from R&D remains in the R&D sphere – e.g. in universities or corporate research units – and hence is not transformed into innovations. Another is that the paradox can be explained by the sectoral allocation of R&D investments. A third is that the internationalization of production has proceeded further than that of R&D, so that R&D carried out in Sweden bears fruit, as innovations, elsewhere, sometimes in the subsidiaries of Swedish multinational enterprises (MNEs) (Edquist, 2002, Sections 4.6 and 4.3). However, we still lack a thorough discussion of the validity of these propositions or of the relations among them.

Against this background, we aim to analyse the Swedish NSI. In doing so, we follow the structure and model table of contents presented in the introduction to this book. Among many other things, we scrutinize whether there is support for the paradox and, if so, how it may be explained. Specifically related to the paradox, we revisit and reformulate the paradox in Section 3 through an analysis of detailed and comparative data from the second and third Community Innovation Surveys (CIS2/3). To assess the grounds for competing explanations of the paradox, a detailed analysis of activities possibly influencing innovation processes in Sweden – also presented in the introductory chapter – follows in Section 4.

2 MAIN HISTORICAL TRENDS

Two main traits characterize the evolution of the Swedish NSI. First, the natural resource base in Sweden – i.e. forests and minerals – and the economic history of Sweden from the industrial revolution onwards have both strongly influenced the present anatomy of the Swedish NSI. Second, the general pattern of economic development can be summarized in terms of 'the combination of exports based on refined and processed materials on the one hand and the multinational engineering firms on the other' (Edquist and Lundvall, 1993, p. 272). As for the resulting character of the NSI, attention should be drawn to the decisive role played by a 'small number of multinational firms in the engineering industry' (ibid.).[2]

In the latter half of the nineteenth century, Sweden was primarily agrarian. Its exports were dominated by products from agriculture and the mining and forest industries (iron and sawn lumber). After the mid-nineteenth century, though, new production processes allowed the export of more refined products from these industries – machinery products and pulp and paper, respectively. The engineering industry subsequently expanded significantly in terms of both employment and export shares, rising from 3 per cent of total exports in 1880 to 10.5 per cent in 1910–11, and reaching over 20 per cent in 1950. Among OECD countries, the share of manufacturing exports held by engineering industries in Sweden during the 1950s was surpassed only by the USA (ibid., p. 271).

Sweden was thus a late but rapidly industrializing country, developing a strong specialization in mechanical engineering technologies related to the extraction and processing of raw materials. Significantly, its major innovations in machinery products during the late nineteenth century were 'all closely related to the export-oriented process industries' (ibid.). Later product innovations that became the basis of multinational firms were also concentrated in engineering firms, although the base widened to include both mechanical and electromechanical technologies.

The Swedish economy has historically been strongly specialized in low-growth sectors (Jacobsson and Philipson, 1996). Before the 1990s, the more knowledge-intensive growth sectors, often referred to as high-technology (i.e. R&D-intensive) production sectors, were relatively underdeveloped (Ohlsson and Vinell, 1987). A study of Sweden's production structure in manufacturing for the period from 1975 to 1991 showed that Sweden actually had a declining proportion of production in the R&D-intensive growth industries – from 100 per cent of the OECD average in 1975 to 76 per cent in 1991 (Edquist and Texier, 1996, p. 110). One consequence of this negative specialization in growth sectors was an exceptionally strong decline of employment in manufacturing (ibid., pp. 113–17).

Sweden joined the EU in 1995 in the hope that increased exposure to international demand would lead to diversification and renewed growth, recognizing that the 'home market' could no longer provide a sufficient basis for growth and the development of new technologies and industries (Benner, 1997, pp. 187–8). Initially, this strategy of exploiting the economies of scale offered by international markets did not bring about diversification, but instead tended to consolidate the pre-existing production structure and established technological trajectories (Carlsson, 1996).

The 1990s witnessed some positive changes in Sweden's sectoral production structure. The general increase in service sector employment relative to manufacturing employment during 1980–94 was marked by a modest increase in the share of employment held by knowledge-intensive service industries (Nutek, 2000, pp. 41–3). Also, from 1980 to 1996 and especially in the latter part of the period, Sweden significantly increased its export specialization in high-technology manufacturing, while losing market shares in medium–high-technology manufacturing (ibid., pp. 47–52).

In 1997, a statistical study of the Swedish NSI, based on a comparison of seven countries (Finland, Germany, Japan, Netherlands, Sweden, the UK and the USA), found that Sweden ranked fourth in terms of the share of manufacturing employment held by high-technology sectors. Furthermore, Sweden ranked fifth in terms of the share of the total labour force employed in high-technology manufacturing (Nutek, 1998, Figure 3.8). Swedish production of high-technology products had also increased from 8.8 per cent of all manufacturing production in 1993 to 12.5 per cent in 1996, owing largely to rapid growth in two high-technology sectors in which Sweden was already specialized – telecommunications equipment and pharmaceutical products (ibid., Table 3.2). These developments improved Sweden's international ranking as a high-technology exporter (Braunerhjelm and Thulin, 2004, Table 1).

To the extent that Sweden's high-technology manufacturing industries expanded their exports of domestically produced goods, international demand acted as a spur to continued technological development, not only within the exporting firms, but also among their domestic suppliers. However, Swedish MNEs – and particularly those specialized in high technology – were simultaneously pursuing a strategy of exploiting international economies of scale through foreign direct investment (FDI), partly in order to avoid high domestic production costs (Braunerhjelm, 2004, p. 18, Figure 16).

3 INNOVATION INTENSITY

3.1 Introduction

The Swedish paradox refers to a mismatch between very high values on indicators of inputs into innovation and low values on output indicators. Here we revisit the alleged paradox and try to reformulate it in more specific terms, based on CIS data and using a comparative research design.[3] First, we identify the strengths and weaknesses of the NSI via comparisons with other countries. We focus on some of the small open European economies included in this book, i.e. the other Nordic countries, the Netherlands and Ireland. Second, to capture the dynamics, we compare the indicators over time for Sweden, using CIS data from two periods, i.e. 1994–96 (CIS2) and 1998–2000 (CIS3).[4] Third, we compare different sectors (manufacturing, knowledge-intensive business sectors (KIBS), finance, trade) and size classes (large firms versus small and medium-sized enterprises (SMEs)).

3.2 The Swedish Paradox Revisited

Revisiting the validity of the paradox in the light of new data presented in a separate paper (Bitard et al., 2005), we can confirm that R&D intensity and innovation intensity (as input measures) of Swedish firms are very high compared to the other small industrialized, European countries (Denmark, Finland, Ireland, the Netherlands and Norway). In 1994–96 Swedish firms invested 4.0 per cent of their turnover in R&D, compared to the group average of 2.3 per cent. Sweden ranked first and none of the other countries invested above average. Intriguingly, the Swedish figure was 38 per cent higher than the figure for the country ranked second (i.e. Denmark).

A complementary but wider input indicator is innovation intensity.[5] For this indicator, too, Sweden ranks first. The Swedish figure in 1994–96 was 6.7 per cent compared to the average of 4.1 per cent, and it was similarly high during 1998–2000.[6] This pattern holds not only for all firms, but also for the manufacturing sector, which is of specific interest.

We conclude that the input component of the Swedish paradox can be extended to all innovation expenditures, and not only to R&D expenditures. Indeed, the difference between Sweden and the other countries was even larger for innovation intensity than for R&D intensity.

At a disaggregated level, however, there is an interesting exception to this overall picture. For SMEs, Sweden ranks only second with regard to innovation intensity, far surpassed by Denmark: Swedish SMEs spent 2.7 per cent of their turnover on innovation, whereas the Danish ones spent 4.9 per cent – i.e. the Danish firms spent 81 per cent more. While in most countries SMEs spend less on innovation than large firms, Sweden had the greatest difference in this respect. This difference was three times more than that in Finland, with the second-greatest gap, where large firms spent 2.5 times more than SMEs.

On the output side, we revisit the paradox by analysing the proportion of innovating firms, the share of all firms that have introduced new processes, and the share of firms having introduced product innovations.

First, the proportion of innovating firms measures the share of firms that have introduced either a product or a process innovation. For this indicator, Sweden (all Swedish firms) ranked only fourth for both periods with a performance only slightly above average. Sweden was followed by Norway and Finland for the 1994–96 period, and by Norway only in the 1998–2000 period. However, when the data are disaggregated into manufacturing, KIBS, finance and trade, Swedish firms perform much better in the service sectors of finance and trade than in manufacturing.

Second, focusing on the share of all firms that have introduced new processes during a three-year period, Sweden's performance was 14 per cent below the average, and Sweden was ranked fourth (out of six) for the first period, and fifth (out of five) in the second period.[7] Hence Sweden is at the bottom in comparison, even though differences among the five countries were rather small. Worryingly, the Swedish position deteriorated over time between the two periods. However, Swedish firms performed somewhat better in services than in manufacturing. It is interesting that previous studies have shown that in the past, Sweden – at least Swedish engineering industry – has been very advanced with regard to the introduction of new process innovations (Edquist and Jacobsson, 1988).[8] However, judging from the CIS data, this no longer seems to be true.

Third, we have analysed four indicators related to product innovations. The indicator 'introduction of new-to-the-firm products' measures the share of firms that during a three-year period introduced products that were new to them (but not to the 'world'). On this indicator, Sweden ranked fourth (out of six) for 1994–96 and fourth (out of five) for 1998–2000.

As a contrast, the indicator 'introduction of new-to-the-market products' measures the share of firms that during a three-year period introduced products that were new to the market (i.e. new to the 'world'). On this indicator, Sweden ranked fourth (out of five), with only Norway behind. Interestingly, on both indicators Swedish firms performed better in comparison to other countries in services, but poorly in comparison to other countries in manufacturing.

The indicator 'turnover due to new-to-the-firm products' is the turnover due to new-to-the-firm products introduced during a certain period, divided by total turnover at the end of the period. On this indicator, Sweden performs very well, ranking first among the five countries compared. Hence the performance is much better in this respect than with regard to the proportion of all firms that innovate in new-to-the-firm products.

The indicator 'turnover due to new-to-the-market products' is the ratio of turnover due to new products or significantly improved products (goods or services) introduced during the period 1998–2000, divided by the total turnover in 2000. On this indicator, Sweden is somewhat below the average, ranking third (out of four). Thus Swedish firms perform relatively worse with regard to creation than to imitation.

It is also interesting that the performance on this indicator is much better for small firms than for large ones, i.e. small firms are much more creative than large ones, as compared to the other countries. Hence the overall performance of all firms – which is, on the average, worse with regard to creation than to imitation – can be explained by the domination of large firms in the Swedish NSI.

3.3 Conclusions

Comparatively speaking, the input indicators for Swedish firms are very high. On the output side all indicators are quite low compared to the other countries – with only one exception: turnover due to new-to-the-firm products.[9]

The comparison made here is with four or five other small industrialized countries in Europe and the result should be tested through further comparisons with more countries. Even so, we have reformulated the paradox in more specific terms than previously discussed in research and policy literature. Our overall conclusion is that the Swedish NSI is not as capable as some other small industrialized countries of transforming the very large resources invested in R&D and innovation activities on the input side into correspondingly large outputs of product and process innovations on the output side. The productivity (or efficiency) of the Swedish NSI is, in this sense, simply not high. (On the surface this conclusion may seem inconsistent with the EU's *Innovation Scoreboard* – where Sweden ranks very high on the summary indicator. However, this indicator is based on both input and output indication and does not make a distinction between them.) Hence the existence of the Swedish paradox is confirmed on the basis of the different, broader and more detailed indicators based on CIS2 and CIS3.[10] More specifically, the results suggest that the underlying problem may reside with the large firms that dominate the NSI, and their underperformance in innovation outputs.

The conclusions of the analysis in Section 3 will be discussed in considerably more detail in Section 7.1.

4 ACTIVITIES THAT INFLUENCE INNOVATION

Having confirmed, extended and specified the Swedish paradox in the previous section, we will now conduct a detailed analysis of the activities possibly influencing innovation processes. Among other things, this will contribute to assessing the validity of the three hypotheses that have been advanced to explain the paradox (see the introduction to this chapter). We follow a set of authors who have stressed the need to go beyond the structural components of an NSI and concentrate on the activities or functions of the system (Johnson, 1998; Rickne, 2000; Liu and White, 2001; Johnson and Jacobsson, 2003; Edquist, 2005; Bergek et al., 2008).

In this book we specifically take an approach based on *activities*. Edquist (2005) has compiled a general set of activities that may serve as a starting point for our analysis. These activities were presented in Box 1.2 of the introductory chapter of this book. This list is only 'provisional'. Thus our analysis does not claim to analyse all vital activities – or all aspects of these activities. Further, it does not rank the activities in importance, or reveal a master plan for redesigning the Swedish NSI. We hope simply to reflect tentatively on the extent to which innovation patterns in Sweden – and specifically the paradox – can be related to the activities of the system.

4.1 Knowledge Inputs to Innovation

4.1.1 R&D activities[11]
Measuring the volume of R&D input by national R&D expenditures as a proportion of gross domestic product (GDP), Sweden figures in the very top among OECD countries together with Israel, spending more than 1.8 times the OECD average and more than twice the EU average on R&D

(Jacobsson and Rickne, 2004). Sweden and Finland are the only European countries that have displayed a catch-up *vis-à-vis* the USA on this indicator since 1991 (European Commission, 2003). Sweden has strongly increased its R&D spending, from a level of 2.3 per cent of GDP in 1981 to 4.3 per cent in 2001 (Marklund et al., 2004).[12] However, even though the growth rate is clearly positive with 2.2 per cent average annual growth from 1995 and onwards, several other countries have a stronger growth rate (e.g. Greece, Finland, Portugal) (European Commission, 2003, Figure 2.1.8).

Sweden's scientific output, as measured by publication, is high, accounting for 1.75 per cent of world publications, and placing it at rank 14 in spite of being a small country (ibid.). In addition, the citation rate, indicating quality, is relatively high, though it has recently declined in some biotechnology-related fields (Sandström and Norgren, 2003). Sweden's scientific productivity is not above that of many other OECD countries (ibid.).[13] However, the technological output as measured by patents is well above the EU average, *vis-à-vis* both actual numbers (rating as number 8) and growth rate, and Sweden is listed among the five fastest growing EU countries as regards patenting in the EU. As to the world's share of US patents, Sweden ranks seventh but shows a moderate growth compared to other European countries (ibid., Table 1.6).

Sweden's relative scientific specialization resembles that of Finland and Denmark, and lies within life sciences, food science and agriculture, environmental sciences, civil engineering and materials science (European Commission, 2003). In most of these fields the citation impact is above average, being especially strong within pharmacology and clinical medicine. The scientific profile is dominated by biomedically related fields, where clinical medicine and health science, biomedicine and pharmacology and basic life science account for 56 per cent of the publications. Only the UK, the Netherlands and the USA have a comparable focus on these fields. Notable in comparison with other OECD countries is also a relatively small focus on chemistry as well as on physics and astronomy. As 'the "age of the atom" is being overtaken by the "age of the molecule" and, more recently, the gene' (ibid., 2003, p. 290), this may mean that Sweden is taking a promising direction. However, the fields of computer science, mathematics and statistics together account for only 3.2 per cent of the publications in this period (ibid., Figures 5.2.12–13).

In contrast, Sweden's technological specialization (as measured by patenting across major technology fields) lies in general in mechanics and process industries with relative strengths in pharmaceuticals, telecoms, materials and analysis-control, and weaknesses in biotechnology, audio-visual, IT and semiconductors. Notably, patenting growth rates are well above average in all fields except biotechnology and materials (ibid.).

Even considering the time lag issue, this mismatch between the scientific and technological profiles may partly explain the low innovation output discussed in Section 3.

The Swedish organization of R&D, whereby the business sector accounts for a major share of the activity, is different from many other OECD countries where firms are less prominent in R&D, but similar to that of the USA, Ireland, Belgium, Korea and Japan (Jacobsson and Rickne, 2004). The business sector has strengthened its dominance over the last two decades, driving the growth of R&D activity. In contrast, expenditure on R&D in the higher education sector has remained fairly constant since the beginning of the 1980s and the government sector had only a slow increase until the end of the 1990s. Within the business enterprise sector, the large firms – with 500 employees or more – account for 83 per cent of R&D.[14] While the contribution of the service sector to R&D was still relatively small in 2001, it was above the EU average and its growth exceeded that of manufacturing. In non-business R&D, the higher education sector assumes a major role, while government research bodies and private non-profit organizations are relatively small actors compared to other countries. As regards sources of R&D financing, the share from corporate sources is large – considerably above the EU average – and comes second only to Japan. Interestingly, it has increased over time, at the expense of public sector financing,[15] and was 72 per cent of the total financing in 2001 (Jacobsson and Rickne, 2004).

In brief, Sweden has clear strengths regarding both input and output of R&D. As pointed out in Section 3, Sweden's innovative firms are now increasingly located in services, and we have seen in this section that Swedish R&D is also characterized by strong dominance of the business enterprise sector – particularly by large firms – and relatively high rates of growth within the service sector. Coupled with these positive traits, concerns are, however, raised regarding a potential mismatch between the scientific profile and the technological profile, potentially explaining that there is a problem in transferring scientific knowledge into industrial needs in Sweden.

4.1.2 Competence building

In 1994, total Swedish spending on education as a proportion of GDP was the highest in the world (OECD, 1998, p. 37), and in 1999 Sweden remained one of the leading OECD countries, with a share of 6.7 per cent just slightly below the leader, Korea, at 6.8 per cent, and well above the average of 5.8 per cent (OECD, 2002, p. 170, Table B2.1a). Sweden also allocates a comparatively high proportion of educational expenditure to tertiary education. In 1999, Sweden spent 2.1 per cent of GDP on tertiary education, compared to an OECD average of 1.2 per cent (ibid., p. 78, Table B3.1). In 2003, this level of expenditure remained essentially unchanged, and Sweden ranked fifth among 25 OECD countries (Högskoleverket, 2003, p. 22).

Consequently, the Swedish labour force has a comparatively high level of educational attainment, with a rate of university graduation above the OECD average (OECD, 2002, p. 54, Table A31b). About 30 per cent of an age cohort graduates from higher education (Högskoleverket, 2003, p. 28). An OECD comparison of the EU-15, along with the USA, New Zealand, Australia and Canada, ranked Sweden third in terms of the proportion of the adult population participating in education and training in 2001. The Swedish participation rate of about 55 per cent in 2001 was surpassed only by Finland (about 56 per cent) and Denmark (around 58 per cent) (OECD, 2002, p. 249, Chart C4.2). In another EU-15 comparison of workplace-based education in 2000, Sweden ranked fifth, with a participation rate of 42 per cent – well above the average of roughly 33 per cent (Aspgren, 2002, pp. 105–6, Figure 5.7).

Recently, Sweden has expanded its higher education system, developing towards a mass rather than an elite system, predominantly academic rather than vocational in orientation (Sohlman, 1996; 1999). The engineering shortages of the past have been overcome, with graduation rates of natural scientists and engineers (NSEs) becoming comparable to those of competitor countries (Aspgren, 2002, p. 102; Jacobsson et al., 2001; Sohlman, 1996, p. 71). A recent international comparison of the proportion of NSEs within the total population holding tertiary qualifications shows Sweden in third place, surpassed by only Germany and Korea (Marklund et al., 2004, p. 47, Chart 13.2). The Swedish educational system remains entirely under Swedish control, and is still largely dominated by the public sector (although private schools are currently growing rapidly). Swedish higher education, however, has strengthened its internationalization since joining the EU in 1995. There is now a fairly even balance between foreign students at Swedish universities and Swedish students abroad (Högskoleverket, 2003, pp. 13–14).

Henrekson and Rosenberg (2001) address some rigidities of teaching in their critique of Swedish higher education. Such rigidities include barriers to competition among universities for both faculty and students, and barriers to competition among faculty within universities, both rendering universities unresponsive to shifting market demands. Historically, low remuneration paid to teaching faculty for high performance in specializations under strong demand, separation of undergraduate teaching and research, and fixed programmes of study providing students with little latitude for choosing courses have all combined to make the Swedish system of tertiary education rather slow to respond to changing markets (ibid., pp. 223–6).

There has been considerable improvement in these areas since the early 1990s, which ushered in decentralization reforms in both tertiary and non-tertiary education (Bauer et al., 1999; Lundahl, 2002). In tertiary education, these reforms were meant to make the system more market-responsive and enhance international competitiveness. Although decentralization has been achieved, it is still unclear whether it has translated into greater competitiveness. Arguably, the reforms have enhanced systemic flexibility at the level of competition *among* universities, but not yet sufficiently stimulated competition *within* them. At the same time, it appears that many Swedish universities and colleges have not yet reorganized themselves to take full advantage of greater freedoms in internal decision making (Alskling, 2001).

To summarize, the Swedish education system scores high in international comparisons of both inputs and outputs, and has improved its flexibility. The fact that most graduates now work in knowledge-intensive services rather than manufacturing (Marklund et al., 2004, p. 17), may help to explain why many of Sweden's innovative firms are now located in services rather than manufacturing.

4.2 Demand-side Factors

Historically, several new industries and technologies in Sweden have been closely tied to new domestic demand, with national procurement initiatives providing initial markets for several 'state-sponsored development blocs' (Glimstedt, 2000, p. 207). Public technology procurement (PTP) has, in earlier times, been an important innovation policy instrument (Edquist and Hommen, 2000).

However, since Sweden joined the EU in 1995, its public agencies have faced greater institutional obstacles in undertaking PTP initiatives under the EC Directives on Public Procurement (Edquist et al., 2000). Sweden's accession to the EU was accompanied by a wave of liberalization reforms that resulted in the dismantling of many state agencies and the privatization of many state-owned companies. There is still some scope for the use of PTP as a demand-side instrument for innovation policy, using the Swedish public sector's comparatively large size and high-quality standards as points of leverage (Marklund et al., 2004, p. 9). However, most PTP projects now under way are mainly characterized by incremental innovation within existing industries. The 24SJU (24SEVEN) project of the Swedish Agency for Administrative Development, in which public administrations will procure information and communication technology solutions to make basic services available 24 hours per day, 7 days per week, provides one example (Karlberg, 2004; Kleja, 2004).

Product market regulation has shaped several important Swedish industries (Glimstedt, 2000, pp. 184–202). As 'institutions specific for each technological system' (Carlsson and Jacobsson, 1997, p. 288), standards have been particularly important in, e.g., mobile telecommunications (ibid., 284–9; Glimstedt, 2001, p. 49). Standard setting contributed to Ericsson's (and Nokia's) current leadership in mobile telecommunications equipment through 'early identification of new technological opportunities' (Carlsson and Jacobsson, 1997, pp. 284–9). However, standard setting has become increasingly internationalized, and private actors, especially producers, have become dominant in influencing the development of standards (Hommen and Manninen, 2003; Hommen, 2003).

Recent Swedish innovation policy has replaced purely demand-side measures with public–private partnerships (PPP) combining demand- and supply-side measures. For instance, the Swedish Agency for Economic and Regional Growth (Nutek) programme 'Design for Environment in SMEs' was based on 'networks of firms involving research institutes, universities, and in some cases customers of the participating SMEs, based on industry-specific supply chains, or on specific product development' (Fukasaku, 1998, p. 124).

In summary, Sweden's accession to the EU led to a shift in Swedish innovation policy, from a strategy of utilizing domestic demand to one of relying upon international demand to stimulate industrial and technological development. Positive effects include gains in high-technology exports and new opportunities for MNEs (see Section 2). However, PTP and standard setting have decreased in importance. These observations may help explain why the Swedish NSI currently performs better in turnover due to products 'new to the firm' rather than products 'new to the market'.

4.3 Provision of Constituents

4.3.1 Provision of organizations[16]

The birth rate of new firms is comparatively low in Sweden. This observation is worrying, since new firms are an important mechanism of industrial renewal. Even though 60 000–75 000 firms were established yearly during the 1990s, the population of new firms was still only 7.4 per cent of all companies in 2001.[17] In a large international survey, Sweden ranked only 33rd out of 41 countries in terms of the share of individuals engaging in firm formation (GEM, 2003). But in a study of new technology-based firms (NTBFs) established between 1975 and 1998, the accumulated population numbered almost 1400 in 1998 (Rickne and Jacobsson, 1999),[18] and their relative share has increased over time. Although firm formation has maintained a constant level during the last decade, there has been a steady increase in science or technology-based spin-offs from universities and companies.

An unusually high proportion of new firms endures: the three-year survival rate was an impressive 55 per cent in 1998 (ITPS, 2003b). Regarding stability, one study showed that 63 per cent of the high-tech spin-offs established in 1996 passed the four-year survival limit, comparing favourably with other Nordic countries (Nås et al., 2004). However, two-thirds of the new firms are one-person companies (ITPS, 2002), and most other firms also remain small. One study reveals that out of firms surviving three years (1998–2001), 40 per cent show some growth (ITPS, 2003b) but only a few grow substantially. Also, less than one-third of the spin-off firms created in 1996 had created any employment expansion in the following four years.

Although many large, international companies have been created in Sweden, few of them were created during the last three decades. Among the newer established firms that are growing, some do so on their own account, but growth frequently seems to be enhanced by becoming part of a larger corporate structure through acquisition (Lindholm, 1994). Through mechanisms such as subcontracting components and subsystems, acquisitions and spin-offs, large companies play an important role in creating and developing innovative new firms.

On balance, Sweden lags in creation of new firms and their contribution to industrial renewal. High survival rates are enlarging the population of firms and the formation rate of high-tech firms is increasing. However, the relative lack of growth may partly explain Sweden's lack of innovation, as discussed in Section 3. The shift towards more service firms can be linked to the finding that Sweden's innovative firms are now increasingly located in services rather than manufacturing, and that the highest rates of 'new-to-the-market' product innovation occur in knowledge-intensive services.

4.3.2 Networking, interactive learning and knowledge integration

Empirical data indicate that innovative collaboration and networking seem to develop organically among Swedish actors and between Swedish and foreign actors. Swedish research often involves collaborations between researchers in firms and in universities or institutes (private or public research organizations), resulting, for example, in joint publications or patents (Sandström, 2002). Out of all Swedish publications, 27 per cent are co-published with a national partner and 39 per cent with a foreign partner (European Commission, 2003, Figure 5.4.2). The importance of spatial closeness is stressed where there is a preference for Nordic partners, but there are many non-Nordic foreign partnerships. Sweden's rate of university participation in research joint ventures with US actors is – despite Sweden's relative smallness – among the six highest in Europe (ibid., Figure 3.3.11). Naturally, patterns of R&D collaboration vary by sector, and science-based sectors such as biotechnology display very high intensities.

University–industry relations are frequent and important in some sectors. One study showed that 93 per cent of the Swedish biotech firms reported university cooperation (VINNOVA, 2001). However, Swedish industrial actors finance fewer activities in universities or research institutes than do firms in other EU countries (European Commission, 2003, Figures 3.1.4–5). Also, a need for improved technology transfer is stressed by the finding that in East Gothia the main partners of firms pursuing product innovations are other firms (suppliers and customers), not universities (Edquist et al., 2000).

Swedish firms frequently enter into licensing, joint development, marketing or distribution, outsourcing agreements, etc. A survey of collaboration in product development, covering all manufacturing firms in East Gothia, found that 70 per cent of all product-innovating firms relied on partnerships (ibid.). This tendency can be illustrated by, for example, the field of biocompatible materials, where innovating firms rely heavily on other actors, and a large variety of partners – national and foreign – supply technological competences, financing, market guidance, etc. (Rickne, 2000). Types of partners and resource exchanges vary substantially across sectors – with, for example, biotech entailing mainly technology development but also market-oriented relations (Alm, 2004).

These findings contrast starkly with evidence from CIS3,[19] where the proportion of cooperating enterprises was shown to be rather low in Sweden (around 30 per cent in 1998–2000) compared to other European countries.[20] The consistent pattern across countries was that a much higher share of large firms cooperate for innovation. In Sweden, two-thirds of large firms cooperated, but only one-third of SMEs. Comparatively, Swedish firms displayed low cooperation in all sectors except KIBS.

These competing observations, based on different data, all find support in the character of the Swedish system. The rather high degree of vertical integration in Sweden implies a lower degree of cooperation and fewer market-based sourcing solutions, as indicated by CIS. But the history and ownership structure of Swedish industry, as well as path dependences involving technological trajectories, resource inertia and variety creation (Glete, 1989) point to a system with extensive networking, as in the East Gothia study. Even so, there is a need to enhance collaboration and learning over organizational borders. Today, private initiatives such as industry associations and bridging organizations, as well as government schemes of various kinds – for example the Innovation Bridge Foundations and

VINNOVA – continue to provide arenas for meetings, coordinate suppliers, or spur university–industry relations by making such cooperation a prerequisite for financing.

4.3.3 Provision of institutions

Here we focus on institutions such as science and technology (S&T) employment rules, corporatist arrangements, intellectual property rights (IPR) laws, competition rules and trade agreements.

Andersson et al. (2002) and Henrekson and Rosenberg (2001) point to insufficient incentives for academic entrepreneurship, with consequently poor performance in commercializing research results via NTBFs (Lindholm-Dahlstrand, 1997a; 1997b; Rickne and Jacobsson, 1999) including university-based start-ups (Olofsson and Wahlbin, 1993; Rickne and Jacobsson, 1996; Marklund, 2001). The Swedish labour market featured low returns on human capital from the 1960s to the 1980s (Edin and Topel, 1997; Fredriksson, 1997). In 1995, Sweden had the lowest wages for experienced teachers among leading OECD countries (OECD, 1995). Rigid pay scales and poor remuneration for high performance and specialization in areas of high demand were long persistent in academia (Etzkowitz et al., 2000; Stankiewicz, 1986, p. 90). However, flexibility has increased greatly during the past two decades.

Sweden's postwar social-democratic welfare state favoured large firms and strong trade unions (Esping-Andersen, 1990). Sweden also developed corporatist economic policy making, based on tripartite cooperation (Ruin, 1974). Initially, the 'core institution' governing economic growth and industrial change was 'labour market regulation' (Benner, 1997, p. 202). Later, public companies, investment planning and R&D policy assumed more importance, and by the 1990s policy aimed at low inflation and labour market flexibility (ibid., pp. 205–13). However, corporatist arrangements remained intact (Edquist and Lundvall, 1993, p. 291). Unions were thus rewarded for cooperation with employers, supporting production-based learning within firms and collaborative learning within industries (Glimstedt, 1995; 2000). However, extensive social security has been confined to large manufacturing firms and the public sector, encouraging a lock-in that can lower the impact of public investments in R&D and education (Andersson et al., 2002, pp. 45–6).

Since 1949, the 'university teachers' exemption' has granted faculty at Swedish universities complete ownership of research results. Arguments for the university teachers' exemption stress that it minimizes bureaucracy and does not preclude voluntary agreements between universities and their scientific staff (Sellenthin, 2004). An alternative arrangement with university involvement would also require more effective technology transfer services (Rosenberg and Hagen, 2003, pp. 25–6). Critics argue that this law does not mitigate costs, uncertainties and risks of commercialization (Brulin et al., 2000). Critics also point to a weak incentive structure with negative effects on both universities (Henrekson and Rosenberg, 2001, p. 225) and faculty (Etzkowitz et al., 2002). There is also evidence of 'anti-entrepreneurial peer pressure' within university departments (SOU, 1996, p. 70). These conditions may have contributed to the underdevelopment of NTBFs in Sweden and may help to explain the low innovation expenditure of SMEs compared to large firms. Some Swedish universities have therefore recently introduced extensive infrastructures for enhancing commercialization.

Sweden's EU accession in 1995 implied liberalization and internationalization. Deregulation of the capital market had already occurred in the 1980s. In the 1990s followed sweeping reforms in domestic air travel (1992), telecommunications (1993), banking, finance, postal services (1993) and electricity (1996). A central aim was to create new entrepreneurial arenas and innovation opportunities, in both Sweden and the EU. Since Sweden joined the EU, moreover, the ownership of Swedish MNEs has become increasingly internationalized (Andersson et al., 2002, pp. 28–9).

To sum up, EU membership made it difficult to pursue 'demand-side' innovation policy (Edquist, 2002, pp. 40–42), as argued in Section 4.2. Liberalization also spelled an end to 'state-sponsored development blocs' (Glimstedt, 2000, p. 207). Both S&T employment relations and IPR law and legislation can be linked to Sweden's continuing underproduction of NTBFs. Conversely, aspects of both corporatist arrangements and competition and trade policy seem to have perpetuated the dominance of large firms and reinforced established technological trajectories. These factors help to account for the much higher innovation expenditure of large firms, relative to SMEs in Sweden, and Sweden's generally poor performance with regard to the introduction of new-to-the-market products.

4.4 Support Services for Innovating Firms

4.4.1 Incubating activities
Sweden's division of labour in initiating, financing and operating science parks and incubators varies greatly and includes government-supported non-profit units, university-driven units, PPPs and private initiatives in corporate incubators.[21] Incubation is seen as a potent policy tool, and university-related incubators have most often been initiated and financed by public money. Recently, a national technology-based incubator programme aiming to operate on a long-term basis and include financial support services has been designed on the initiative of government actors.

Following the example of US and UK science park establishments in the 1970s, Sweden's incubation activities began in 1983 with the Ideon Science Park in Lund (Bengtsson, 2003) and an additional 15 parks were established between 1983 and 1989. However, the positive results were not as strong or direct as anticipated (Ferguson, 1998; Lindelöf and Löfsten, 1999; Löfsten and Lindelöf, 2002). This led to a systematic review, in which science parks were highlighted as only one instrument in an innovation environment (VINNOVA, 2002). Today, some incubators are stable and successful, while many still struggle.

The Swedish universities have incorporated 'technology transfer offices' similar to those at Stanford University or Massachusetts Institute of Technology only since the mid-1990s. Today, most large universities have some form of unit for handling patent and licensing issues, and promoting entrepreneurial and cooperative activities. Searches for entrepreneurial opportunities are undertaken through, for example, venture competitions or innovation prizes. However, there is much more to be done. In 1998, Sweden officially assigned universities a 'third mission' of diffusing knowledge for societal use, but few means are devoted to it by governmental or other bodies. Academic researchers own the right to their inventions, but other supports for commercial activity – i.e. incentives, suitable career structure, time, financial resources, role models and experience – are often missing.

While deficiencies in incubation may help to explain the relatively low innovation expenditure of Swedish SMEs relative to large firms, there have been dramatic enhancements since the beginning of the 1990s. Policy actors and the bridging organizations they have formed, as well as universities and private firms, have played important roles.

4.4.2 Financing

Since the joint effort by government and a merchant bank to create the first venture capital (VC) firm in 1973, the Swedish VC industry has experienced waves of increase and decline (Berggren, 2002; Isaksson et al., 2004).[22] In the early 1980s a promising stock market and the formation of the over-the-counter (OTC) list in Sweden encouraged both private actors and government funds to enter the industry. However, a shakeout followed, due to high interest rates, a weakening stock market and a promising real-estate market. This resulted in a shift to majority investments and late-stage financing. The 1990s saw a moderate growth, and the valuable experience cultivated by the long-term surviving VC firms was important when the situation evolved into a significant expansion of the industry in the latter part of the decade. This was a response to the increase in the number of high-tech firms, the growth in the stock market, and input from both private savings and pension funds (VINNOVA, 2002), and was consistent with European patterns. However, the global downturn affected the Swedish VC market in 2000 and a severe decline has followed.

Based on the description above, Sweden has often been pointed out as having an impressive level of VC activity. It is indeed true that there is an increasing number of actors on the VC market, and that the percentage of GDP devoted to VC is well above the EU average (Eurostat, 2003). In fact, the number of actors tripled (from 50 to 150 firms) between 1998 and 2002, at the same time as the funds managed quadrupled (from 45 to SEK190 billion). However, as Sweden started from a low figure, she is still in somewhat of a catch-up situation. In fact, many developed EU countries have been ahead of Sweden for many years, and Sweden has yet to develop a fully competent VC market with experienced actors and sufficient institutional support (Karaömerlioglu and Jacobsson, 2000; European Commission, 2003). While an upsurge has certainly put Sweden on the map and been important for firm formation and growth, the industry can still be characterized as relatively immature, in terms of institutional structure, phase of financing and sectoral focus.

Thus, first, there are some misgivings about the institutional structure underlying the VC market. In effect, in a comparative study analysing the regulatory environment for VC, Sweden was ranked below average in Europe (EVAC, 2004). Positive features mentioned are the fund structure, the company tax rate, the ease of registering a new company, and the regulation for reorganization and bankruptcy of a company. More negative aspects include the strict regulation of mergers, the lack of a special tax rate for SMEs, the income and capital gain taxes for individuals, the lack of tax incentives for individuals, and the lack of fiscal incentives for interfirm cooperation (see also SVCA, 2002).

Second, as regards the phases of development which are VC-financed, a relatively large share of the funds has been allocated over time to late-stage development. Indeed, surveys show that 30–50 per cent of the funds managed by Swedish VC firms are invested in any of the phases from seed to expansion, and the rest in buyouts (EVAC, 2001). While the heavy lagging seed financing has displayed an upsurge since 1998, later figures have disputed this trend (Nutek, 2004). Interestingly, although the government aimed to increase the volume of seed capital through the establishment of two large investment organizations in 1992 (Atle and Bure), these bodies subsequently refocused on later-stage financing (Isaksson and Cornelius, 1998). While the lack of early-stage financing to some extent seems to be handled by the entrepreneur's own sources, bank loans are mainly an option for more mature buyouts and neither source is sufficient (Nutek, 2004).

Third, another worry concerns the sectoral focus, where only 28 per cent of the total equity capital in Sweden is allocated to high-tech sectors, as compared to the EU average of 38 per cent and the astonishing US figure of 79 per cent (European Commission, 2003).

All in all, even though several EU countries have long been ahead, Sweden is in a good situation, with growing financial options for firm formation and expansion. However, there is clearly a quandary in Sweden as regards maturity of the VC market and the involvement of all the types of actors necessary for a smooth sequence of financing and the provision of resources to high-technology firms. Much has been done towards the development of the VC industry in Sweden, but it still requires improvement, and its current state may help to explain why the innovation intensity of Swedish SMEs is not exceptionally high. In particular, the fact that there is a relative shortage of seed and early-stage financing and the lack of high-tech focus may possibly contribute to explaining the Swedish paradox. One positive sign is a visible internationalization of the VC market. Although domestic actors dominate the financing of innovation and VC firms located in Sweden invest mostly in Swedish firms (82 per cent) or in other Nordic companies (13 per cent) (Nutek, 2004), foreign organizations are nevertheless involved in every fifth investment and the financing process has become more internationalized.

4.4.3 Provision of consultancy services

Nearly all Sweden's private consultancies are located in the KIBS sector.[23] VINNOVA's recent comparison of nine countries – Austria, Germany, Denmark, Finland, France, Italy, Norway, Sweden and the USA (Marklund et al., 2004, Figure 4.4) – shows that Sweden's KIBS sector is not especially large. Sweden ranks seventh in the proportion of total services belonging to KIBS, sixth in the percentage of the total labour force employed in KIBS, and sixth in the percentage of total population employed in KIBS (ibid.). The sector has recently expanded rapidly in Sweden, with high employment growth from 1981 to 1991, returning to more moderate rates in 1991–2000 (ibid., Table 5.4). This development was part of a more general change in sectoral employment patterns, whereby increasing employment in knowledge-intensive services, combined with stable employment in other services, contributed to a net increase in private sector services until 1985. Thereafter, private sector employment in knowledge-intensive services continued to increase, while other private sector services, as well as public sector services, stagnated (ibid., p. 17).

KIBS has clearly become important for innovation processes, due to the reorganization of other sectors. Thus KIBS has for some time accounted for a very large share of the employment of all Sweden's qualified NSEs. In 1996, 41 per cent of all NSEs employed in private or public organizations were employed by manufacturing firms, and nearly as many were employed by firms in KIBS (Nutek, 1998, p. 133).[24] Moreover, a majority of the NSEs employed in KIBS were employed in small firms, bolstering the innovation capacity of SMEs. In manufacturing, especially in high-technology and medium–high-technology industries, there has been (and remains) a strong positive association between firm size and NSE employment. Large firms in these industries accounted for two-thirds of the net increase in NSE employment in manufacturing over the period 1993–96 (ibid., p. 137). In services, though, a different pattern prevails. In KIBS, 63 per cent of the net increase in NSE employment took place in SMEs, and was fairly evenly divided between small firms and medium-sized firms, with 34 and 29 percentage points respectively (ibid.).

Sweden's KIBS sector also exhibits a high level of innovative activity itself. An analysis of CIS2 data for Sweden has shown that a high proportion of all innovating firms, well above the service sector average of 36 per cent, were found in the financial intermediation and KIBS sectors, where the shares of innovating firms were 59 per cent and 51 per cent, respectively (Nählinder and Hommen, 2002, p. 11, Table 2). KIBS firms were also especially strong investors in human resource development related to innovation, with the proportion of all innovative KIBs firms investing in innovation-related training standing at 67 per cent – far more than in any other of the service sectors covered by CIS2 (ibid., p. 12). A more recent analysis of independent survey data has confirmed these findings and provided a more detailed profile of innovation in Sweden's KIBS sector (Nählinder, 2003). According to this survey's results, 82 per cent of Swedish KIBS firms exhibit a high level of knowledge intensity in terms of the employment of qualified personnel (ibid., p. 14), and some 82 per cent of this population of firms engaged in some form of innovation during the period from 2000 to 2002 (ibid., p. 15). This figure is much higher than the corresponding figure of 51 per cent arrived at by the Swedish CIS2, and is arguably more reliable, given that CIS2 in Sweden provided poor and uneven coverage of the service sectors.

To summarize, the recent expansion of Sweden's KIBS sector, together with the centrality of KIBS firms to many innovation processes, and their typically high levels of knowledge intensity, may help to explain why Swedish firms are currently more innovative in some service sectors, particularly finance and trade, as compared to manufacturing. These observations may also help to explain why the Swedish NSI also performs well in new-to-the-market products within such service sectors.

4.5 Summary of the Main Activities Influencing Innovation

In our discussion of the nine activities influencing innovation processes we have, at times, related the arguments to the Swedish paradox. In the introduction, we mentioned three hypothetical explanations for the paradox: (1) there are obstacles to technology transfer from the R&D sphere to the commercial sphere; (2) sectoral allocation of R&D is problematic; and (3) internationalization of production means that the results of Swedish R&D is increasingly exploited abroad. We have found support for the first hypothesis under Section 4.1 (Knowledge inputs to innovation), Section 4.3 (Provision of constituents) and Section 4.4 (Support services for innovating firms). We have also found some support for the second hypothesis under Section 4.3 (Provision of constituents). However, we have found no support for the third hypothesis, which will be revisited in Section 6.

5 CONSEQUENCES OF INNOVATION

In this section, we address the consequences of innovation. We focus on productivity at the micro-, meso- and macro-levels. First, we assess the relation between innovation expenditure and turnover growth. Subsequently, we examine the relation between turnover and growth of value-added. Then we consider evolution of labour productivity and sectoral changes in value-added. Finally, we assess changes in sectoral value-added.

5.1 Micro-level

At the micro-level, we find a weak,[25] but significant,[26] association for manufacturing firms between turnover increases of at least 10 per cent and engagement in innovation (as indicated by the level of innovation expenditure), for both the 1994–96 and 1998–2000 periods. Thus, the most successful firms (as measured by turnover increases) are likely to be those that have invested in innovation. However, it is problematic to identify causality here. It might be that the most successful firms are more likely to invest in innovation.

Turning to the sectoral level, we hypothesize that sectors with the highest shares of turnover due to new products during the 1998–2000 period were also those with the highest growth of value-added in the following year. Changes in labour productivity between 2000 and 2001, derived from the STAN dataset, show how productively labour is used to generate value-added. We couple this measure with the share of turnover due to new products during 1998–2000 according to CIS3.[27]

The result of the correlation test indicates a negative and significant correlation between the two variables.[28] This suggests that, in a given sector, the higher the share of turnover due to new products during 1998–2000, the lower the growth of value-added was likely to be between 2000 and 2001.

This result rests on a small sample and must be regarded with caution. It may be partially explained by the spectacular drop in value-added of the 'machinery and equipment' sector between 2000 and 2001.[29] However, it could also be evidence that 'successful' innovative sectors – as measured by the share of turnover due to new products – also experience the smallest increases in value-added. This is counter-intuitive since, statistically, innovation's impact on value-added seems to be negative.

5.2 Meso-level

To assess structural changes, we examine the last decade. We assess the average share of sectoral value-added as related to value-added created by the whole economy, as well as the variation of these shares relative to the grand total between 1991 and 2001 (see Table 7.1).[30]

There were significant structural changes during the period. The weight of the manufacturing sectors increased relative to service sectors in the Swedish economy, representing an average share of 21 per cent versus 69 per cent respectively of the total value-added between 1991 and 2001. 'Total manufacturing' grew by 9.17 per cent whereas 'total services' grew only by 3.2 per cent. However, 'computer and related activities' experienced a spectacular 192 per cent increase, its relative share in the total value-added standing at 1.65 per cent. There was a concurrent decline of traditional low-tech sectors. Both 'construction' and 'agriculture, hunting, forestry and fishing' dropped by about 34 percentage points. The latter represented an average proportion of total value-added of 2.40 per cent, and the former an average share of 4.71 per cent.

Comparing these sectoral differences in value-added with R&D expenditures between 1993 and 2001, we note that the sector with the most dramatic growth in value-added – i.e. 'computer and related activities' – has also undergone the strongest growth in R&D expenditures (from index 20 in 1995 to nearly index 120 in 2001). It experienced a 100 percentage point growth of R&D investment (see Figure 7.1).

5.3 Macro-level

Comparing evolution of labour productivity, measured as GDP per hour worked between 1979 and 2001 (see Figure 7.2) with other countries, Sweden neither catches up with nor lags behind the USA. Sweden has remained at a high level, slightly above 80 per cent of the US level of labour productivity. As related to the other countries, Sweden remained at fourth ranking almost all through the period, despite remarkable catching up in most of the other countries. By 1997, Ireland had overtaken Sweden, and has performed better ever since, reaching the third rank and approaching becoming second-best (replacing the Netherlands). At this macro-level, it

Table 7.1 Industrial structure of Sweden – average share of the value-added of the different sectors in the grand total (%) and variation of the share in the grand total (%), 1991–2001

Sectors	Average share 1991–2001 (%)	Variation 1991–2001 (%)
Agriculture, hunting, forestry and fishing	2.40	−34.26
Mining and quarrying	0.31	−30.99
Total manufacturing	20.79	9.17
Food products, beverages & tobacco	1.83	−4.51
Textiles, textile products, leather & footwear	0.29	−21.13
Wood & products of wood & cork	0.93	−11.93
Pulp, paper, paper products, printing & publishing	3.30	18.04
Chemical, rubber, plastics & fuel products	2.77	33.72
Other non-metallic mineral products	0.51	−18.85
Basic metals & fabricated metal products	2.76	10.99
Machinery & equipment	5.15	−3.62
Transport equipment	2.70	25.80
Manufacturing n.e.c.	0.55	15.31
Construction	4.71	−33.84
Total services	68.79	3.20
Wholesale and retail trade; restaurants and hotels	12.15	0.75
Finance, insurance, real-estate and business services	23.60	14.38
Financial intermediation	4.03	−25.96
Computer and related activities	1.65	191.83

Source: Own calculation and presentation based on STAN database for Industrial Analysis (OECD, 2004).

is difficult to investigate clear relations between different kinds of innovations and performance; therefore, we have to rely more on lower levels of aggregation for these purposes.

5.4 Conclusions

In summary, the most successful Swedish firms are likely to be those investing most in innovation. However, the most innovative sectors are also

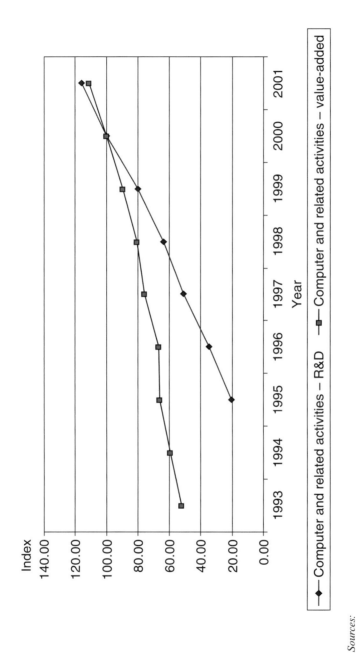

Sources:
[1] Own calculations ANBERD – R&D Expenditure in Industry (ISIC Rev. 3), Vol. 2003, release 01.
[2] Using the STAN database for Industrial Analysis, Vol. 2004, release 04.

Figure 7.1 R&D expenditure[1] and value-added[2] for the sector computer and related activities in Sweden (year 2000=100)

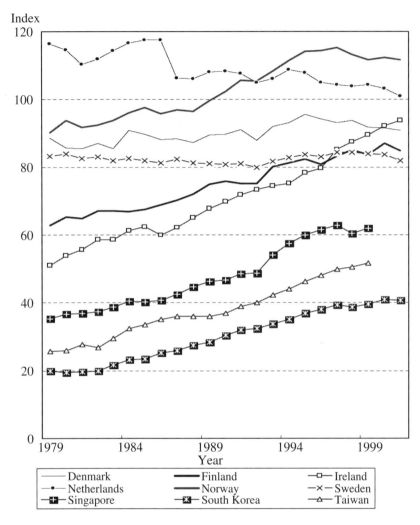

Notes: The Geary–Khamis method is used when benchmarking PPPs for GDP. It is used as an aggregation method to weight and sum the commodity-group parities to arrive at PPPs for each category of expenditure up to the level of GDP.

Source: Appendix Figure A2.3.

Figure 7.2   GDP per hour worked, in 1990 PPP US$, USA=100, 1979–2001

those experiencing the smallest increases in value-added. Sweden has had moderate success in evolution of labour productivity. However, the example of 'Computer and related activities' illustrates innovation's positive impact on firms' value-added.

6 GLOBALIZATION

Multinational enterprises (MNEs) have played a central role in the Swedish NSI, accounting for as much as 70 per cent of the total private sector R&D in the later twentieth century (Braunerhjelm, 1998). As shown in Section 4, the dominance of domestic MNEs has contributed to the Swedish paradox by diminishing commercialization of research results and maintaining a disproportionately high allocation of R&D resources to low- and medium-technology sectors with little potential for growth. Many of Sweden's large firms have long been highly internationalized in production and sales; more recently, ownership has also been internationalized (Edquist and Lundvall, 1993, pp. 291–2). Foreign ownership and relocation of head offices has created great concern (Andersson, 1998) and offshoring of more sophisticated forms of production threatens the innovative capacity of supplier industries (Metall, 1998). As suggested in Section 2, the latter trend may also eventually undermine the increases in high-technology manufacturing exports that Sweden achieved during the 1990s.

In the mid-1990s, Sweden's high-technology manufacturing MNEs had not yet begun to make the majority of their R&D investments abroad, and they are not likely to do so within the near future (Nutek, 1998, pp. 113–18, Figures 6.11 and 6.13). Moreover, foreign subsidiaries still rely strongly upon exports from Sweden, and Sweden continues to have a positive trade balance in high-technology products (Marklund et al., 2004, pp. 13 and 32, Figure 9.3). However, Swedish high-tech MNEs have begun to substitute outward FDI for exports based on domestic production (Braunerhjelm, 2004). Further, although these firms continue to invest strongly in R&D within Sweden, an increasing share of their production is located abroad (ITPS, 2003a), and their contribution to GDP continues to decline (Marklund et al., 2004, p. 13). It is clear that the internationalization of production in Sweden has proceeded further than the internationalization of R&D, and that 'multinational industrial groups find Sweden considerably more attractive for R&D activities than for production' (ibid., p. 32). Thus there is substantial support for the hypothesis that the Swedish paradox can be at least partly explained by globalization, in the sense that R&D carried out in Sweden increasingly bears fruit in terms of innovations in other countries.

7 STRENGTHS AND WEAKNESSES OF THE SYSTEM AND INNOVATION POLICIES

Section 7 is based on our previous analysis. In Section 7.1, we concentrate on the strengths and weaknesses of the NSI and in Section 7.2 we focus on policies recently pursued. In Section 7.3, we address innovation policy implications for the future, based on the preceding discussion.

7.1 Strengths and Weaknesses

From Section 3, we conclude that the Swedish NSI is strong on the input side and rather weak on the output side, i.e. the Swedish paradox is confirmed. One exception to the overall pattern of strength on the input side is that innovation expenditures of SMEs were not exceptionally high. Sweden has the greatest difference between large firms and SMEs in this respect. If high innovation expenditures is considered to be a weakness, we have thus also identified a weakness at the input end of the paradox. However, if a high innovation output can be achieved with a low input, it can also be considered to be a strength (see the discussion of small firms' performance below).

On the output side, Swedish firms were not particularly innovative according to an indicator measuring process and product innovation combined. However, they were more innovative in some service sectors than in manufacturing; manufacturing was weaker than some other parts of the system in this respect.

Performance was poorer for process innovations than for new (to the firm) product innovations. This weakness is surprising in the light of previous studies, covering earlier periods. Judging from the CIS data, a new weakness in process innovations seems to have emerged during the 1990s.

As regards introduction of new-to-the-firm products, Sweden performed badly by one measure (proportion of firms carrying out product innovations) and well by another (turnover due to new-to-the-firm products).[31] On the output side, the latter is the only indicator for which the Swedish NSI performs well. Hence the two indicators on new-to-the-firm products point in different directions. However, with Sweden's very high R&D and innovation intensities, this performance should have been better. On both indicators, Swedish firms performed somewhat better in some service sectors than in manufacturing.

For new-to-the-market products, Sweden performed very poorly on both available indicators (proportion of all firms carrying out new-to-the-market product innovations and proportion of turnover due to new-to-the-market products). The paradox is certainly strong in this respect.

It can also be noted that the performance on this indicator (new-to-the-market product innovations) is much better for small firms than for large ones, i.e. small firms are much more 'creative' than large ones in comparison with the other countries in the sample. Hence the overall performance of all firms – which is, on the average, worse with regard to creation than to imitation – can be explained by the domination of large firms in the Swedish NSI. We have seen above that small firms spend considerably less than large ones on innovation, and that (as expected) they perform rather badly with regard to the number of innovative firms, but that they perform well above the average with regard to turnover due to new (to the market) products. This is a great strength of small firms within the Swedish NSI.

Taken together, the results on the four last indicators discussed can be interpreted in the following way. As compared to input efforts, Swedish firms performed well with regard to one of the indicators capturing new-to-the-firm products, but badly on the other one. Swedish firms performed weakly with regard to both indicators capturing new-to-the-market products. More specifically, Swedish firms are reasonably good at imitating products that have already been introduced elsewhere by other firms, but they are less good at innovations that are brand new (new to the world). In broader terms, this means that the Swedish NSI is not creative in a profound way. It is locked into producing products that are not unique.

Turning to the activities – or determinants of innovation processes – analysed in Section 4, Sweden is strong with regard to R&D and competence building. However, the generation of organizations causes concern. The volume of new firm formation is simply too low. Connected to this is a VC market whose growth has finally taken off, but which has not yet supported early stages and high-tech ventures sufficiently.

Other support services for innovating firms have been weak in the past but are now improving. Incubation support has been established in recent decades, through diverse actors and initiatives, and is now better coordinated. With the rapid expansion of KIBS, consultancy services are plentiful.

As regards networking, a high degree of vertical integration may imply a lower degree of market-based sourcing solutions, as indicated by data from the CISs, but in fact other studies point to a system with extensive networking, even though strengthening is needed in relation to, for example, university–industry collaboration.

Demand-side activities, generally, are underdeveloped, having been largely reduced to seeking global markets through internationalization and restructuring domestic markets through liberalization.

Many problems of the Swedish NSI relate to institutions. Rigidities in S&T employment and uncertainties related to IPR legislation may have contributed to low rates of new firm creation. The relative success of large firms has been supported by corporatist organization and competition and trade rules, but these institutions may also have hindered technological renewal by impeding the creation of new firms.

Large firms remain central to the NSI, and, as shown in Section 6, they have also been the primary agents of globalization through outward FDI. As a result, much of the return on Sweden's R&D investment is captured abroad, rather than domestically.

7.2 Summary and Evaluation of the Innovation Policy Pursued

We now address Swedish innovation policies pursued during the last two to three decades. We define innovation policy as all actions by public organizations related to the nine activities discussed in Section 4.[32]

7.2.1 Knowledge inputs to innovation

The total R&D expenditures are high in the Swedish NSI. However, while the business sector is strong in this respect, the public sector is weaker. The public funding has also been distributed more widely among an increased population of higher education organizations whose numbers have been swelled by the creation of many new regional universities and university colleges. Hence, established research universities may have experienced a real decline in public research funding (Sörlin and Thörnqvist, 2000). However, such losses cannot have been very large, since 'new' universities and university colleges account for only about 5 per cent of the national research budget.

Sweden has had a persistent underproduction, relative to other economically advanced OECD countries, of university graduates in natural sciences and engineering subjects, particularly in disciplines related to high-technology industries, such as electronics and computer science (IVA, 1986). During the 1990s, therefore, Sweden greatly expanded its higher education system, focusing especially on increasing enrolments in natural sciences and engineering, and eventually reaching a level of NSEs graduation comparable to that of the USA (Jacobsson et al., 2001).

7.2.2 Demand-side factors

Sweden's relatively poor innovation output may partly be explained by the lack of market formation, where traditional instruments like regulation or PTP have recently had little scope, as compared to earlier decades.

Historically, Sweden's policy of 'armed neutrality' has meant that the military has been an important actor in the development of 'indigenous military technology' (Edquist and Lundvall, 1993, p. 281). After the fall of the Berlin Wall, it no longer plays that role. Other influential public agencies in Sweden have included state-owned authorities for infrastructure in areas such as power, transport and communications. During the mid-twentieth century, procurement contracts between the state power authority, Vattenfall, and ASEA (now merged into Asea Brown Bovery, ABB) led to ASEA's early development of high-voltage direct current transmission technology (Fridlund, 2000a). From 1954 to 1980, Televerket, the telecommunications authority, fostered Ericsson as a major supplier of telecommunications equipment, and later facilitated Ericsson's entry into mobile telecommunications (Fridlund, 2000b; Hommen and Manninen, 2003). PTP by the Swedish Railway authority, SJ, supported the development of the X2000 high-speed train during the 1980s by the transport division of ASEA (Edquist et al., 2000).

Sweden's accession to the EU has made it awkward to utilize many of the policy instruments formerly used by public organizations to stimulate the development of new technologies from the demand side. PTP is now seldom pursued. Similarly, technological standard setting (see Section 4.2) is now carried out primarily by private sector actors. In addition, large firms have also become less suitable partners for national 'innovation policy' due to the effects of globalization.

7.2.3 Provision of constituents

When it comes to public organizations related to innovation, there have been frequent restructurings. In the late 1960s, there occurred an 'industrial policy offensive', characterized by 'an emphasis on state ownership and public support to industrial renewal' (Benner, 1997, p. 221). It included large public subsidies to sunset industries such as textiles and shipyards. For example, the support to the shipyard industry amounted to as much as 0.5 per cent of Sweden's GDP for a ten-year period. It had no lasting results. Hence the industrial policy offensive eventually failed as industrial policy *per se* (Arvidsson et al., 2007, pp. 36 and 101–2). Failing support to ailing industrial sectors served as a lesson that everyone in Sweden seems to have accepted. No one now advocates public support for established industries that are not competitive. However, as we shall see in Section 7.3, the negative attitude towards public support to specific sectors of production changed in 2004.

However, the industrial policy offensive marked an important turning point for technology policy in another respect. It led to the creation of the Swedish Board for Technical Development – later transformed into Nutek – and the initiation of a number of large-scale projects involving public and private sector cooperation in the development of new technologies in fields such as nuclear energy, telecommunications and military aircraft (Benner, 1997, pp. 121–3).

In 2001 important changes were made in the organizational set-up of innovation policy in Sweden.[33] Nutek was divided into two parts, one still named Nutek and the other called the Swedish Agency for Innovation Systems (VINNOVA). VINNOVA's mission is to promote sustainable growth by developing effective systems of innovation and funding problem-oriented research. The name is rather unusual, since national policy organizations are seldom named after an academic theory or approach. Renamings of relevant public activities and organizations from the 1960s to the early years of the twenty-first century also reflect a changing policy emphasis: from industrial policy, to technology policy and then to innovation policy.

One important institutional measure has been to charge the universities with a third mission, which in 1998 was explicitly stated in the new regulation of universities as the task of engaging with the surrounding society, disseminating research information outside of academia and facilitating societal access to relevant information about research results (SOU, 1998, pp. 128 and 153–4). This reform was largely, though not exclusively, directed towards the commercialization of university-based research, through the promotion of various forms of university–industry collaboration. However, this third task is not regarded as at all as important as the 'original' tasks (teaching and research), for example, in academic appointments.[34]

7.2.4 Support services for innovating firms

The main policy initiatives taken in recent years to provide support services to innovating firms, particularly NTBFs, have been concerned with academic–industry relations, in areas such as public R&D expenditures, technology transfer initiatives (including the third mission), and public support for the financing of innovation. Higher education reforms (see Section 4.1.2) have figured prominently in this context, as have efforts to develop a VC industry in Sweden.

In addition to the third mission, a number of other reforms in the area of academic–industry relations have been implemented in recent years. From the early 1980s onwards, several Swedish universities have sought to build up an infrastructure for the exploitation of university patents and other research results. Between 1983 and 1997, 17 science parks were established in Sweden with government assistance, and since 1993 universities have been allowed to set up wholly owned companies for the commercialization of their research (Henrekson and Rosenberg, 2001, p. 212).

Increased public support for the financing of innovation has complemented the above-mentioned reforms of higher education. Nutek has continued its activities in this area, and since 1994 the Swedish government has also established seven Innovation Bridging Foundations in major university regions. Their mandate is to support the commercialization of (largely university-based) R&D by assisting inventors with patenting and aiding the start-up of SMEs – by, for example, locating appropriate VC financing (Henrekson and Rosenberg, 2001, p. 212).[35]

Numerous government schemes – as many as 140 in 1998 – have been introduced to increase the proportion of Swedish VC investment allocated to seed and early-stage financing for NTBFs, albeit with rather modest success (Landell et al., 1998). Public actors were essential for the formation of the VC industry, establishing the first VC firm, encouraging regional formation, and supplying most of the monetary resources during the 1970s and a significant share in the 1980s. Also, in the surge of VC formation in 1982–84 many smaller funds were formed by pension funds, insurance companies and real-estate companies. Policy changes were crucial to these developments. For example, regulatory reforms allowed government pension funds to make equity investments (Karaömerlioglu and Jacobsson, 2000). In addition, the creation of the OTC list in 1982 opened up the stock market as an exit route (CEBR, 2001). Recently, the Innovation Bridging Foundations have become an increasingly important tool. As discussed in Section 4.4.2, there is still a relative shortage of seed and early-stage financing. However, efforts to increase the overall size of Sweden's VC industry have met with considerable success. During the late 1990s, Sweden became the third-ranking EU country in terms of the amount of VC relative to GDP (Isaksson, 1999; Karaömerlioglu and Jacobsson, 2000).

7.3 Future Innovation Policy

We now turn to suggesting policies for the future development of the Swedish NSI. Above, we have identified weaknesses representing problems and unexploited opportunities that should be subject to policy interventions or changes.[36]

We have confirmed the existence of the Swedish paradox, enlarging the input side to cover not only R&D but also all other innovation expenditures. We have also specified it on the output side. Further, we have found support for all three hypotheses that have been advanced to explain the paradox. In Section 4, on activities that influence innovation, we found support for the first hypothesis, concerning obstacles to technology transfer from the R&D sphere to the commercial sphere, in relation to Section 4.1 (Knowledge inputs to innovation), Section 4.3 (Provision of constituents) and Section 4.4 (Support services for innovating firms). We also found support for the second hypothesis, which points to a problematic sectoral allocation of R&D, under Section 4.3 (Provision of constituents). The third hypothesis, according to which internationalization of production means that Swedish R&D is increasingly exploited abroad, was supported by our assessment of globalization in Section 6.

The dominance of incumbent large manufacturing firms (MNEs) is a common element in all these explanations. We are therefore persuaded that the underlying problem concerns the apparent inability of these large firms to translate innovation inputs into outputs – at least not in a way that secures that the return on Sweden's R&D investment is captured domestically, rather than abroad.

Regarding obstacles to technology transfer from the R&D to the commercial sphere, most recent policies have concentrated on creating incentives and infrastructures for improving university-to-industry technology transfer. Given that corporate sources account for 72 per cent of R&D funding, it would be logical to address the overwhelming domination of business sector R&D by large firms in more detail. This is especially so since small firms are more efficient innovators than large ones – comparing inputs and outputs, i.e. productivity – and also perform better in product innovation. Innovation expenditures and resources are much lower for SMEs than for large firms. At the same time, large firms are becoming less suitable partners for a national innovation policy because of ongoing globalization.

Hence there are strong reasons to increase R&D and innovation expenditures and efforts in SMEs in advanced sectors. The recently started VINNOVA programme entitled 'Do Research and Grow' (Forska och Väx) may be instrumental here. Regional clusters and collaboration in strategic R&D and innovation including SMEs should also be strengthened. One thing that could be done is to facilitate the spin-off of new firms from large firms, in cases where the latter are not commercializing results from R&D and innovation efforts to a sufficient extent. These instruments would lead to the establishment of more new innovation-based firms.

With respect to the problematic sectoral allocation of R&D, policy makers have generally ignored the institutionally induced lock-in of R&D resources and results to large firms in traditional sectors. Public agencies have even supported R&D in traditional sectors to a large extent, such as research in relation to forest-based industries, and provided direct subsidies to the textile and shipyard industries mentioned earlier. Further, many of the reforms introduced in recent decades have actually exacerbated this problem, reinforcing existing sectoral and technological specialization patterns. Therefore there are reasons to stimulate the development of new knowledge-intensive industries, by encouraging large firms to diversify into them, by assisting the birth and growth of new innovation-based firms in new sectors and by attracting foreign firms in advanced sectors of production. One infrastructural mode of doing so would be to make more strategic use of public funding for R&D (Edquist, 2002, pp. 53–4).[37]

In 2004 the Swedish Ministry of Industry abandonded the dogmatic resistance to formulating policies at – and providing public support to – a sectoral level. This was actually a crucial paradigm shift in Swedish innovation policy which replaced the dogmatic rejection of sector-oriented policies based on the failures in textiles and shipyards mentioned in Section 7.2.3. The Ministry formulated six sectoral policy initiatives for the following sectors: Aerospace, Motor Vehicles, Metals, Information Technology/Telecom, Forest and Wood and Pharmaceuticals/Biotech/Medical Technology.

It is a major step forward that the policy is formulated at the sectoral level. However, it is (still) a problem that the list of sectors includes a large part of industry – and, accordingly, also established and traditional sectors that can be expected to finance their own future development. Less policy effort and fewer public resources should be allocated to well-established, 'traditional' sectors, and stronger, more focused interventions should be pursued in radically new areas of technical development (Edquist, 2002, pp. 53–4; Arvidsson et al., 2007, pp. 9–18). In other words, public R&D and innovation efforts should be more effectively targeted to sectors of production that are new and where uncertainty is large.[38] Such a strategy can be seen as an attempt to balance previous policy measures – or, rather, mistakes – in Sweden. These mistakes have contributed to a lock-in effect that has actually supported the maintenance of the existing production structure. Examples are substantial support to ailing industries through subsidies, currency devaluations in the 1970s and 1980s, and public R&D support to traditional industries.

Complementary measures could be developed on the largely neglected demand side of innovation policy by, for example, following the EU's recent 'rediscovery' of PTP as a policy instrument for stimulating private sector innovation. Sweden's current lack of attention to the demand side is reflected by the country's poor performance in new-to-the-market product innovations, with the exception of a few service sectors. These exceptions should provide models for new thinking on, and initiatives in, demand-side policies, including new forms of PPPs and new combinations of supply- and demand-side measures.

Regarding the internationalization of production by MNEs and the resulting failure to capture returns on R&D investment within the domestic economy, Sweden faces a quandary. On one hand, outward FDI has meant declining benefits from Sweden's historical specialization in low- and medium-tech sectors and industries dominated by very large and increasingly internationalized firms. On the other, it has also meant the development of Sweden into a global centre for R&D activities and services – a potential source of comparative advantage which, however, remains underutilized. Public policy cannot intervene very much in the internal affairs of large firms in order to exploit this source of opportunity. Instead, it should try to build upon and complement their valuable contributions to the NSI, including the creation of a strong labour market for NSEs and other R&D personnel and expression of sophisticated, 'leading-edge' demand in relation to domestic supplier industries.

Such innovation policies should include elements of 'attraction policies'. These are a matter of how MNEs can be influenced to locate high-productivity activities (such as R&D) within the borders of Sweden (Arvidsson et al., ch. 8). However, there are certainly dilemmas associated with pursuing such policies in the present era of globalization. It can be questioned whether the state in a small country, for example, should subsidize R&D activities of large, foreign-owned MNEs. At the same time, public support to (R&D in) Swedish innovation-based SMEs can also mean that the pay-off for Sweden disappears if the firms move early to other countries, perhaps because they get larger subsidies there (Borrás et al., 2008).

What should be addressed is the industrial ecology surrounding the large international firms in an effort to replicate the virtuous relationships between KIBS firms and the large service sector firms whose unbundling created them (Nählinder, 2005). Much could be done to help achieve such a balance. For instance, supplier firms that already benefit from collaboration with MNEs should be encouraged to interact with a broader range of customers. Interfirm networks of innovation in Sweden have a strong 'vertical' character, due to domination and control by a few large firms, and could be greatly enhanced by measures to support collaboration and learning over organizational borders. Increasing collaboration with customers through diversification should markedly improve Swedish firms' poor performance with respect to product innovations – both those that are new to the market and those that are new to the firm.

In addition to indicating some new policy directions, sketched above, our analysis also recommends continued support for some initiatives already under way. Efforts to stimulate translation of research results from universities into innovations in firms should be strengthened by pushing the third mission, and improving both financing and additional support services for innovating firms, particularly those formed to exploit academic research results. Increasing the presence of this type of firm should help to ameliorate low innovation expenditures by SMEs in Sweden. The innovation gap between Sweden and other countries is greatest in manufacturing, and calls for more policy efforts targeted towards manufacturing. For instance, policy should try to make process innovation a preferable alternative to relocating production abroad.

However, there is still also a general need to stimulate product innovation, since such innovation is the main engine for renewing the production structure of any NSI. Here, newer and smaller firms seem to be more creative than older and larger ones, and should therefore be the main focus. Efforts to alter the production structure towards stronger representation of high-technology sectors should also be continued, with emphasis on entry into new knowledge fields and creation of new sectoral innovation systems. A shift towards a more knowledge-intensive structure of production would increase productivity, economic growth and employment. However, this can only be achieved by combining many of the policy measures discussed here.

With regard to practically all the issues addressed in this chapter, much more data could – and should – be created and collected, and the analysis should be made much more profound in many respects. This chapter has only scratched the surface and calls for more thorough analysis of many issues. One such issue concerns the role of MNEs and their networks within the Swedish NSI. How do these actors operate in relation to others? How do they obtain knowledge from, and establish ties with, private and public research organizations? Do they act differently abroad than they do in Sweden?

In addition, the NSI's strengths and weaknesses will also change over time, and policies will have to be adjusted. This task requires continuous and in-depth analyses, to which considerable resources should be committed. Therefore our most important policy proposal is that a collective analytical effort is needed to create a knowledge basis for innovation policy. Learning for policy and through policy is crucial.

ACKNOWLEDGEMENTS

Preparation of this chapter was jointly financed by VINNOVA, Sweden's Agency for Innovation Systems, and ITPS (the Swedish Institute for Growth Policy Studies). For comments on earlier drafts of this work, we are grateful to a number of colleagues, including Yu-Ling Luo, Nola Hewitt-Dundas Stephen Roper, Jan Fagerberg, Astrid Kander, Olof Ejermo, Bo Carlsson and Mats Benner.

NOTES

1. This publication of 1998, internally published in 1996, was written in 1994, based on a publication from 1992 – which, in its turn, was a translation of a chapter in an appendix to the final study of the Swedish Productivity Delegation from 1991 (Edquist and McKelvey, 1991).
2. See Sections 6 and 7 on the dominance of large firms in the Swedish NSI.

3. The CIS data referred to here is presented in Section 4 of the Appendix. A series of 15 tables provides detailed comparative data on the countries mentioned here.
4. This also means that we will give priority to indicators that are available for both periods.
5. This indicator includes not only R&D but also acquisition of machinery, equipment and knowledge, training, market introduction of innovations, design and other preparations for production or distribution.
6. In Bitard et al. (2005), Section 1, footnote 1, we pointed out, however, that the data seem to be uncertain for innovation intensity in 1998–2000.
7. Our data measured mainly technological process innovations and did not include organizational process innovations.
8. There it was shown that Swedish manufacturing firms were among the world leaders in the 1970s and 1980s with regard to the diffusion of computer-controlled process technologies (numerically controlled machine tools, industrial robots and flexible manufacturing systems) in the engineering industry.
9. This could indicate that the new (to the firm) product innovations, on average, account for large volumes of sales, which is certainly a great strength of the Swedish NSI.
10. In addition, the input component of the Swedish paradox can be extended to all innovation expenditures, not only R&D expenditures. Further, the difference between Sweden and the other countries with regard to this indicator was even larger for innovation intensity than for R&D intensity. In other words, the paradox can be reformulated along these lines: on the input side we could use innovation intensity instead of R&D intensity – or both.
11. The analysis in this section partly supplements the discussion in Section 3. There the discussion was focused upon R&D performed by firms. Here both private and public R&D are discussed. The sources used are also different between Section 3 and Section 4.1.1.
12. While we have no reason to doubt the high R&D expenditure in the business sector (accounting for approximately 75 per cent of total R&D expenditure), a recent study shows that there are some measurement problems involved in assessing non-business R&D, making this part of the R&D volume somewhat overestimated (Jacobsson and Rickne, 2004). This means that although Sweden does have a very high R&D expenditure, the figure of 4.3 per cent may be somewhat overestimated.
13. As measured by the number of publications per input unit (e.g. the number of R&D personnel or researchers).
14. Although this may not appear much higher than the EU average (77.9 per cent), there are large differences across countries (Finland, 71.8 per cent; Denmark, 60.6 per cent).
15. The common trend of reduction of defence budgets at the beginning of the 1990s has naturally had a strong influence on public R&D expenditures.
16. This subsection focuses on new firms. Other organizations, especially those that support innovation, will be discussed in Sections 4.4 and 7.
17. There is a relatively large stock of small firms in Sweden, but not a high formation rate of new firms.
18. NTBFs, or high-tech firms, are those with a clear scientifically or technologically innovative character (Rickne, 2000).
19. CIS asked whether or not the firm had cooperated on innovation activities, and if so with what kind of partner.
20. Note that the study by Edquist et al. (2000) relates to product innovations only, while the CIS survey also refers to process innovations.
21. Science parks and incubators are, of course, two different things. In Sweden, however, most science parks have deliberately incorporated incubator functions, either formally or informally, and few incubators are found outside science parks.
22. The venture capital and private equity industry (here termed the VC industry) involves the support of unlisted companies, both economically and with active owner involvement.
23. The following discussion focuses on private consultancy services, and therefore on the KIBS sector. Public consultancy services have been addressed in Section 4.4.1 and will also be dealt with in Section 7.

24. This source defines KIBS as including 'business service firms, R&D firms and firms engaged in wholesale trade with machinery and equipment' (Nutek, 1998, p. 133).
25. Phi and Cramer's V was equal to 0.115 in the first period and 0.148 in the second period. The values range between 0 (no correlation) and 1 (perfect correlation).
26. At the 5 per cent level of confidence.
27. A ten-sector decomposition was chosen, including the following sectors: Food products, beverages and tobacco; Textiles, textile products, leather and footwear; Wood and products of wood and cork; Chemical, rubber, plastics and fuel products; Other non-metallic mineral products; Machinery and equipment; Transport equipment; Manufacturing n.e.c. (not elsewhere classified); Financial intermediation and Computer and related activities.
28. Cf. Pearson coefficient = −0.673, significant at the 5 per cent confidence level.
29. At the same time, the number of hours worked in this sector remained constant.
30. Based on STAN database for Industrial Analysis (OECD, 2004).
31. As mentioned in Section 3, this could indicate that the new (to the firm) product innovations, on average, account for large volumes of sales, which is certainly a great strength of the Swedish NSI.
32. The discussion here is structured according to the areas of activity discussed in Section 4. It will concentrate on outlining broad, general trends in policy, since it is beyond the scope and possibility of this subsection to mention all the specific policy measures that have been taken. Instances of specific policy measures will only be referred to as examples used for illustration and explanation.
33. With regard to the provision of organizations, public efforts to encourage the formation of new firms was discussed under Section 4.4 (Support services for innovating firms) above.
34. Institutional reforms, such as deregulation and privatization measures, have been mentioned in Section 4.2 (Demand-side factors). The same is true for policies for supporting networking and collaboration between organizations.
35. In late 2004, the seven Innovation Bridge Foundations were reorganized into one national organization with regional branches.
36. We have also identified 'strengths'. These should not be subject to policy or policy changes (since private actors or prevailing policies already secure a good performance).
37. Anyone reflecting on this realizes that most policies – including publicly funded R&D – are problem-oriented and selective rather than neutral. Of course, firm strategies are also a matter of selection between alternatives. For both public and private actors such choices are extremely difficult, but cannot be avoided. (These arguments are developed in the concluding chapter in this book.)
38. As shown in the concluding chapter of this book, such policies have been pursued in many of the ten countries addressed.

REFERENCES

Alm, H. (2004), 'External relations in the product development process: a study of biotechnology firms in Sweden', Linköping, Linköping University.

Alskling, B. (2001), 'In search of new models of institutional governance: some Swedish experiences', *Tertiary Education and Management*, **7**, 197–210.

Andersson, T. (1998), 'Internationalization of research and development – causes and consequences for a small economy', *Economics of Innovation and New Technology*, **7**(1), 71–91.

Andersson, T., O. Asplund and M. Henrekson (2002), *Betydelsen av innovationssystem: utmaningar för samhället och för politiken*, Stockholm: VINNOVA (in Swedish).

Arvidsson, G., H. Bergström, C. Edquist, D. Högberg and B. Jönsson (2007), *Medicin för Sverige – Nytt liv i en framtidsbransch*, Stockholm: SNS (in Swedish).
Aspgren, M. (2002), 'Är vi lika bra som andra? Sveriges kompetensförsörjning i ett internationellt perspektiv', in K. Abrahamsson, L. Abrahamsson, T. Björkman, P.-E. Ellström and J. Johansson, *Utbildning, kompetens och arbete*, Lund: Studentlitteratur (in Swedish).
Bauer, M., B. Askling, S.G. Marton and F. Marton (1999), *Transforming Universities: Changing Patterns of Governance, Structure and Learning in Swedish Higher Education*, London and Philadelphia: Jessica Kingsley.
Bengtsson, L. (2003), 'Forskningparkens betydelse för företagens och regionens utveckling och tillväxt – en studie av Ideon 1983–2003', Lund: School of Economics, Lund University (in Swedish).
Benner, M. (1997), *The Politics of Growth and Economic Regulation in Sweden, 1930–1994*, Lund: Arkiv förlag.
Bergek, A., S. Jacobsson, B. Carlsson, S. Lindmark and A. Rickne (2008), 'Analyzing the functional dynamics of technological innovation systems', *Research Policy* (forthcoming).
Berggren, T. (2002), 'The development of the venture capital industry in Sweden', http://www.vencap.se/article_view.asp?ArticleID=31.
Bitard, P., C. Edquist, L. Hommen and A. Rickne (2005), 'A CIS-based analysis of intensity and characteristics of innovation in the Swedish national system of innovation', Division of Innovation, Lund Institute of Technology, Lund University, January (mimeo).
Borrás, S., C. Chaminade and C. Edquist (2008), 'The challenges of globalisation: strategic choices for innovation policy', in G. Marklund, N. Vorontas and C. Wessner (eds), *The Innovation Imperative: Globalisation and National Competitiveness*, Cheltenham, UK and Northampton, MA, USA: Edward Elgar (forthcoming).
Braunerhjelm, P. (1998), 'Varför leder inte ökade FOU-satsningar till mer högteknologisk export?', *Ekonomiska Samfundets Tidskrift*, **51**(2), 113–22 (in Swedish).
Braunerhjelm, P. (2004), 'Knowledge capital and economic growth: Sweden as an emblematic example', paper presented at the Conference 'New Science, New Industry', Rome, October.
Braunerhjelm, P. and P. Thulin (2004), 'Can countries create comparative advantages? R&D expenditures, market size and institutions in 19 OECD countries, 1981–1999', Working Paper, Centre for Business and Policy Studies (SNS). Stockholm: SNS.
Brulin, G., P.-E. Ellström and L. Svensson (2000), 'Interactive knowledge formation: a challenge for Swedish research and higher education', Paper presented at HSS-03, 14–16 May, Ronneby, Sweden.
Carlsson, B. (1996), 'Innovation and success in Sweden: technological systems', in J. de la Mothe and G. Paquet (eds), *Evolutionary Economics and the New International Political Economy*, London: Pinter.
Carlsson, B. and S. Jacobsson (1997), 'Diversity creation and technological systems: a technology policy perspective', in C. Edquist (ed.), *Systems of Innovation: Technologies, Organisations and Institutions*, London: Pinter Publishers/Cassell Academic, pp. 266–94.
CEBR (2001), *Seed Capital in the Nordic Countries: Best Practice*, Copenhagen: Centre for Economic and Business Research.

Edin, P.-A. and R. Topel (1997), 'Wage policy and restructuring – the Swedish labor market since 1960', in R.B. Freemen, R. Topel and R. Swedenborg (eds), *The Welfare State in Transition*, Chicago: University of Chicago Press, pp. 155–201.
Edquist, C. (2002), *Innovationspolitik för Sverige: mål, skäl, problem och åtgärder*, Stockholm: VINNOVA (in Swedish).
Edquist, C. (2005), 'Systems of innovation – perspectives and challenges', in J. Fagerberg, D.C. Mowery and R.R. Nelson (eds), *The Oxford Handbook of Innovation*, Oxford: Oxford University Press, pp. 181–208.
Edquist, C. and L. Hommen (2000), 'Public technology procurement and innovation theory', in C. Edquist, L. Hommen and L. Tsipouri (eds), *Public Technology Procurement and Innovation*, Boston, Dordrecht and London: Kluwer Academic Publishers, pp. 5–70.
Edquist, C. and S. Jacobsson (1988), *Flexible Automation: The Global Diffusion of New Technology in the Engineering Industry*, Oxford: Basil Blackwell.
Edquist, C. and B.-Å. Lundvall (1993), 'Comparing the Danish and Swedish national systems of innovation', in R.R. Nelson (ed.), *National Systems of Innovation: A Comparative Study*, Oxford: Oxford University Press, pp. 265–98.
Edquist, C. and M. McKelvey (1991), 'The diffusion of new product technologies and productivity growth in Swedish industry', Consortium on Competitiveness and Cooperation (CCC) Working Paper No. 91-15, 1992, Center for Research in Management, Berkeley, CA: University of California.
Edquist, C. and M. McKelvey (1998), 'High R&D intensity without high tech products: a Swedish paradox?', in K. Nielsen and B. Johnson (eds), *Institutions and Economic Change: New Perspectives on Markets, Firms and Technology*, Cheltenham, UK and Northampton, MA, USA: Edward Elgar, pp. 131–49.
Edquist, C. and F. Texier (1996), 'The perverted growth pattern of Swedish industry – current situation and policy implications', in O. Kuusi (ed.), *Innovation Systems and Competitiveness*, Helsinki: Taloustieto Oy.
Edquist, C., M.-L. Eriksson and H. Sjögren (2000), 'Collaboration in product innovation in the East Gothia regional system of innovation, *Enterprise and Innovation Management Studies*, 1(1), 37–56.
Edquist, C., P. Hammarqvist and L. Hommen (2000), 'Public technology procurement in Sweden: the X2000 high speed train', in C. Edquist, L. Hommen and L. Tsipouri (eds), *Public Technology Procurement and Innovation*, Boston, MA, Dordrecht and London: Kluwer Academic Publishers, pp. 79–98.
Edquist, C., L. Hommen and L. Tsipouri (2000), 'Policy implications', in C. Edquist, L. Hommen and L. Tsipouri (eds), *Public Technology Procurement and Innovation*, Boston, MA, Dordrecht and London: Kluwer Academic Publishers, pp. 301–11.
Esping-Andersen, G. (1990), *The Three Worlds of Welfare Capitalism*, Princeton, NJ: Princeton University Press.
Etzkowitz, H., P. Asplund and N. Nordman (2002), 'The university and regional renewal: emergence of an entrepreneurial paradigm in the US and Sweden', in G. Törnquist and S. Sörlin, *The Wealth of Knowledge: Universities and the New Economy*, Stockholm: SNS Förlag.
European Commission (2003), *3rd European Report on Science and Technology Indicators: Towards a Knowledge-Based Economy*, EUR 20025 EN, Brussels: European Commission.
Eurostat (2003), *Long term indicators*, http://epp.eurostat.ec.europa.eu/portal/page?_pageid=1996,45323734&_dad=portal&_schema=PORTAL&screen=

welcomeref&close=/yearlies&language=en&product=EU_yearlies&root=EU_ yearlies&scrollto=0, accessed 21 June 2007.
EVAC (2001), 'Survey of the economic and social impact of management buyouts and buyins in Europe', Brussels: European Private Equity and Venture Capital Association.
EVAC (2004), 'Benchmarking European tax and legal environments: indicators of tax and legal environments favouring the development of private equity and venture capital and entrepreneurship in Europe', May, Brussels: European Private Equity and Venture Capital Association.
Ferguson, R. (1998), *What's in a Location? Science Parks and the Support of NTBFs*, PhD dissertation, Uppsala: SLU.
Fredriksson, P. (1997), 'Economic incentives and the demand for higher education', *Scandinavian Journal of Economics*, **99**(1), 129–42.
Fridlund, M. (2000a), 'Procuring products and power: developing international competitiveness in Swedish electrotechnology and electric power', in C. Edquist, L. Hommen and L. Tsipouri (eds), *Public Technology Procurement and Innovation*, Boston, MA, Dordrecht and London: Kluwer Academic Publishers, pp. 99–120.
Fridlund, M. (2000b), 'Switching relations and trajectories: the development of the AXE Swedish switching technology', in C. Edquist, L. Hommen and L. Tsipouri (eds), *Public Technology Procurement and Innovation*, Boston, MA, Dordrecht and London: Kluwer Academic Publishers, pp. 143–66.
Fukasaku, Y. (1998), 'Public/private partnerships for developing environmental technology', *OECD STI Review*, **23**, 105–30.
GEM (2003), *Global Report* 2003.
Glete J. (1989), 'Long-term firm growth and ownership organisation – a study of business histories', *Journal of Economic Behavior & Organisation*, **12**(3), 329–52.
Glimstedt, H. (1995), 'Non-Fordist routes to modernization: production, innovation and the political construction of markets in the Swedish automobile industry before 1960', *Business and Economic History*, **24**(1), 243–53.
Glimstedt, H. (2000), 'Creative cross-fertilization and uneven Americanization of Swedish industry: sources of innovation in post-war motor vehicles and electrical manufacturing', in J. Zeitlin and G. Herrigel (eds), *Americanization and its Limits*, Oxford: Oxford University Press, pp. 180–208.
Glimstedt, H. (2001), 'Competitive dynamics of technological standardization: the case of third generation cellular communications', *Industry and Innovation*, **8**(1), 49–78.
Henrekson, M. and N. Rosenberg (2001), 'Designing efficient institutions for science-based entrepreneurship: lessons from the US and Sweden', *Journal of Technology Transfer*, **26**(3), 207–31.
Högskoleverket (2003), 'Swedish universities and university colleges – short version of annual report, 2003', Stockholm: Högskoleverket.
Hommen, L. (2003), 'Third generation mobile telecommunications – UMTS', in C. Edquist (ed.), *The Fixed Internet and Mobile Telecommunications Sectoral System of Innovation: Equipment Production, Access Provision and Content Provision*, Cheltenham, UK and Northampton, MA, USA: Edward Elgar, pp. 129–61.
Hommen, L. and E. Manninen (2003), 'Second generation mobile telecommunications – GSM', in C. Edquist (ed.), *The Fixed Internet and Mobile*

Telecommunications Sectoral System of Innovation: Equipment Production, Access Provision and Content Provision, Cheltenham, UK and Northampton, MA, USA: Edward Elgar, pp. 71–128.

Isaksson, A. (1999), *Effekter av venture Capital i Sverige (Effects of Venture Capital in Sweden)*, B1999:3, Stockholm: Nutek Förlag (in Swedish).

Isaksson, A. and B. Cornelius (1998), 'Venture capital incentives: a two country comparison', tenth Nordic Conference on Small Business Research, Växjö University, Sweden, 14–16 June.

Isaksson, A., B. Cornelius, S. Junghagen and H. Landström (2004), 'Institutional theory and contracting in venture capital: the Swedish experience', *Venture Capital*, **6**(1), 47–71.

ITPS (2002), *Newly-started enterprises in Sweden 2000 and 2001*, S2002:008, Östersund: Swedish Institute for Growth Policy Studies.

ITPS (2003a), *Svenskägda koncerner med verksamhet i utlandet 2001*, SOS 2003:006, Östersund: Swedish Institute for Growth Policy Studies (in Swedish).

ITPS (2003b), *Uppföljning av 1998 års nystartade företag – tre år efter start*, S2003:005, Östersund: Swedish Institute for Growth Policy Studies (in Swedish).

IVA (1986), *Ingenjörer för framtiden (Engineers for the Future)*, Ingenjörsakademien meddelande 249 (Engineering Academy Communication 249), Stockholm: IVA (in Swedish).

Jacobsson, S. and J. Philipson (1996), 'Sweden's technological profile', *Technovation*, **26**(5), 245–53.

Jacobsson, S., C. Sjöberg and M. Wahlstöm (2001), 'Alternative specifications of the institutional constraints to economic growth – or why is there a shortage of electronic engineers and computer scientists in Sweden?', *Technology Analysis and Strategic Management*, **13**(2), 179–93.

Jacobsson, S. and A. Rickne (2004), 'How large is the Swedish "academic" sector really? A critical analysis of the use of science and technology indicators', *Research Policy*, Special Issue in Honor of Keith Pavitt: What do we know about innovation?, **33**(3), 1355–72.

Johnson, A. (1998), 'Functions in innovation systems approaches', mimeo, Department of Industrial Dynamics, Chalmers University of Technology, Gothenburg, Sweden.

Johnson, A. and S. Jacobsson (2003), 'The emergence of a growth industry: a comparative analysis of the German, Dutch and Swedish wind turbine industries', in S. Metcalfe and U. Cantner (eds), *Transformation and Development: Schumpeterian Perspectives*, Heidelberg: Physica/Springer.

Karaömerlioglu, D.C. and S. Jacobsson (2000), 'The Swedish venture capital industry: an infant, adolescent or grown-up?', *Venture Capital*, **2**, 61–88.

Karlberg, L.-A. (2004), 'Tre IT-bolag fick ramavtal med ESV', *Ny Teknik*, http://nyteknik.se/art/35569, accessed 21 June 2007.

Kleja, M. (2004), 'IT-ministrar i bråk om 24-timmarsmyndigheten', *Ny Teknik*, http://nyteknik.se/art/36700, accessed 21 June 2007 (in Swedish).

Landell, E. et al. (1998), *Entreprenörsfonder: riskkapital till växande småföretag (Entrepreneurs' Funds: Venture Capital for Growing Small Firms)*, Stockholm: The Federation of Swedish Industries and Nutek (in Swedish).

Lindelöf, H. and H. Löfsten (1999), 'Teknik- och forskningsparker' ('Technology- and research-parks'), WP 99:108, Gothenburg: IMIT, Chalmers University of Technology (in Swedish).

Lindholm, Å. (1994), 'The economics of technology-related ownership changes: a

study of innovativeness and growth through acquisition and spin-offs', Chalmers University of Technology, Gothenburg, Sweden.
Lindholm-Dahlstrand, Å. (1997a), 'Growth and inventiveness in technology-based spin-off firms', *Research Policy*, **26**(3), 331–44.
Lindholm-Dahlstrand, Å. (1997b), 'Entrepreneurial spin-off enterprises in Göteborg, Sweden', *European Planning Studies*, **5**(5), 659–73.
Liu, X. and S. White (2001), 'Comparing innovation systems: a framework and application to China's transitional context', *Research Policy*, **30**, 1091–114.
Löfsten, H. and H. Lindelöf (2002), 'Science parks and the growth of new technology-based firms – academic–industry links, innovation and markets', *Research Policy*, **31**, 859–76.
Lundahl, L. (2002), 'Sweden: decentralization, deregulation, quasi-markets – and then what?', *Journal of Educational Policy*, **17**(6), 687–97.
Marklund, G. (2001), *International Benchmarking av det Svenska FoU-Systemet*, mimeo, Stockholm: VINNOVA (in Swedish).
Marklund, G., R. Nilsson, P. Sandgren, J. Granat Thorsland and J. Ullström (2004), *The Swedish National Innovation System, 1970–2003: A Quantitative International Benchmarking Analysis*, VINNOVA Analysis VA 2004:1. Stockholm: VINNOVA.
Metall (1998), '15 storföretag 1997/98 – Tema Östeuropa', Stockholm: Metall.
Nählinder, J. (2003), 'Syselsättningseffeker av innovationer i kunskapsintensiva tjänsteföretag', Working Paper No. 266, Linköping: Department of Technology and Social Change, Linköping University (in Swedish).
Nählinder, J. (2005), *Innovation and Employment in Services: The Case of Knowledge Intensive Business Services in Sweden*, Linköping: Department of Technology and Social Change, Linköping University.
Nählinder, J. and L. Hommen (2002), 'Employment and innovation in services: knowledge intensive business services in Sweden', paper read at the final meeting and conference of the AITEG project, 18–19 April, Clore Management Centre, Birkbeck College, London.
Nås, S., T. Sandven, T. Eriksson, J. Andersson, B. Tegsjö, O. Lehtoranta and M. Virtaharju (2004), *High-Tech Spin-Offs in the Nordic Countries*, STEP Report 22-2003. SINTEF.
Nutek (1998), *The Swedish National Innovation System: A Quantitative Study*, Stockholm: Nuteks Förlag.
Nutek (2000), 'Internationella jämförelser för näringslivets tillväxt: Tillväxt-indikatorn', Rapport 2000:17, Stockholm: Nuteks förlag (in Swedish).
Nutek (2004), *Riskkapitalbolagens aktiviteter – första kvartalet 2004 (Venture Capital Firms' Activities – First Quarter 2004)*, R 2004:07, Stockholm: Nutek (in Swedish).
OECD (1995), *Education at a Glance*, Paris: OECD.
OECD (1998), *Human Capital Investment*, Paris: OECD.
OECD (2002), *Education at a Glance*, Paris: OECD.
OECD (2003), *Main Science and Technology Indicators*, 2003:1, Paris: OECD.
OECD (2004), *Economic Surveys, Sweden*, Paris: OECD.
Ohlsson, L. and C. Vinell (1987), *Tillväxtens drivkrafter: en studie av industriers framtidsvillkor*, Stockholm: Industriförbundets Förlag (in Swedish).
Olofsson, C. and C. Wahlbin (1993), *Teknikbaserade företag från högskolan*, Stockholm: Institute for Management Innovation and Technology (IMIT) (in Swedish).
Rickne, A. (2000), *New Technology-Based Firms and Industrial Dynamics – Evidence*

from the Technological System of Biomaterials in Sweden, Ohio and Massachusetts, PhD thesis, Gothenburg: Department of Industrial Dynamics, Chalmers University of Technology.

Rickne, A. and S. Jacobsson (1996), 'New technology-based firms – an exploratory study of technology exploitation and industrial renewal', *International Journal of Technology Management*, 11(3/4), 238–57.

Rickne, A. and S. Jacobsson (1999), 'New technology-based firms in Sweden: a study of their impact on industrial renewal', *Economics of Innovation and New Technology*, 8(2), 197–223.

Rosenberg, N. and H.-O. Hagen (2003), *The Responsiveness of the Universities*, ITPS Report A2003:019, Östersund: Swedish Institute for Growth Policy Studies.

Ruin, O. (1974), 'Participatory democracy and corporativism: the case of Sweden', *Scandinavian Political Studies Yearbook*, 9, 171–84.

Sandström, A. and L. Norgren (2003), *Swedish Biotechnology*, VA 2003:02, Stockholm: VINNOVA.

Sandström, U. (2002), *Det nya forskningslandskapet*, Stockholm: Sister (in Swedish).

Sellenthin, M.O. (2004), 'Who should own university research? An exploratory study of the impact of patent rights regimes in Sweden and Germany on the incentives to patent research results', ITPS Report A2004:013, Östersund: Swedish Institute for Growth Policy Studies.

Sohlman, Å. (1996), *Framtidens utbildning: Sverige i internationell konkurrens*, Stockholm: SNS Förlag (in Swedish).

Sohlman, Å. (1999), 'Sweden – a learning society? The educational system's performance in international perspective', *Swedish Economic Policy Review*, 6(2), 403–44.

Sörlin, S. and G. Thörnqvist (2000), *Kunskap för Välstånd: Universiteten och omvandlingen av Sverige (The Wealth of Knowledge: Universities and the New Economy in Sweden)*, Stockholm: SNS Förlag (in Swedish).

Stankiewicz, R. (1986), *Academics and Entrepreneurs: Developing university–industry relations*, London: Frances Pinter.

SOU (1996), *Samverkan mellan högskolan och näringslivet, Huvudbetänkande av NYFOR*, Stockholm: Fritzes (in Swedish).

SOU (1998), *Forskningspolitik: Slutbetänkade av kommitén för översyn av den svenska forskningspolitiken (Forskning 2000)*, Stockholm: Fritzes (in Swedish).

SVCA (2002), *Svenska Riskkapitalföreningens Vitbok och agenda*, http://www.mcit.se/svca/article_view.asp?ArticleID=32, accessed 21 June 2007 (in Swedish).

VINNOVA (2001), *The Swedish Biotechnology Innovation System*, VF: 2001:2, Stockholm: VINNOVA.

VINNOVA (2002), *Nationellt Inkubatorprogram*, VP 2002:2, Stockholm: VINNOVA (in Swedish).

8. Low innovation intensity, high growth and specialized trajectories: Norway

Terje Grønning, Svein Erik Moen and Dorothy Sutherland Olsen

1 INTRODUCTION

The Norwegian economy is one of the major puzzles within studies of economic growth and welfare. The country ranks high on indicators for economic output and standard of living, but low on innovation output indicators. In this chapter, we explore the functioning of the Norwegian national system of innovation (NSI) with this main puzzle in mind. The account builds on official statistics, published survey results, secondary literature, and, in connection with Section 4.4, interviews with two firms, three ministries and nine different support organizations for incubation, funding and policy.

The chapter follows the same structure as other chapters in this volume: an examination of the main traits of the NSI and of the propensity to innovate, analyses of activities within the NSI, of the system's degree of openness and of policy traits and concerns. In order to identify and describe the main traits of the economy, we include an explicit focus on technological trajectories (Pavitt, 1984; Archibugi, 2001). On the one hand, a large segment of the economy is related to extraction of natural resources and is populated mainly by the scale-intensive and supplier-based trajectories. On the other hand, a limited number of firms within the science-based trajectory constitute an alternative segment where one part is linked to extraction of natural resources through supplies and services, but where another part is relatively independent of those activities. As a third segment, there is an innovation-intensive trajectory of specialized suppliers with strong linkages to the scale-intensive and supplier-dominated trajectories in the form of supplies to these segments. The framework is further presented and discussed below in Sections 2 and 3.

Our investigation into the relationship between Norway's low innovation output and high economic output includes the more obvious factors, such as the influence of the oil and gas revenues, but also tries to take more specific mechanisms into account. Affluence has come as a result of increasing specialization in low-tech resource extraction in combination with the existence of innovation-intensive technological trajectories within sectors such as mechanical engineering, engineering consultancy and suppliers to the aquaculture sectors. This development has partly been due to technology innovation policies, which have been generous in terms of affording infrastructure financing on a broad basis. The current and future competitive global situation, together with reduction of access to natural resources, may force Norwegian government and business alike into serious prioritization. Future policies may be forced from their current broad and general orientation into either a portfolio of instruments catering more explicitly to fostering diversity and multiple knowledge-intensive activities, or a more targeted and competitive policy with fewer focus areas.

2 MAIN HISTORICAL TRENDS

The geography of Norway facilitated an industrialization process that relied heavily on natural endowments. Thus it was timber, fisheries and shipping that became the first industrial basis of the economy in the late nineteenth century (Hodne and Grytten, 1992, pp. 30–45; Sejersted, 1993). In the period between 1905 and 1920, the wider foundations of the modern economy were laid in the form of private and government initiatives for co-locating heavy industry and power plants at large waterfalls. Innovations developed by the first companies of this kind were highly knowledge-intensive and considered as technological breakthroughs. This industrial build-up was heavily supported by a large amount of foreign direct investment (FDI) from Europe.

During its occupation of Norway in the Second World War, Germany started numerous projects related to railways, energy and heavy industry, and the Norwegian government completed many of these projects after the war. The government furthermore planned an overall postwar modernization by way of rebuilding and improving infrastructure. Public programmes for regional planning and industrial development in this era included the Plan for rebuilding Northern Norway (1951) and establishment of the Regional Development Fund (1961).

Employment in mining and manufacturing industries rose until 1974. From 1974 to 1992, employment levels in industries such as iron, steel, ferro-alloys and paper manufacturing suffered most among those that competed on the international market. Machine production, however, experienced rising employment due to offshore industry supplies. The service sector, as in most other industrialized countries, experienced considerable growth. Today, the most important sectors based on natural resources extraction are, in addition to fisheries and aquaculture, the sectors centred on oil and gas extraction, which were developed from the early 1970s onwards (see Box 8.1). By the beginning of the 2000s, oil and gas extraction accounted for close to a quarter of gross domestic product (GDP) (Baygan, 2003), and in terms of all export revenues it accounted for 42.5 per cent (Figure 8.1). Oil and gas extraction has become increasingly supported by adjoining sectors which are specialized in high-technology supply ships, engineering, administration,

---

BOX 8.1 NORWEGIAN OIL AND GAS

In 1969 oil and gas potential was discovered below the North Sea. Much of the capital and technology had to be transferred from abroad, but at the same time the Norwegian government wanted to establish and maintain indigenous oil extraction companies. The government went into alliances with foreign oil companies, giving concessions for some of the oil drilling. In return, the oil companies had to locate some of their R&D in Norway, something that was crucial in order to build up a domestic knowledge base. Norwegian solutions were provided by the Norwegian sources within certain fields, such as H-3 platforms and the so-called Condeep constructions. In addition, Norwegian ship-owners strategically ordered supply ships and drilling rigs at Norwegian shipping yards. One estimate is that Norwegian oil companies with operations in the North Sea placed 46 per cent of all contracts with domestic suppliers between 1979 and 1993. By the mid-1990s approximately 11–12 per cent of the total employment in the oil and gas industry were within approximately 150 companies within the offshore engineering consulting industry. The oil extraction activities thus created new opportunities for the supplier industry and the service sector. A central debate in recent years among both scholars and policy makers has been related to the Norwegian 'oil dependency'. One view is that it is necessary to redirect development before the natural endowments are exhausted around 2050, since future wealth depends on whether or not alternative and competitive sectors are developed.

Sources: Cumbers (2000); Howie and Lipka (1993, as cited in Cumbers, 2000, p. 245); Olsen and Sejersted (1997) and Skogli (1998).

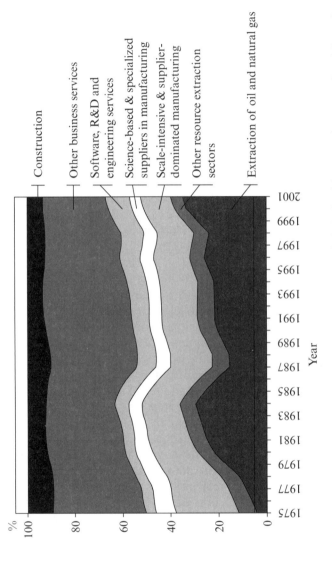

Note: Based on three-year centred averages. Other resource extracting sectors include electricity, gas and water supply; mining and quarrying (except oil and natural gas) and fisheries/aquaculture.

Source: OECD STAN database.

Figure 8.1 Value-added in the Norwegian business sector, 1975–2001

finance and information and communication technology (ICT) services. Regarding the last, the worldwide trend is that ICT services relevant to oil extraction have become increasingly important within both exploration and production (Pinder, 2001), and several specialized Norwegian ICT firms are at the international forefront (UFD, 2005a).

State ownership within business was as of the mid-1990s the third highest in Europe next to Finland and Italy (Bøhren and Ødegaard, 2003), and has historically been evident in the form of two distinct waves, where the first coincided with the postwar industrial build-up and consolidation in the 1945–70s period. Strong state involvement was seen as necessary if Norway was to become a serious actor within selected heavy industries (Wicken, 2000). In addition to direct ownership, the state also provided a very hands-on type of coordination by way of nurturing a large research institute sector servicing industry (Benner, 2003, p. 138), stimulating entrepreneurship, entering licensing agreements and ensuring privileged access to capital and special considerations in incomes policy. An additional, although passing, wave of increased state ownership later occurred and focused mainly on banking and finances due to the 1990s crisis within this sector (Huber and Stephens, 1998, p. 368).

Large firms in the core sectors are important, but that is not to say that small and medium-sized enterprises (SMEs) play a minor role. Referring to sheer numbers, there were as of 2006 merely 610 firms with more than 250 employees, whereas there were 2076 with 100–249 employees and 81 313 firms with 5–99 employees (Statistics Norway, 2007). Smaller proprietor-managed firms in agriculture, trade and manufacturing gave vital input to the Norwegian political economy during the 1800s, and many later became central within sectors such as furniture, engineering and machinery. The SMEs were (and still are) often family-owned and dependent on local financing. Their business activities were then gradually supplemented, as reviewed above, with a few larger enterprises exploiting natural resources. Thus, similar to Finland and different from Sweden, big business did not gain overwhelming societal influence (Huber and Stephens, 1998, p. 368; Sejersted, 1993).

3 INNOVATION INTENSITY

According to the Community Innovation Survey (CIS) there was a slight negative trend regarding innovation intensity in Norway during 1994–96 and during 1997–2001 (see Appendix Table A4.1). This negative development can be seen in most other countries that participated in the CIS study. Still, the overall assessment is that Norwegian firms on average are not particularly innovative. Only about 30 per cent of Norwegian firms can be classified as innovative, and the share of innovating firms remained unaltered in 1997–2001. Denmark and Sweden have a share of innovating firms that is considerably higher (approximately 40 per cent, which is in line with the average score in the survey). It is also worth mentioning that the share of Norwegian firms that have introduced products that are new to the market is quite low, which indicates that many of the innovations are diffusion-based, i.e. adoptions of innovations made by others.

One feature of innovation intensity in Norway is the difference between SMEs and large firms, with the latter being considerably more innovative (see Appendix Table A4.2). There may be several reasons for this difference. Large enterprises (LEs) often have more financial and knowledge resources, and also usually have a broader range of products and more processes than smaller firms. However, while the share of innovating Norwegian LEs is on the same level as in the other CIS countries, the Norwegian SMEs distinguish themselves from other European SMEs by having a low share of innovators.

Among the Norwegian SMEs, the average innovation expenditures were 2.3 per cent of turnover in 1997 and 1.4 per cent in 2001. But the LEs also spent less on innovation in 2001 as seen in relation to turnover. Looking at research and development (R&D) in isolation, R&D intensity declined from 1.7 per cent of turnover in 1997 to 0.9 per cent in 2001 for large firms, and from 1.4 per cent to 0.9 per cent on average. This decline may be partly explained by the general state of the market around 2000–2001, and can be seen in most other European countries as well, with the exception of Sweden. Among Norwegian SMEs there was an increase from 1997 to 2001 of firms that had introduced products new to the market. The large firms' share in this respect was more than twice that of the SMEs. Compared with Denmark, Ireland, the Netherlands and Sweden, both Norwegian SMEs and LEs have a lower share of innovating firms (see Appendix Tables A4.4 and A4.10).

Among the knowledge-intensive business services (KIBS) there was an increase in R&D intensity, simultaneous with a slight decline in innovation intensity from an average of 5.9 per cent of turnover in 1997 to 5.4 per cent in 2001. Innovation intensity among manufacturing firms declined from 2.7 per cent in 1997 to 2.1 per cent in 2001. In the same period, innovation intensity also decreased among manufacturers of basic metals and fabricated metal products.

One cause for decline in innovation intensity during the time period in question may simply have been the 'relationship between technological change and the low oil-price regime which prevailed towards the end of the 1990s' (Pinder, 2001, p. 580). In other words, many oil companies restrained their exploration and production activity due to overall low prices for oil (ibid.). Another contributing cause may have been the general downturn within high-tech investments following the 2000 crisis within ICTs and biotechnology (Grønning et al., 2006). More extensive explanations have also been offered, such as the time-lag hypothesis suggesting that Norway is an extreme case of high economic performance today being based on the high innovation input of yesterday (i.e. during the 1970s), the structural weakness hypothesis stating the accumulated effects of erroneous choices, and the hypothesis stating that the innovation indicators presently in use may actually be unable to capture the particularities of the Norwegian economy in an adequate way (Aanstad et al., 2005; Moen, 2005).[1]

Coupled to this last indicator (ir)relevancy hypothesis we may pursue still another type of explanation in line with the main argument of this chapter, which is to show that there is considerable heterogeneity both within the country and within sectors when it comes to innovation intensity. We are in this respect inspired by the technological trajectory perspective pioneered by Pavitt (1984; see also Archibugi, 2001; Srholec and Grønning, 2006), and see the oil and gas extraction sector as part of a scale-intensive trajectory and fisheries/aquaculture as part of a supplier-based trajectory. The scale-intensive trajectory has been associated with learning by doing in connection with process technology, internal non-R&D inputs, and external technology sourcing from specialized suppliers, and the supplier-based trajectory with dependence on external inputs when it comes to technology sourcing (Pavitt, 1984). Whereas both of these trajectories in themselves score relatively modestly on the various indicators for innovation and R&D intensity (see Tables 8.1 and 8.2), the neighbouring specialized supplier trajectory is innovation- and R&D-intensive, and large parts of the specialized supplier trajectory's output consists of supplies to the trajectories dominated by the primary sector. Indeed, even parts of the science-based trajectory in Norway are tightly linked to the less innovation-intensive trajectories by way of R&D output to these trajectories, e.g. vaccines and other equipment for aquaculture and high-tech equipment, as well as R&D services for oil and gas exploration. As for value-added, when extraction of oil and natural gas as well as fisheries and aquaculture are included in the analysis, the value-added proportion between different trajectories is as shown in Figure 8.1. There has been an abrupt increase of oil and gas extraction value-added up until 1984, and then again from 1987 onwards. Fisheries and aquaculture have been of more limited importance when it comes to value-added (Figure 8.1), but have been extremely important for employment (Table 8.1).

Table 8.2 shows the main findings for innovation and R&D intensity, as well as the nature of innovativeness and interaction patterns. Of special interest are oil and gas extraction as seen in relation to other sectors. One

Table 8.1 Characteristics of selected sectors in Norway, 2001

	Share in the business sector (%)				Market structure		
	Value-added	Employment	Export	Internal R&D expenditure	% of small firms	Capital intensity	Export intensity
Average	n.a.	n.a.	n.a.	n.a.	40	1 949	18
Median	n.a.	n.a.	n.a.	n.a.	41	600	13
Extraction of oil and natural gas	35.0	2.5	45.9	6.0	3	20 178	85
Scale-intensive and supplier-dominated manufacturing							
Food	3.0	5.1	3.4	2.9	28	819	16
Wood	0.7	1.5	0.3	0.4	55	583	8
Paper	0.8	0.9	2.0	1.3	9	2 289	44
Basic metals	1.4	1.4	5.8	3.2	7	2 644	53
Fabricated metal	1.0	1.8	0.5	1.1	65	368	9
Furniture	0.5	1.2	0.4	1.0	51	391	11
Science-based and specialized suppliers in manufacturing							
Chemicals	1.5	1.5	5.6	8.4	9	2 939	34
Machinery, N.e.c.	1.4	2.4	2.1	7.4	43	351	22
Computers	0.0	0.1	0.4	0.5	51	2 832	11
Electrical machinery	0.5	0.8	0.6	3.1	38	612	16
Radio, TV, phones	0.4	0.7	0.9	13.8	20	390	19
Instruments	0.4	0.8	0.5	4.1	30	188	16

Software, R&D and engineering services					
Software	2.3	3.7	0.8	14.9	55	192	6
R&D services	0.5	1.1	0.2	0.9	23	433	12
Engineering	7.2	15.0	2.9	5.7	53	120	12
Other business services						
Water transport	3.9	4.8	12.4	0.3	19	2 328	77
Other resource extracting sectors						
Fish	1.1	1.5	1.1	2.2	93	1 515	30
Electricity & water	3.5	1.6	1.3	0.7	30	12 473	20

Notes: Export intensity includes both intermediate goods as well as capital goods.

Source: OECD STAN database.

Table 8.2 Innovativeness and nature of innovation process in selected sectors in Norway, 2001

	Innovativeness			Nature of innovation process				Interaction			
	Overall innovativeness	Product innovation	Product imitation	Process vs. product innovation	Appropriability by patents	Intensity on R&D inputs	Intensity on non-R&D inputs	Interaction with suppliers	Interaction with customers	Interaction with science	Cooperation agreements
Average	41	5	9	0.6	23	5.4	1.6	21	41	5	48
Median	40	2	6	0.5	20	1.6	0.9	20	40	4	47
Extraction of oil and natural gas	43	1	3	0.4	73	1.3	0.0	23	35	14	58
Scale-intensive and supplier-dominated manufacturing											
Food	35	2	6	0.7	8	0.6	0.6	22	17	5	30
Wood	25	2	2	0.7	18	0.9	1.0	29	28	5	44
Paper	31	1	11	0.7	18	1.3	0.3	38	56	0	69
Basic metals	48	1	11	1.0	37	1.4	0.4	21	40	10	55
Fabricated metal	29	1	6	0.5	24	2.2	1.0	20	39	4	40
Furniture	42	7	12	0.5	38	2.4	1.0	29	48	7	38
Science-based and specialized suppliers in manufacturing											
Chemicals	65	2	8	0.5	56	1.7	0.4	23	46	6	68
Machinery, N.e.c.	45	8	19	0.5	46	4.7	1.2	4	46	3	46

Computers	100	9	43	0.4	50	5.1	0.0	50	100	13	75
Electrical machinery	46	4	16	0.5	20	5.4	0.8	21	63	5	35
Radio, TV, phones	61	19	26	0.4	40	27.3	2.0	18	65	5	72
Instruments	60	8	31	0.5	40	7.1	1.8	17	50	4	60
Software, R&D and engineering services											
Software	63	7	12	0.1	12	14.2	3.4	18	52	1	33
R&D services	100	53	31	0.9	33	72.1	0.9	0	67	17	50
Enginering	34	7	4	0.4	20	6.2	1.5	13	40	3	54
Other business services											
Water transport	10	4	1	0.7	14	0.5	1.0	23	32	7	55
Other sectors											
Fish	38	15	2	0.8	5	7.4	1.0	27	41	18	53
Electricity & water	28	2	2	0.9	1	0.4	0.3	18	10	7	65

Source: CIS3 data.

could have expected machinery and instruments to rank even higher due to their highly specialized nature as suppliers to natural resources and transportation sectors. However, the figures for machinery product imitation, as well as product imitation and innovation for instruments, stand out (Table 8.2, Innovativeness column).

4 ACTIVITIES THAT INFLUENCE INNOVATION

In the following we examine four groups of NSI activities: knowledge inputs to innovation, demand-side factors, provision of constituents and support services for innovating firms. It would be an exaggeration to state that all activities within the Norwegian system revolve around the resource-extracting sectors. Other sectors, including public sector services, occupy an important place within the Norwegian economy. However, the section focuses on the business sector with the aim of assessing the relative role of resource-extracting scale-intensive and supplier-based trajectories versus other trajectories when it comes to the specific types of activities.

4.1 Knowledge Inputs to Innovation

4.1.1 R&D activities
During 1991–2003 Norway had low R&D expenditure in relation to GDP compared to the OECD average. The share was 1.75 per cent in 2003, while OECD average spending was 2.24 per cent (OECD, 2005). NOK29 517 billion was spent on R&D in Norway: the business sector financed 49 per cent, the public sector financed 42 per cent, while the rest was financed by other domestic and foreign sources. Regarding expenditures, the business sector's share of R&D expenditures in Norway was 57.5 per cent in 2003, compared to the OECD average of 67.3 per cent. The public sector and the higher education sector accounted for around 42.5 per cent, which is higher than the OECD average (NFR, 2005; OECD, 2005).

When we separate the public sector (higher education and research institutes) from the private sector (business), we find that R&D expenditures are relatively different in character. The private sector spends more within ICT, offshore technology and materials technology. It also spends a large amount within 'other fields' (medicine, social sciences, human sciences and natural sciences). The public sector has by far the majority of its expenses within 'other fields', but also within 'marine R&D' (NFR, 2003, p. 35).

The research institute sector is large in Norway, and 23 per cent of all Norwegian R&D expenditures in Norway are spent there. There were, as of 2005, 63 research institutes and approximately 50 more 'other institutes with R&D'. The former accounted for 85 per cent of R&D expenditures in the institute sector and received some basic funding from the Norwegian Research Council or from the ministries. The industrial research institutes accounted for the largest group. The Foundation for Scientific and Industrial Research at the Norwegian Institute of Technology was by far the largest unit with approximately 1800 employees, and was heavily involved within research areas such as petroleum, gas and marine research (NFR, 2005, pp. 34–5).

The relatively low Norwegian expenditure on R&D in relation to GDP is closely related to the structure of the Norwegian economy, in which there is – as reviewed – a large proportion of SMEs without much tradition of research activity. And most LEs have – also as reviewed above – their basis in the exploitation of natural endowments. These are as such low- and medium–low-technology firms with relatively low levels of R&D expenditure. Thus Norway looks weak in the OECD R&D comparisons (see Appendix Table A2.1). Another important factor in considering the low provision of R&D is the large income from the oil and gas sector contributing to a high GDP, which in turn makes the Norwegian R&D effort look very low (Nås and Hauknes, 2004). By accounting for R&D intensity in the business sector adjusted for variations in industry structure based on averages from 1999 to 2002, Norway's relative ranking moves up substantially in terms of business R&D intensity, while Korea and Finland drop to a lower level (OECD, 2006, pp. 57–9). Also, measuring the provision of R&D in relation to population size from 1991 to 2003 shows that Norway has spent more than the OECD average (OECD, 2005).

Norway's share of the world's total output of scientific articles in the period 1998–2002 was nearly 0.6 per cent. Based on Appendix Figure A3.1, the average annual growth was 2.86 per cent, which is a relatively low growth rate. Publications come from two major areas, with medical sciences accounting for the largest share (52 per cent) and natural sciences the second largest (24 per cent). As for the latter, the historical changes in publication output from 1982 to 2001 show a remarkable growth within geosciences. This development is related to the oil exploration activities. Ecology and plant and animal sciences are also subjects of high specialization, reflecting the activities taking place within fisheries and aquaculture. Subjects such as engineering, physics, chemistry and computer science have a relatively low specialization level.

Researchers in Norway increasingly collaborate with foreign researchers. Collaboration between Norwegian researchers on scientific publications increased from 23 per cent in 1985 to 53 per cent in 2004. However, this tendency is apparent in all countries, reflecting the increasing internationalization of research in general (NFR, 2005, pp. 112–13).

294   *Slow growth countries*

Norway ranks low when it comes to patenting activities in international and comparative terms, with only a 0.2 per cent share of the total US Patent and Trademark Office grants (as compared to 0.3 for Denmark, 0.6 for Finland, 0.8 for the Netherlands and 0.9 for Sweden), and 67.54 European Patent Office patent applications per million inhabitants as compared to, for example, 160.63 for Denmark and 261.35 for Finland (see Appendix Table A3.1). These findings are consistent with other aspects of the low-tech nature of large parts of the Norwegian business sectors. The Norwegian Patent Office registered 69 795 patent applications in the period 1991–2002. Applications of foreign origin constituted around 80 per cent, highly represented by the USA, Germany and Sweden. There was a difference between patent applications of Norwegian and foreign origin with regard to technological fields. During 1995–2002, most of the foreign patent applications were within chemicals and pharmaceuticals. Most of the Norwegian applications were related to shipbuilding, machine tools, pumps/turbines, oil rigs, drilling techniques and instruments within medical technology (NFR, 2003, pp. 181–6).

In sum, the importance of the oil and gas extraction sector as part of a scale-intensive trajectory and fisheries/aquaculture as part of a supplier-based trajectory is reflected in Norway's revealed technological advantage (see Appendix Figure A3.2). Oil and gas have experienced a dramatic increase during the whole period, while transportation services (basically shipping) have become more competitive since 1993 after experiencing a downturn during 1980–86. Medical electronics have also very recently become more competitive. However, other sectors related to natural resources such as metals, wood and paper have experienced a decrease in terms of technological advantage since 1980.

4.1.2 Competence building

As of 2004, the share of 20–24-year-olds having enrolled in higher education was over 95 per cent, compared to the EU-15 average of 76 per cent. In 2004 there were 209 000 students in Norway within higher education. In 2002, 31 per cent of the population between 25 and 64 years in Norway had higher education,[2] which was nearly the same as in Sweden, Denmark and Finland, but much higher than the OECD average (23 per cent). These numbers indicate that Norway has an advantage compared to the OECD average in terms of competence in the labour force. However, looking at the numbers of 20–29-year-olds with science and technology education in general, and also the number of people with PhDs within these fields, Norway ranks relatively low compared to the EU-15 average, as well as to Sweden, Ireland and Denmark (NFR, 2005, pp. 230–31). The quality of education within mathematics has also been questioned: Norway ranks statistically below the OECD average when it comes to mathematics performance (OECD, 2004b). This has been taken seriously by the government, which in 2005 started to consider measures for increasing recruitment to science and mathematics studies (Aanstad et al., 2005).

The size, importance and the technical complexity of the Norwegian oil and gas system is reflected in the data on people with higher education. As of 2003, the largest share of people with higher education in the labour force was found within oil, gas and mining, where 16 per cent of all employees had a higher education, as compared to less than 5 per cent in manufacturing. It is also within oil, gas and mining that we find the highest share of people educated within technology and natural sciences (NFR, 2005, pp. 83–4).

4.2 Demand-side Factors

Three types of activities directly or indirectly aimed at increasing demand and creating new markets are of special interest. First, publicly owned organizations have used procurement to stimulate innovation. Second, businesses have been active as front-runners in the implementation of ICTs in the case of both banking and oil and gas. Third, Norwegian firms are globally at the forefront when it comes to demand-related quality assurance regarding shipbuilding and oil platforms, and energy production in general.

In 2003 public procurement accounted for 15 per cent of Norwegian GDP (OECD, 2004a). A procurement programme established in 1986 aimed at stimulating the development of new products or processes for which a government department has a requirement. The aim was principally to speed up the purchase and implementation of new products and services in government ministries and give industry a 'pilot' customer and reference (Remøe et al., 2004). During the 1990s public procurement policy had been directed towards achieving the greatest possible efficiency and lowest prices, unlike in Sweden, where public procurement had been used more actively to stimulate technology development. During the second half of the 1990s there was an increasing awareness of public procurement as a potential means of increasing innovation, providing customer references, setting high requirements and generally professionalizing firms (NHD, 1997). According to an evaluation report, the existing guidelines on procurement still concentrate on efficiency, whereas neither innovation nor stimulation of markets is mentioned (NHD, 2004, p. 50).

Formerly state-owned organizations that have been privatized engage actively in procurement, which is stimulating innovation. Perhaps the best-known example is Statoil's role in the development of the oil industry (Rothwell, 1994). Statoil deliberately created a local market for engineering suppliers by engaging with them in innovative collaborations. The legacy of this policy, combined with the goodwill agreements with foreign firms, is a robust engineering sector with sufficient technological capabilities to win global contracts. On a similar theme, the DEMO 2000 programme was started in order to remove market barriers and improve competition for suppliers and services to oil producers. It has aimed at trying out new technology in pilot projects involving producers, suppliers and research institutes (Hansen et al., 2005). Also in 2001, the OG21 project was initiated by government and business with the aim of improving research, demonstration and commercialization of new technology for use in the oil-related businesses.

When it comes to other businesses, a group of competing banks, software suppliers and the retail industry collaborated during the 1980s in an attempt to stimulate use of electronic payments systems. This initiative gave Norway a standardized electronic platform on which to develop new payment products and Internet banking, and in 2002 Norway had the highest worldwide use of electronic debit cards for consumer purchases (Bank of Norway, 2003). In 2002 the e-Norway programme was initiated by the Ministry of Trade and Industry with aims including 'creating value through enhanced innovation and competitiveness in Norwegian industry' (NHD, 2003a, p. 6). A programme entitled BIT has been aimed at improving the profitability and competitive ability of firms by developing common ICT solutions adapted to specific sectors of industry, and has received favourable evaluation results (Kallerud et al., 2006, p. 43).

As for quality assurance, the Norwegian organization Det Norske Veritas is one of the main organizations setting quality standards for, for example, oil-rig building and shipping industry worldwide.

4.3 Provision of Constituents

4.3.1 Provision of organizations

This section deals with provision of organizations in two different senses. The first is the establishment of new business organizations and the second the provision of innovation policy organizations. Norway performs particularly well when it comes to the creation of entirely new business organizations. Birth rates are, according to Eurostat (2007), higher than in, for example, Sweden, Finland, Denmark and the Netherlands. Norway also holds a strong potential in this area, since the country consistently ranks very high on the ratio of the active population engaged in start-ups. In addition, Norway ranks well ahead of Sweden, Denmark and Finland when it comes to start-ups representing something new in the domestic market (Kolvereid and Alsos, 2003, p. 13).

Sweden excels, however, when it comes to survival rate, at least during the 2000–2001 period. This phenomenon is obviously connected to the differences in death rates, where Sweden is clearly lower, with 5.55 percentage points compared to 8.30 in Norway (Eurostat, 2007).

The figures above concern all kinds of firms. A survey by Matson (2005) gives an indication of Norway's performance when it comes to the creation and bankruptcy rates of new technology-based firms (NTBFs) during 1990–2001. Within his sample of 3055 firms, computer and R&D services had 10.7 and 13.0 per cent bankruptcy rates respectively, compared to between *circa* 20 and 30 per cent for mining, manufacturing, construction and transportation and 9.4 per cent for oil and gas exploitation (Matson, 2005, p. 70). According to another study (Nås et al., 2003a; 2003b) only 11.2 per cent of all spin-off firms in Norway were spin-offs from high-technology industries. This figure compares to 17.5 per cent for Sweden and Denmark, and 17.3 per cent for Finland. Not surprisingly, Norway holds (together with Denmark) a considerably lower share of spin-offs from high-technology manufacturing, mainly due to the overall higher presence of such industry in Sweden and Finland.

Regarding agencies and organizations for innovation policy, the development has been towards consolidation in larger units. Until 1993, Norway had a funding system spread across several research councils according to broad disciplinary bases. These research councils were then merged into one, the Research Council of Norway, which was further reorganized in 2004, based on the recommendations of a large-scale evaluation exercise (Kuhlman and Arnold, 2001). Simultaneously, the trend has been from smaller research programmes and funding of individual research projects to formation of larger and more comprehensive research programmes. In addition, Norwegian business may participate in regional development or start-up support programmes. In 2004 these programmes were merged into Innovation Norway, together with the Trade Council of Norway and two other support organizations.

4.3.2 Networking, interactive learning and knowledge integration

Collaboration between firms is important for innovation in the Norwegian NSI. Forty-one per cent of all the surveyed firms collaborated actively with others in the period 1999–2001 (cf. CIS3). Suppliers were the most frequent partners (around 70 per cent of the surveyed firms). However, 39 per cent of the firms reported that customers were the most important actors in collaboration, while competitors were the least important. Twenty-eight per cent of the firms reported that they collaborated with universities and state/scientific colleges, and 33 per cent collaborated with public or private research institutes. The level of innovation cooperation was thus very high in Norway compared to the EU-15 average.[3] Innovation cooperation with national partners in Norway was considerably higher than the EU-15 average. In Norway, suppliers, customers and consultants were the most important partners. Thirty-three per cent of the firms were engaged in innovation cooperation as compared to 23 per cent in the EU. Regarding international collaboration and the nationality of the collaborating partner firms, 29 and 34 per cent collaborated with firms within Scandinavia and within EU/EFTA respectively. Most of the Norwegian firms that collaborated with partners in the USA were large firms with activities related to oil and gas, shipping and production of chemicals. Smaller firms within other sectors such as aquaculture and furniture did not report any such collaboration (Statistics Norway, 2001). It should however be noted that collaboration between firms on the one hand, and universities, scientific colleges and research institutes on the other, had relatively low importance for innovation processes. Correspondingly, interactions within the firm were the most important source of information during innovation processes (ibid.).

When it comes to R&D collaboration between firms in Norway, approximately 50 per cent of all firms with R&D activities had a formal relation of R&D collaboration with another firm in 2003. This was especially apparent in aquaculture, extraction of oil and gas, metal products, communication equipment and medical instruments (NFR, 2005, pp. 104–5).

Networking programmes set up by national or regional authorities have included the BUNT (Business Development Using New Technology) programme in the early 1990s, targeting firms' abilities to find and use new technology developed in other companies or research institutions, a follow-up FRAM programme (est. 1992) supporting basic learning in SMEs, and a programme for mobilization of R&D-related innovation in SMEs (Aslesen, 2004, p. 31). Looking at oil and gas in particular, the Norwegian Oil and Gas Partners network was established jointly by the business community and government in 1997, and had 160 partner companies as of 2004 (INTSOK, 2005).

4.3.3 Provision of institutions

This section contains a review of institutional arrangements related to the activities reviewed elsewhere in this Section 4. The review contains only the most significant developments in recent years, but some issues will be pursued in further detail in Section 7 in connection with the overview of innovation policies.

In connection with knowledge inputs to innovation, such as provision of R&D (see Section 4.1.1), the institutional framework has undergone a series of changes. In 1993, the goodwill R&D agreements with foreign oil companies were abolished, making it a necessity to sustain this kind of research on a more competitive basis. In 2002, amendments to the Act on Universities and Colleges (UFD, 2002) gave universities and colleges formal responsibility for assisting in the process of making research results available for society. In more concrete terms, the traditional 'professorial exemption rule' in the act was dissolved. Employees at higher education organizations traditionally had rights connected with the inventions and discoveries that were made, and the change in this respect was part of an intended shift towards higher participation by higher education organizations in the commercialization process by way of being entitled to the intellectual property rights (Askevold et al., 2003, p. 13). The organizations have in this respect also been faced with new tasks and responsibilities, and as a consequence several technology transfer units have been established.

The White Paper 'Commitment to Research' (UFD, 2005a) states that Norway's future goals include raising the number of researchers per 1000 employees. In concrete terms the declining number of new recruits to science and technology subjects is to be countered first and foremost by improving the conditions for science teachers at secondary and tertiary levels. Furthermore, recruitment to researcher posts is to be improved by giving special premiums for science and technology doctoral candidates (UFD, 2005b, p. 30), and making the postgraduate researcher career opportunities and working conditions in general more attractive through improved working conditions (UFD, 2005a, p. 7).

Furthermore, and also related to the institutional framework of competence building (see Section 4.1.2), a new Act on Universities and Colleges passed in 2002 (UFD, 2002) included changes in the control and management structures of universities and colleges. In the ensuing debate many have seen this as a threat to academic freedom and internal democracy, and substantial freedom when it comes to deciding governance structures has been reinstated (Aanstad et al., 2005).

Regarding demand-side factors (cf. Section 4.2), competition legislation and its agency were coordinated into a new competition authority in 1994 in order to enforce both national and European competition law. Norway is as of 2007 not an EU member, but adheres to EU directives through signing the Agreement on the European Economic Area. The OECD has, however, pointed to prevailing problems with the Norwegian market and recommends that greater efforts should be made to improve competition even further (OECD, 2004a).

One of the main traits of the Norwegian economy is, apart from the Norwegian state having ownership interests in parts of industry, that the Norwegian firms are governed by institutionalized employee representation on the board. There are no signs of deinstitutionalization on this point. Regarding corporate governance, there have been changes in the wake of scandals such as Enron abroad as well as within Norway in the early 2000s. Main actors in the financial market agreed on a code for corporate governance in 2005. In parallel, another institution of domestic origin and based on long-standing egalitarian norms is the proposed minimum ratios of females on corporate boards. The demand for qualified females has incidentally led to the establishment of databases listing women eligible for board service.

One unresolved issue of perhaps even greater implications for organizations and their innovativeness is the overall framework for corporate ownership. As mentioned in Section 2, ownership patterns of Norwegian corporations differ from those in the rest of Europe, starting with high state ownership, much as in Finland and Italy. Figures from a 1993–97 study of listed companies show that the public owns 14 per cent as compared to 8 per cent in the other European countries in the study. Overseas owners constitute 32 per cent as compared to 21 per cent, owners such as pension funds and so on constitute 21 per cent as compared to 25 per cent, industrial owners 25 per cent as compared to 18 per cent, and personal ownership is restricted to a mere 8 per cent as compared to 28 per cent (Bøhren and Ødegaard, 2003). This low level of direct ownership is rather unusual, and has been explained by referring to more long-term and normatively based institutions such as taxation aimed at equalization of income and assets. Gradual change in this respect started, however, at the end of the 1990s with periodic tax alleviations for savings in bonds (ibid.). There have also been examples in recent years of firms and business persons first successful in, for example, mechanical engineering, aquaculture and retail sales later investing in other sectors than their own.

Regarding institutions related to support services for innovating firms (Section 4.4), we find the most dramatic developments within financing. The surpluses from oil revenues have been first and foremost accumulated in the form of the Petroleum Fund established in 1990 and starting operations in 1996. The fund is an integral part of Norwegian fiscal policy and allows Norway to finance a budget deficit which should, over time, not surpass the 'real return' on the Petroleum Fund, estimated at 4 per cent of the fund value. The objective is thus also to secure present oil and gas revenues for future generations (OECD, 2004a, pp. 11 and 21). Conditions are attached to this capital stipulating that it has to be invested abroad in order to avoid artificial overstimulation of the domestic economy. This policy has been adjusted in recent years with the establishment of a separate Fund for Research and Innovation (FRI) in 1999 with a capital of NOK3 billion, as well as funds for seed capital. The FRI capital base has been increased, with the 2005 White Paper mentioned above (UFD, 2005a), to NOK50 billion starting from 2006 (Aanstad et al., 2005). Needless to say, increased funding may in turn have implications for the scale of R&D provision already reviewed above. One may speculate that large-scale research programmes catering mainly to the science-based trajectory by focusing on biotechnology (2002–11) and nanotechnology (2002–8 with planned extension through 2011) might not have been possible without these means.

Also related to financing, albeit on a more detailed level, is the institutionalization of R&D support in the form of tax deductions to private enterprises with sales less than NOK80 million and fewer than 100 employees. This was instituted in January 2002 and became a popular opportunity among SMEs. The programme was based on predecessors in Austria and Spain, and nearly 60 per cent of the applicants in the 2004–6 period have been firms with fewer than ten employees (Kallerud et al., 2006, pp. 35–7). From 2007 onwards the criteria have become somewhat stricter, including a maximum level of deductions when it comes to hourly expenses as well as annual hours worked per person (Munch, 2007). Nevertheless the character of the policy instrument as being reserved for rather small firms and as working more on an egalitarian rather than competitive basis remains intact.

4.4 Support Services for Innovating Firms

4.4.1 Incubating activities

Incubation activities are based on experiences gained in Europe and the USA, although the recommended size of the average unit has been modified in order to fit with Norwegian conditions. The most important public actor within incubation is the Industrial Development Corporation of Norway (SIVA) (est. 1968), which was, as of 2002, the joint owner of 10 research parks, 15 knowledge parks, 34 business parks, 8 R&D companies, 18 singular incubators and 12 seed capital/venture investment companies (SIVA, 2002). There are in addition 14 other science or research parks in Norway, a majority of which were established in the late 1990s and have mixed public and private ownership. These are in general located adjacent to one of the universities or colleges.

The creation and maintenance of incubators is relatively recent in Norway. Evaluations cite examples of successes, but point out that the business aims are vague and not directed towards any particular market (Havnes, 2003), and that research output in the form of, for example, patenting so far has been limited (Askevold et al., 2003). From initially being mainly property developers, the incubator hosts now see themselves more as providing various services for innovators. One aspect which may differ from other countries is that the publicly funded incubators usually have a dual aim of creating economic growth through innovation while at the same time promoting regional development (SIVA, 2004). Thus although research park and incubation activities are in a sense by definition high-tech, and hence enhance a science-based trajectory, this dual function emphasizes the much broader aim of the initiative in Norway.

4.4.2 Financing

In this section we briefly describe in sequence the provision of both venture and seed capital, and the activities of private investors. Note that we focus on financing of innovation in start-up firms, and not on financing of innovation in larger, existing firms.

It has been pointed out that there is no real lack of capital in the Norwegian market (NHD, 2001; FD, 2004). But venture capital (VC) raised in 2001 was an amount equivalent to 0.152 per cent of GDP, which is lower than in most other European countries (EVCA, 2002). The composition of the investor population is mainly public funds and a few large private investors, and in this respect it differs from European VC. Norwegian investors are investing well below the European average in the seed capital phase, and during 2000–2002 almost 60 per cent has been invested in the buyout stage (NVCA, 2001; EVCA, 2002). During the same period less than 30 per cent of total VC investments were in seed and start-up firms and projects (Baygan, 2003, p. 8). An alliance of public and private partners was created in 1997 aimed at providing seed capital for new businesses. However, in practice the capital raised was invested in later stages than is typical for seed capital (Sydnes and Halvorsen, 2003).

Innovation Norway is responsible for a system of grants and loans for developing ideas, and for the creation and development of new firms. Innovation Norway and its predecessors had a fairly constant level of resources available for new start-ups in the period from around 1980 and up until 1990. In the period 1994–2003 there was a general trend of reductions in funding (from NOK4 billion to NOK2 billion). This reduction was particularly dramatic in urban areas, where funding fell by over 60 per cent (Innovation Norway, 2004). Reasons for the reductions aimed at entrepreneurialism are not entirely clear, but may have been related to the general decentralization of regional funding to regional and municipal councils (KRD, 2003), to increases in R&D funding (Aanstad et al., 2005, p. 18), or to the lack of coherence in implementing innovation policies (Remøe et al., 2004, p. 92). This reduction may contribute to the funding problems experienced by entrepreneurs, particularly urban NTBFs.

According to CIS3 data, 15 per cent of innovative companies experience problems regarding financing of innovations, an increase from 12 per cent in 1997. The only factor that more firms (17 per cent) rate as an obstacle to innovation is cost. According to CIS data, under 1 per cent of the firms in the 2001 survey used VC to finance innovation, and the vast majority of innovation was financed using internal resources. Borch et al. (2002) also show that over 60 per cent of the entrepreneurs involved in starting up technology-based companies saw obtaining finance as their greatest challenge. It is, however, not clear if this problem is due to lack of finance or to lack of communication between investors and entrepreneurs.

4.4.3 Provision of consultancy services

Consultancy and advisory services for innovating firms are available in Norway from both public and private organizations. In recent years there has been an expansion in the number of KIBS, while the publicly funded services have increased their emphasis on the support of SMEs, particularly in rural areas.

If we first look at the development of KIBS, we see a marked growth in both number of firms and number of employees. Stambøl (2005) states that around 6 per cent of the Norwegian workforce were employed in KIBS in 1994.[4] By 1999 this figure had risen to around 12 per cent. This is similar to the growth experienced in Sweden in the same period. As of the mid-1990s more than one-tenth of total employment related directly to oil and gas extraction worked in smaller firms providing knowledge-intensive consultancy services to the resource-extraction firms themselves (Skogli, 1998).

It is difficult to compare exactly with other countries, but in the period between 1998 and 2001, Norway had a greater increase in the number of firms in the service sector overall than the other Scandinavian countries. During the 1990s there was an increase in the education levels of those employed in KIBS, and they now have on average 1.5 more years of education than the average for the Norwegian workforce (Stambøl, 2005, p. 93). It should be mentioned that there has been an increase in the externalization, or outsourcing, of certain large firm business functions such as ICT services, and this may be partly responsible for the growth in KIBS.

CIS3 data indicate that the KIBS sector is the most innovative in Norway, since 48.9 per cent of the firms are innovating and 7.8 per cent of turnover is from new-to-the-market products or services. KIBS are also active in assimilating products developed by others, with 46.3 per cent introducing products new to the firm accounting for 23.2 per cent of turnover.

The major public actors are Innovation Norway, providing local consultancy and advisory services aimed at SMEs especially in rural areas, SIVA, providing similar services for firms in their incubators, and business parks and the semi-public research institutes (see Sections 4.1.1 and 4.4.1 for more details). Additionally there are private foundations receiving public support, but having more specific aims than public consultancy services. The Technological Institute offers technological expertise, advisory services, training, technology transfer programmes as well as laboratory testing and certifying services to SMEs, and the advisory institute in Northern Norway offers advice and contract research for that specific region. The Public Advisory Service for Inventors is a public agency offering advice and scholarships for inventors. The office supports patent applications and the building of prototypes (Aanstad et al., 2005).

Studies of knowledge-intensive services in the software and aquaculture industries have shown that although innovative firms use KIBS actively, the internal activities play a more important role in innovation (Broch and Isaksen, 2004; Aslesen, 2004). This finding matches CIS3 data on sources of innovation, where 46 per cent of innovative firms cite internal sources as the most important for innovation.

4.5 Summary of the Main Activities Influencing Innovation

The provision of R&D is without doubt an important activity within the Norwegian system. SMEs are not particularly active in this regard, which comes as no surprise. A particular feature of R&D provision is the close relationships between the large firms within resource extraction and mechanical engineering firms on the one hand and the semi-public research institutes on the other hand. This particular type of relationship also explains the relatively low participation of KIBS.

Regarding knowledge inputs to innovation, the importance of the natural-resource-extraction-related trajectories is reflected in Norway's revealed technological advantage. At the same time, there are more recent and increasingly significant features related to other 'challenging' trajectories, such as the science-based trajectory. The latter is highly visible through large sums of R&D expenditure, as well as high competence levels. The leading firms still need science and technology graduates, as do the specialized supplying research institutes and firms. Seen from the needs of the dominant trajectories, the low number of graduates is alarming. This stagnation in competence development may also pose a problem for the successful development of alternative strongholds within the science-based trajectory.

The distribution of activities in the case of demand aspects has always been heavily tilted towards the segment of the economy not directly related to natural resources extraction. This is most visible in the form of programmes for ICT diffusion. However, when it comes to quality requirement organizations, Norway has been and continues to be prominent by supplying some of the leading organizations for certification of equipment within natural-resources-related sectors.

The Norwegian system shows dynamism in terms of organization turnover as well as support services. While organizational longevity could ideally be at least at the level of Sweden's, the recruitment to entrepreneurial activities seems to be adequate. The level of support services in terms of incubators, financing and consulting seems to be high. On an institutional level the review shows adjustment to supranational trends (such as, e.g., the change of university-related legislation and implementation of corporate governance codes) blended with domestic initiatives (such as continued strong employee representation and gender-based corporate board quotas).

5 CONSEQUENCES OF INNOVATION

The Norwegian NSI, as repeatedly stated, scores relatively low on innovation and R&D intensity in the CIS data (see Sections 3 and 4.1.1), while relative GDP and productivity growth are high. Economic growth in Norway during 1980–2003 has been higher than the average for the OECD countries (see Appendix Table A2.3). In the 1992–2002 period, average employment growth was twice as high as within the EU, with a simultaneous reduction in unemployment. Norway has had a steady, positive development in productivity during 1980–2001 (see Appendix Table A2.2 for the 1995–2003 period). This positive development also goes for labour productivity/total employment, labour productivity/hours worked, and total factor productivity. Moreover, Norway ranks as number one on the Human Development Index (see Appendix Table A1.2) and relatively high on literacy indicators (see Appendix Table A1.3).

One explanation is that this prosperity is predominantly due to revenues from the energy sector, in particular oil and gas, but also from hydroelectricity in earlier times. For example, looking at the structural non-oil budget balance during the 1980–2004 period, there are 20 years of negative figures. In 1992, the balance was down to nearly minus 10 per cent (OECD, 2004a, p. 27). It would thus be safe to conclude that the fluctuations of Norwegian GDP in recent decades are closely related to fluctuations in energy production. Benner (2003) claims that Norway's lack of innovation intensity is due to a lack of incentives for innovation. It is argued that the country 'seems stuck in its traditional growth paradigm, which at the moment is more than sufficient to support a full employment labour market and a universal social policy regime' (ibid., p. 140).

This can, however, only constitute part of the explanation, since compared to most other 'oil-dependent' economies the Norwegian economy is far more heterogeneous, includes a large public sector and sustains a significantly higher standard of living for the general population. Complementary (and also partial) explanations for the puzzle are thus, first, that while CIS takes into account many types of firms and sectors, the highly profitable oil and gas firms constitute only a small fraction of these firms. In addition, oil and gas drilling is not represented statistically as manufacturing, but rather as a primary industry together with fisheries and aquaculture. Second, while Norwegian innovation output on average is low due to, among other things, the low-tech character of many Norwegian firms, there are indeed competitive and innovative niches such as supply services to oil drilling, medical technology and environmental technology.

With these reservations in mind, we sum up some overall observations. According to the Norwegian part of CIS3, the positive effects of the firms' innovative activities in Norway are related to improved products and services. Sixteen per cent reported that they increased their market share due to innovations. Twenty-seven per cent of firms reported that in the 1999–2001 period the most important effect of innovation was better-quality products. This was particularly common among producers of ICT equipment and other electronic goods. Twenty-three per cent of the firms reported a broader range of goods and services as being the most important effect, and this was particularly noticeable for those producing agricultural and forestry machinery. Sixteen per cent of the firms saw increased market share as the most important consequence. Within aquaculture, oil and gas extraction, and also within construction, the most important innovation effects were improved flexibility in production, increased production capacity and reductions in labour costs. Process improvements were the most important in the service sector as well. There are no noticeable differences between the effects on small or large firms, but a larger number of the small firms do stress the importance of improved production flexibility. However, this was the opposite in the case of the service sector. Ørstavik (2000) followed up CIS2, and the relative significance of successful innovations was related to increased competence and technology more than to increased turnover. This was in particular a dominant trend within ICT. With regard to turnover, around 50 per cent of the firms reported that innovations had a positive effect (ibid.).

As mentioned in Section 3, various hypotheses have been proposed to explain the low innovation intensity of the Norwegian firms (Aanstad et al., 2005; Moen, 2005). One is that there is a long time lag between innovation indicators and economic performance. Strong macroeconomic performance in the present could therefore partly be explained by choices and activities of the 1970s and 1980s when Norway invested strongly in the oil- and gas-related efforts as well as within the marine sciences (Aanstad et al., 2005, p. 20).

A second type of explanation is that the Norwegian economy suffers from structural weaknesses resulting from erroneous choices and strategies within both business and government. According to this view there is currently an aversion to, or perhaps even inability to, foster high-tech innovation intensive sectors (Moen, 2005, p. 7). This practice differs significantly from other countries such as Ireland, Finland and Singapore, and Norway should emulate these countries and channel more of its surplus resources into high-tech efforts in order to become a fully fledged member of the knowledge economy (Moen, 2002).

The third type of explanation criticizes this 'high-tech bias' (Aanstad et al., 2005, p. 21) in modern policies consisting of 'the mind-set seeing high-tech sectors as the future focus and loci of the economic activity in the advanced economies' (ibid.). This view is to a great extent supported by our study. It questions the relevance of benchmarking the Norwegian experience against such a biased perspective, and rather suggests that the innovation indicators presently in use may be unable to capture the particularities of the Norwegian economy. Norway is specialized in the low-tech industrial range, and has little activity in the high-tech sectors except for in some niches (see Appendix Table A2.1). There is, as reviewed above, a large number of small companies that invest little in R&D. Manufacturing concentrates, also as reviewed above, much of its efforts on process innovations, but 'process innovation may be equally or even more profitable in oil and gas, metals etc. compared to new to firm or new to market products' (Aanstad et al., 2005, p. 20). It could thus be argued that the indicators such as those used within the European Innovation Scoreboard (European Commission, 2005) provide a misleading picture of actual innovation performance and dynamism.

6 GLOBALIZATION

Norway's degree of openness in terms of import and export ratios has been steadily declining since the 1980s. As for 2000, the country's position is number 13 on the list, and this compares to Ireland as number 1, Sweden as number 6, Finland as number 7, and Denmark as number 8 (see Appendix Table A2.6). As for share of high-technology products within exports, Norway does not score particularly high either. The 1999 level of 3.91 per cent is well below Finland, Sweden and Denmark (see Appendix Table A2.6). Again it could be noted that these figures are special in Norway's case, since they refer to the average and not to particular niches such as oil drilling equipment or aquaculture equipment, where we would expect considerably higher figures.

There has been a gradual increase in the amount of FDI in the Norwegian stock market since the 1980s, and as of 2003 it accounts for over 20 per cent. In comparative terms, however, this figure is far below Sweden (47.5 per cent), Denmark (36.1 per cent) and Finland (28.6 per cent). It could however be added that Norway ranked as number 2 on UNCTAD's Inward FDI Potential Index, 2000–2002 (p. 15) after the USA (number 1), but before, for example, Singapore (4), Ireland (7), Sweden (10), the Netherlands (11), Hong Kong (12), Finland (13), Korea (18), Denmark (19) and Taiwan (21) (UNCTAD, 2004, p. 15). Value-added to Norwegian manufacturing industry by foreign interests increased from under 10 per cent in 1991 to over 25 per cent in 2000. In 1999, the value-added by these foreign affiliates was spread over several sectors, with the main ones being pharmaceuticals (89 per cent), electrical machinery (47 per cent), chemical products (38 per cent) and petroleum products (32 per cent).

As for outward FDI stocks as a percentage of GDP, the increase has been considerably steeper, whereas the comparative situation is that Norway, with its 18.4 per cent as of 2003, is far behind Sweden (62.7 per cent), Finland (42.4 per cent) and Denmark (36.6 per cent) (see Appendix Table A2.6). Here one could add that the Petroleum Fund instituted in 1990 and starting its overseas investment activities in 1996 (see Section 4.3.3) is bound to contribute to changes of these figures and rankings provided the institutional framework for the fund remain.

Regarding knowledge inputs to innovation, Norway ranks very high on outward student mobility. In 2001, 6.9 per cent of the students studied at universities outside Norway. Foreign students studying in Norway have increased from around 3 per cent to over 4 per cent during the 1998–2001 period. Present policies and immigration rules, however, are not designed to encourage these students to remain in Norway (KUF, 2000). Graversen et al. (2003) suggest that the main flow when it comes to migration among the Scandinavian countries is not knowledge workers, but rather trades people, and that most of the migration between the Scandinavian countries is short-term.

Regarding support services for innovating firms, more specifically financing, there has been an increase in the number of transnational syndicates in the private equity market, mostly resulting in Norwegian capital being invested abroad. VC raised from foreign sources amounted to 4 per cent in 2001, 6 per cent in 2002 and 26 per cent in 2003 (EVCA, 2003), whereas the comparable figure for most other European countries was approximately 50 per cent in the 1997–2001 period (Baygan, 2003, p. 11).

7 STRENGTHS AND WEAKNESSES OF THE SYSTEM AND INNOVATION POLICIES

7.1 Strengths and Weaknesses

Norway's high competence level and industrial specialization pattern are predominantly related to resource extraction and to transportation-related services (mainly shipping). It has recently been argued that the NSI heavily depends on these trajectories, and is being 'locked into' their needs (Narula, 2002). The trend within scientific specialization also reflects such a view. There is a correlation between specialization in some scientific fields and the country's industrial strengths. Furthermore, most of Norwegian patenting is related to activities taking place in sectors such as shipbuilding, machine tools, pumps/turbines, oil rigs and drilling techniques. Norway also shows a competitive publication advantage in geosciences related to oil exploration. These are technologies applied in sectors characterized as low- and medium–low-technology fields, which have been dominating the structure of the Norwegian manufacturing industry for a long period of time. Here, Norway seems to do very well.

Strengths are also found for example in relatively high numbers of start-up firms and entrepreneurship, high enrolment and graduation rates within the education system, a developed banking sector, as well as a regional system of providing advice and grants to entrepreneurs. Both public and private actors contribute to seed capital funds, and there is a growing number of increasingly professional venture capitalists. Public and private actors cooperate in providing incubator facilities with good regional links and contact with most of the academic environments. Finally, the level of innovation collaboration is very high in Norway.[5]

As for weaknesses, we have found a slight negative trend in the propensity to innovate during 1994–97 as well as during 1998–2001. Norway also shows a relatively low innovation intensity and R&D intensity compared to the other Nordic countries. However, we know that in relation to population size, Norway's provision of R&D is close to the OECD average. A further reason for the low propensity to innovate is the existence of many SMEs without much tradition of R&D. In addition, some large firms are within low- and low–medium-tech industries.

Related to high-technology fields, we have seen that the country already performs low in physics and computer science publication output. In higher education, a low performance in mathematics threatens the country's ability to educate sufficient numbers of scientists and engineers.

Other weaknesses include the investment climate and structural rigidity. Investments in innovation (both VC and seed funding) are concentrated around the expansion phase of new firm developments, and there is limited availability of risk capital, particularly when it comes to larger projects in urban areas. This pattern of investment has contributed to a situation where there is only a small number of new high growth firms and few NTBFs. The Norwegian incubators are very small by European standards, and many are only now moving their focus from property development to innovation. At this stage, many of the incubators still have vague aims and are not directed towards particular sectors. In comparison with other countries, little has been done in the area of competitive incentives in order to encourage investment or to facilitate the early-phase development of new firms.

The prominent position of the semi-public research institutes is of an ambiguous nature. On the one hand they may serve pragmatic buffer and systemic 'lubrication' functions (Nerdrum and Gulbrandsen, 2006) and constitute a strength, while they may at the same time serve as a disincentive to firms developing competitive in-house or firm-to-firm collaborative R&D activities.

7.2 Summary and Evaluation of the Innovation Policy Pursued

Wicken (2000) and Remøe et al. (2004, pp. 10–20) suggest that Norwegian innovation policies have evolved from a '1st generation innovation policy' (Lengrand et al., 2002) during *circa* 1946 to the 1970s, where innovation was conceived of as a linear progression from basic science to applied technology, through two distinct stages of a '2nd generation policy' during the 1980s and 1990s focused on the interactive aspects of the innovation process.

The first stage of the second-generation era (1978–92) had a strong technology focus where new technology came to be recognized as the most important factor in creating economic growth. Targeted areas were ICT, biotechnology, materials technology, aquaculture and offshore technology. The second stage of the second-generation era must be seen in relation to the economic context of the very end of the 1980s and the beginning of the 1990s, when Norway had problems with increasing unemployment, lower oil prices, stagnation within the farmed salmon market and problems with the development of high-technology industries. Accordingly, priority was no longer to be given to selected sectors. Although there were some main areas of focus (marine research, ICT, medicine and health, and energy and environment), the new policy aimed at improving the general performance in all firms with innovation potential. The numerous R&D and innovation promotion organizations were collected into two major organizations, at the same time as more funds and grants for development of the industries became available. In a White Paper on R&D policy (KUF, 1998), the central objective was to reach the average OECD level (3 per cent) of R&D expenditures in relation to GDP before 2005. This goal was to be reached by means of a considerable escalation in public expenditure, partly through the establishment of a research and innovation fund. At the same time, the private sector was to be stimulated to invest in R&D.

During the early 2000s, there have been attempts at implementing a broader '3rd generation policy' (Remøe et al., 2004) in the form of an outline for a 'holistic innovation policy' (NHD, 2003b), followed up by a new R&D policy (UFD, 2005a) building directly on the initiative of several years before (KUF, 1998). The priority areas of the 'holistic innovation policy' are the development and maintenance of educational organizations that produce and disseminate relevant knowledge on a high international level, better competence in natural sciences and mathematics, strengthening of lifelong learning and the capacity of firms to apply knowledge in practice, and increase of knowledge flows between industry and milieus of knowledge and competence – regionally, nationally and internationally (Remøe et al., 2004, p. 68). Furthermore, 'research, development, and commercialization' included aims such as working to get Norway to the OECD gross expenditure on R&D average by 2005, improving quality and internationalization in Norwegian research, stimulating increased research in the private sector through a tax deduction scheme, encouraging commercialization of results of research and promoting better interaction between knowledge organizations and private industrial actors. In concrete and post-White Paper terms, a programme for centres for research-based innovation started in 2006, supplementing the centres for excellence and centres of expertise programmes that had started in 2003. A programme for 'industrial PhDs' based on a Danish predecessor was also suggested and is as of 2006 in the process of being implemented (Kallerud et al., 2006, p. 29).[6]

However, and as Remøe et al. (2004, p. 20) point out, it is questionable whether the holistic policy constitutes a departure from previous policies. It is a third-generation policy mainly in the sense that innovation concerns now permeate other policy areas, while the policy is very vague when it comes to more concrete measures regarding any changes of framework conditions and infrastructure. The relationship between the political system and demand aspects, which is another key point of a third-generation policy (Arnold, 2004), are absent from the policy proposition.

The 2005 'Commitment to Research' (UFD, 2005a) is in essence centred on the goal of achieving 3 per cent R&D expenditure as seen in relation to GDP, with the new target date set to 2010. Basic priorities include internationalization, increased focus on natural sciences when it comes to basic research, and research-based innovation within both the public and the private sectors. The four thematic areas continued more or less from the late 1990s policy are energy and environment, health, marine activities and food, and the three technological areas of nano-/materials technology, biotechnology and ICT are in addition specified as supporting the overall targeted areas.

In this rather ambitious policy, the relationship between maintaining a broad orientation simultaneous with specialization is explained as a two-level relationship. The country must on one level be able to understand and make use of a broad range of research results from other countries, although the prerequisites for it being in the lead within these fields are not present. In addition to this, Norway should exploit national advantages and take responsibility within research areas where the country has such prerequisites (UFD, 2005a, pp. 24 and 28–9). In these respects, the 2005 'Commitment to Research' (UFD, 2005a) demonstrates a strong belief in research-driven innovation.[7]

7.3 Future Innovation Policy

The Norwegian government as well as business community is at a watershed when it comes to the formulation and implementation of innovation policy for the coming years. The main issues of this watershed are twofold. First, there is the issue of deciding between a broad versus a more targeted approach in a technical sense. In addition to general upgrading, the government has proposed targeted areas for further development. In the case of general upgrading as the main and underlying philosophy, there is a need to further formulate the conditions under which general upgrading is to take place. In other words, if the rationale is that heterogeneity leads to increased output and that a subsequent core strategy is to foster heterogeneity by way of having broad and multiple targets, this must be stated in an explicit way. In the case of targeted areas, one might want to ask whether they are too broad and ambitious for a small country such as Norway. One also might want to ask whether the technical focus areas are the right ones in the sense that they are only partly connected to past areas of strength, and instead mainly preoccupied with the fostering of new and hitherto largely unknown areas.

Second, there may also be an issue of prioritizing between innovation and growth issues in a narrow sense, and broader issues such as regional, environmental and welfare development. Of course, these two areas of concern need not constitute a contradiction in essence. It may perfectly well be possible to combine, for example, regional development with growth and innovation. However, the supposed symbiosis between innovation on the one hand, and regional, environmental and welfare development on the other hand, is, in our interpretation, more assumed than proven. In such a situation, there is a high risk that the development of a critical mass will suffer, since there is a risk of compromises in the form of broad distribution of funds rather than serious pursuit of actual and 'deserved' needs.

Indeed, one might be able to argue that the Norwegian economy can afford to maintain a broad approach in both of the senses referred to above, at least in the near to medium term. We are, however, convinced that it will become necessary to face some unpleasant prioritizing tasks. We sum up the main alternative options as follows:

1. Increased attention to the possibilities of further and advanced exploitation of the existing trajectories, such as oil and gas extraction sector as part of a scale-intensive trajectory and fisheries/aquaculture as part of a supplier-based trajectory.
2. Examining the realism and need for specific percentage 'push' goals when it comes to R&D expenditure, and rather devising sectoral 'pull' goals.
3. Abolition of the general and integrated approach consisting of including multiple aims within one and the same policy, and instead devising competitive instruments where such symbiosis may (or may not) be part of the end result.

First, current policy seems to be 'high-tech biased' (see Section 5) in the sense that it is overly concerned with the need to develop a set of so-called new technologies as the basis for the knowledge society in Norway, and thereby trying to fill purported 'gaps' between the low-tech status of the Norwegian system and high-tech areas. This may distract attention from innovation potential in the predominantly low-tech areas. Policy might instead have greater promise when focusing more on innovation in low-tech sectors in line with the argument posed by von Tunzelmann and Acha (2004). In the case of Norway, the low-tech and the medium–low-tech sectors are certainly such fields.

Second, current R&D and innovation policy is, as reviewed, mainly centred on the specific numerical goal of reaching a certain level as compared to the OECD average. Apart from representing a partial reversion to the linear way of perceiving the innovation process, such a strategy may be side-stepping the real problems addressed in our chapter. Indeed, financing is necessary and welcomed by the public and private sectors alike, but to design the entire R&D policy around this goal reflects a push-type philosophy. Instead, differentiated intermediate goals should be set where each targeted area or sector may be envisioned as reaching certain levels of expenditure at certain dates, but where these goals may be adjusted or even abandoned along the way according to performance criteria. Whether the end aggregate level is higher than, lower than, or exactly 3 per cent of GDP thus becomes less relevant than the content of the expenditure. Such an alternative strategy may have an additional strength in the sense that it will be forced to deal with the business financing part of the 3 per cent goal in a more direct as well as pragmatic way than as within current policy. Current policy expects business to partake in reaching the 3 per cent goal by being responsible for *circa* two-thirds of the expenditures, something that must be confronted as a rather unrealistic goal in view of past history and current structure of Norwegian business and its R&D expenditure patterns.

Third and similarly, a general and integrated approach consisting in including multiple aims, such as innovation and growth concerns integrated with regional development concerns, within one and the same policy may be admirable in itself. It is not entirely clear, however, that integration of aims is the most efficient means of achieving optimal results. Measures directed at enhancing innovation output should be designed and directed primarily towards this objective on competitive terms in order to avoid the financing of 'mediocre' projects and firms. Such an alternative and competitive approach may indeed have one or several weaknesses, including the administrative costs. We are, however, convinced that drastic measures are needed within this realm if there is to be significant progress.

ACKNOWLEDGEMENTS

We would very much like to thank the interview respondents for their time and help, Ingunn Steinsland Kristiansen and Martin Srholec for their crucial assistance regarding data preparation in Section 3, the Research Council of Norway for funding (grant no. 148191/510) and Charles Edquist, Jan Fagerberg, Leif Hommen, Svein Olav Nås, Annika Rickne, Kung Wang and Olav Wicken for comments on early drafts.

LIST OF INTERVIEWEES

Marianne Botten, Project Manager, SIVA.
Henning Qvale, Manager of Qubator AS, previously manager of Kampus Kjeller Science Park.
Lars Monrad Krohn, Manager of training programme for entrepreneurs Grunderskolen and adviser at Oslo Research Park.

NOTES

1. An extended discussion of the various hypotheses on innovation intensity is given in Section 5.
2. Higher education is defined as four years or more with education at the university or college level.
3. Cf. also Kallerud et al. (2006, p. 58) for recent figures confirming this situation with special reference to SMEs.
4. Stambøl (2005) bases his figures on the accumulation of NACE codes 642, 72, 671–672, 73 and 74. Other comparative measurements (Eurostat, 2005) state the share of employment as considerably lower due to a much more narrow conception of the KIBS concept, and here the Norwegian level is stated as higher than in the Netherlands, but lower than in Sweden, Finland and Denmark.
5. It may be noted that a crucial and long-standing part of networking programmes has indeed been to create incentives for collaboration. We are indebted to Svein Olav Nås for this observation.
6. For more detailed descriptions of recent and contemporary policy instruments, see Remøe et al. (2004), Aanstad et al. (2005) and Kallerud et al. (2006).
7. At the time of completing this manuscript (2007) the government is in the process of preparing a White Paper on innovation, which may or may not include initiatives of another character when compared to those described here.

REFERENCES

Aanstad, S., P. Koch and A. Kaloudis (2005), *European Trend Chart on Innovation: Annual Innovation Policy Trends and Appraisal Report Norway 2004–2005*, Brussels: European Commission.

Archibugi, D. (2001), 'Pavitt's taxonomy sixteen years on: a review article', *Economics of Innovation and New Technology*, **10**, 415–25.

Arnold, E. (2004), 'Evaluating research and innovation policy: a systems world needs systems evaluations', *Research Evaluation*, **13**(1), 3–17.

Askevold, E.O., T. Halvorsen, T. Laudal and J.M. Steineke (2003), 'Norske forskningsparker: mot en bedre organisert nyskaping?', RF-2003/174, Stavanger: Rogalandsforskning.

Aslesen, H.W. (2004), 'Knowledge intensive service activities and innovation in the Norwegian aquaculture industry: part project report from the OECD KISA study', STEP Report 05, 2004, Oslo: STEP.

Bank of Norway (2003), *Årsrapport om Betalingsformidling 2003*, Oslo: Norges Bank.

Baygan, G. (2003), 'Venture capital review Norway', STI Working Paper 2003/17, Industry Issues, Paris: OECD.

Benner, M. (2003), 'The Scandinavian challenge: the future of advanced welfare states in the knowledge economy', *Acta Sociologica*, **46**(2), 132–49.

Bøhren, Ø. and B.A. Ødegaard (2003), 'Norsk eierskap: Særtrekk og sære trekk', *Praktisk økonomi og finans*, **20**, 3–17.

Borch, O.J. et al. (2002), 'Kapitalmarkedet for nyetablerte foretak: en studie av etterspørsels- og tilbudssiden', KPB-rapport 1/2002, Oslo: KPB.

Broch, M. and A. Isaksen (2004), 'Knowledge intensive service activities and innovation in the Norwegian software industry: part project report from the OECD KISA study', STEP Report 03/2004, Oslo: STEP.

Cumbers, A. (2000), 'The national state as mediator of regional development outcomes in a global era: a comparative analysis from the UK and Norway', *European Urban and Regional Studies*, **7**(3), 237–52.

Eurostat (2005), *Employment in Knowledge-intensive Service Sectors: Share of Total Employment (%)*, Brussels: Eurostat.

Eurostat (2007), *Business Demography – Survival Rate*, Brussels: Eurostat.

EVCA (2002), *European Venture Capital Association Year Book 2002*, Brussels: EVCA.

EVCA (2003), *European Venture Capital Association Year Book 2003*, Brussels: EVCA.

European Commission (2005), *European Innovation Scoreboard 2005*, Brussels: European Commission.

FD (2004), *Kapitaltilgang og økonomisk utvikling: rapport fra ekspertgruppen som har vurdert Norges kapitalstyrke*, Oslo: FD.

Graversen, E. et al. (2003), 'Migration between the Nordic countries: what do register data tell us about the knowledge flows?', STEP Report 10/2003, Oslo: STEP.

Grønning, T., E. Dobos, M. Knell, D.S. Olsen and B.K. Veistein (2006), 'Norway', in C.M. Enzing (ed.), *Innovation in Pharmaceutical Biotechnology: Comparing National Innovation Systems at the Sectoral Level*, Paris: OECD, pp. 86–95.

Hansen, T.B., T. Karlsson and H. Godø (2005), 'Evaluation of the DEMO 2000 program', NIFU STEP Report 7/2005, Oslo: NIFU STEP.

Havnes, P.-A. (2003), 'Ni norske inkubatorer: midtveis gjennomgang av første pulje i SIVAs program', AgderForskning FoU Report 6/2003, Kristiansand: AgderForskning.

Hodne, F. and O.H. Grytten (1992), *Norsk økonomi 1900–1990*, Oslo: Tano.

Huber, E. and J.D. Stephens (1998), 'Internationalization and the social democratic model: crisis and future prospects', *Comparative Political Studies*, **31**(3), 353–97.

Innovation Norway (2004), 'Tilsagn til nyskaping, og totalt 1994–2003', unpublished report on resources allocated to establishment of new firms, advice on entrepreneurship etc. per region in the period 1994–2003, Oslo: Innovation Norway.

INTSOK (2005), INTSOK's home page http://www.intsok.no/, accessed in May 2005.

Kallerud, E., J. Hauknes and P. Koch (2006), *European Trend Chart on Innovation: Annual Innovation Policy Trends and Appraisal Report Norway 2006*, Brussels: European Commission.

Kolvereid, L. and G.A. Alsos (2003), *Entreprenørskap i Norge 2002: Global Entrepreneurship Monitor*, Bodø: Handelshøgskolen i Bodø.

KRD (2003), 'Storbymeldingen: om utvikling av storbypolitikk', St.meld. nr. 31 (2002–2003), Oslo: KRD.

KUF (1998), 'Forskning ved et tidsskille', St meld nr 39 (1998–99), Oslo: KUF.

KUF (2000), 'Frihet med ansvar: om høgre utdanning og forskning i Norge', NOU 2000/14, Oslo: KUF.

Kuhlman, S. and E. Arnold (2001), 'RCN in the Norwegian Research and Innovation System', Report to Ministry of Church, Education and Research Affairs, Oslo: KUF.

Lengrand, L. and Associés, PREST and ANRT (2002), 'Innovation tomorrow: innovation policy and the regulatory framework: making innovation an integral part of the broader structural agenda', Innovation Paper No. 28, European Commission Directorate-General for Enterprise, Brussels: European Commission.

Matson, E. (2005), 'Deregistrering og videreføring av nye teknologifirma', *Beta*, **2**, 63–77.
Moen, E. (2002), 'Globalisering og industripolitiske strategier: en sammenligning mellom Finland og Norge', Makt- og demokratiutredningens rapportserie, Oslo: Makt- og demokratiutredningen.
Moen, E. (2005), '"Næringsnøytralitet" eller næringsavvikling? Norsk oljeøkonomi, næringsutvikling og næringspolitikk i et politisk-institusjonelt perspektiv', BI Discussion Paper 1/2005, Oslo: BI.
Munch, A.T. (2007), 'SkatteFUNN begrenses', *Bladet Forskning*, **15**(1), 10.
Narula, R. (2002), 'Innovation systems and "inertia" in R&D location: Norwegian firms and the role of systemic lock-in', *Research Policy*, **31**, 795–816.
Nås, S.O., T. Sandven et al. (2003a), 'High-tech spin-offs in the Nordic countries', STEP Report 23/2003, Oslo: STEP.
Nås, S.O., T. Sandven et al. (2003b), 'High-tech spin-offs in the Nordic countries: statistical annex', STEP report 22/2003, Oslo: STEP.
Nås, S.O. and J. Hauknes (2004), 'Er det FoU vi skal leve av?', *Forskningspolitikk*, **27**(2).
Nerdrum, L. and M. Gulbrandsen (2006), 'The research institutes in the Norwegian innovation system', Paper to TIK-Centre Innovation Seminar, Oslo: University of Oslo.
NFR (2003), *Det Norske Forsknings- og innovasjonssystemet – statistikk og indikatorer 2003*, Oslo: Norges Forskningsråd.
NFR (2005), *Det Norske Forsknings- og innovasjonssystemet – statistikk og indikatorer 2005*, Oslo: Norges Forskningsråd.
NHD (1997), 'Offentlige Anskaffelser', NOU 1997 (21), Oslo: NHD.
NHD (2001), 'Best i test? Referansetesting av rammevilkår for verdiskaping i næringslivet', NOU 2001 (29), Oslo: NHD.
NHD (2003a), *eNorway Status Report, January 2003*, Oslo: NHD.
NHD (2003b), *Plan fra idé til verdi: regjeringens plan for en helhetlig innovasjonspolitikk*, Oslo: NHD.
NHD (2004), 'Evaluering av regelverket for offentlige anskaffelser', Report H2004-019, Oslo: Asplan Viak.
NVCA (2001), *Activity in the Norwegian VC-market 2001*, Oslo: NVCA.
OECD (2004a), *Economic Survey Norway 2004*, Paris: OECD.
OECD (2004b), *Learning for Tomorrow's World: First Results from PISA 2003*, Paris: OECD.
OECD (2005), *Main Science and Technology Indicators*, Paris: OECD.
OECD (2006), *Economic Policy Reforms: Going for Growth, Structural Policy Indicators and Priorities in OECD Countries*, Paris: OECD.
Olsen, O.-E. and F. Sejersted (eds) (1997), *Oljevirksomheten som teknologiutviklingsprosjekt*, Oslo: AdNotam Gyldendal.
Ørstavik, F. (2000), 'Innovasjoner – suksesser. Identifiserte innovasjoner 3 år etter', STEP report 11/ 2000, Oslo: STEP.
Pavitt, K. (1984), 'Sectoral patterns of innovation: towards a taxonomy and a theory', *Research Policy*, **13**, 343–74.
Pinder, D. (2001), 'Offshore oil and gas: global resource knowledge and technological change', *Ocean & Coastal Management*, **44**, 579–600.
Remøe, S.O., M. Fraas, A. Kaloudis, Å. Mariussen, R. Røste, F. Ørstavik and S. Aanstad (2004), 'Governance of the Norwegian innovation policy system: contribution to the MONIT collaborative project for the OECD', NIFU STEP Report 6/2004, Oslo: NIFU STEP.

Rothwell, R. (1994), 'Issues in user–producer relations in the innovation process: the role of government', *International Journal of Technology Management*, **9**(5/6/7), 629–49.

Sejersted, F. (1993), *Demokratisk kapitalisme*, Oslo: Universitetsforlaget.

SIVA (2002), *Inkubator April 2002*, Oslo: SIVA.

SIVA (2004), *Norske og internasjonale inkubatorer: Inkubator undersøkelser i andre land sammenholdt med norske forhold*, Oslo: SIVA.

Skogli, E. (1998), 'Offshore engineering, consulting and innovation', STEP WP A-04-1998, Oslo: STEP.

Srholec, M. and T. Grønning (2006), 'Sectoral patterns of innovation in Norway', paper to TIK-Centre Innovation Seminar, Oslo: University of Oslo.

Stambøl, L.S. (2005), *Urban and Regional Labour Market Mobility in Norway*, Oslo: Statistics Norway.

Statistics Norway (1995), *Historical Statistics 1994*, Oslo: Statistics Norway.

Statistics Norway (2001), *Innovation Statistics in the Business Enterprise Sector 2001*, Oslo: Statistics Norway.

Statistics Norway (2007), *470 000 bedrifter i Norge ved årsskiftet*, Oslo: Statistics Norway.

Sydnes, T. and K. Halvorsen (2003), *Evaluering av Såkornordningen*, Oslo: NHD.

UFD (2002), *Om lov om endringer i lov 12. mai 1995 nr. 22 om universiteter og høgskoler og lov 2. juli 1999 nr. 64 om helsepersonell: ot.prp. nr. 40 (2001–2002)*, Oslo: UFD.

UFD (2005a), *Vilje til forskning: st.meld. nr. 20 (2004–2005)*, Oslo: UFD.

UFD (2005b), *Realfag, naturligvis! Strategi for styrking av realfagene 2002–2007*, Oslo: UFD.

UNCTAD (2004), *World Investment Report 2004*, Geneva: UNCTAD.

von Tunzelmann, N. and V. Acha (2004), 'Innovation in "low-tech" industries', in J. Fagerberg, D.C. Mowery and R. Nelson (eds), *The Oxford Handbook of Innovation*, Oxford and New York: Oxford University Press.

Wicken, O. (2000), 'Forskning, næringsliv og politikk: en historisk fremstilling av norsk næringslivsforskning og politikk', Senter for teknologi, innovasjon og kultur arbeidsnotat 6/2000, Oslo: University of Oslo.

9. Challenged leadership or renewed vitality? The Netherlands

Bart Verspagen

1 INTRODUCTION

The Netherlands national system of innovation (NSI) has deep roots in the history of the country. After having been, once upon a time, the world economic leader, the Netherlands has been forced to follow other countries in terms of technological developments, but it has done so with its own specific way of adapting to global developments. The result, by the end of the 1960s, was an NSI that operated at a high level of performance, hosting a number of global companies that played dominant roles in their industries at the world level.

However, much has changed in the world economy since this period. Global competition has intensified, and the Netherlands system has felt this pressure from abroad. In 2002, foreign direct investment outflows from the Netherlands equalled 8 per cent of gross domestic product (GDP), and inflows 6 per cent. These figures are higher than those for any of the other European countries in this study (Sweden is second, with 4 per cent outflow, 5 per cent inflow). In addition, European integration has affected the Netherlands system. As will be shown in this chapter, the Netherlands NSI has been severely challenged by this globalization process, and a few years into the twenty-first century, one has to conclude that the system is losing momentum. Innovation performance indicators show a persistent downward trend, especially so in relative (to other countries) terms. The source of this relative decline also seems evident from the data. Although the public parts of the NSI remain strong, business innovation efforts have stagnated.

Innovation policy makers are well aware of the problems that face the Netherlands NSI. But whether they will be able to turn the tide remains doubtful. Available government budgets for innovation policy are small. Moreover, as will be argued in the final section of this chapter, some of the analysis underlying the problem analysis may be challenged by adopting an explicit systems of innovation (SI) perspective.

2 MAIN HISTORICAL TRENDS

The Netherlands NSI inherited two important bequests from its pre-twentieth-century history. These are the broad institutional history of corporatism, and a specialization pattern that is strongly biased in favour of services and agriculture and the latter's related industries. In their roots, both of these factors go back to the hegemony of the Dutch Republic in the seventeenth century (Schot, 1995; Van Zanden and Van Riel, 2000).

In terms of its industrial structure, the merchant trade-based system of the seventeenth and eighteenth centuries obviously entailed a strong specialization in the service industries. But this was not limited to producer services, such as banking; it also included a large government sector and a range of consumer services such as education and domestic servants (Van Zanden and Van Riel, 2000, p. 78). In addition, the agricultural sector had traditionally shown high productivity, especially in specialized segments such as horticulture and dairy cattle.

The success of the merchant trade-based economic system of the Republic was to a significant extent based on institutional foundations whereby society was ruled by a large number of 'organizations', working in parallel at all levels of society, and with a great degree of independence. Examples of these institutions were the federal nature of the Republic, in which the provinces had a great deal of power through a system of representation, and the influence of the *polders*, which were responsible for water management. Even the international trading companies that held monopolies on trade with a specific part of the world, such as the Verenigde Oostindische Compagnie, were part of this system. They had significant powers and held diplomatic relationships with foreign governments, and were themselves ruled by a complicated system of representation.

Van Zanden and Van Riel (2000) argue that although this institutional fabric greatly contributed to the economic success of the Republic (for example, it laid the basis for an efficient capital market), its rigidities were also ultimately responsible for its demise. Eventually, a period of economic stagnation set in, and when the Industrial Revolution emerged in the UK, this turned into an age of industrial retardation (Griffiths, 1979). Industrialization was slow and late in the Netherlands, and it took until the 1860s before economic growth took off again (Van Zanden and Van Riel, 2000, ch. 8).

The renewed process of economic growth was fed by a specific pattern of industrialization, which was based on diffusion of technologies developed abroad (albeit with a specific Dutch style) and a high degree of dependence on foreign markets (Schot, 1995). The country's historical involvement in international trade and specialization towards agriculture became the basis of a new system of commercial capitalism, strongly dependent on exports and companies that operated globally. During this period, 'innovation-based entrepreneurship was primarily the same as good business sense' (ibid., p. 237).

Van Zanden and Van Riel (2000) argue that along with this new age of economic growth, there emerged a period of neo-corporatism, which cast its shadow into the twentieth century, when the Netherlands society became strongly segregated into a number of religious and non-religious 'pillars', each with their own societal organizations (see Box 9.1). Although differences among these pillars tended to be pronounced, in the end there was always a spirit of coalition forming, and the governance system tended to be based on compromise and unanimity (the so-called *poldermodel*; see Box 9.1).

BOX 9.1   THE NEO-CORPORATIST *POLDERMODEL* IN THE NETHERLANDS

The popular term *poldermodel* is used in the Netherlands to represent the typical tendency for consensus-seeking decision making in which all strata of society are represented and consulted. The term itself refers to the typical landscape of grassland and ditches that prevails in the artificially made land areas in North and South Holland (the two western coastal provinces in the Netherlands). These *polder* areas are traditionally administered by organizations in which representation plays an important role, but the term is now more symbolic than a direct reference to the polders.

In general terms, the *poldermodel* refers to the tendency in the Netherlands to seek consensus and compromises, rather than polarization and polemic debates. At a central government level, the *poldermodel* is characterized by harmonic relations between the 'social partners', i.e. employers and trade unions, leading to low wage growth and a low number of strikes over the last decades. Government itself is always a coalition government, in which often more than two parties participate. At lower levels of government, the so-called 'Public Industry Organizations' (*Publieke Bedrijfsorganisaties*) are a typical expression of the *poldermodel*. These are organizations in which employers and trade unions cooperate, and which are responsible for a range of affairs within a specific sector of the economy. Their activities range from promotion of product quality and employee education to the execution of specific pieces of government regulation.

> The *poldermodel* has its origin in the Dutch Republic in the sixteenth century (see Section 2), when it also incorporated tolerance of immigrants as a main ingredient, but it has evolved significantly since then. In the first half of the twentieth century, society in the Netherlands became 'pillarized' into religious and political streams. In this process, all kinds of organizations affecting daily life (such as sports associations, radio stations, trade unions, political parties, etc.) were split into separate organizations for Roman Catholics, Protestants (of various persuasions), and social democrats. Although it may sound paradoxical, this high degree of compartmentalization in fact strengthened the desire for consensus seeking. After the Second World War, when religion became less important, the tendency towards 'pillarization' became much less pronounced, although traces of this system remain (e.g. in the system for public television, and in the two large federations of trade unions).
>
> In the twenty-first century, the *poldermodel* came under pressure when criticism of the 'purple coalition' (social democrats, left liberals and right liberals) that had governed the country for most of the 1990s became ubiquitous. The populist politician Pim Fortuyn evidently rose to power on a wave of criticism of this government, as well as anti-immigration sentiments. His assassination, a week before elections for parliament in 2002, polarized politics even more. Since then, the term *poldermodel* has significant negative connotations among a large part of the population – e.g. a tendency to strike powerless compromises.

The wave of industrialization continued into the twentieth century, when the Netherlands partially caught up with the early industrializing nations of Europe. New branches of industrial specialization emerged, especially in the chemical industry, but the old comparative advantages in agriculture and its related industries, as well as services, also remained (Schot and Van Lente, 2003). A particular feature of this long period of industrialization was the emergence of a limited number of very large corporations, in an economic system that was otherwise dominated by small and medium-sized companies. Philips Electronics, Royal Shell and AKZO are examples of this trend. These firms, which were founded relatively early (late nineteenth and early twentieth century), became, often through a process of mergers and acquisitions, dominant world players in their respective markets.

3 INNOVATION INTENSITY

Since the early 1980s, the propensity to innovate in Dutch manufacturing has been declining. For services, data only exist since the early 1990s, and indicate a relatively flat trend.[1] These data are displayed in Figure 9.1. For manufacturing, the observations for product innovations for 1992 seem rather low, which might be due to some statistical artefact (although the worldwide recession in this period may also play a role). Disregarding this year, the overall trend from 1983 onwards is remarkably smooth, and clearly points to the declining propensity to innovate in manufacturing. The difference between 1983 and 2000 in the percentage of all firms in manufacturing that innovate is almost 10 percentage points. Roughly half of this decline is observed in the last half of the 1990s.

Despite this relative decline, innovation rates are still high in the Netherlands as compared to other (European) countries in this study. The percentage of innovators in manufacturing in 2000 is higher than in any

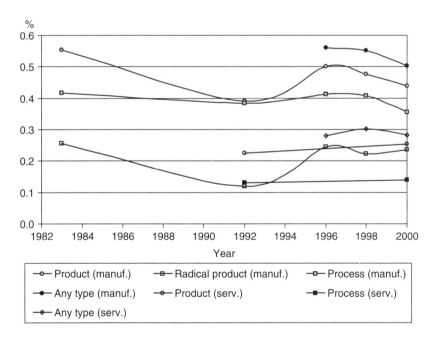

Source: Calculations based on Kleinknecht (1987), Kleinknecht et al. (1990), and Statistics Netherlands, Statline.

Figure 9.1 The propensity to innovate in the Netherlands innovation system

other country (Denmark is second, with only a minimal difference compared to the Netherlands). Also, when viewed for specific subsectors, the rates of innovators in the Netherlands are always high (see Appendix Table A4.4).

Nevertheless, the decline of the propensity to innovate in manufacturing is indicative of a trend that has worried policy and opinion makers ever since the mid-1980s. The 'public feeling' is that the Netherlands used to be (in the 1960s and 1970s) a country at the frontier of technological development, but that this position is eroding. Both the perceived competition from abroad (e.g. 'Asian tigers' such as Korea) and domestic factors have played a role in this debate. Several of these factors (e.g. education, public–private research and development (R&D) cooperation) will be discussed in Section 4 below. Another interesting suggestion has been made by Kleinknecht (1998), who argued that the policy of limited wage growth was responsible for the declining innovation performance. In his 'evolutionary' reasoning, high real-wage growth puts strong selective pressures on firms, which are then forced to implement innovations in order to remain competitive. Because wage policy is considered as one of the main achievements of the neo-corporatist *poldermodel* of consensus decision making, Kleinknecht's argument has met with much disfavour. However, although there is little direct evidence to support the negative relationship between low wage growth and innovativeness, it seems obvious that competitiveness based on low wage growth is often an aspect of a defensive innovation strategy.

Figure 9.2 documents a broader range of indicators of the propensity to innovate in four sectors of the Netherlands economy. Clockwise, starting from the top, these diagrams represent total innovation costs as a proportion of turnover, R&D (internal and outsourced) as a fraction of turnover,[2] the share of new (to the firm) products in total turnover, the share of radical (new-to-the-market) products in turnover, firms with product innovations (new to the firm) as a proportion of all firms, firms with radical (new-to-the-market) product innovations as a proportion of all firms, firms with process innovations as a proportion of all firms and firms with any type of innovation as a proportion of all firms. The maximum value on the axes (i.e. one) indicates the maximum value over the four broad sectors.

There is a clear distinction between the two sectors at the top (manufacturing and knowledge-intensive business services (KIBS)[3]) and the two at the bottom (finance and trade). Manufacturing and KIBS show a 'complete' innovation profile, i.e. they rank relatively high on all eight indicators. Manufacturing has the highest value on process innovators and on the share of new (to-the-firm) products in turnover. KIBS has the highest value on all other indicators. This clearly indicates the importance of innovation in services. In the Netherlands economy, with its strong emphasis on services, knowledge-intensive business services are an important sector for

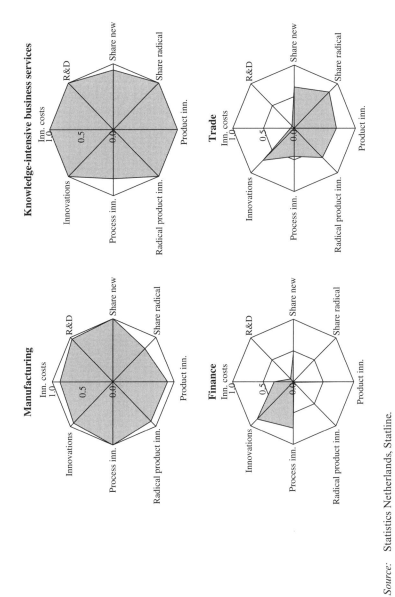

Source: Statistics Netherlands, Statline.

Figure 9.2 Innovation profiles of four broad sectors

innovation in all respects. To focus only on manufacturing would leave out an important part of innovation activities.

The sectors trade and finance show a much more 'concentrated' picture. For example, innovation in finance and trade is an activity that consumes relatively little costs (as compared to KIBS and manufacturing). In finance, radical product innovation is completely absent. In trade, process innovation is relatively weak (trade is also the sector in which performance in the Netherlands is relatively weak as compared to other countries in the study). In general, scores on all indicators are lower than for KIBS and manufacturing.

The numbers presented so far hide any effects of the specific nature of the Netherlands firm size distribution. It was already indicated in Section 2 that the Netherlands has a rather peculiar firm size distribution. Figure 9.3 breaks

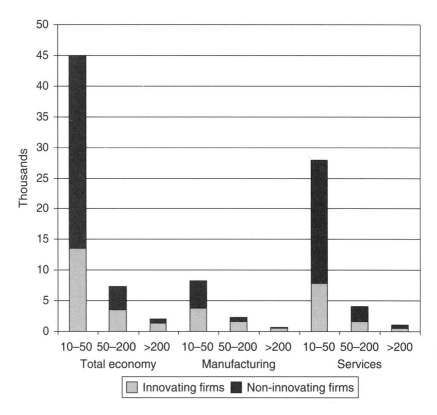

Source: Calculations based on data in Statistics Netherlands (2002).

Figure 9.3 Innovation and firm size in the Netherlands system

down the total number of firms in the economy and the total number of innovating enterprises into broad size classes. Small firms (10–50 employees) are by far the largest group, but, expectedly, they show the lowest degree of innovativeness. This is true for most other countries, and small and medium-sized enterprises (SMEs) in the Netherlands still rank highest in terms of innovation rates in the sample of countries in the present study. Medium-sized and large firms are an order of magnitude less frequent, and show higher propensities to innovate. However, despite their smaller number, large firms accounted for 76 per cent of total innovation expenditures in the Netherlands in 2000 (the percentage for small firms is 11). Within the group of large firms, the role of five large multinational firms stands out. For example, towards the end of the 1990s, these firms accounted for 45 per cent of all business R&D in the Netherlands (Tijssen et al., 2000).

Table 9.1 gives more details with regard to the sectoral concentration of innovation in the Netherlands. Besides the traditional 'high-tech' sectors (e.g. pharmaceuticals, computer services), one also finds a number of typical strongholds of the Netherlands economy, especially in services (e.g. finance) and manufacturing (food products). Together, these sectors account for three-quarters of total business innovation expenditures in the Netherlands.

Finally, with regard to non-technological innovation, the Netherlands data for 2000 show that change in the strategic goals of the enterprise is the most dominant form of this. In all four main sectors documented above (manufacturing, KIBS, finance and trade), this is the most frequently observed form of non-technological innovation. When we break down manufacturing into 13 sub-branches, this form of non-technological

Table 9.1 Top ten of sectors with highest total innovation expenditures, 2000

Sector	Innovation expenditures, 2000 (€ million)
1. Electrical machinery	1968
2. Finance	1254
3. Transport & communication	615
4. Machinery	608
5. Pharmaceuticals	571
6. Mining	529
7. Food products	458
8. Wholesale trade	408
9. Basic chemicals	394
10. Computer services	370

Source: Statistics Netherlands (2002).

328   *Slow growth countries*

innovation ranks highest eight times (organizational change follows, with three occurrences). It might, however, be the case that this is a specific feature of the economic recession that had already set in in 2000, forcing firms to redefine their strategic goals.

4   ACTIVITIES THAT INFLUENCE INNOVATION

4.1   Knowledge Inputs to Innovation

4.1.1   R&D activities

R&D intensity (defined as R&D as a fraction of GDP) has been remarkably constant at an approximate 2 per cent over the last three decades, as shown in Figure 9.4. Firms account for slightly more than half of total

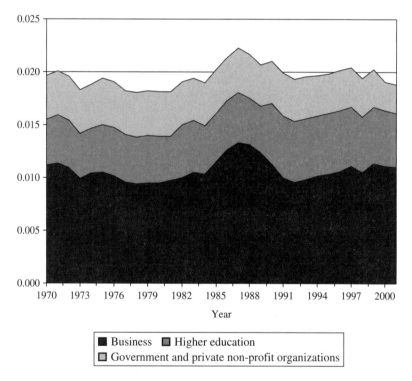

Source: Calculations on data from Statistics Netherlands, Statline.

Figure 9.4   R&D spending as a fraction of GDP, broken down by institutional sector

R&D over this entire period. Of the other half, higher education accounts for the largest part, just leading over the government sector (the latter includes private non-profit organizations).

In a world where many countries have increased their R&D spending, the 30-year flat trend in the Netherlands has led to a relative decline in the country's R&D score relative to the rest of the world. Whereas the Netherlands ranked among the most R&D-intensive countries in the world in the 1960s and 1970s, it has fallen back to the EU average in the 1990s. In terms of the countries in this study, the Netherlands R&D intensity only ranks higher than that of Ireland (at 1.2 per cent). Tijssen et al. (2003) argue that especially the slow growth of business R&D is responsible for this fallback.

As mentioned above, business R&D activities in the Netherlands are dominated by a few large multinational companies. As a result, globalization is an important force determining the pattern of R&D spending. Ever since the 1980s, worries have been expressed about the possibility that firms may locate their R&D abroad. In particular, Asia (and recently, more specifically, China) has been mentioned as a region to which firms might like to move their R&D. Philips Electronics, which is the largest R&D spender in the Netherlands system, currently spends 35 per cent of its total R&D in the Netherlands.[4] In the first half of the 1990s, this was still 45 per cent.[5] More evidence on the globalization of R&D will be provided in Section 6 below.

Higher education R&D in the Netherlands, despite budget cuts in higher education that have been going on ever since the 1980s, has been roughly constant as a share of GDP since the 1970s.[6] The Netherlands has three technical universities (in Delft, Eindhoven and Twente), one agricultural university (in Wageningen), and nine general universities. Besides R&D in universities, a small amount of R&D takes place in institutes for higher vocational training. The university funding system is largely based on 'general university funds', which are distributed on the basis of the number of students (graduate and undergraduate) and characteristics of the field of study (e.g. whether high investments in equipment are necessary). In 2000, these funds were responsible for paying 52 per cent of all university researchers. Contract research was responsible for 27 per cent of university researchers.[7]

The remaining part of higher education R&D, 21 per cent, is financed by the research councils, of which the Netherlands Organization for Scientific Research (NWO in the Dutch abbreviation) is the most influential one. Recent developments in science policy attribute more influence to the research councils, due to an attempt by policy makers to gain more influence over the direction of university research. Whereas the spending of general university funds is at the discretion of the universities themselves, research council funds can be focused on specific fields of science. Thus funds are being shifted from general university funds, as well as from contract research paid by the government, to the research councils.

University research in the Netherlands traditionally stands at a high-quality level. Tijssen et al. (2003) show that, in a group of 30 countries, the Netherlands is the country with the third-highest citation impact in the period 1998–2001 (the first and second were the USA and Switzerland, respectively). Similar conclusions are reached by the European Commission (2003) and Wang et al. (2003). But Tijssen et al. (2003) also show that the growth of publications by university researchers is relatively slow as compared to other countries, and stresses that the high citation impact is largely based on research undertaken several years before the measurement period 1998–2001; hence the potential impact of recent budget cuts is not yet visible.

Regarding the specialization pattern of science in the Netherlands, the exact conclusions depend both on the set of reference countries, and on the breakdown into disciplines. Thus, for example, the conclusions reached in Wang et al. (2003) and the European Commission (2003) differ. However, both studies seem to agree that the medical sciences, environmental sciences, astronomy and physics are strongholds of the Netherlands science system. The two recent Nobel prizes awarded to scientists from the Netherlands ('t Hooft in physics and Crutzen in chemistry, the latter related to environmental issues) seem to confirm this, although both Nobel laureates have been working abroad for large parts of their careers. Interestingly, Wang et al. (2003) and the European Commission (2003) reach opposite conclusions on the engineering sciences. Wang et al. (2003), who use the countries in this study as a reference point, and hence include Asian countries but exclude large EU countries, conclude that the Netherlands is underspecialized in engineering, while the European Commission (2003) concludes the opposite in a comparison with EU member states.

Finally, R&D in the government sector has been a traditional stronghold of the Netherlands R&D system. Traditionally, this institutional sector has undertaken applied research, and hence acted as an intermediate between the more fundamentally oriented research in the university system and business R&D. In this sector, which in the statistical definition used above also includes a minor share of private non-profit institutions, the Netherlands Organization for Applied Scientific Research (TNO in the Dutch abbreviation) is the largest player. Besides TNO, one also finds a number of specialized institutes, dealing, for example, with water-related civil engineering, energy and aerospace. TNO is a collection of institutes and technology-oriented companies and in some respects resembles the

German Fraunhofer Institutes. TNO is concentrated around the city of Apeldoorn in the east of the country, but also has locations at the technical universities and Delft.

TNO was traditionally funded almost completely by the Netherlands government, which provided both general funds aimed at keeping up the knowledge infrastructure of TNO, and specifically targeted funds aimed at fields that were considered to be of special interest. Since the 1980s, a trend has been set in which TNO has been forced to be more dependent on generating income from the market. In 2003, two-thirds of its €553 million turnover was generated from market sources.[8]

4.1.2 Competence building

The Netherlands system of education includes three levels of vocational training (not including universities), and, within the secondary educational level, three levels of preparatory education. The structure of the system has been stable for the last decades, with the exceptions of some changes in the university system and major changes in the lowest levels of vocational training and preparatory education. The latter refers to a recent merger of these two types of education, although separate tracks remain within the unified level. Formally, the lowest vocational training is now no longer considered as an end level, but in practice it often remains so.

Budget cuts have been common in the Netherlands educational system for decades, and this has led to problems in many schools with regard to housing, educational materials and the attractiveness of teaching as a job (low salaries). As a result of these budget cuts, the Netherlands now ranks about half a percentage point below the average of countries reported in the OECD publication *Education at a Glance* (2003) with regard to spending on primary and secondary education as a percentage of GDP, and about average with regard to spending in tertiary education as a percentage of GDP.

One of the worries about the Netherlands educational system is that it produces too few graduates in the technical sciences, and therefore does not provide enough relevant input into the science and technology system (Ministry of Economic Affairs, 2003). Figure 9.5 provides some data on this phenomenon. The indicator of revealed comparative advantage in the figures is calculated on the number of graduates in tertiary education broken down by field of study. A negative (positive) value points to a relatively low (high) number of graduates in the Netherlands relative to the OECD average. The top panel shows that in the university part of higher education, the Netherlands is specialized in health and welfare. On the negative specialization side, one finds a larger range of fields, including, indeed, all the technical and 'hard science' fields. Especially computing, life

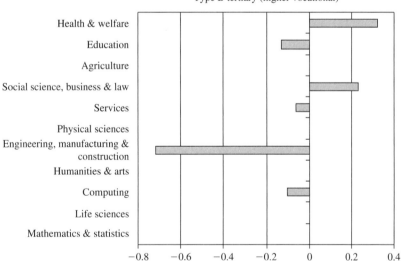

Source: Calculations on data from OECD (2003).

Figure 9.5 Revealed comparative advantages in tertiary education, Netherlands versus OECD, 2001

sciences, and mathematics and statistics score rather negatively in the university system. Concentrating on the higher vocational part of higher education, the shortfall in terms of the engineering sciences becomes even more obvious: this category is the overall strongest negative specialization of the Netherlands system.[9]

4.2 Demand-side Factors

Schot (1995) describes how during the post-Second World War rise of the 'consumption society', consumer associations played a large role in the formation of new markets. As a prime example, he mentions the automobile association ANWB, which provided an exemplary pattern of automobile use for its members. The role of these organizations is totally in line with the neo-corporatist nature of the Netherlands society. Examples of these types of organizations can still be found. The Hobby Computer Club is an organization that attracted a large membership during the 1980s, and, through its yearly budget market, computer magazine with product reviews, and sub-associations aimed at specific computer brands or interest groups (such as genealogy or model trains), had a large influence on early adopters of information and communication technologies (ICT) equipment. However, no contemporary research on these associations has been undertaken, and therefore little can be said about the specific influence of these organizations in the current Netherlands NSI.

Innovation policy in the Netherlands has charted a careful course away from anything that could be interpreted as industrial policy, at least nominally. The origin of this tendency lies in the 1970s, when government showed strong commitment to a number of specific sectors, such as shipbuilding. The support given to these sectors turned out to be a big failure, because, despite the large amount of money invested by the government, firms kept on going bankrupt, not being able to keep up with the (Asian) competition. The ensuing 'RSV debacle' caused politicians to avoid any type of industrial policy until the present day.[10]

Nevertheless, the innovation policy efforts of the Ministry of Economic Affairs have usually included a range of policy instruments aimed at two specific new markets: ICT and biotechnology (life sciences). Both were targeted as important key technologies for the future, and a purely market-based development was considered to be inadequate.

In biotechnology, the policy initiatives have mainly taken a supply-side-oriented perspective (Ministry of Economic Affairs, 2004), but in ICT, policies have taken a broader approach, also involving users. One of the main efforts in this area was the action plan for Electronic Highways undertaken by the Ministry of Economic Affairs in the 1990s (see also the follow-up note in Ministry of Economic Affairs, 1999). The main aim was to increase the use of ICT, and especially electronic communication and the Internet by small and medium-sized firms. Demonstration projects, subsidies and specific technological development were all part of the action plan. More recently, broadband communication has been the target of the new policy initiative *Kenniswijk* ('Urban Knowledge Area'). This project awarded funds (on a competitive basis) to two cities (Amsterdam and Eindhoven) for developing broadband services to the home. Project proposals had to involve a range of parties, including users, suppliers, local governments, social welfare organizations, educational institutes, etc. The two chosen projects were started with high expectations, but the burst of the dot-com bubble has significantly affected the willingness of private parties (firms) to contribute.

Control of regulations with regard to quality and safety in crucial sectors such as food and health is carried out in the Netherlands by the relevant ministries, which have special departments that operate somewhat independently of other parts of government. For example, in the field of food safety, the Food and Consumer Product Safety Authority takes care of control and implementation of regulations. This institute also has a department dealing with animal diseases. The Ministry of Health similarly has a number of institutes that deal with various aspects of regulation and quality control in healthcare (e.g. medical inspection, registration of medicines, etc.).[11]

An interesting case where regulation, quality control and R&D are combined in one institute is the RIVM (National Institute for Health and the Environment). In this institution, R&D is combined with the development of methodologies and models used for regulatory policy (e.g. environmental norms with regard to the emission of certain substances). RIVM is formally a part of the Ministry of Health, but operates independently.[12]

International standards and norms (such as the ISO norm and the European EN45000 series of norms) are the basis for standardization and quality control in the Netherlands system of production. Specific national norms are almost extinct now. Implementation of these norms and accreditation of organizations that want to subscribe to the norms is in the hands of a private foundation (Raad voor Accreditatie). In line with the corporatist structure of the Netherlands system, the government supports this 'private initiative to support public interests' (Tweede Kamer, 2003).

In the field of electrical norms and standards, KEMA is a private organization that takes care of the implementation of norms and standards. This company was started in 1927 as the test-house of the national electrical company, and was for a long time part of the public utilities sector. Now it has been privatized and is also active in the field of (international) consulting.[13]

4.3 Provision of Constituents

4.3.1 Provision of organizations

The Netherlands NSI is relatively densely populated with actors of various types. The neo-corporatist nature of society reveals itself in the vague boundaries between publicly and privately provided organizations dealing with innovation. A prime example is the agriculture and related industries sector. Research in this sector is concentrated in a complex set of interacting institutes, for which the newly established Wageningen University and Research (WUR) Institute is a centre of gravity. This institute includes both the research part of Wageningen University and a number of other research institutes in the agro-field that were previously working under the heading of a sub-department of the Ministry of Agriculture. WUR now encompasses a network of specialized institutes, dealing with specific aspects of the industry. In terms of the number of employees, WUR is larger than any other university in the Netherlands.[14]

Not only pure research, but also quality control and veterinary aspects are addressed by WUR. Often, the institutes work in close cooperation with farmers' associations, both at a practical level (e.g. demonstration projects and new product development) and at the level of strategic planning (e.g. programmes aimed at developing new specializations for a region). Semi-public organizations such as the 'sector council' National Council for Agricultural Research (Nationale Raad voor het Landbouwkundig Onderzoek), which is financed by the Ministry of Agriculture, but acts independently, also play a role in setting out the strategic directions of innovation.

In this sector, one also finds private organizations that are essentially legitimized as collective organizations. The Aalsmeer flower auction, as well as other agricultural auctions, provide examples. One of the largest banks in the country, the Rabobank, also had its origins in this type of organization. The flower auction is quite active in terms of using ICT equipment to help its primary function, i.e. auctioning flowers. One of its primary innovative efforts has been aimed at allowing sellers and buyers to participate in the auctions at a distance, i.e. over an electronic connection. Since reaction speed is crucial in the auctions, the speed of electronic connections is of the utmost importance in such a project. But the auction's innovative activities have also been aimed more broadly at the agro-logistic chain.

In the field of public research organizations, recent government policy has been aimed at increasing the interaction between public and private parties. Section 4.1 has already described how this led to an increasing proprotion of market funding for TNO and other public research organizations. A new initiative that came out of this policy orientation was the establishment of so-called Technology Top Institutes (TTIs). These are newly founded institutes that perform research aimed at specific fields of interest to firms, and in which firms and public research organizations cooperate. Proposals for these TTIs were submitted by universities in cooperation with firms. A competitive process awarded funds to the most promising of these proposals. Currently, four TTIs exist, in polymer science (connected to Eindhoven University), food science (connected to Wageningen University), 'telematics' (connected to Twente University) and metallurgy (connected to Delft University).

4.3.2 Networking, interactive learning and knowledge integration
According to the 1998–2000 Community Innovation Survey (CIS3), 24 per cent of all innovating firms in the Netherlands innovated in partnership. This percentage is nearly equal between manufacturing and services. Individual sectors with particularly high rates of innovation partnerships were oil refining and the chemical industry, basic metals (in manufacturing), architects and engineering consultants and environmental services (in services). Firms within the same group, clients or suppliers, and competitors are the most important partners of those firms that innovate in partnership. Large innovating firms have a much higher propensity to cooperate (46 per cent) than small firms (21 per cent) or medium-sized firms (27 per cent).

The Ministry of Economic Affairs (2003) considers the low rate of cooperation between the private sector and the public sector as one of the weak points of the Netherlands system. This view is primarily based on an EU comparison of the percentage of higher education R&D financed by business enterprise sources. The European Commission (2003, Table 2.4.9) shows that this percentage is 5.1 in the Netherlands versus 6.9 in the European Union. But Tijssen et al. (2003, Fig. 2.16) show that this might be the result of the strong attractiveness for firms of the public research organizations in the Netherlands. The share of business financing in total R&D in the government and university sector stands at 10 per cent in the period 1995–2000, which is second in the EU only to Belgium (11 per cent). Tijssen et al. (2003, Fig. 4.3) also show that the share of co-publications between privately and publicly employed authors in total publications in the Netherlands is second highest in the EU (Belgium is again first).

Despite this mixed evidence, the Ministry of Economic Affairs has made public–private interaction for innovation one of its focal points for policy. The TTIs, where public and private parties perform joint research, have already been mentioned. An older form of public–private interaction is the so-called Innovation Research Programmes, which are research programmes aimed at a specific technology field (such as genomics or precision engineering). Broad outlines of these programmes are selected for funding, after which specific projects involving cooperation between private and public parties are awarded funds. In its most recent innovation policy document, the Ministry of Economic Affairs (2003) also announced several new measures aimed at increasing the level of public–private interaction, including a new policy instrument for fostering cooperation.

4.3.3 Provision of institutions

The institutional context of the Netherlands NSI is to an important extent determined by the policy environment created by the Ministry of Economic Affairs, as well as, to a lesser extent, the Ministry of Education, Science and Culture, the Ministry of Health and the Ministry of Agriculture. Innovation policy in the Netherlands has a relatively rich and long tradition, and the set of policy instruments is varied and wide.

For the last decade (broadly), the policy of the Ministry of Economic Affairs has embraced the idea of SI, and tried to develop specific policies consistent with this theoretical notion (e.g. Ministry of Economic Affairs, 2003). Nevertheless, traditional economic theory also still has an influence on policy thinking in this field. The debate where this tension is perhaps most obvious is the one on the generic versus specific nature of innovation policy. Argued from a traditional economics point of view, the idea of a (sector) specific policy is often connected to the practice of 'picking winners', and this is considered to be a bad idea (because of a lack of information on the side of policy makers) (see AWT, 2003). Using the point of view of SI, policy makers argue that policies need to be addressed to specific forms of systemic failures in specific parts of the system, and this often requires instruments aimed at only a small part of economic activity.

In practice, elements of both views have been incorporated in the policy programme. The most important generic policy instrument is the R&D tax credit measure, which is aimed mostly at small firms. Sectorally specific policies (subsidy schemes) exist in various fields, such as biotech and ICT (see also Section 4.3). Arrangements for subsidies and tax credits in the field of innovation are carried out by SenterNovem, an agency working under the Ministry of Economic Affairs. Cooperation and interaction are also an important element of policies inspired by the system view on innovation. Finally, also inspired by the systems idea, the Innovation Centres, recently renamed Synthens, play a role in diffusing technical and related knowledge to firms, especially small and medium-sized ones.

The involvement of a large range of public and private institutions in innovation policy sometimes raises problems of coordination. The biotechnology sector exemplifies this trend. A range of subsidy programmes has been, and still is, available for firms in this sector (Ministry of Economic

Affairs, 2004). But at the same time, in view of public opinion that many innovations in biotechnology (such as genetic modification) are dangerous for the environment, government has kept a relatively tight legal framework for experimentation in this sector. This has led to tension between far-reaching ambitions and the firms that may realize them, and the space that the Netherlands society (including non-governmental organizations such as Greenpeace as well as political parties) is willing to provide. For example, one of the first genetically modified mammals, the bull Herman, was bred in 1991 by the Pharming company (a Leiden University spin-off) with the aim of producing lactoferrine in cow milk. But this company later moved its activities in this field to Belgium, allegedly because of an excessively strict legal environment and negative public opinion in the Netherlands. Another example is the activities of the AVEBE company, which is a world leader in agricultural seeds, but which has problems getting permission for field experiments using genetically modified seeds.

The most important new element in the institutional set-up of the Netherlands NSI is the introduction of the so-called Innovation Platform in 2003. The Innovation Platform is modelled after the Finnish example (see Chapter 10, this volume), and is chaired by the Prime Minister. Its goal is to 'troubleshoot' the Netherlands NSI, and to provide indications of potential solutions. The platform meets several times per year, and has a bureau attached to it that organizes sub-committees involving people who are not actually members of the platform. Members of the platform are the relevant ministers, captains of industry, scientists and labour union representatives. So far, the actions of the platform have not been very visible, but among the topics that it has addressed have been issues of generic versus specific policy, and the importance of making it easier for foreign knowledge workers to enter the Netherlands (in response to the perceived labour shortages that were discussed in Section 4.1.2 above).

Another important discussion on institutions regards the role of intellectual property rights, especially in the public sector. Following the discussion on the Bayh–Dole Act in the USA, it has been suggested that university patenting is important for public–private knowledge transfer. Although theory in this field is far from clear in prescribing stronger university patenting as a remedy for a lack of interaction between universities and firms (e.g. Verspagen, 2004), the Ministry of Economic Affairs (Buijink, 2004) seems to be convinced that stronger patenting by universities in the Netherlands is necessary. The research council STW (formally part of NWO), which is responsible for the engineering sciences, already has an active policy of claiming property rights (patents) on all research undertaken using its funds, but such a policy does not seem to be in place yet for the broader activities of NWO (the general research council).

4.4 Support Services for Innovating Firms

4.4.1 Incubating activities

Although policies aimed at incubation activities have traditionally been part of the activities of universities and the government, the number of technology-based start-up companies is considered too low by the Ministry of Economic Affairs (2003). GEM (2003) indeed positions the Netherlands as a country with a low entrepreneurial nature. Both total entrepreneurial activity (or the number of new businesses per 100 adult population) and firm entrepreneurial activity (a composite index measuring entrepreneurship in existing firms) rank the Netherlands in the lowest group. Reliable and comparable data on specific technology-based entrepreneurship are not available.

The Ministry of Economic Affairs has launched a new programme called TechnoPartner to stimulate technology-based start-ups. According to Buijink (2004), an important reason for the low rate of technology-based start-ups lies with the public knowledge organizations in the Netherlands, which do not receive enough incentives to generate spin-offs, and do not have the necessary know-how and procedures installed. But little or no research exists to substantiate these claims, and the role of institutions like TNO, from which a large number of technology-based ventures originate, are ignored in this discussion. The TechnoPartner programme intends to provide seed capital (in the form of loans), subsidies for exploitation of public knowledge, and a platform for exchanging information about technology-based entrepreneurship.

Under the influence of this policy attention to technology-based start-ups, several universities have already initiated programmes to stimulate entrepreneurship. The three universities of technology are in the process of coordinating their activities with each other.[15] An important element of the so-called 3TU agreement is to give more attention to entrepreneurship in the course programmes for engineers and to start an institution called 3TULab exclusively aimed at screening for, generating and supporting technology-based start-ups. These activities are part of a larger range of activities on the boundaries of public and private initiatives, such as the Biopartner network (aimed at providing support to starting companies in the biotech field), and the LiveWire initiative by Synthens (the association of Innovation Centres in the Netherlands), etc.

An interesting element of the Netherlands system that remains largely unexplored until today is the notion of 'extrapreneurship' (Hulsink, 2003), i.e. the spin-off of new companies from existing ones. A prime example of this process is the highly successful firm ASML, which produces equipment for manufacturing electronic chips. This firm emerged as a spin-off from Philips Electronics in a rather problematic process involving conflicts between the ASML extrapreneurs and Philips management. Whether there is any unexploited potential for more of these corporate spin-offs remains an unexplored question. It is interesting in this respect that Philips Electronics is now implementing an active policy of 'open innovation' in its Eindhoven lab. This entails an open policy of on-campus cooperation with other firms, and stimulating entrepreneurship by its own employees.

4.4.2 Financing

In the CIS3 survey, referring to the period 1998–2000, 26 per cent of all innovating firms in the Netherlands system reported that a lack of finance for innovative projects had a weak negative influence on their innovative performance, 20 per cent reported that it contributed negatively, and 8 per cent reported that it was a strong negative factor. As compared to other possible innovation problems, this is a relatively low score (for example, inflexible organizational structures and a lack of qualified personnel score higher). This suggests that a lack of finance for innovative projects is not a large problem in the Netherlands system, at least not for established firms (as opposed to start-ups).

However, the problem of lack of finance for innovation is unequally distributed over sectors. In manufacturing, pharmaceuticals and basic metals a relatively high percentage of innovators report that a lack of finance is a problem that has a strong negative effect on innovative performance. Services, legal and economic consultancies, and architects and engineering services are sectors where finance is a considered to be a large problem.

An important element of the financial system geared to innovative activity is the provision of venture capital funds. As compared to the (European) countries in this study, the availability of venture capital in the Netherlands seems to be higher than in most other countries (except Sweden). The European Commission (2003, Table 3.5.2) presents quantitative data on venture capital investment as a percentage of GDP in 2001. The Netherlands value is 0.43 per cent, which is right on the EU average. Venture capital amounts to 49.2 per cent of total invested equity capital in the Netherlands, which is again right on the European average. Where the Netherlands is below the European average is on the growth rate of venture capital over the period 1995–2001 (15 per cent on an annual basis in the Netherlands versus 25 per cent on the European average). Also, venture capital in the Netherlands is aimed more at the expansion phase (as opposed to the seed and start-up phases). In summary, this suggests a relatively conservative attitude of venture capitalists in the Netherlands. The Ministry of Economic Affairs (2003) indeed considers this one of the weak points of the Netherlands NSI (especially the lack of venture capital available for the early stages of innovation).

4.4.3 Provision of consultancy services

The importance of consultancy services as a source of knowledge used in innovation can again be assessed using the CIS3 data. In the Netherlands survey, firms were asked to rate 14 different sources of knowledge used in the innovation process on a three-point scale (somewhat important – important – very important). Five of the 14 listed sources refer to organizations outside the firm's own direct environment. One of these five sources refers to 'private research bureaux', which is the closest to consultancies. The other four sources are public research institutes, universities, innovation centres and industrial organizations.

The outcome of the survey is that, overall, the importance of these five types of organizations as sources of knowledge is much greater than that of sources from within the firm itself or the direct environment of the firm (competitors, clients, suppliers or firms in the same corporate group). However, within the group of five, private research organizations are a relatively important source of knowledge. The most important sources of knowledge are industrial organizations (22 per cent of all innovating firms list this as an important or very important source) and private consultants (18 per cent). Public research organizations (11 per cent), universities (6 per cent) and innovation centres (6 per cent) rank well behind these two sources.

There is, however, a difference between manufacturing and services in this respect. Public research institutes are rated as somewhat more important (12 per cent) than private consultants (11 per cent) in manufacturing, Individual sectors for which private research consultants were relatively important are pharmaceuticals, finance, legal and economic consultants, and environmental services. It is likely that the importance of various (external) sources of knowledge differs between size classes, but, unfortunately, no information about this aspect has been made public by Statistics Netherlands.

4.5 Summary of the Main Activities Influencing Innovation

Concluding, one may say that the Netherlands NSI is relatively well developed. Although the Netherlands is a small country, the diversity and institutional range of its NSI compares well with that of larger countries, such as Germany or France. This is related to the historical phenomenon of the Netherlands being a relatively advanced country from the technology point of view.

Among the specific assets of the Netherlands system is first of all its relatively rich public (and semi-public) sector related to innovation. University research in the Netherlands is of high quality (as indicated by citation counts), although budget cuts have negatively influenced this part of the system. Public research organizations span a wide range of research topics in the Netherlands, and are responsible for a significant part of R&D expenditures. Increasing interaction with private firms (in terms of getting funds from the market) has been an important policy-initiated trend in this sector over the past decades.

Still, one gets the impression that the role of innovation and R&D in firms has been declining over the last 30 years, although perhaps only at a relative rate (i.e. compared to other countries). The stagnation of R&D intensity in the Netherlands system is to an important extent related to the (lack of) activities of firms in this respect, and the propensity to innovate in firms seems to be on a downward trend.

In summary, this appears to be a system that is relatively advanced from a structural point of view, but which is losing momentum in terms of performance. The Netherlands system is also one in which the specific pattern of specialization (in services, agro-food and selected manufacturing sectors) plays an important role, and in which, thus, a standard prescription of increasing R&D inputs would not necessarily be the best option. Because of the open nature of the economy, the role of globalization is also likely to play a large role (see Section 6 below).

5 CONSEQUENCES OF INNOVATION

Unemployment has traditionally been the economic issue on which governments in the Netherlands are evaluated, and wage politics has been the main policy instrument in this field. While this seemed to be a successful policy during the late 1990s, when unemployment was at a record low, the worldwide recession since 2001 has hit the Netherlands economy particularly hard. In the debate following this crisis, the link between productivity and innovation has become the central issue.

Productivity in the Netherlands is high. Among the countries in this study, output per hour worked is higher only in Norway and Ireland. In terms of GDP per capita, the Netherlands ranks lower only than Norway, Ireland and Denmark (see especially Table A2.2 in the Appendix). Overall, the country occupies rank 5 on the Human Development Index of the World Bank in 2002. Obviously, from a global perspective, this is rather high, but in the sample of countries in this study, two countries rank even higher (Norway and Sweden).

Figure 9.6 displays the productivity trends in the Netherlands economy over the period 1950–2001. Three different indicators have been used: GDP per capita (i.e. production divided by total population, as a crude indicator

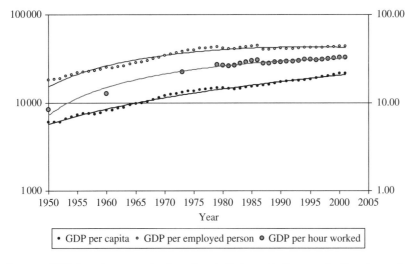

Source: GGDC total economy database; lines indicate quadratic regression lines (see main text); GDP per capita and per person employed are plotted on the left axis, GDP per hour worked is plotted on the right axis; both vertical axes have logarithmic scales, i.e. a straight line indicates a fixed growth rate over time.

Figure 9.6   Productivity trends in the Netherlands innovation system, 1950–2002

of economic wealth), GDP per worker (a rough measure of labour productivity) and GDP per worked hour (a more precise indicator of labour productivity). In each of the three cases, GDP has been measured in 1990 US$ in Gheary–Kamis purchasing power parities (a procedure aimed at making the numbers comparable between countries). The dots in the figure indicate actual observations. The lines indicate regression trend lines, based on a quadratic trend (various other functional forms for the trend lines were tested, but a quadratic trend yields the highest R^2 in all cases).[16]

The most obvious feature of the graph is the fact that productivity (in all three definitions) is slowing down over the 50-year period (the quadratic trend lines only fit the upward-sloping part of the hill-shaped parabola). This is consistent with the general pattern of a productivity slowdown in the OECD (developed economies) over this period. There are differences, however, in the extent of this slowdown among the three indicators. The indicator that shows least slowdown (i.e. flattest curvature) is GDP per capita. GDP per hour worked shows the strongest slowdown. Towards the end of the period (2001), this indicator approaches the maximum of the quadratic trend – i.e. a zero growth rate of GDP per hour worked has been approached.

Indeed, in the group of 25 (advanced) countries of the GGDC (Groningen Growth & Development Centre) database, the Netherlands shows a relatively low growth rate of GDP per hour worked. Over the period 1995–2001, only four of the 25 countries show slower growth,[17] and over the period 1990–2001 only a single country.[18] The situation is much the same when GDP per worker is considered, but totally different when looked at in terms of GDP per capita. In the latter case, the Netherlands occupies rank nine (1990–2001) or eight (1995–2001) in the growth rating of the 25 countries. The explanation for this paradoxical result is that employment in the Netherlands grew rapidly over the 1990s. Indeed, unemployment was the most central economic problem in the 1980s, as in many of the other EU countries. Over the 1990s, however, employment growth took off rapidly, including a much larger participation of women and part-timers in the labour market. Thus the relatively rapid growth of GDP per capita was achieved by working more and harder, rather than by productivity growth.

With the data available here, a more elaborate analysis addressing the relationship between innovation and economic performance is only possible at a broad cross-sectional level of 20 observations. For these sectors, we can calculate a number of CIS3 innovation indicators (share of innovators in a sector, share of product innovators, share of process innovators, share of radical product innovators, share of new products in turnover, share of products new to the market in turnover), as well as indicators of productivity growth (growth of value-added per hour worked) and structural change (change in share in total value-added, and share in total hours worked).[19]

Structural change in the Netherlands economy over the period 1990–2001 is rather low. Over this period, only five of the 20 sectors saw a change of their share in total value-added of one percentage point or more. These sectors were chemicals (−1 percentage point), electrical machinery (−1 percentage point), finance (+2 percentage points), computer services (+1 percentage point) and other services (including business services, +4 percentage points).

A simple correlation analysis of the available variables reveals no significant correlations at all between the innovation variables on the one hand, and the economic performance variables on the other hand. The productivity variables never generate any significant correlations, while some of the structural change variables correlate negatively with innovation (higher innovation goes hand in hand with shrinking shares of the sector in the total economy).

The absence of correlation between innovation and performance does not necessarily mean that innovation does not increase economic performance. The causal relation between innovation and the economy may simply be too complex, and too much interwoven with the general business cycles that greatly influenced economic developments in the 1990s, for our simple analysis to capture the essentials. At the firm level, Statistics Netherlands (2003) did observe differences between firms with and without innovations. Firms with innovations showed 2 percentage points higher turnover growth than firms without innovations, and 3 percentage points higher employment growth.

At the micro-level, the CIS3 shows that the most important goals for firms in their innovation processes were related to demand factors. For the total economy, 47 per cent of innovating firms claimed that innovation had strongly contributed to product quality; 39 per cent said that enlarging the product range had been a strong pay-off of innovation. Innovation goals related to processes (e.g. increasing production flexibility or lowering costs) scored much lower as important outcomes of the innovation process. In manufacturing, demand factors tend to be even more important than in the total economy.

6 GLOBALIZATION

The Netherlands is a small open economy that has participated from the beginning in the process of European unification. It is therefore likely to be strongly affected by globalization. One of the traditional debates to which this process has led is the extent of relocation of firms' activities to foreign countries, both in terms of production (with a direct impact on employment), and with regard to R&D. The latter has a more direct impact on the NSI, and has led to worries about a loss of competitiveness in the knowledge field, with indirect effects in terms of a loss of knowledge externalities and high-quality employment.

The impact of globalization on the general Netherlands economy is well illustrated using a range of indicators. According to data in Dunn & Bradstreet's Linkages database for 1998, the Netherlands attracts foreign multinational firms from a wide range of countries, and also sees its own firms active in foreign countries. An entropy index for the internationalization of large companies in Denmark, Finland, Hong Kong, Ireland, Korea, the Netherlands, Norway, Singapore, Sweden and Taiwan ranks the Netherlands on top of the list of internationalization, both inward (i.e. 'receiving' companies from the other countries), and outward ('sending' companies abroad).[20]

Knell (2003), on the basis of the OECD Activities of Foreign Affiliates database, calculates several more indicators of (inward) internationalization.

These show that in 1999, 28 per cent of manufacturing value-added is produced by foreign affiliates in the Netherlands. The top sectors in this respect are coke and petroleum products (52 per cent) and chemicals (49 per cent). One-third of manufacturing business R&D in 1999 was performed by foreign affiliates in the Netherlands. This share is significantly higher in selected industries, such as chemicals (61 per cent), pharmaceuticals (65 per cent) and instruments (70 per cent).

Evidence on the impact of globalization on business R&D for a sample of multinational enterprises (MNEs) that covers a significant part of total business R&D in the Netherlands is available in Goedegebuure (2003). Here, a sample of 35 firms is considered for 1996–99 and a sample of 15 firms for the longer period 1990–2001. The sample of 35 firms can be divided into four sub-groups of varying degrees of internationalization of R&D. Although no information on the identity of firms in the sample is provided, it is evident that Philips Electronics is not part of the sample (its R&D expenditures are too large compared to the total for sub-groups in the sample). Data on the four sub-groups are presented in Table 9.2.

The table shows that different patterns in the globalization of R&D exist: we find a range of strategies, from keeping R&D totally concentrated in the Netherlands, to relocating R&D abroad, to foreign MNEs who locate part of their R&D in the Netherlands. The group of companies that is mostly responsible for the public concern on relocating R&D is the third group. Over the period 1996–99, this group was responsible for a quarter of total Netherlands business R&D expenditures, but this excludes Philips, a company that shows the same behaviour. Over the 1990s, the share of R&D done abroad in this group of companies increased rapidly from 10 per cent to up to 44 per cent in the early twenty-first century.

Obviously, however, the impact of globalization on the NSI is much broader than just R&D. At the basis of the system lies the impact of globalization on human resources. Historically, the Netherlands has relied greatly on inward migration of skilled labour, for example during the period of the Reformation, when many Huguenots moved from France to the Netherlands. During the liberation war with Spain (sixteenth century), many fled from the economically leading southern part of the Netherlands, which remained under Spanish, and hence, Catholic rule, to the North, where the Spanish king had lost influence and the Reformists were ruling. On this basis, the Dutch Republic built its hegemony of the seas and world economy. It is only recently that the Netherlands attitude towards inward migration has become much more conservative.

Specifically with regard to highly skilled (knowledge) workers, the European Commission (2003, Table 4.4.5) concludes that the share of

Table 9.2 *Internationalization of business R&D by MNEs in the Netherlands*

	R&D in NL, average 1996–99	Share of NL R&D (%)	R&D worldwide average, 1996–99 (€ million)	Share abroad, 1996–99 (%)	Share abroad, 1998–2001 (%)	Share abroad, 1994–97 (%)	Share abroad, 1990–93 (%)
		Sample of 35 firms				Sample of 15 firms	
R&D concentrated in NL	516	14	516		0	0	0
R&D in NL stable	719	19	1434	50	50–60	50–60	50–60
R&D moving out of NL	930	25	4676	80	21–44	16–24	10
Foreign MNEs active in NL	273	7	10694	97	97–99	97–99	97–99

Sources: Goedegebuure (2003) and calculations on his data; data of Statistics Netherlands (Statline).

non-natives in the total working population, in highly skilled workers and in workers engaged in science and technology activities, lies just below the EU average (data refer to 2000).[21] Of the 4 per cent foreign-born science and technology workers active in the Netherlands in 2000, the largest part (2.3 per cent) are from the EU. Within the EU, about as many Netherlands citizens work abroad in science and technology as foreigners working in the Netherlands in this field.

7 STRENGTHS AND WEAKNESSES OF THE SYSTEM AND INNOVATION POLICIES

7.1 Strengths and Weaknesses

The Ministry of Economic Affairs (2003) provides an analysis of the Netherlands NSI in terms of five strong points and six weak points. Because this analysis is the starting point of the formulation of innovation policy for the near future, we take this list of strengths and weaknesses as the guidance for our discussion. The five strong points are:

1. High quality of scientific research
2. Good performance in patenting
3. High share of private funds in public research organizations
4. The diffusion of, and access to, ICT stands at a high level
5. The number of knowledge workers is high.

The six weak points are:

1. A backlog in business R&D intensity
2. A potential future shortage of knowledge workers, especially in the sciences and engineering subjects
3. Too little innovative entrepreneurship
4. Too little commercial usage of the results of scientific research
5. Weak points in the interaction between the private and public knowledge infrastructure: little private funding in university research, and little interaction between SMEs and public research
6. The financing of innovation is problematic.

Although these strengths and weaknesses tend to be formulated in terms of indicator benchmarks rather than true analysis, they do contain a number of useful ideas. But there is also room for disagreement, especially with regard to the role of the public knowledge infrastructure.

The Netherlands NSI does indeed have a high-quality basis in its university research (strong point 1). Partly, the renown of this university system resides in precisely those fields that do not have an immediate economic application. At the same time, it is another particular strong point (number 3) of the system that it has a strongly developed set of public research institutes, which operate at a high-quality level, and interact extensively with private firms. The way in which firms in the Netherlands system are fed with ideas and knowledge from publicly financed research must be seen from the joint perspective of universities and public research organizations. Together, these two types of actors have a specific role to play in the system.

The main question that remains from an analysis of the strong and weak points of the Netherlands NSI is why business R&D is so weak. Two obvious factors in explaining this are the strong impact of large multinational firms, and the specialization pattern of the Netherlands. With regard to the latter, the Netherlands has traditionally been locked into a sectoral structure in which manufacturing plays a relatively minor role. Successive waves of industrialization since the 1860s have, of course, brought industrial production to the Netherlands on a large scale, and specific strongholds (chemicals and electronics) have emerged. But the Netherlands has never wholeheartedly become an industrial economy. Agriculture (and related industries) and services are the historical sectors in which the Netherlands economy has been, and remains, specialized. Electronics and chemicals have been able to influence the overall business R&D intensity for a while, but with the emergence of a post-industrial society, these sectors have proven not to be the main drivers of industrial R&D that one might have thought them to be. Large electronics and chemicals firms will remain dominant within business R&D in the Netherlands for a while, but one should not expect strong growth of business R&D, either from them or from the other established sectors in the Netherlands. Growth of business R&D in the Netherlands will require a targeted policy, taking into account where the 'new economy' will have its focus in terms of technological development.

7.2 Summary and Evaluation of the Innovation Policy Pursued

Innovation policy in the Netherlands has a relatively rich tradition. It encompasses a large range of activities within the NSI, and has recently, at least partly, adopted a systems perspective for its design and evaluation. This development has been inspired both by the SI literature, and by Porter's idea of clusters (e.g. Porter, 1985). Innovation policy makers of the Ministry of Economic Affairs have initiated SI analysis of the Netherlands economy, they have written on the policy implications of the SI idea (addressing, among other things, systemic failure), and they have played a leading role in the OECD work aimed at measuring and benchmarking SI. Analyses aimed at identifying and examining Porterian clusters in the Netherlands system have also been undertaken by the Ministry of Economic Affairs.

At the same time, ideas from traditional economic analysis on market failure and how to address it continue to play a role in innovation policy. In this respect, the Ministry of Economic Affairs, as the prime ministry responsible for innovation policy, still takes its role as a guardian of the competitiveness level of Netherlands firms very seriously. This implies that, for example, specific subsidy schemes and export promotion are also still closely connected to innovation policy. More recently, competition policy has become an aim of the Ministry of Economic Affairs, and the relation between competition and innovation is now one of the focal points. Although the theory of SI does not specifically promote market power as a means of raising innovation performance, it seems clear that too much emphasis on competition as a way to stimulate innovation may be counterproductive.

A large problem with innovation policy has been the cutting of budgets. Universities, public research organizations, subsidies schemes, etc. have all been cut severely by successive governments over the past decades. An indication of this is the virtual lack of growth of publicly financed R&D as a percentage of GDP over the period since 1970.

7.3 Future Innovation Policy

The Ministry of Economic Affairs (2003) has recently announced an action programme for stimulating the knowledge economy in the Netherlands. This programme has three focal points:

1. Stimulating the innovation environment
2. Stimulating more firms to innovate
3. Focusing more on public research efforts in an attempt to raise pay-offs of public research.

Specific measures have been proposed in order to achieve these three issues. Examples of these measures are an increase of the available budget for the R&D tax credit by €100 million, a new policy instrument aimed at intensifying R&D cooperation, various measures to stimulate students to follow innovation-relevant study programmes, the TechnoPartner initiative (see Section 4.4), and a focus on ICT, life sciences and nanotechnology.

Although the proposed set of policy initiatives looks elaborate, the available budget for these measures is not very large. The Netherlands government has implemented severe budget cuts in an effort to maintain its obligations under the Stabilization Pact. This has also implied that little extra money is available for innovation policy, although the government has made this a focal point of its overall policy. Whether it is possible to implement an effective innovation policy with so little extra money is questionable at least.

In addition to this, on the basis of the above analysis, one may argue that the proposed policies overlook an important aspect of the Netherlands NSI, to the extent that they propose measures to stimulate interaction between university research and private firms. The evaluation of the 'bridge function' of public research organizations between fundamental science and business innovation efforts was one explicit point in the proposed plans. This evaluation had, in the meantime, been undertaken (by the so-called Committee Wijffels), and the conclusion, in broad terms, was that more efforts are necessary to increase incentives for public research organizations to interact with the market. However, this has been the aim of policy for the last decade at least, and it is unclear how continuation of this trend will really make a difference. Applying such logic to the university system may indeed lead to an erosion of the strong points of this system – i.e. its high-quality fundamental research may be substituted for more applied work that is already undertaken in the public research organization sector.

From the NSI perspective adopted here, it might make (more) sense to look at the interaction between the university system and the public research organizations. At least three issues stand out here as possible alternative directions for policy development. First, although some degree of cooperation between public research organizations and universities has been achieved, for example in the TTIs discussed in Section 4.3, this may still be intensified. A closer coordination between high-quality (fundamental) university research and applied research of, for example, TNO is an attractive option to increase the linkages between the market and fundamental science in the Netherlands system.

Second, the system of research councils in the Netherlands is still mainly aimed at universities. The research council STW, aimed at the engineering sciences, is an exception to this, because it involves 'users' in its projects. But more opportunities for public research organizations to get their research financed in the broad research council system, in cooperation with universities, would benefit the interaction between different types of research. Obviously, this would have to involve a substantial increase of the budgets of the research councils; otherwise, such an operation would lead to a (renewed) budget cut for universities.

Third and finally, one may think about strategic choices on the topics addressed by public research organizations in the Netherlands. The specialized institutes in this sector are all aimed at relatively old specializations, such as civil waterworks and shipbuilding. A thorough evaluation of how much these topics contribute to the current innovation strengths of the Netherlands might be in order, and could lead to a shift in focus of efforts in this field. 'New' topics such as ICT and life sciences are exclusively addressed by new organizations like the TTIs, but it might be beneficial to have an old-fashioned, publicly financed research organization active in this field.

NOTES

1. The first innovation survey in the Netherlands was done for the year 1983 (Kleinknecht, 1987); a second one was done for the period 1988, but did not include any questions on product or process innovations (Kleinknecht et al., 1990); the CIS1 results were reported in Brouwer and Kleinknecht (1994). The methods of these early surveys were crude as compared to the recent ones (also the samples were smaller), so that the margins of error were probably higher in the beginning. The CIS2, CIS'2½' (an intermediate survey between the ones coordinated by Eurostat) and CIS3 were done by Statistics Netherlands and are available in their online database Statline.
2. R&D is also a part of total innovation costs, i.e. the first indicator.
3. The Netherlands Innovation surveys (CIS2 to CIS3) do not include engineering services, which is normally included in KIBS.
4. http://www.cpb.nl/nl/data/rd/.
5. According to data collected at the time by the Ministry of Economic Affairs.
6. In some cases, these cuts were relative (i.e. spending grew less than GDP), and in some cases the cuts were absolute (i.e. the actual budget of specific educational institutes was cut).
7. All data on source of financing of university researchers are drawn from Tijssen et al. (2003), Figure 2.31.
8. Information in this paragraph is taken from various TNO annual reports and the TNO website.
9. Absent bars in the bottom panel of Figure 9.5 indicate missing values in the OECD database, allegedly because these categories are not relevant in the Netherlands system.
10. Rijn Schelde Verolme (RSV) was the name of a shipyard in which government invested heavily, but which, after a long time, went bankrupt.
11. See www.vwa.nl for details.
12. See www.rivm.nl.
13. See www.kema.nl.
14. Tijssen et al. (2003), Table 2.29; although this table lists WUR and LEI separately, LEI is now formally part of WUR (LEI website, August 2004).
15. See www.3tue.nl.
16. A linear trend, a logarithmic trend, a power trend and an exponential trend.
17. The four countries are Canada, Italy, New Zealand and Spain.
18. New Zealand.
19. All data on the economic indicators are taken from the GGDC 60-sector industrial database; all innovation data are taken from Statistics Netherlands, Statline.
20. The indicator can be calculated in two ways: including or excluding domestic companies in the country for which the indicator is calculated. With inward and outward

internationalization, this yields four possible indicators. On three of these, the Netherlands ranks first; on the fourth (outward, including domestic companies), the Netherlands ranks second (Sweden is first).
21. It is unclear, however, how the data in various countries are influenced by relationships with former colonies of various countries (in the Netherlands, Surinam).

REFERENCES

AWT (Adviesraad voor het Wetenschaps- en Technologiebeleid) (2003), 'Backing Winners', Van Generiek technologiebeleid naar actief innovatiebeleid, Den Haag, advies no. 53.
Brouwer, E. and A.H. Kleinknecht (1994), *Innovatie in de Nederlandse Industrie en Dienstverlening (1992)*, Den Haag: Ministerie van Economische Zaken.
Buijink, C. (2004), Presentation at the TechnoPartner Workshop, 18 May.
European Commission (2003), *3rd European Report on Science and Technology Indicators*, Brussels: European Commission, DG Research.
GEM (2003), *Global Entrepreneurship Monitor 2003, Executive Report*.
Goedegebuure, R. (2003), 'Internationalisering van Innovatie: een verkenning van twee gegevensbronnen', in *Kennis en Economie 2002*, Voorburg/Heerlen: Centraal Bureau voor de Statistiek, pp. 273–84.
Griffiths, R.T. (1979), *Industrial Retardation in the Netherlands 1830–1850*, Den Haag.
Hulsink, W. (2003), 'Extrapreneurship, Nieuwe combinaties van ondernemers, ideeënleveranciers, investeerders en incubators', mimeo, Erasmus Universiteit Rotterdam.
Knell, M. (2003), 'On measuring structural change in small European and Asian economies', mimeo, University of Oslo, TIK.
Kleinknecht, A. (1987), *Industriele innovatie in Nederland*, Assen and Maastricht: Van Gorcum.
Kleinknecht, A.H. (1998), 'Mythen in de Polder', inaugural lecture, Delft University.
Kleinknecht, A.H., J.O.N. Reijnen and J. Verweij (1990), *Innovatie in de Nederlandse Industrie en Dienstverlening*, Den Haag: Ministerie van Economische Zaken.
Ministry of Economic Affairs (1999), *De Digitale Delta*, The Hague, various ministries.
Ministry of Economic Affairs (2003), *Innovatie brief*.
Ministry of Economic Affairs (2004), *Actieplan Life Sciences*, The Hague.
OECD (2003), *Education at a Glance*, OECD: Paris.
Porter, M. (1985), *Competitive Advantage*, New York: Free Press.
Schot, J.W. (1995), 'Innoveren in Nederland', in H.W. Lintsen, M.S.C. Bakker, E. Homburg, D. Van Lente, J.W. Schot and G.P.J. Verbong (eds), *Techniek in Nederland. De Wording van een Moderne Samenleving 1800–1890*, Stichting Historie der techniek/Walburg Pers.
Schot, J.W. and D. Van Lente (2003), 'Techniek, industrialisatie en de betwiste modernisering van Nederland', in J.W. Schot, H.W. Lintsen, A. Rip and A.A.A. de la Bruhèze (eds), *Techniek in Nederland in de twintigste eeuw. VII. Techniek en modernisering. Balans van de twintigste eeuw*, Stichting Historie der techniek/Walburg Pers, pp. 257–83.

Statistics Netherlands (2003), *Kennis en Economie 2002*, Voorburg: Centraal Bureau voor de Statistiek.

Tijssen, R., H. Hollanders, T. Van Leeuwen and B. Verspagen (2000), *Wetenschaps- en Technologie Indicatoren 2000*, Leiden and Maastricht: CWTS/Merit, http://www.nowt.nl.

Tijssen, R., H. Hollanders, T. Van Leeuwen and T. Nederhof (2003), *Wetenschaps- en Technologie Indicatoren 2003*, Leiden and Maastricht: CWTS/Merit, http://www.nowt.nl.

Tweede Kamer (2003), *Certificatie en accreditatie in het kader van het overheids- beleid, Tweede Kamer der Staten Generaal*, vergaderjaar 2003/2004, stuk 29304.

Van Zanden, J.L. and A. Van Riel (2000), *Nederland 1780–1914. Staat, Instituties and Economische Ontwikkeling*, Amsterdam: Balans.

Verspagen, B. (2004), 'University research, intellectual property rights and the Dutch innovation system', Ecis, Eindhoven University of Technology, paper commissioned by the Ministry of Finance.

Wang, K., M.-H. Tsai, C.-C. Liu, I.Y.-L. Luo, A. Balaguer, S.-C. Hung, F.-S. Wu, M.-Y. Hsu and Y.-Y. Chu (2003), *Intensities of Scientific Performance: Publication and Citation at a Macro and Sectoral Level of All the Nine Countries* (version of 20 November 2003), STIC, Taipei.

10. Not just Nokia: Finland
Ville Kaitila and Markku Kotilainen

1 INTRODUCTION

Industrial development in Finland can be divided into three phases: (1) a factor-driven economy from the mid-1800s to the early 1900s; (2) an investment-driven economy from the end of the Second World War to the 1980s; and (3) an innovation-driven economy since the late 1980s. The innovation-driven economy developed as a result of increasing science and technology (S&T) content in production. Gradually, enhancing this development also became a political target. The birth of the Finnish S&T policy goes back to the 1960s and 1970s. This policy framework served as an important basis for the development of an explicit Finnish national system of innovation (NSI), which reached more or less its present form in the 1990s.

Early in the development of the NSI the key actors were public. An important driving force behind the development of the NSI was the idea of strengthening economic competitiveness and diversifying the production structure. Since the early 1990s, private firms have become more important, especially in the information and communication technology (ICT) sector. Finland experienced a severe depression in the early 1990s, and the recovery from it was to a large extent due to fast growth in the ICT sector (see e.g. Paija, 2001, for an analysis of the evolution of the Finnish ICT cluster).

Lately, innovative activity in Finland has been dominated by the electronics industry as reflected in the success of this sector, and particularly of Nokia. Application of ICT in other sectors is less widespread in Finland than, for example, in the USA. Knowledge-intensive business services (KIBS) are, however, also innovative to some degree in Finland. The two other pillars of the manufacturing industry (besides the electronics industry) are the manufacturing of paper and the manufacturing of machinery and equipment. The former is good in process innovations; the latter, in product innovations. There are, thus, also innovative manufacturing sectors and firms in the country other than just the electronics industry and Nokia.

Research and development (R&D) and competence building have always been crucial elements in the system. Provision of organizations has to a large extent been based on the realization of these activities. Furthermore, networking between different firms and other organizations has been a key factor in the success of the Finnish NSI.

The future challenges of the Finnish NSI include the widening of innovative activities from the ICT sector to other manufacturing industries and to service sectors. In addition to technical innovations, the role of organizational innovations should be strengthened, and technical and organizational innovations should be integrated more than they currently are.

2 MAIN HISTORICAL TRENDS

Through the centuries, promotion of education and learning has been the most important way of enhancing innovativeness. In Finland, as in many other countries, basic education was initially provided by the Church for those who were able and interested in learning. The first university in Finland was founded in Turku in 1640. Before that, Finns received their higher education mainly in Germany and after that in Sweden. The university was moved from Turku to Helsinki in 1827. The current Helsinki University of Technology was founded in 1849 and the Helsinki School of Economics in 1911.

Legal property rights and freedom of economic activity were important factors for the development of economic incentives. In the promotion of agriculture, the land reform of the late eighteenth and early nineteenth centuries was crucial. The land reform turned the farms into private property, which could be inherited. Before that, land was owned by the state (the crown) and rented by farmers either directly or against taxes. Private ownership existed in trade and industry but it was quite strictly regulated until the end of the nineteenth century.

The industrialization of the country took place fairly recently by West European standards. It was heavily dominated by the development of wood processing. Sawmills were developed and pulp and paper factories started to expand. Foreign entrepreneurs were important in the early development of some manufacturing industries. Imports of machinery and consumer goods as well as journeys by specialists abroad were important in introducing foreign innovations (see Myllyntaus, 1992). This has been a general pattern in all Nordic countries.

Finland's Economic Society, founded in 1797, was one concrete example of non-governmental initiatives to promote innovativeness and economic activity. The leaders of the Society consisted of the most important members of the Finnish intellectual elite. In the beginning, the main task of the Society was to fight human and animal diseases.

Finland became independent in 1917, in the aftermath of the Russian Revolution. The country had been a part of Sweden for over 600 years until 1809, and after that an autonomous grand duchy of Russia.

The twentieth century saw the state actively participating in the deepening and widening of the industrial base. State-owned companies were founded and some previously privately owned and bankrupted ones were bought by the state. In the 1920s, the state continued to support agriculture and to develop the transport and communication infrastructure as well as the education system. Obligatory basic education was established in 1922, which is early by international standards. In the 1920s and 1930s, about 10 per cent of the eligible age cohort studied in secondary schools.

After the Second World War, at least two important interrelated tendencies marked Finnish society: (1) rapid economic development and (2) the build-up of a welfare state and free education. They both also supported innovativeness in the Finnish economy.

The birth of an explicit science and technology policy goes back to the 1960s and 1970s. According to Lemola (2002), this was later than in many other OECD countries. Several factors contributed to this development. These included: (1) the need to strengthen the competitiveness of the economy and to diversify the production structure; (2) the widening system of higher education (combined with the better-educated post-Second World War baby-boom generation); and (3) closer integration with the OECD and its science and technology policy.

The Finnish approach to S&T policy, both the policy doctrines and the institutional and organizational models, was largely adopted and imitated from several OECD countries, especially from Sweden, the UK and the USA. In the 1980s, Japan provided an inspirational benchmark case. As Japan's economic performance dwindled at the turn of the 1990s, so did its appeal as a model case. The most important single aim of Finnish science policy since the late 1960s, and of S&T policy since the 1970s, has been the growth of R&D expenditure as a proportion of gross domestic product (GDP) (Lemola, 2002).

A ministerial committee on science, the Science Policy Council (later Science and Technology Policy Council (STPC)) was established in 1963. It is a high-level political body for the formulation of S&T policy guidelines. It is also a coordinator of the ministries that deal with different aspects of these matters.

A reform of research councils, and the formation of the Academy of Finland in 1969–71 marked an important step in science policy and in research funding. The Academy founded new research posts and started funding project research. It became a central organization in research funding but also an important actor in research policy.

An important instrument in the promotion of industrial R&D, the Finnish National Fund for Research and Development (Sitra) was founded in 1967. The Ministry of Trade and Industry (MTI) started to finance firms' R&D in 1968. However, the oil crisis of the early 1970s slowed down the development of S&T policy. Economic resources were not allocated to R&D as generously as before. Planning optimism also declined in the face of these exogenous shocks, which could not be avoided by policy measures.

In the late 1970s and early 1980s, there was a resurgence of development optimism and, at the same time, a need to invest more in R&D. Most OECD countries, including Finland, started to fund large national, cooperative programmes for the development of new technologies, primarily information technology (IT), materials technology and biotechnology (Lemola, 2002).

In Finland, the role of new technology in economic growth and employment creation was highly appreciated, and emphasis on new technology became a new core for the S&T policy. A new organization, Tekes (the National Technology Agency), was founded in 1983 to promote technological R&D and diffusion of technology in Finland. National technology programmes were developed to give Tekes a framework for controlling and promoting R&D. The first programmes concentrated on IT.

As in many other countries, several organizations were founded in Finland at the national as well as regional levels to support technology transfer, diffusion and commercialization. Nationwide networks of technology parks and centres of expertise were set up. Technology parks initiated spin-off projects and business incubators. Technology transfer companies were established to commercialize the results generated in universities and research institutes.

In its *Review 1990*, the STPC took the NSI concept as a starting point for its policy and gave it the following definition:

> An NSI means a whole set of factors influencing the development and utilisation of new knowledge and know-how. The concept allows these factors and their development needs to be examined in aggregate. In addition, it offers a framework for analysing interrelationships between different factors. These relationships are relevant to general development capability and they have proved to be essential for the creation of new innovations. (Science and Technology Policy Council of Finland, 1990)

3 INNOVATION INTENSITY

According to the summary innovation index of the European Innovation Scoreboard 2004, Finland was second in innovativeness among EU countries (after Sweden). Japan was ahead of these countries but the USA was ranked after Finland. The index is a composite indicator, consisting of 12–20 individual indicators depending on the country (European Commission, 2004).

In terms of innovation intensity in the Community Innovation Survey (CIS), Finland ranked second after Sweden within a group of six countries that included Denmark, Ireland, the Netherlands and Norway (see Appendix Tables A4.1–A4.15). The Finnish manufacturing sector also ranked second, again after Sweden, in intensity of innovativeness and R&D. The share of innovative firms was the third highest after the Netherlands and Denmark. In this respect, there was a clear improvement between 1996 and 2000 (based on comparing CIS2 and CIS3).

Within the above-mentioned group of six countries, Finland ranked first in the share of new-to-the-market product innovations. The position was the same irrespective of the size or the sector. There was a slight improvement between 1996 and 2000.

Finland ranked third, again after Denmark and the Netherlands, in the share of new-to-the-firm product innovations and process innovations. In this respect, also, there has been a clear improvement in Finland between 1996 and 2000. There was no substantial difference between small and large enterprises. The trade sector had the second-highest ranking in terms of product and process innovations. KIBS and the manufacturing sectors also had quite good rankings.

In what follows, the propensity to innovate is measured in different sectors using nine indicators from CIS3. There are no data available for financial services. The results for Finland are presented in Table 10.1. We highlight in bold the three highest-ranking sectors in each indicator.

Radio, TV and communication equipment is among the three best performers in all except one case (introduction of new processes, where it is fourth). This industry has the highest ranking in five out of eight quantitative indicators. Thus the propensity to innovate is the highest in the manufacturing of communication equipment. Its strengths lie especially in the share of new products, the share of firms introducing new-to-the-firm products, the share of firms introducing new-to-the-market products and the share of innovative firms. In the case of non-technological innovations, this sector concentrates on organizational innovations. The communication equipment industry produces mainly mobile phones and mobile phone networks. The largest firm in the sector is Nokia. For a discussion of the Finnish ICT cluster and its history, see Paija (2001).

The second-best sector is KIBS, which is among the three highest-ranking sectors in four out of eight indicators. It is second in R&D intensity, share of new products and share of new-to-the-market products, and third in innovation intensity. On the other indicators, KIBS also performs reasonably well, except in the introduction of new processes, where it is

Table 10.1 Propensity to innovate in Finnish firms by sectors

Sector	ININT2K	RDINT2K	TURNIN	TURNMAR	NPDT	NTTM	INPCS	INNO	MODNTIA	% of GDP in 2003
All enterprises	3.75	2.73	9.02	5.98	35.06	28.08	23.50	40.45	3	63.6
SMEs	3.21	2.87	8.49	5.89	32.22	26.17	20.89	37.39	3	–
Large enterprises	3.83	2.72	12.29	6.52	52.50	39.87	39.51	59.28	3	–
Food, beverages & tobacco	1.37	0.92	5.03	1.98	25.85	16.35	25.08	32.34	5	1.8
Textiles, clothing & footwear	2.35	2.30	11.14	7.81	50.80	38.39	19.56	51.50	5	0.4
Wood & cork products	0.51	0.26	4.24	1.39	17.34	11.67	28.82	33.15	2	1.0
Pulp, paper, paper board	4.48	0.81	4.57	2.47	33.99	22.53	**39.74**	47.62	3	3.0
Printing, publishing	2.21	1.17	4.73	3.37	24.66	21.25	28.08	36.34	3	1.3
Oil & chemical products	5.16	4.66	8.04	3.52	52.53	37.60	**37.01**	**62.40**	3	1.6
Rubber & plastic products	0.88	0.79	7.84	5.57	37.89	34.73	34.16	50.28	1	0.8
Non-metallic mineral products	4.03	1.05	6.08	4.95	28.16	23.34	31.06	35.53	2	0.8
Metal processing	0.70	0.63	3.51	2.05	46.02	24.41	**48.75**	52.92	3	1.0

Sector									
Metal products	3.54	1.79	9.15	3.44	32.70	24.19	24.57	39.48	3	1.6
Machinery & equipment	4.12	3.74	13.64	**10.17**	48.52	**41.17**	26.86	50.33	3	2.5
Office machines & computers	2.31	1.97	**14.15**	9.15	**58.32**	40.77	31.20	**62.19**	3	0.6
Radio, TV & communication equipment	**9.87**	**7.39**	**35.91**	**25.88**	70.07	**54.53**	36.20	**76.08**	3	4.6
Medical & optical equipment	**12.93**	**10.49**	13.12	6.85	**55.00**	38.07	27.90	55.00	2	0.6
Production of cars & lorries	3.60	2.37	7.99	7.99	18.91	18.91	15.59	22.14	5	0.3
Other transport equipment	0.90	0.71	8.00	5.32	25.91	22.89	27.54	32.78	3	0.6
Furniture, recycling	2.58	0.97	10.24	7.11	46.78	**43.57**	34.20	51.83	5	0.5
KIBS	**8.43**	**7.67**	**15.66**	**10.33**	49.00	38.14	22.81	51.13	3	≈2.5
Finance	–	–	–	–	–	–	–	–	–	3.2
Trade	1.37	2.02	11.12	8.70	39.19	33.34	16.68	40.77	3	10.6
Other	1.14	0.86	3.48	2.53	18.62	15.56	16.74	24.69	3	≈24.3

Notes: ININT2K: Innovation intensity in 2000; RDINT2K: R&D intensity in 2000; TURNIN: Share of new products in turnover; TURNMAR: Share of new-to-the-market products in turnover; NPDT: Share of innovative firms who introduced new products; NTTM: Share of innovative firms who introduced new-to-the-market products; INPCS: Share of innovative firms who introduced new processes; INNO: Innovative firms; MODNT1A: Main non-technological innovation of innovative firms (1: Strategy; 2: Management; 3: Organization; 4: Marketing; 5: Aesthetic). SMEs = Small and medium-sized enterprises. GDP shares are estimates for 2003.

Sources: Community Innovation Survey (CIS), Number 3; Statistics Finland; Eurostat.

below average. The main way of innovating non-technologically is by introducing new organizations.

Medical and optical equipment is a small sector in Finland. It ranks among the three best-performing industries on three measures: innovation intensity, R&D intensity and the share of innovative firms introducing new-to-the-firm products. In the first two, this sector has the leading position. The main non-technical innovation is the implementation of advanced management techniques.

Office machines and computers has a small output share too. It is among the three best-performing sectors in the share of new products in turnover, the share of innovative firms introducing new products and the share of innovative firms. The main focus in this sector is clearly in the diffusion of innovations. The performance of this sector is below average in innovation intensity and R&D intensity. The main non-technological innovation is implementation of new organizational structures.

Machinery and equipment, and oil and chemical products are among the three best performers on two measures. The machinery and equipment sector occupies this position in the share of new-to-the-market products and in the share of innovative firms that introduced such products. The sector invests heavily in new innovations. These include paper processing machines, different kinds of engines, forest harvesters and lifting automation. The sector is also above average on all other indicators. Organizational innovations are the main way to innovate non-technologically. The oil and chemicals sector is among the three best performers in introducing process innovations and in the share of innovative firms. Organizational innovations are also the main non-technological innovation in this sector.

Pulp and paper is among the three best performers in the introduction of new process innovations. This sector performs below average on six indicators. In the case of the share of innovative firms it is somewhat above average. In this industry products remains relatively unchanged for quite long periods of time. Innovations thus occur mainly in processes. Non-technological innovations occur mainly in organizations. The paper industry has important links to the rest of the economy. The paper cluster includes, for example, forestry, wood industry, production of paper chemicals as well as parts of the transportation and energy sectors. Innovations in these sectors are often linked to each other.

Metal processing performs the best in the introduction of new processes. The products are often fairly standardized. Non-technological innovations are mainly organizational. Furniture and recycling performs relatively well in the share of firms who introduced new-to-the-market products. It is below average in innovation and R&D intensities. Non-technological innovations occur mainly as aesthetic changes.

The rest of the sectors are not among the three best performers on any of the indicators. For example, the foodstuffs industry is below average in all criteria except in introducing new processes, where it is slightly above average. This is again understandable, as the majority of foodstuffs stay the same for quite long periods of time. Innovations occur more often in processes, but even in this case the sector is not among the best performers. About 24 per cent of value-added is produced in firms included in 'other sectors'. According to the survey, these are not particularly innovative.

According to the European Innovation Scoreboard 2004 (European Commission, 2004), the Finnish electrical and optical equipment sector ranks at the top in innovativeness among the EU countries. In Finland, this sector consists mainly of the production of telecommunication equipment. The index number was 0.62 in Finland, 0.46 in Belgium, 0.45 in the Netherlands, 0.42 in Sweden and 0.37 in the EU-15 (EU countries' residents on 1 January 1998) on average. Even if the manufacturing of textiles and clothing is not an important sector in Finland, its innovativeness was ranked first among the EU-15 countries with 0.20 points, with an EU-15 average of 0.11. This good performance was partly due to a high number of innovative firms and a high share of new-to-the-market and new-to-the-firm products.

When looking at innovativeness in terms of firm size, large firms performed better than small and medium-sized ones on all except one indicator, namely R&D intensity (see Table 10.1 and Appendix Tables A4.1–A4.15).

Organizational changes were the main way of innovating non-technologically in 12 out of 20 industries; this was the case in small as well as in large enterprises. In four industries it was aesthetic change, in three industries management and in one industry implementation of a new corporate strategy.

Innovations are mostly implemented by firms registered in Finland. However, the largest of them are international (like Nokia, all paper companies and the most important machinery firms). Some of these firms also have innovation activity abroad. Currently, it is still modest when compared to that conducted in Finland.

4 ACTIVITIES THAT INFLUENCE INNOVATION

4.1 Knowledge Inputs to Innovation

4.1.1 R&D activities

Extent of R&D and main actors   In 2005 total R&D expenditure in Finland was 3.5 per cent of GDP. In the early 1980s it had been just slightly above

1 per cent, among the lowest in the OECD. The minimum target for the EU in 2010, set at the Barcelona Summit in March 2002, is 3 per cent of GDP. Finland and Sweden are currently the only EU countries to reach this target.

In the early 1980s, the share of public funding in total R&D expenditure was about 45 per cent and even in the 1990s it was about 40 per cent. In 2004 it had declined to 28 per cent, reflecting fast growth in private R&D. However, public R&D also grew rapidly by international standards during this time.

The Ministry of Education was responsible for 34.5 per cent of the public R&D, Tekes for 18.9 per cent, the Academy of Finland for 11.4 per cent, the Ministry of Social Affairs for 8.5 per cent, the Ministry of Agriculture and Forestry for 6.3 per cent and the MTI for 4.9 per cent. EU funding has increased gradually since Finland's EU membership in 1995 but it is still small. The shares of the business sector and of foreign countries were 67.8 and 3.2 per cent, respectively.

The decline in the share of public R&D funding has raised concern about the long-term viability of innovation activities. R&D in especially new areas has traditionally depended on public funding. There is a danger that there will not be enough new opportunities for innovation based on new developments in publicly funded research. In addition to the electronics industry, biotechnology currently receives substantial amounts of public R&D funding.

The share of the business sector in performing R&D has increased from 55 per cent in the early 1980s to 70 per cent in 2004. The share of the university sector was 20 per cent and that of the rest of the public sector 10 per cent in 2004.[1] The business sector financed 93 per cent of its own R&D. About 5–6 per cent was financed by the public sector, mainly by Tekes.

In 2004, 57 per cent of business sector R&D was performed in the electronics and telecommunications equipment industry. The metal products and machinery industries ranked second with an 8 per cent share, followed by data processing services with a 7 per cent share. This is the main sector where the role of KIBS is reflected. The chemical industry was fourth with a 6 per cent share. R&D had a 5 per cent share.

Large firms dominate Finnish R&D. This is partly due to their central role in production and exports (85 per cent of total merchandise exports). In this respect Finland resembles Sweden (see Chapter 7 in this volume). Nokia alone performs about one-third of total R&D in Finland (see Box 10.1). In biotechnology, firms are small but the R&D performed in them is substantial. Biotechnology firms are currently at the very beginning of their product development work (see Hermans, 2004). In R&D, as in other areas, the role of large companies has been strengthened in the Finnish economy during the past 20 years.

BOX 10.1   NOKIA – A BIG COMPANY IN A SMALL COUNTRY

Nokia is an international firm producing wireless telecommunication equipment and solutions. It produces mobile phones, telecommunication networks, multimedia and enterprise solutions. Until the early 1980s, it was a multi-sector company producing several kinds of goods, such as paper, cables, tyres, rubber boots, TV sets and portable phones (Nokia Mobira). During the 1980s and 1990s, it made a strategic decision to concentrate on mobile telecommunication. This decision appeared to be successful, and the company became a market leader in mobile phones and a top player in networks. Currently, it emphasizes multimedia and wireless enterprise solutions. The headquarters are located in Finland, and the key positions are occupied by Finns. However, the number of foreigners is increasing in top management. The company is currently about 90 per cent foreign-owned. The foreign ownership is widely dispersed between individual and institutional investors.

Nokia is important for the Finnish economy. Its direct share in GDP was about 3.5 per cent in 2003, in exports it was 18 per cent, and in total employment more than 1 per cent. Nokia grew rapidly in the late 1990s. At its peak in 2000, it contributed 1.5 percentage points to the GDP growth rate. However, in the first years of the twenty-first century the company's contribution to growth declined to about zero, following the international slowdown in the ICT sector. In 2003, its contribution recovered to 0.3 percentage points.

However, the direct effects of Nokia do not tell the whole story, because Nokia has important linkages to the rest of the economy and the whole society. In production, there are normal leveraging links, subcontractor links, business links with other firms and institutions, the impact on general and local government tax receipts, etc. As for R&D, Nokia affects and is affected by the research and education institutions (see Figure 10.4).

Public R&D was important for Nokia in the early stages of its strategic reorientation. As the company has grown, the relative importance of public R&D has diminished. In the 1970s the average financing share of the technology office of the MTI (predecessor of Tekes) in Nokia's total R&D expenditure was 7 per cent. In the 1980s the corresponding share of Tekes was 8 per cent and in the 1990s it was 3 per cent. In 2000, this share was just 0.3 per cent.

> Nokia is very important for the Finnish economy and for the NSI, and also reflects the general importance of ICT for the NSI. Other high-tech sectors, such as biotechnology, are much smaller. Nokia as a company is not, however, the only contributor to the Finnish NSI. For example, ICT development conducted by other firms and development work done in paper processing technology and in many areas of the machinery industry are also important.
>
> Sources: Statistics Finland and Nokia's Annual Reports; see also Ali-Yrkkö and Hermans (2002).

The role of foreign firms is very small in business R&D. Big internationally operating Finnish companies are, however, owned to a large extent by foreign portfolio investors. The majority of their R&D is still performed in Finland. R&D conducted abroad is increasing due to wider foreign activities (partly because of mergers and acquisitions), a need to adapt innovations to local conditions, and to some extent also to lower R&D costs abroad. However, the main reason is a need to tap into the best knowledge available, wherever it is located (see Pajarinen and Ylä-Anttila, 1999). In 2003, 23 per cent of the R&D personnel of Finnish-owned manufacturing firms worked abroad (Ali-Yrkkö et al., 2004). Private sector R&D systems are very complex networks. They change flexibly according to the theme. Often, however, established connections are also used in several subsequent projects.

Figure 10.1 depicts the Finnish public NSI and its main actors. At the top there is the Parliament, which decides the total resources to be used. At the Cabinet level, the STPC plays an important role. Among the funding institutions, the Academy of Finland is responsible for funding basic research, while Sitra and Tekes fund applied R&D projects. Sitra also acts as a think-tank in the field.

There are some public and private non-profit research institutes, which perform research with their own financing or are financed by organizations such as the Academy of Finland, Sitra, Tekes and several ministries. This financing is based on competitions. The largest research institutes are the Technical Research Centre (VTT) and the National Research and Development Centre for Welfare and Health (Stakes), which are both public organizations. The research institutes conduct mainly applied research while the universities take care of basic research and education. Some universities have founded separate research units for applied research where they conduct research with outside financing (public or private). Of the total public sector R&D, 40 per cent was technical in 2004 and 18 per cent went into agricultural and forestry research. Natural and medical

Finland 367

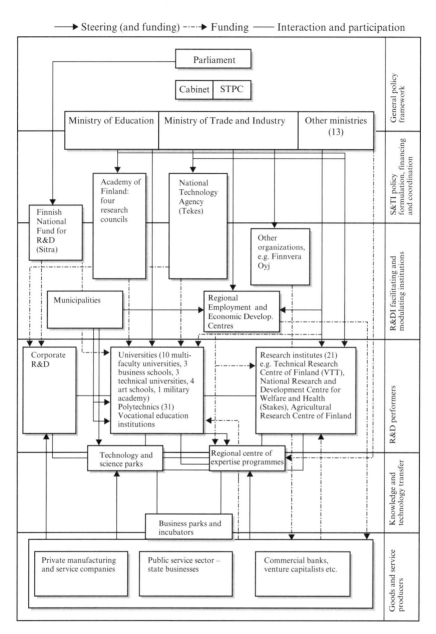

Source: Nieminen and Kaukonen (2001).

Figure 10.1 A policy-centred organizational map of the Finnish system of innovation

sciences both had a 15 per cent share, and 13 per cent was channelled into social and cultural sciences. In technical research, one-third of the R&D was conducted in electrical engineering (Statistics Finland, 2005).

At the regional level, the Employment and Economic Development Centres (T&E Centres) have an important role in the coordination of regional actions. There are 21 universities, 31 polytechnics and 21 research institutes in different parts of the country. Among the technical research institutes, VTT is the largest and most important. Additionally, the official NSI includes science and technology parks, regional centres of expertise programmes and business parks and incubators.

Performance of R&D   In what follows, some aspects of the performance of R&D are described by publication, citation and patent data in relation to the size of the population. Here, Finland performs well with respect to the other EU-15 countries on all measures (Table 10.2; see also Appendix, Table A3.1). With respect to the USA, Finland performs better when measured on scientific publications and European patents. When measured on US patents the USA performs, naturally, much better. In the case of highly cited publications the USA performs slightly better.

In Finland, patenting is most active in the wood and paper industries as well as in electronics (see Appendix Figure A3.2). This is explained by the high shares of these sectors in exports. Patenting is quite strong in industrial process equipment too. Especially in the electronics industry, patenting activity has increased during the 1990s. Due to the high speed of technical change in electronics, the average time the patents are in effect has declined (Balaguer and Tsai, 2004).

Table 10.2   Scientific publications, citation and patent indicators in Finland, EU-15 and the USA in 1996 and 2000, per million population

Publications and patents	1996			2000		
	Finland	EU-15	USA	Finland	EU-15	USA
Scientific publications	1094	682	785	1270	803	909
Highly cited publications				31	19	38
European patents	163	88	101	283	139	144
US patents	92	49	233	130	74	315

Note: Highly cited publications in 1998–2001.

Source: European Commission, CORDIS: Science and Technology Indicators.

The share of applied research in total R&D has increased slightly since the early 1990s. The share of applied private R&D has clearly increased and that of applied public R&D has declined, but by a couple of percentage points less than the private share has increased. The share of universities in total R&D has consequently declined from 22 to 20 per cent.

Public policies and institutions   Organizational changes have been crucial in the implementation of public R&D policy. These changes include the foundation of the STPC and Sitra, the reorganization of the Academy of Finland, the foundation of new universities in the 1960s and 1970s, and the foundation of Tekes in 1983 (see Section 2).

All these were aimed at, among other things, strengthening aggregate R&D input. The founding of the STPC reflected a realization that something must be done in this field and that activities should be coordinated within public administration. Changes in the Academy of Finland and the establishment of new universities were aimed at increasing basic research. The foundation of Sitra and Tekes reflected the need to put more emphasis on applied, mainly technical, R&D. These changes can be regarded as successful. The organizations named here have been functioning well and have contributed to the high share of R&D in GDP.

The large number of small universities in minor cities can, however, be criticized for contributing to inefficiencies in research. This deficiency has to some extent been overcome by nationwide research networks and joint doctoral programmes.

Changes in institutions are closely related to organizational changes. These include university laws, the laws pertaining to the Academy of Finland, Tekes, Sitra etc. All these organizations also have their internal constitutions. One important institutional requirement of Tekes in the provision of R&D financing is that large companies must have small firms as members of their research networks. This measure aims at broadening the R&D base. Another crucial institutional factor behind the fast development of the telecommunication industry was the early liberalization of the Nordic telecom market (see Section 4.2).

4.1.2 Competence building

Organizations – main actors and the division of labour   The general education system is free and mainly public in Finland. Only a few high schools and vocational schools are private, and even they are financed publicly. Studying is limited by the application of quotas. The qualifying criteria are based on former academic results, on exams, or on a combination of these indicators.

The university sector is fully public. It expanded in the 1960s and 1970s to different parts of the country. Currently there are 21 universities, of which 10 are multi-faculty universities, 3 business schools, 3 technical universities, 4 art schools and 1 military academy (Figure 10.1; the Ministry of Education). The system of polytechnic education (tertiary B) was founded in the 1990s by upgrading former upper/post-secondary schools to polytechnics and by founding new ones. Currently, there are 31 polytechnics. Some are private in form but financed from public sources.

Doctoral education is provided by the universities. In several disciplines, national doctoral programmes combine the educational and supervisory resources of different universities. The degrees are, however, awarded by individual universities.

Studying abroad is relatively uncommon. Only a few students take their master's or doctoral degrees abroad, mainly in the USA or in the EU. There are no foreign universities in the country, either. The number of foreign students is also relatively small in Finland at only 2 per cent of all tertiary students.

The financing of studies is mainly based on a general public support system. In addition to direct support there is also a loan-based support system with subsidized interest rates. Doctoral studies are usually financed by grants. Some doctoral students are able to prepare their theses in their jobs (at universities, research institutes or in firms). Additionally, studies are partly financed by parents and by students' own earnings.

The private sector enters the education scene usually around or after graduation as people may receive applied training in their jobs. At the secondary level there is an on-the-job learning system. In 2000, there were 25 900 students in on-the-job learning, of whom half were in basic vocational education and half in advanced vocational education. The aim of the government is that 10 per cent of those who begin their secondary-stage vocational education should start in the job learning system. For the unemployed, the public sector organizes training courses, or people participate in existing vocational training (see e.g. Koski et al., 2006, on Finnish education strategies from the point of view of innovation).

Performance – quantity and quality of output   In 2001, total public expenditure on education was 5.8 per cent of GDP in Finland, down by one percentage point since 1995. The OECD average was 5.6 per cent, up by 0.2 percentage points since 1995. In Finland, almost one-third of public education expenditure was used in higher education, while the figure in the OECD on average was one-quarter. The high share of public expenditure on higher education in Finland reflects the public ownership of the universities and the absence of tuition fees.

In 2002, 75 per cent of people aged 25–64 had at least an upper secondary education, while the corresponding figure for people aged 25–34 was 88 per cent. In the OECD these figures were 67 and 78 per cent, respectively (Table 10.3).

In 2002, 19 per cent of people aged 25–34 had a tertiary type B education and 21 per cent a tertiary type A or more advanced education in Finland. More new entrants to the labour market have a tertiary education in Finland than in the OECD on average. This is because of the higher share of tertiary type B education.

Currently there is a mismatch between education and the needs of firms in Finland, especially in some practical professions, which require vocational training. These include different kinds of construction and metal workers. The main reasons for this mismatch are the low prestige of these professions among young people, and limited knowledge about these kinds of jobs while students are still attending school. In the ICT sector, the number of people with education or practical knowledge is quite large. The slowdown of this sector during the first years of the twenty-first century has diminished the degree of mismatch. In general, though, there is a constant need for those who have a high education in technical, computing or physical sciences.

Participation in continuing education is high in Finland. About 55 per cent of the labour force had this kind of education in 2001. Among 19 OECD countries, this figure was the second highest after Denmark. People with tertiary education have the highest participation rates in continuing education, and those who have lower secondary education have the lowest (Edqvist and Göktepe, 2003; original data from the OECD). Working life is however changing rapidly, with increasing needs for enlarging the quantity and improving the quality of continuing education.

4.2 Demand-side Factors

Free trade and deeper economic integration between Finland and other European countries since the early 1960s has provided Finnish firms with an enlarging export market and forced them into increasing competition both at home and abroad. European integration has also unified industrial standards in many sectors, and further unification is taking place in the internal market. The liberalization of trade has thus led not only to a lowering and/or removal of tariffs and quotas in foreign trade but also to a lowering of technical trade barriers. In the EU, many environmental and product standards are now set at the European level. This development has increased export markets and competition (see also Section 6 on globalization).

Table 10.3 Educational attainment in 2002, % of population

	Pre-primary and primary	Lower secondary	Upper secondary	Post-secondary, non-tertiary	Tertiary, type B		Tertiary, type A, or advanced	
					Aged 25–64	Aged 25–34	Aged 25–64	Aged 25–34
Finland	–	25	42	–	17	19	16	21
Males	–	27	44	–	14	–	16	–
Females	–	24	40	–	20	–	16	–
OECD, mean	14	18	41	3	8	9	16	19
Males	13	17	43	3	7	–	16	–
Females	15	19	49	3	8	–	14	–

Source: OECD (2004).

As a small country, Finland has typically not been able to export its own standards. A notable exception was the NMT (Nordic Mobile Telephony) standard implemented by the Nordic countries. The planning of the NMT network started in 1973 and the official decision to build an NMT 450 network in Finland was taken in 1977. In 1981–82, the network was brought into operation in Finland and the other Nordic countries, thus forming a lead market. This initiative supported the growth of Nordic firms, especially Nokia and Ericsson, as the small size of the national market could otherwise have constrained product development (see e.g. Paija, 2001, and Palmberg and Martikainen, 2005, on the development in Finland). While ICT-related services are not as developed as in some other countries, these too started to grow considerably, especially after 1995. Nokia has also increased markets for Finnish software and other firms as a part of Nokia's R&D (see also Section 4.4.3 on KIBS).

There are cooperation and feedback links between high-tech firms, notably Nokia and its subcontractors, and universities and research laboratories (for example VTT) in addition to links with customer firms (see also Section 4.3.2). Feedback from and quality requirements set by customers lead to innovations and enhanced competitiveness, or – if not met – a gradual decline of the company. In non-high-tech sectors, the Finnish forest cluster, for example, has also increased the quality of its products through such feedback (see Hernesniemi et al., 1996).

Trade associations bring together firms operating in certain foreign markets, helping them make business contacts and disseminating information. The central interest organizations of Finnish industries also promote internationalization and thereby help in 'creating' new markets. There are also public sector actors operating in this field. At the governmental level, the liberalization of trade and investment flows in Europe within the European Free Trade Area and the EU and also globally within the context of GATT/WTO has enlarged the markets for Finnish products. For firms, this has formed 'new markets' or enlarged existing ones. Further, Finnvera grants export guarantees,[2] Tekes promotes networking and technology transfer at both the national and international levels, and Finpro (founded in 1919 as the Finnish Export Association) provides various business support services for internationalization.

The public sector may also have a role as a lead user because of its large size. It can, for example, introduce new electronic systems in administration or public services. According to Palmberg (2000, 2002), public technology procurement (by the main public telecom operator) has had a positive effect on Nokia's growth and internationalization. In the EU, public procurement has to be Union-wide, however, so there should be no possibility of favouring domestic producers.

Although the business-to-business market is important for Nokia, the company's rise has also meant that the global consumer market has become relatively speaking more important for Finland than it was before. Traditionally, Finland has mainly exported paper and wood products, investment goods (machinery) and basic metal products, for which customers are typically other firms. While these products remain important, their share in exports has declined since the beginning of the 1990s. Figure 10.2 shows how the structure of Finnish exports has changed and become more diversified.

4.3 Provision of Constituents

4.3.1 Provision of organizations

To simplify, there was a change in policy thinking in Finland in the 1980s from 'backing-old-industries' towards 'trying-to-pick-the-winners' policies. At the end of that decade, another transition occurred (see Section 1) when the STPC took the NSI concept as a starting point for its policy.

Figure 10.3 shows a stylized picture of how the innovation support system has changed since the 1980s. The figure shows different phases of research, product development and commercialization on the horizontal axis and different actions or measures on the vertical axis. The dotted arrows show how product development advances. For example, universities are involved in basic research and deal with grants to their research personnel. Research institutes do more applied research. The Academy of Finland acts with both universities and research institutes. Looking at the first stylized picture that depicts the situation in the 1980s, we move up and right. Tekes was involved at the product development phase and granted loans to firms. Sitra and Finpro, among others, were involved later in the product life cycle when the product was being commercialized. They were involved in equity investment (Sitra) or services for the firms (Finpro). The 1980s model was a linear innovation model.

In the 1990s, there was a change in policy because the identification and prediction of future growth sectors proved difficult. The policies of the 1990s have been characterized more as 'conditions-providing' horizontal policies that seek to avoid direct interventions in product markets but instead concentrate on rectifying market failures and promoting competition. To simplify, there has occurred a shift from macro-oriented structural policies towards long-term micro-based growth policies (see Georghiou et al., 2003; Ylä-Anttila and Palmberg, 2005). These choices are also present in the European discussion of whether 'national champions' should be supported. Typically, supporting them has not been the best course of action.

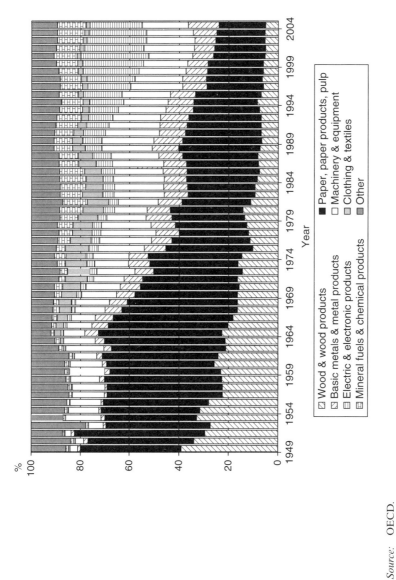

Source: OECD.

Figure 10.2 The structure of Finnish exports of goods in 1949–2004, %

376   *Slow growth countries*

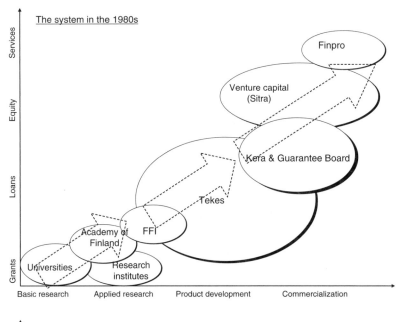

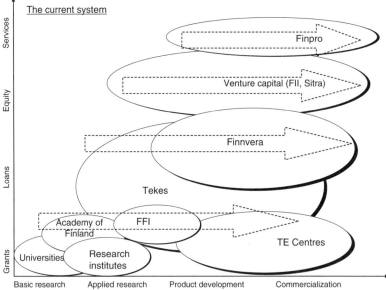

Source: Georghiou et al. (2003).

Figure 10.3   Organizations that support innovations in Finland

As we see from the lower stylized picture in Figure 10.3, there has been a horizontal extension in the innovation support organizations' activities. Kera and the Guarantee Board have merged and are now called Finnvera. FII (Finnish Industry Investment) has joined Sitra as an actor in venture capital and T&E Centres have been created as a new set of organizations. For example, the Academy of Finland and Tekes have increased their cooperation. However, there have not been any real changes in the vertical dimension.

There has been a move from a linear innovation model in the 1980s towards a systemic model where innovation is seen as a process with several interactions and different phases becoming increasingly simultaneous. Indeed, there is increasing need for coordination of the activities of the principal public organizations in the Finnish NSI. A 'Group of Six' has been established, in which all five organizations administrated by the MTI and Sitra take part.[3] There is also a great deal of informal cooperation between the policy agencies (Georghiou et al., 2003).[4]

Still, as Ylä-Anttila and Palmberg (2005) argue, there has been no 'master plan', with the government playing a strong leading role. Rather, there has been an effort to improve the general operating conditions of firms, especially in terms of knowledge development and diffusion, innovation and clustering of industrial activities. Public–private partnerships have been important in this respect.

In the OECD, Finland ranks rather low in the share of entrepreneurs (self-employment as a percentage of all non-agricultural employment) in total employment. However, the share is even lower in Norway, the USA, Denmark and Sweden, among others. Among the more advanced OECD countries, Finland ranks as an average country. The average for 1997 and 2000 showed that entry and exit rates in manufacturing and business services in Finland were both relatively low compared with other OECD countries. Finland was at about the same level as Denmark and the Netherlands, but above Sweden. In business sector services, the entry and exit rates were higher than in manufacturing (OECD Labour Force Statistics; OECD STI Scoreboard, 2003). Entry and exit rates are also affected by the business cycle, however, which may be idiosyncratic.

4.3.2 Networking, interactive learning and knowledge integration
Networking is an important element in the Finnish NSI. From a policy standpoint, the purpose is to increase firm–firm and firm–research institute connections. Consequently, small firms with limited resources can gain from the strengths of other firms. In the 1970s, cooperation between firms and universities was relatively limited, but especially since the 1990s there has been a strong expansion in this respect. A full 70 per cent of Finnish

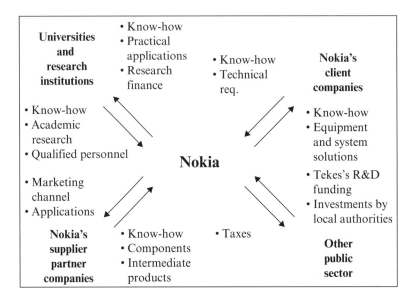

Source: Ali-Yrkkö and Hermans (2002).

Figure 10.4 Nokia in the Finnish system of innovation

innovating firms were cooperating with other firms, universities or public research institutes in 1996 as opposed to just 25 per cent in the EU on average. Finland was followed by Sweden and Denmark with 57 and 53 per cent shares respectively (European Commission, 2001b).

Nokia is the single most important firm in the Finnish NSI. Nokia has extensive R&D and cooperation with firms, universities and research institutes, not just in Finland but also globally (see Figure 10.4). Research cooperation with universities in the field of mobile phones dates back to the 1980s. According to Ali-Yrkkö and Hermans (2002), cooperation has led to a flow of knowledge between the partners. In addition to cooperation in R&D, recruitment from universities to firms and vice versa has served as another important channel of knowledge transfer.

According to Nieminen and Kaukonen (2001), subcontractors, competitors and non-university research institutes are more important partners than technical or multi-faculty universities in innovation-related cooperation. They argue that Finnish firms cooperating with universities are usually high- or medium–high-tech or KIBS firms that have regular in-house R&D activities. Collaboration among Finnish firms is usually connected to product development. Finnish firms that did not cooperate with universities gave their reasons as lack of time and lack of information about cooperation possibilities. The study found that cooperating companies considered commercial utilization of knowledge, acquiring new scientific knowledge and monitoring technological development as the most important goals for university cooperation. Leiponen (2001) shows that networking and feedback from clients are important for KIBS firms.

Tekes has technology programmes where cooperation and networking are encouraged at both the national and international levels as the main instruments for the implementation of innovation policies. Finland relies on foreign connections and cooperation in research, both basic and applied, as well as in development of products into finalized goods. Also, the Finnish home market is too small to support independent commercialization, so international connections are valuable for marketing purposes, especially for small and medium-sized enterprises (SMEs).

Pan-European cooperation is becoming more important in innovation and R&D activities. At the European level, Tekes is involved in EU R&D programmes, EUREKA (a pan-European network for market-oriented and industry-related R&D), COST (European Cooperation in Scientific and Technical Research), ESA (European Space Agency), NI (Nordic Industrial Fund), IRC (Innovation Relay Centre), OPET (Organizations for the Promotion of Energy Technologies) and TAFTIE (Association for Technology Implementation in Europe). Tekes hosts the Finnish Secretariat for EU R&D and it also has cooperation networks at the global level in the USA, Japan and China.

4.3.3 Provision of institutions

Ginarte and Park (1997) constructed an index of patent rights ranging from 0 to 5. Finland scored 2.95 in 1980–90, while the figure in 1990 was 4.52 in the USA, 3.90 in Sweden and 3.71 in Germany, for example. A high figure represents a strong level of protection for patent rights.

As Finland joined the EU in 1995, rules governing competition, foreign trade and intellectual property rights (IPR) have since been harmonized or are being harmonized to comply with Union regulations. The creation of a Community patent will harmonize intellectual property legislation. According to the Innobarometer published by the European Commission in February 2004, Finnish firms are rather pessimistic about any positive effects from the creation of a Community patent for innovation, both in general and in terms of their own interests (Flash Eurobarometer 144).

Some changes in the Finnish IPR legislation have taken effect in the second half of 2005. These changes concern the rights of universities over inventions made under their auspices and the rights of firms over inventions made by their employees. Universities will be given a part of the rights to the inventions made in their facilities. Furthermore, universities have been given the right to set up and own companies.

4.4 Support Services for Innovating Firms

4.4.1 Incubating activities

Since 1980, public sector support for starting up a new firm has increased. This support does not necessarily take the form of direct financial assistance but may consist of different types of services available to entrepreneurs.

Several organizations provide incubation services in Finland. The Finnish Science Park Association, founded in 1988, is a nationwide cooperation network and connects 22 science and technology parks operating in the university cities.[5] These science parks promote enterprises implementing innovative research in businesses, their growth and internationalization. They include 1600 enterprises and other organizations, and 32 000 experts working in information, telecommunications, environmental, energy, health and medical technology, life sciences and material research. The science parks provide training and consulting in founding new firms, business management, financing, IPR and other legal services, marketing and communications.

The Foundation for Finnish Inventions (FFI), founded in 1971, supports and helps private individuals and small entrepreneurs residing in Finland to develop and exploit inventions. It also helps entrepreneurs with new business and product ideas for licensing. The T&E Centres support and advise small and medium-sized enterprises at the various stages of their life cycle and also when a firm is first founded.

4.4.2 Financing

In the past, Finland was on balance a capital-importing country. Financing possibilities for firms were both limited and largely supplied by the banking sector. The situation changed during the 1990s. Subsequently, Finland has become a capital-exporting country with a relatively large current account surplus; a venture capital market has developed and financing in general is now less centred on the banking sector.[6] Public sector decisions have been important in the development of the venture capital market. For large firms, too, the possibilities of obtaining financing from the global financial markets have become more extensive than they were even in the 1980s.

The Finnish financial markets were liberalized during the 1980s and early 1990s. In 1980–2000, the rights of shareholders were strengthened and those of creditors weakened. Overall, a bank-centred financial system shifted from relationship-based debt finance towards increasing dominance by the stock market. The liberalization of financial markets has also led to a considerable increase in the availability of financing. Still, the growth-oriented and innovative SMEs remain financially constrained (see Georghiou et al., 2003; Hyytinen et al., 2001 and 2003). According to Hyytinen and Pajarinen (2003b), innovative firms – especially firms engaged in R&D and firms that own patents and/or intangible assets – run a lower debt ratio than other firms. These firms rely more on inside equity.

According to Hyytinen and Toivanen (2003), capital market imperfections have held back innovation and growth but public policy can complement the capital markets. According to their results, firms in industries that are more dependent on external financing invest relatively more in R&D and are relatively more growth-oriented when they have more government funding (potentially) available. According to Ali-Yrkkö and Pajarinen (2003), public R&D funding has not reduced private R&D expenditure in the Finnish metal and electronics industry.

The share of high-tech industries in total venture capital investment is relatively high in Finland, 57.5 per cent against 45.4 per cent in the EU-15 countries on average.[7] In 1998–2001, Finland was close to the OECD average in this respect. According to CIS3, the share of early-stage venture capital in GDP was 0.087 per cent in Finland, which is second only to Sweden in the EU and 0.037 per cent in the EU-15 on average. However, this ratio was 0.218 in the USA.[8]

There are a number of public and semi-public organizations as well as private funds providing financing for innovation processes (see Figure 10.5). Tekes helps companies by providing R&D funding and expert services. Sitra provides funding for technology start-ups, regional growth companies and commercialization activities. Sitra also provides support for the technological and business-related innovations of new and existing companies with focus on technological enterprise, technology transfer and business development.

The Foundation for Finnish Inventions (FII), owned by the state and administered by the MTI, encourages a more efficient functioning of the venture capital market by investing actively in new venture capital and private equity funds in Finland. It also invests in seed and growth-stage enterprises together with private investors, promotes regional venture capital investments and uses direct investments to enable major investments in corporate development, corporate restructuring and the launch of new industrial projects. The FFI supports the inventions of private persons and small-scale entrepreneurs. Additional sources of financing are the Finnish Work Environment Fund and the Academy of Finland.

According to the Finnish Venture Capital Association, Finnish private equity companies invested a total of €328 million, equivalent to 0.23 per cent of GDP, in 435 investment projects in 2003. Firms at their seed phase received €11 million. The total capital under management of the

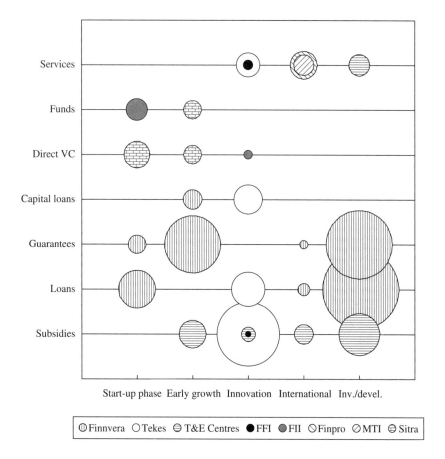

Note: Inv./devel.: Investments/development; FFI: Foundation for Finnish Inventions; FII: Finnish Industry Investment; MTI: Ministry of Trade and Industry.

Source: Georghiou et al. (2003).

Figure 10.5 Service expenditures and the amount of financing granted in Finland

Finnish private equity companies totalled €3.1 billion (equivalent to about 2.2 per cent of GDP) at the end of the year.

4.4.3 Provision of consultancy services

The supply of business services in Finland has been growing (see Figure 10.6). Out of a total employment of 2.4 million in 2005, computer and related activities (NACE 72) employed 42 900 people, R&D (NACE 73)

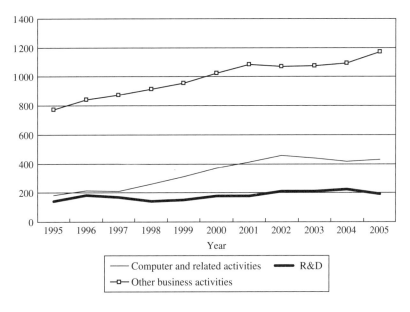

Note: The sectors are NACE 72, 73 and 74.1–74.5.

Source: Statistics Finland.

Figure 10.6 KIBS employment in Finland, 100 people

employed 19 300 and other business activities (NACE 74.1–5) employed 117 200 people. There has been a total increase of 63 per cent since 1995. The share of these sectors in total employment has risen from 5.2 per cent in 1995 to 7.5 per cent in 2005. Nokia has somewhat increased overall demand for software and R&D services. Schienstock and Hämäläinen (2001) identify the development of KIBS as a major challenge for Finnish innovation policy.

Among others, Finpro, the T&E Centres, Tekes and FFI provide different kinds of consultancy services to firms. According to Georghiou et al. (2003), Tekes's services are used at the early phase of firms' growth cycle or innovation process, while Finpro's services are used at a later stage. Also the age of the firm positively affects the probability of using the services of the T&E Centres. Fintra,[9] Finnish industry and universities provide training in international business management, leadership, sales and marketing, communication skills, and international business operations.

Private consultancy services complement those provided by the public sector, but the latter, often with larger resources, may also reduce private sector business opportunities in this regard. This limits growth in KIBS in the private sector and may help to explain their relative underdevelopment compared with the other small European countries.

4.5 Summary of the Main Activities Influencing Innovation

The different activities presented above can be classified into three groups according to their past relevance for innovations in Finland. The most relevant activities include: provision of R&D, competence building, provision of organizations and networking, interactive learning and knowledge integration. Activities of medium relevance have been: formation of new markets, provision of institutions (through international organizations and agreements), demand articulation of quality requirements and financing. The least relevant activities have been incubation activities and provision of consultancy services. The relevance of these activities has, however, varied over time.

Provision of R&D and competence building are the most traditional and most important activities in the Finnish NSI. Provision of organizations was a crucial element in the birth of an explicit NSI and organizations which still forms the basic channels for different kinds of activities. As a small country Finland has benefited greatly from open-minded and intensive networking in R&D and learning.

Formation of new markets has been especially important for the electronics industry, as evidenced by the historical importance of the NMT standard for the development of mobile phone technology. Provision of institutions has been closely related to the formation of organizations and of new markets (especially memberships in the OECD, EFTA and EU). IPR and competition legislation etc. are currently gaining some importance. Demand articulation of quality requirements is central in the development of, for example, paper processing machines and production automation, where customers' opinions are crucial guidelines in the tailoring of the technology. Financing is a key activity for many organizations, especially financing of R&D. Venture capitalism is mainly private, and it is still relatively underdeveloped in Finland. There are also important public venture capitalists such as FII, Finnvera and Sitra.

Incubation activities and provision of consultancy services are relatively underdeveloped in Finland. There is a network of incubators but until lately they have mainly had only local importance. Development work in biotechnology has been based within these kinds of organizations but the results are yet to come. There are important consulting firms in the country, some of which are foreign-owned. Many of them are concentrated on administrative development and market research. In the forest industries and in the development of wood processing technology there is an international Finnish consulting firm, Jaakko Pöyry Ltd.

There are some important relations between different activities in: (1) R&D and competence building at universities; (2) research in different organizations (R&D and its financing); and (3) networks of research institutes and universities, firms (including subcontractors), consultants and customers in applied R&D. Concerning relations between the main functions, activities, organizations and institutions, two are especially important: (1) the relation between activities and organizations, and (2) the relations between institutions and organizations. Institutions determine and regulate the organizations as well as the activities they are responsible for (the independent role of institutions is not very prominent, however).

Innovation policy has had a strong influence on three activities: R&D, competence building and networking. The first two are basic activities while the third is used in the mobilization of the scarce resources of a small country.

Globalization has had a strong effect on Finnish firms since the early 1990s. Firms have become international and they have grown in size. Globalization has affected even rather small firms through export and import. Subcontracting firms, in particular, have also invested abroad. Foreign portfolio investments in Finnish companies are currently important. Foreign firms have bought Finnish companies, especially in the construction and service sectors, and in the production of machinery. This has had some, but not large, effects on the activities of the Finnish NSI. Finnish firms have some R&D abroad, such as Nokia in, for example, the USA, Hungary or China. In big companies, the system of innovation is globalized within the firm. Attracted by Finnish R&D know-how some foreign firms also have small R&D units in Finland. Globalization can also be seen in the increasing number of international research networks. Consultancy and financing are becoming more internationalized too (see Section 6 for further discussion on globalization).

5 CONSEQUENCES OF INNOVATION

Successful product or process innovations will raise productivity and output and thereby incomes. Consequently, the simplest way of evaluating the success of innovations is to analyse the development of productivity.

Finland has been catching up with the USA fairly steadily since 1960 in terms of GDP per hour worked (labour productivity) although it has not quite kept up with US growth during the past few years (see Figure 10.7). Finland has also caught up with, for example, Sweden (see also the Appendix).

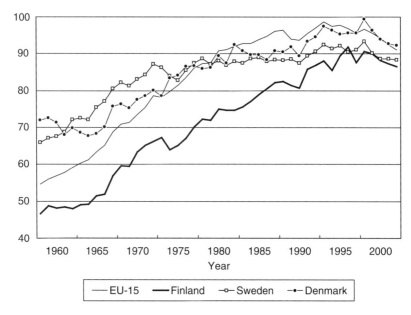

Note: Measured in nominal purchasing-power-adjusted terms.

Sources: Eurostat; Groningen Growth and Development Centre and The Conference Board, Total Economy Database, http://www.ggdc.net.

Figure 10.7 GDP per hour worked in Finland, Sweden, Denmark and average of EU-15, USA=100

In the 1975–2003 period, total employment in Finland grew by just over 1 per cent and in manufacturing it declined by 24 per cent. Meanwhile, total hours worked declined by 8 per cent and by 30 per cent respectively. Consequently, GDP growth is due to continuing rapid growth in labour productivity. Fixed investment has been considerably lower since the beginning of the 1990s than before, and Finland now has one of the lowest investment-to-GDP ratios in the EU.

As can be seen in Table 10.4, productivity growth in the traditional sectors was relatively rapid up until about the mid-1990s, by which time the labour shedding in the depression had more or less come to an end. The increase in R&D expenditure has only occurred since the early 1990s, and its effects can be seen especially in productivity growth in the manufacturing of telecommunications equipment, thus contributing significantly to the catch-up in labour productivity seen in Figure 10.7.

As we have seen, R&D expenditure has been dominated by telecommunication equipment and therein to a large extent by one company, which

Table 10.4 Average annual real growth rate of value added per hour worked in selected industries in Finland, and share in total value added in 2003 in current prices, %

Sector	1975–85	1985–95	1995–2003	Share in 2003
Total economy	3.2	3.2	2.4	100.0
Total manufacturing	4.9	6.0	4.3	22.6
Food products, beverages	3.1	5.7	3.1	1.7
Textiles, leather & footwear	4.4	4.1	1.8	0.4
Wood & products of wood	6.4	5.4	4.8	1.0
Pulp, paper & paper products	7.7	6.8	3.3	2.9
Chemical, rubber, plastics & fuel products	5.7	5.1	1.7	2.4
Other non-metallic mineral products	3.8	4.6	1.4	0.7
Basic metals & fabricated metal products	5.9	6.9	0.7	2.5
Machinery & equipment	5.3	4.2	0.8	2.4
Electrical & optical equipment	3.7	9.2	13.5	5.7
Transport equipment	0.6	4.0	1.1	0.9
Wholesale & retail trade; repairs	2.5	2.8	2.5	10.4
Transport & storage	1.8	3.8	1.5	7.3
Post & telecommunications	3.3	6.2	11.2	3.5
Financial intermediation	3.9	1.8	7.4	3.1
Real estate, renting & business activities	0.6	0.3	−0.9	18.3
Business sector services	2.7	3.0	2.0	44.0

Note: The base year is 2000.

Source: STAN database and own calculations.

may be a structural weakness in the Finnish economy. In any case, the growth of productivity in the production of telecommunications equipment has been spectacular, as can be seen in Figure 10.8. Rapid productivity growth in telecommunications equipment started in 1992. Obviously, higher R&D investment is linked with this development.

As already noted, productivity growth in basic metals industries and machinery and equipment has been dismal in 1995–2003. Productivity growth has also slowed in the paper and chemicals industries. This can

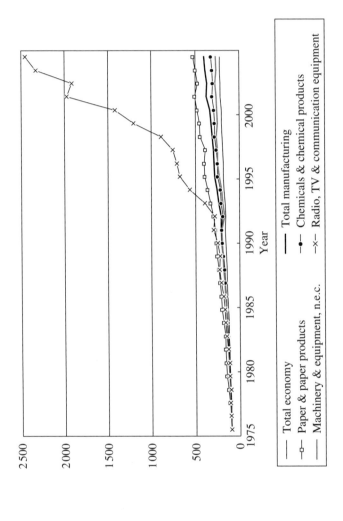

Note: The base year is 2000.

Source: STAN database and own calculations.

Figure 10.8 Labour productivity: value-added per hour worked in Finland in 1975–2003, 1975 = 100

partly be explained by the finalization of a catching-up phase in many sectors of Finnish manufacturing. On the other hand, the difference in productivity growth between ICT-producing and ICT-using sectors is very large. According to calculations by Estevão (2004), growth in labour productivity in ICT-using sectors in Finland was the lowest in the EU after Luxembourg in the 1995–2001 period. The difference from the EU average was 1.5 percentage points and from the USA 4.4 percentage points per annum.

In addition to productivity developments we should also look at price developments in order to have a better understanding of the whole situation.[10] The manufacturing of electronic and optical equipment that is dominated by mobile phones is a special case among Finnish export sectors, in the sense that export prices there have been declining quite rapidly since 1997 (see Figure 10.9). Consequently, the rapid growth in productivity is partly 'deflated' by the declining prices for these products. Declining prices are due to rapid development of technology, productivity and competition. This is typical for much of consumer electronics. Price developments in machinery, transport equipment, chemical products and paper have been much more favourable.

6 GLOBALIZATION

Large export firms became much more internationalized during the course of the 1990s than they were even in the 1980s. This was partly due to the full liberalization of foreign ownership in firms via direct portfolio investments in the stock market after 1992 when foreign ownership started to grow rapidly. At the end of 2004, almost 50 per cent of total market capitalization of the firms listed on the Helsinki stock exchange was under foreign ownership. Nokia accounted for 34 per cent of the total market capitalization. About 85 per cent of Nokia's shares were under foreign ownership.

The stock of Finnish direct investment abroad is about 50 per cent larger than the foreign direct investment (FDI) stock in Finland. The ratio of the inward FDI stock to GDP was 27 per cent in 2004, which is about the same as in the EU-15 countries on average. In Finland, foreign-owned firms accounted for about 30 per cent of total exports in 1999. In Sweden the corresponding figure was about 35 per cent (Ali-Yrkkö et al., 2004).

Sweden is the largest source of FDI in Finland when measured by the number of firms and personnel and the investment stock. At the end of 2004, Swedish companies accounted for 56 per cent of the total FDI stock in Finland. The Netherlands, the UK, Germany, Denmark, Luxembourg

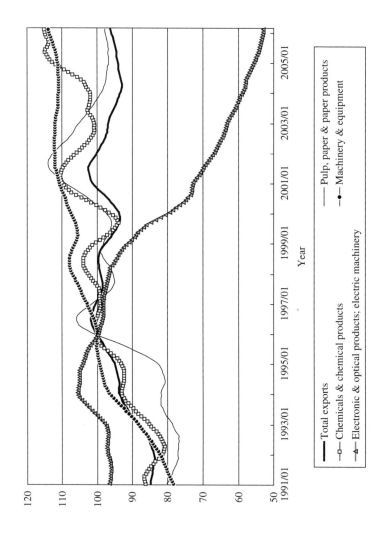

Source: Statistics Finland.

Figure 10.9 Finnish export prices by sectors, 12-month moving averages of the index values, 1995 = 100

and the USA are also important investors. Together with Sweden, these countries account for 87 per cent of the FDI stock in Finland.

In 2002, foreign-owned firms accounted for 13 per cent of employment in Finland. Measured this way, foreign ownership was largest in the chemicals industry (37 per cent of employment). The corresponding figure was 29 per cent in the electronics industry and 19 per cent in the food industry. In manufacturing as a whole it was 17 per cent.

The growth of the ICT sector and innovative activities therein have increased FDI inflows. For example Sweden's L.M. Ericsson has been present in Finland since 1918 but the firm has also had R&D in mobile networks in Finland since the early 1980s. Success in ICT-related technology has therefore increased FDI into Finland in this sector. Indeed, a high technological level has been an important reason for FDI in Finland (see Pajarinen and Ylä-Anttila, 1999; Lehto, 2004). R&D activity has thus usually been acquired through a merger or an acquisition.

It is not clear how foreign ownership has affected Finnish companies. Informal evidence seems to indicate that foreign ownership has had a positive effect on R&D and firms' long-run growth prospects. Pajarinen and Ylä-Anttila (2001) show that there is some evidence that most of the firms acquired by foreign companies have benefited from the technological know-how or managerial capabilities of the new owners. Furthermore, being part of a larger multinational corporation has also opened new marketing channels and thus formed new markets from the point of view of the purchased Finnish firms (Georghiou et al., 2003, Section 3.2).

Some non-commercial foreign organizations also exert control over Finnish innovations. This is the case in foreign-led R&D projects where Finnish participants are minor contributors. It is also the case with international financing organizations (such as the EU), which exercise control over R&D when they are funding it. The role of international R&D projects has increased recently along with growing EU funding (see Section 4.1.1).

Finnish firms have internationalized their activities intensively since the early 1990s. Different sectors have become internationalized at different speeds, with electronics and forest industries as leaders. Different measures of internationalization are presented in Figure 10.10.

When measured by the share of foreign employment, the forest industry is the most internationalized manufacturing sector. More than 50 per cent of the employees are abroad. The most important firms in this sector are Stora Enso, UPM-Kymmene and the Metsäliitto Group. Recently, this sector has become rapidly internationalized due to mergers, which have also caused the large share of foreign R&D.

In the chemicals industry internationalization has mainly affected production. The share of foreign R&D is still low. In the metals, engineering

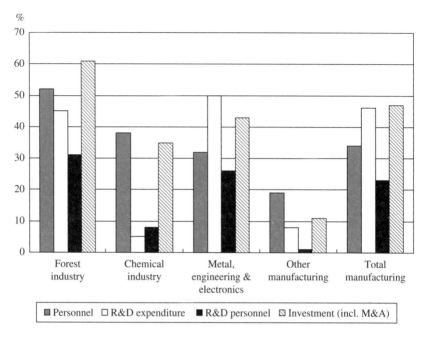

Note: The calculations are based on statistics by Statistics Finland, Bank of Finland and the Confederation of Finnish Industries.

Source: Ali-Yrkkö et al. (2004).

Figure 10.10 Foreign shares of personnel, R&D expenditure, R&D personnel and investment (including mergers and acquisitions) in Finland in 2003

and electronics industry one-third of employees work abroad. Its internationalization is mainly due to electronics, especially Nokia and its subcontractors. The high share of foreign R&D is also due to Nokia, which has established several greenfield R&D units abroad. These are mainly responsible for applied development work such as adjusting mobile phones to use the Chinese language. Electronics accounts for about 95 per cent of foreign R&D expenditure and 90 per cent of foreign R&D personnel (Ali-Yrkkö et al., 2004).

According to recent studies, Finland is a very competitive location for R&D in terms of quality as well as costs (see Pajarinen and Ylä-Anttila, 2001; Lovio, 2002, 2004). Furthermore, according to Van Beers et al. (2004), foreign-owned companies cooperate as much with Finnish science and technology institutions as do domestic firms. This has not been the case in, for example, the Netherlands, where foreign-owned firms have been much less likely to cooperate with the local institutions.

Increase in foreign R&D is thus not a sign of decreasing competitiveness of Finnish R&D and of relocation abroad, but instead a sign of expansion abroad, mainly through mergers and applied development work. In the future, the continuing internationalization of firms will also mean that the share of foreign R&D will grow.

Movement of researchers to and from Finland has been modest. The northern location of the country, a strange language and high taxation are disincentives for foreigners to work there. However, highly paid foreign personnel can have tax exemptions thanks to specific legislation, the purpose of which is to ease the transfer of employees of international firms to Finland. The possibility of studying in English at Finnish universities has to some extent attracted foreign students to the country. Many Finnish researchers in turn are so rooted in Finland through family ties and lifestyle factors that they are not very eager to emigrate. However, researcher flows in both directions exist and may increase in the future.

The Finnish NSI is in principle internationally oriented. As a small country, Finland tries to exploit international networks. This is seen in academic as well as in applied research, in the context of universities, research institutes and international organizations such as the EU, OECD, etc.

In 2001, affiliates under foreign control accounted for about 17 per cent of firms' turnover and 13 per cent of R&D expenditure in Finland; i.e. their importance was relatively small. These figures were even higher in the USA, not to mention the Netherlands, Sweden or Ireland. Also in 1997–99, the foreign ownership of domestic inventions, or the share of patent applications to the European Patent Office (EPO) owned by foreign residents in total patents invented domestically, was lower in Finland than in the USA or even the EU after intra-EU cooperation had been netted out. On the other hand, the domestic ownership of inventions made abroad, or the share of patent applications to the EPO invented abroad in total patents owned by domestic residents, was relatively high in Finland and also higher than in the USA (OECD STI Scoreboard, 2003).

Globalization has not brought about a large-scale relocation of innovations and innovative activities outside the NSI. They are still controlled by domestic actors. Some diffusion of technology through foreign units is unavoidable, however. Foreign production units also increase productivity in foreign countries. This is not necessarily a substitute for, but instead a complement to, domestic productivity increase.

7 STRENGTHS AND WEAKNESSES OF THE SYSTEM AND INNOVATION POLICIES

7.1 Strengths and Weaknesses

The strengths of the Finnish NSI include a high level of basic education (see the so-called PISA studies), a high share of the population with tertiary education, high public and business investment in R&D, high innovativeness of firms (especially a high share of firms with new-to-the-market products), extensive high-tech patenting, high Internet penetration and rather strong networking among the different sectors. Recently Finland has made progress in increasing the share of tertiary education, in raising business R&D, in increasing US patenting and in rising high-tech value-added.

The propensity to innovate is the highest in the manufacturing of communications technology in Finland. No other EU-15 country has higher innovativeness in that sector (European Commission, 2004). After the electronics industry, the second-best performing sector in innovativeness, according to CIS3, is KIBS. This sector is important for the diffusion of innovations to the rest of the economy. Even though KIBS performs reasonably well, its future development is a major challenge for the application of new technology in Finland. It is also relatively underdeveloped compared with other Nordic countries or the Netherlands.

The paper industry has maintained its important position although its share in production and exports has declined, reflecting the faster increase in output of the electronics industry. This industry is good at introducing new process innovations, but does not perform well otherwise in innovation. The machinery and equipment sector invests heavily in new product innovations in paper processing machines, different kinds of engines, forest harvesters and lifting automation.

One of the weaknesses of the Finnish NSI is the small population. It is a problem in diversification of resources and in use of scale economies. This is reflected for example in the small size of many universities, in strong dependence on one sector (electronics) and in fact on one firm (Nokia). Other weaknesses include unsatisfactory application of ICT outside the electronics industry, low employment in medium–high-tech manufacturing and a small number of innovative small and medium-sized firms. SMEs perform worse than large firms with respect to all other measures of innovativeness of CIS3, except with respect to R&D intensity. When compared to the SMEs in other countries covered in this volume, the Finnish SMEs have, however, on average an intermediate ranking.

Even though the share of firms that have introduced new-to-the-market products has been high in Finland, globalization creates extra challenges for product innovations (see Table 10.1 and Appendix Tables A4.9–A4.15). Several products have reached their peak production in Finland and new products should be invented and commercialized to maintain production and employment of the domestic open sector.

7.2 Summary and Evaluation of the Innovation Policy Pursued

The key organizations of the Finnish NSI were established gradually in the 1960s, 1970s and 1980s. The system as a whole assumed its present form in the 1990s. Finland was the first country to adopt the NSI concept as a basic foundation for its S&T policy (Miettinen, 2002, p. 12).

The depression of the early 1990s urgently required an increase in exports. The overheating of the economy and the consequent large current account deficit, as well as a drastic decline of the Soviet/Russian market in 1991, meant that new types of products had to be invented in order to win market shares in Western markets. It was not possible to increase exports of forest industry products to a great extent, and the potential of the machine and metal product industry was not large enough either. Many Western economies were also in a recession during the early 1990s, which meant weak demand for Finnish exports.

High unit labour costs due to overly rapid wage increases in the 1980s and misfortune in the export markets led to a marked depreciation of the Finnish currency. The depreciation restored the competitiveness of the traditional export sectors. Luckily, the innovation and product development work done in the electronics industry also started to bear fruit at the right time. A sufficient amount of research was conducted and there were enough educated engineers to put the ideas into practice. This was to a large extent done in one firm, Nokia, which together with its subcontractors contributed markedly to the recovery of the Finnish economy. At its peak, in 2000, the electronics industry contributed about 1.5 percentage points to the Finnish GDP growth rate. High-tech exports increased accordingly in relation to their imports. Finland has thus performed well in terms of manufacturing and exporting high-tech products. However, it has not performed equally well in using these products. While the contribution of ICT use to productivity growth has been two-thirds in the USA, it has been just one-third in Finland (Koski et al., 2002).

Total R&D expenditure is currently about 3.5 per cent of GDP, which is clearly above the EU target of 3 per cent set for 2010. In addition to Finland, Sweden is currently the only other EU country to exceed this target. In the early 1990s, the share of public funding (including the university sector) in total R&D expenditure was 40 per cent. Currently it is less than 30 per cent. This reflects fast growth in private R&D. A substantial part of the development work is done in one sector, namely the electronics and telecommunications equipment industry. This tendency has also raised concern about the long-term viability of innovation activities. R&D investments especially in new areas have traditionally been dependent on public funding. There is a danger that there will not be enough new openings in innovation activities. This fear has already been reflected in the demands of different political parties. In addition to the electronics industry, biotechnology has also received substantial amounts of public R&D funding, for example from Sitra.

7.3 Future Innovation Policy

Long lists of challenges for the Finnish NSI can be identified (see, e.g., Schienstock and Hämäläinen, 2002; Sitra, 2002; Georghiou et al., 2003). On the basis of the material presented above, we stress the following development needs:

1. techno-organizational restructuring and competence building at the firm level (especially in SMEs);
2. modernization of low-tech industries (for example application of ICT more widely in manufacturing and services);
3. development of KIBS;
4. supporting inter-organizational network formation (especially between high-tech and low-tech firms and between the KIBS and other firms);
5. strengthening basic research;
6. adapting the education system more to the needs of the firms;
7. fostering inter-regional cooperation (small size of the regions);
8. more financing to young, small, growth-oriented R&D-intensive firms and more tolerance towards failure and risk;
9. more emphasis on user perspectives in innovation projects.

Techno-organizational restructuring and competence building at the firm level refers to a wider use of social and organizational innovations. SMEs, in particular, have neglected these. The SMEs have not invested much in other kinds of innovations, either. Finnish innovation activity is rather concentrated in large firms and is very technically oriented. However, the importance of social and organizational innovations is growing. The ageing of the population is one trend that requires more social innovations. Within firms, more emphasis should be placed on organizational innovations and staff training. One advantage of these kinds of innovations is that they cannot be easily imitated.

There is a great deal of ICT production in Finland, but in ICT use the country's success is modest. This is reflected in low productivity growth of many industries and services. Because the requirements of industries differ substantially from each other, there should be more tailor-made innovations, technical as well as organizational. The problems lie to a large extent in transforming research into marketable products. This problem is also emphasized by Georghiou et al. (2003). One way to go further in this respect is to strengthen the cooperation between high-tech and low-tech firms. Also public support is needed because low-tech firms seldom have enough own innovative resources. One crucial factor is also increasing the quality of upper-secondary-level vocational education. Low-tech firms are often very dependent on workers with practical engineering skills. Tertiary B education should also be directed towards a better match with the needs of the firms.

The Finnish service sector is poorly linked to the NSI. ICT is not intensively applied in the sector. The KIBS sector is underdeveloped. Its share in the Finnish GDP was only 2.5 per cent in 2000, which was about the EU-15 average. In the UK the corresponding share was 4.5 per cent. More attention should be given to the development of KIBS and on their links to industry, universities, public services and foreign firms.

Schienstock and Hämäläinen (2001) identify the development of KIBS as one of the major challenges for Finnish innovation policy. They emphasize the need to support the development of KIBS in areas where the existing demand is not satisfied (marketing, management, design, organization). They recommend strengthening the role of KIBS in the Finnish economy and internationally by supporting their growth strategies, including the application of new ICT-based organizational forms, improved human resources management and interfirm networking. They also advocate supporting the development of research and training capacities in universities and polytechnics specialized in the service sector, promoting cooperation between private and public KIBS providers, strengthening the role of private KIBS firms in cooperative networks and creating a more service-oriented culture in Finland.

Finnish companies have quite extensive and intensive innovation networks. There is, however, a need to improve some things in this area. Local and sectoral organizations should be used more in finding and reaching potential network partners. More emphasis should be given to creating networks between high- and low-tech companies and KIBS firms should be integrated into the networks more frequently. The network approach should be used more extensively to create service innovations.

Resources for basic research have continued to increase, but to a lesser extent than in applied research. More emphasis and resources should be allocated to basic research because many processes have become more complicated and less controllable and they need input from basic research. Cooperation with other European researchers in basic research is quite intensive in Finland. However, contacts with the USA and Japan should be intensified. Interdisciplinary research and teaching should be promoted.

The education system should be adapted so that it stresses more the ability to learn new things. Because working life is changing rapidly, less permanent knowledge can be taught and students should be prepared for continuing education. Teaching should thus be more problem-oriented. Also closer links to firms should be established while children are at school. The education system should be more flexible and diversified to take into account the differing needs and differing capabilities of students. The idea of continuous learning should be implemented more fully. The system of polytechnics should be developed so that the roles of polytechnics and universities are more clearly defined, enabling the polytechnics to becomer more specialized.

Improving the education system is also important from the point of view of preventing social segmentation and social exclusion. This problem has become severe in the context of rapid structural change, which is accelerated by the application of high technology in production. It has both saved labour and shifted labour demand towards high-skilled labour. A part of the labour force is thus in danger of being excluded from the labour market. This risk requires several kinds of policies, including re-education of people to new professions and to be able to use ICT in their work as well as in their leisure time.

In spite of the internationalization of firms, the Finnish NSI is still under national control. It has a strong sectoral emphasis. There is also a regional dimension in the policy, but the problem in Finland is that the regions are small. Competition between regions is thus often not the most efficient route to take. Strengthening regional innovation policy means, among other things, more interregional cooperation, as well as between regions and the national level so that a critical mass for innovations is created. This applies, for example, to cooperation between universities of different regions.

Georghiou et al. (2003) emphasize the development of financing for especially young, small, growth-oriented, R&D-intensive and high-tech firms. These firms have problems with financing although R&D financing is functioning quite well in general. This should also mean more tolerance towards risk. The private venture capital sector is still relatively underdeveloped.

Putting more emphasis on user perspectives in R&D is a challenge in several fields. They include the application of ICT in the whole society and development of mobile phones and other wireless solutions to take the user needs more into account. A user perspective is also important in traditional industries, for example in metal and engineering, where the products are becoming more tailor-made.

The continuing globalization of the economy in production as well as research activity is a challenge for the Finnish NSI, as well as for other small open economies. An increasing share of innovation activity in large companies will be carried out abroad. The internationalization of research will also shift a part of control outside the country. The challenge for the Finnish NSI is to utilize national innovation activity to enhance domestic welfare, in accordance with market economy rules, thereby taking advantage of the international division of labour and international networking.

ACKNOWLEDGEMENTS

We would like to thank Charles Edqvist, Birgitte Gregersen, Leif Hommen, Pekka Ylä-Anttila and the participants of the Oslo, Taipei, Lund and Seoul workshops of the project for helpful comments on this chapter.

NOTES

1. Universities are public in Finland, but they also receive some private research funding.
2. Finnvera finances export promotion and supports the internationalization of businesses. It was formed in 1999 by merging Kera (the Regional Development Fund established in 1971) and Takuukeskus (the Finnish Guarantee Board). Finnvera is Finland's official Export Credit Agency.
3. Tekes, Finnvera, Finpro, the T&E Centres and FII.
4. They have, among other things, an Internet portal (www.enterprisefinland.fi) on the setting up and running a business in Finland.
5. Koskenlinna (2004) finds that in the summer of 2004 there were slightly over 100 incubators in Finland.
6. The Finnish venture capital market remains less developed than, for example, the Swedish (see Hyytinen and Pajarinen, 2003a). In 1999, seed and start-up venture capital investment were 0.56 per cent of GDP in Finland relative to 0.38 per cent in the EU-15 area on average. Sweden, the Netherlands and Belgium exceeded the Finnish figure (European Commission, 2001b).
7. Computer-related fields, electronics, biotechnology, medical/health, industrial automation and financial services. Total venture capital is defined as the sum of early-stage capital (seed and start-up) plus expansion capital.
8. Sources: European Innovation Scoreboard: Technical Paper No. 1, 'Indicators and Definitions', 11 November 2003, European Commission, using CIS3; OECD STI Scoreboard.
9. Fintra was established under the name of Vientikoulutussäätiö in 1962 by the Finnish government.
10. The price indices used in different countries affect the results. Edqvist (2005) discusses this issue in the context of Sweden and argues that in Sweden, where the ICT production has

been important for total productivity growth, growth has been overestimated because of the value-added price deflators used there. This may also concern Finland, Ireland and South Korea.

REFERENCES

Ali-Yrkkö, J. and R. Hermans (2002), 'Nokia in the Finnish Innovation System', ETLA Discussion Papers No. 811.
Ali-Yrkkö, J. and M. Pajarinen (2003), 'Julkinen t&k-rahoitus ja sen vaikutus yrityksiin – analyysi metalli- ja elektroniikkateollisuudesta', ETLA Discussion Papers No. 846 (in Finnish).
Ali-Yrkkö, J., M. Lindström, M. Pajarinen and P. Ylä-Anttila (2004), 'Suomen asema globaalissa kilpailussa – yritysten sijaintipäätöksiin vaikuttavat tekijät' (Finland in Global Competition – Determinants of Firms' Locational Decisions), ETLA Discussion Papers No. 927 (in Finnish).
Balaguer, A. and M.-H. Tsai (2004), 'Technological specialization in small and open economies', STIC Working Paper Series, Taipei (in Chinese).
Edquist, H. (2005), 'The Swedish ICT miracle – myth or reality?', *Information Economics and Policy*, **17**, 275–301.
Edqvist, C. and D. Göktepe (2003), 'Indicators on education and competence building', paper presented at the Taipei workshop in the SF project, November.
Estevão, M.M. (2004), 'Why is productivity growth in the euro area so sluggish?', IMF Working Paper WP/04/200.
European Commission (2001a), 'Innovation policy in Europe 2001 – European trend chart on innovation', Innovation Papers No. 17.
European Commission (2001b), 'Towards a European research area – key figures 2001', special edition, Indicators for benchmarking of national research policies.
European Commission (2004), 'European Innovation Scoreboard 2004 – comparative analysis of innovation performance', Commission Staff Paper, Brussels, 19.11.2004, SEC(2004) 1475.
European Commission, Science and Technology Indicators. CORDIS database.
Georghiou, L., K. Smith, O. Toivanen and P. Ylä-Anttila (2003), 'Evaluation of the Finnish innovation support system', Ministry of Trade and Industry of Finland, Publications 5/2003.
Ginarte, J.C. and W.G. Park (1997), 'Determinants of patent rights: a cross-national study', *Research Policy*, **26**, 283–301.
Hermans, R. (2004), *International Megatrends and Growth Prospects of the Finnish Bio-technology Industry*, ETLA, A 40, Helsinki.
Hernesniemi, H., M. Lammi and P. Ylä-Anttila (1996), *Advantage Finland – The Future of Finnish Industries*, ETLA, B 113, Helsinki.
Hyytinen, A. and M. Pajarinen (2003a), 'Financial systems and venture capital in Nordic countries: a comparative study', in A. Hyytinen and M. Pajarinen (eds), *Financial Systems and Firm Performance: Theoretical and Empirical Perspectives*, ETLA, B 200, Helsinki, pp. 19–64.
Hyytinen, A. and M. Pajarinen (2003b), 'Small business finance in Finland – a descriptive study', in A. Hyytinen and M. Pajarinen (eds), *Financial Systems and Firm Performance: Theoretical and Empirical Perspectives*, ETLA, B 200, Helsinki, pp. 203–47.

Hyytinen, A. and O. Toivanen (2003), 'Do financial constraints hold back innovation and growth? Evidence on the role of public policy', ETLA Discussion Papers No. 820.
Hyytinen, A., I. Kuosa and T. Takalo (2001), 'Law or finance: evidence from Finland', ETLA Discussion Papers No. 775.
Hyytinen, A., P. Rouvinen, O. Toivanen and P. Ylä-Anttila (2003), 'Does financial development matter for innovation and economic growth? Implications for public policy', in A. Hyytinen and M. Pajarinen (eds), *Financial Systems and Firm Performance: Theoretical and Empirical Perspectives*, ETLA, B 200, Helsinki, pp. 379–456.
Koskenlinna, M. (2004), *Välittäjäorganisaatiot Suomessa – rakenteelliset haasteet*, Publications of Ministry of Trade and Industry of Finland.
Koski, H., P. Rouvinen and P. Ylä-Anttila (2002), *Tieto ja talous – Mitä 'uudesta taloudesta' jäi?*, Helsinki: Edita.
Koski, H., L. Leijola, C. Palmberg and P. Ylä-Anttila (2006), 'Innovation and education strategies and policies in Finland', in C.J. Dahlman, J. Routti and P. Ylä-Anttila (eds), *Finland as a Knowledge Economy: Elements of Success and Lessons Learned*, Knowledge for Development Program, Washington, DC: The World Bank, pp. 39–62.
Lehto, E. (2004), 'Motives to restructure industries – Finnish evidence of cross-border and domestic mergers and acquisitions', Labour Institute for Economic Research, Working paper No. 195, Helsinki.
Leiponen, A. (2001), *Knowledge Services in the Innovation System*, ETLA, B 185, Helsinki.
Lemola, T. (2002), 'Convergence of national science and technology policies: the case of Finland', *Research Policy*, 31, 1481–90.
Lovio, R. (2002), 'Suomalaisten monikansallisten yritysten t&k-toiminnan kansainvälistyminen – perustietoja ja kysymyksenasetteluja' (Internationalization of the R&D activities of Finnish multinational enterprises – current facts and research questions), ETLA Discussion Papers No. 804 (in Finnish).
Lovio, R. (2004), 'Internationalization of R&D activities of Finnish corporations – recent facts and management and policy issues', in J. Ali-Yrkkö, R. Lovio and P. Ylä-Anttila (eds), *Multinational Enterprises in the Finnish Innovation System*, ETLA, B 208, Helsinki, pp. 39–74.
Miettinen, R. (2002), *National Innovation System: Scientific Concept or Political Rhetoric?* Helsinki: Edita.
Myllyntaus, T. (1992), 'Technology transfer and the contextual filter in the Finnish setting – transfer channels and mechanisms in a historical perspective', in S. Vuori and P. Ylä-Anttila (eds), *Mastering Technology Diffusion: The Finnish Experience*, ETLA, B 82, Helsinki: The Research Institute of the Finnish Economy, pp. 195–251.
Nieminen, M. and E. Kaukonen (2001), 'Universities and R&D networking in a knowledge-based economy: a glance at Finnish developments', SITRA Reports No. 11.
OECD (2003 and 2004), *Education at a Glance*, Paris: OECD.
Paija, L. (2001), 'The ICT cluster in Finland – can we explain it?', in L. Paija (ed.), *Finnish ICT Cluster in the Digital Economy*, Helsinki: Taloustieto Oy.
Pajarinen, M. and P. Ylä-Anttila (eds) (1999), *Cross-border R&D in a Small Country: The Case of Finland*, Helsinki: Taloustieto Oy.

Pajarinen, M. and P. Ylä-Anttila (2001), *Maat kilpailevat investoinneista – teknologia vetää sijoituksia Suomeen*, ETLA, B 173, Helsinki.
Palmberg, C. (2000), 'Industrial transformation through public technology procurement? The case of Nokia and the Finnish telecommunications industry', in C. Edquist, L. Hommen and L. Tsipouri (eds), *Public Technology Procurement and Innovation*, Dordrecht, Boston and London: Kluwer Academic Publishers.
Palmberg, C. (2002), 'Technological systems and competent procurers – the transformation of Nokia and the Finnish telecom industry revisited', *Telecommunications Policy*, **26**, 129–48.
Palmberg, C. and O. Martikainen (2005), 'The GSM standard and Nokia as an incubating entrant', *Innovation: Management, Policy & Practice*, **7**(1), 61–78.
Schienstock, G. and T. Hämäläinen (2001), 'Transformation of the Finnish innovation system: a network approach', SITRA Reports No. 7.
Science and Technology Policy Council of Finland (1990), *Review 1990: Guidelines for Science and Technology Policy in the 1990s*, Helsinki.
Statistics Finland (2005), *R&D Activity in Finland 1991–2004: Selected Statistics*, http://www.stat.fi/tk/yr/ttt-ko.html.
Van Beers, C., E. Berghäll and T. Poot (2004), 'Foreign direct investment and science and technology infrastructure in small countries: evidence from Finland and the Netherlands', VATT (Government Institute for Economic Research) Discussion Papers No. 357.
Ylä-Anttila, P. and C. Palmberg (2005), 'The specificities of Finnish industrial policy – challenges and initiatives at the turn of the century', ETLA Discussion Papers No. 973.

11. An NSI in transition? Denmark

**Jesper Lindgaard Christensen,
Birgitte Gregersen, Björn Johnson,
Bengt-Åke Lundvall and Mark Tomlinson**

1 INTRODUCTION

How has a small high-income country with high wages, high taxes, a large public sector, an export specialization in low-tech products (with a few exceptions) and a relatively low proportion of people with a higher education in science and technology been able to adjust to changing international market pressures and stay competitive and wealthy for decades? In particular, two interdependent explanatory factors have been put forward in recent studies of the Danish national system of innovation (NSI) (Lundvall, 2002).

The first explanatory factor is the Danish welfare state model. Since the 1960s, Denmark has emphasized social cohesion and a relatively equal income distribution based on comprehensive redistribution mechanisms. Since the 1930s the country has had strong trade unions and a strong political presence of the Social Democratic Party even in periods when that Party did not form the government. A central institution in the formulation and implementation of economic policies has been the corporatist system of interactions between the state, the trade unions and the employers. This has created a labour market with a high degree of 'flexicurity', combining high flexibility for employers to hire and fire with relatively high degree of income security for the employees (Madsen, 2006). A crucial related aspect of the social cohesion model is the high labour market participation rate for women in combination with an extended public service scheme for child and elder care.

The second and related explanatory factor has to do with the 'mode of innovation' dominated by small and medium-sized enterprises (SMEs) continuously making incremental innovations based on learning by doing, learning by using and learning by interacting, especially with customers and suppliers. One exception to this general picture is the traditional scale-intensive agro-industrial sector with a high degree of standardization. This sector has stayed relatively competitive due to high efficiency in the processing industries, heavy EU subsidies to primary production, forceful marketing and efficient distribution channels. Another exception is pharmaceuticals – a science-based industry with a high level of patent activities.

In a study of Danish competitiveness, Maskell (2004) shows that many small countries pursue a specialization strategy based on low-tech goods, but that the Danish case also has specific features. In particular, informal institutions such as the negotiated economy, egalitarian social values and the role played by established trust relations in easing the exchange of information are pointed out as significant elements (Lundvall, 2002; Maskell, 2004). This type of 'village economy', stable macroeconomic conditions, and an advanced public service sector are important keys to understanding how Danish industry has remained relatively competitive without substantial inputs of formal research and development (R&D).

However, in recent years the two fundamental pillars of the Danish NSI have come under increasing pressure. First, the social cohesion model is under political pressure from neoliberal tendencies common to most of the Western world and also from an increasingly introverted approach to tackling immigration issues. Second, ongoing globalization implies changes in the international division of labour and, consequently, the prevailing mode of innovation.

The current Danish government has established two different committees to come up with solutions to these challenges. One committee has investigated how to change the income redistribution mechanisms of the Danish welfare state and the other – with the prime minister in the forefront – has discussed challenges of globalization, with a special focus on how to stimulate innovation and adaptive capabilities of the labour market. However, a closer look at the evolution of the existing strongholds of the Danish NSI shows that they have relied, to a large extent, on the unique combination of a welfare state emphasizing social cohesion and a mode of innovation based on interactive learning and international trade. The following sections provide further documentation for this statement.

2 MAIN HISTORICAL TRENDS

2.1   From Agriculture to Agro-industry

In the nineteenth century, the Danish economy was predominantly agrarian and linked to the world economy mainly through its exports of grain to the UK. In response to the entry of new competitive grain producers from Russia and the USA, Danish agriculture was reoriented from grain to animal products and also gradually toward more processed products.

The rapid diffusion of technical innovations through the Danish economy since the late nineteenth century occurred in connection with an important institutional innovation – the introduction of cooperatively organized dairies and slaughterhouses. The cooperative ownership model proved very efficient in stimulating a rapid technology diffusion based on education, training and a widespread consultancy system targeting both the primary producers and the agro-industry.

The history of the transformation of the agro-industrial complex is reflected both in the pattern of specialization of Danish technical competence and in the organization of innovation. Over the years, the agro-industrial sector has been especially profitable in developing, producing and marketing selective food products of a uniform and relatively high standard. On the one hand, this standardization process initiated a strong focus on process innovations emphasizing automation and reductions of direct labour costs in dairies and slaughterhouses. This tendency still prevails, with the effect that increasing numbers of jobs have recently moved from Denmark to other EU countries with lower wages. On the other hand, the process innovations created a demand for innovations in machinery and equipment for milk and meat processing that has fostered Danish strongholds within these technological fields. Gradually, a cluster of firms within the different parts of the value chain for pork production has emerged. One of the main players in this cluster, Danish Crown, is the second largest slaughterhouse in the world, processing around 20 million pigs per year. Various waves of concentration have influenced the Danish slaughterhouse sector, and today Danish Crown totally dominates the Danish scene with an export value close to DKK30 billion (approx. €3.7 billion) per year – equivalent to 56 per cent of the Danish total export in agriculture (www.danishcrown.dk). In 2004, around 15 per cent of all Danish exports were agricultural products. The Danish export of agro-industrial equipment is currently about DKK11 billion per year.

The long-term changes in the overall employment structure in Denmark have followed the general pattern of most OECD countries during the twentieth century – from agriculture to manufacturing to service sectors. In the Danish case, the strong expansion of public service sector employment took place from the mid-1960s to the early 1980s. After that period, employment growth in the public service sector has stagnated while employment in the private service sector has continued to increase.

2.2 Contemporary Clusters of Competence

In 2001, The Danish National Agency for Enterprise and Housing identified 20 existing and nine potential (emerging) Danish clusters of competence listed in Table 11.1.[1] The 29 identified clusters of competence include not only traditional Danish strongholds within agriculture (especially pork and dairy products) and the maritime sector (shipping companies, shipyards), but also the development of other Danish clusters of competence that have historical roots in the agriculture and the maritime areas. There is, for instance, a direct link between communication technology and sea transport, a link between refrigeration technology and food production and food transport, a link between insulin and pork production and probably also a link between stainless steel and dairy products.

It is, however, interesting to note that quite a few of the Danish clusters of competence – wind energy, water supply and wastewater treatment, hearing aids, handicap equipment, and the medical and pharmaceutical industry – are linked to public utilities and public services, such as healthcare. Public technology procurement, regulation and advanced public demand have been important driving forces in both the creation and the further development of these clusters. Additionally, there are direct and indirect links and spin-offs based on public sector demand and regulation within the communication technology and offshore industry clusters. Within the group of potential or emerging clusters of competence, moreover, there are also clear links to public sector demand and regulation in relation to bio-informatics and waste handling. We will return to such 'demand-side factors' in Section 4.2.

2.3 Low-tech Export Specialization, but Changes Under Way

At first sight, a relatively high share of the Danish manufacturing value-added, employment and export appears to be within low-tech industries (especially food products and beverages), although the share is decreasing (see Appendix Table A2.1). As mentioned in the introduction, there are exceptions to the traditional dominance of low- and medium-tech sectors – the most important being the pharmaceuticals and the medical industries. Both sectors have gained important terrain in terms of share of manufacturing value-added and export.

Low or medium R&D intensity does not, however, mean that production is not knowledge-intensive. In fact, production in many of the industries characterizing Denmark's so-called low- and medium-tech production is based upon extensive knowledge inputs related to a high degree of change and flexibility in firms' use of resources, including rapid diffusion of new technologies and frequent incremental product innovation combining a high level of competence in industrial design with advanced organizational techniques and marketing methods. Such knowledge components are not necessarily registered as research activities in the

Table 11.1 Danish clusters of competence, 2001

	Existing clusters of competence	Potential (emerging) clusters
Agriculture	Fur (DK produces 40% of world's mink) Seeds (DK supply app. 50% of world market) Commercial gardening Pork (hog breeding, bacon factories) Dairy products	Øresund Food Network Organic food
Transport and communication	'The blue Denmark' (shipping companies, shipyards, equipment, sea transport) Telecommunication (especially mobile and satellite communication) Transport (inland, goods)	PR/communication
Public utilities and public services (especially healthcare)	Wind energy (windmills, turbines, various components) Water (water supply, wastewater treatment) Hearing aids Disability equipment Medical and pharmaceutical industry	Bio-informatics Waste handling
Mixed	Power electronics Cooling and heating technology, ventilation equipment Stainless steel Offshore industry Textiles and clothing Furniture Business tourism	Sensor technology Ubiquitous computing Movie and video production Toys and games

Notes: The identification of clusters is based on the following four criteria:
1. High OECD market share (above average for Danish export to OECD, 1990–99). This criterion indicates an established cluster.
2. High export growth rate (above 13% p.a. (average)). This criterion indicates a potential cluster.
3. Highly priced products (prices exceed the average world market price by more than 15%). This criterion indicates high quality.
4. High market growth on import markets (OECD import growth exceeds 3.7% p.a.). This criterion indicates a growing market.

Source: Based on Danish Enterprise and Construction Authority (2003), Appendix C.

statistics but nevertheless add high value to the products. Clearly, industrial design is an input of that type. Examples of Danish design, such as Bang & Olufsen hi-fi equipment, Novo's insulin pen, the Wegner chairs, PH lamps, Royal Copenhagen (porcelain), and Georg Jensen silver illustrate how applied industrial design may be a (if not the) fundamental ingredient in international market success for these products. A study of the economic importance of design has estimated that Danish companies annually invest around DKK7 billion in design activities, and that companies investing in design show significantly better bottom-line results than companies not engaged in design activities (Danish Enterprise and Construction Authority, 2003).[2]

A supplementary explanation of the overall performance of the Danish NSI could be that innovations flourish in industries not usually regarded as traditional engines of growth, the so-called 'creative industries'. These include, for example, the music industry, the film industry, leisure, sports and arts. In Denmark, these industries make up a substantial and increasing proportion of the economy. It is estimated that these industries contribute 5 per cent of gross domestic product (GDP), represent 12 per cent of total employment, and account for 16 per cent of total Danish exports (Ministry of Science, Technology and Innovation, 2005a).

3 INNOVATION INTENSITY

Community Innovation Survey 2 and 3 (CIS2 and CIS3) data indicate that Danish firms are relatively innovative, but also that there are differences among sectors. According to the CIS3 data, the sectors with the highest shares of innovative firms are manufacturing and knowledge-intensive business services (KIBS), where respectively 50 per cent and 51 per cent of the firms have introduced product or process innovations. Within finance and trade, the share of innovative firms is lower (respectively 38 per cent and 35 per cent). Compared to other small open economies like Sweden, Finland and the Netherlands, the data (CIS3) indicate that the Danish finance and trade sectors are lagging behind in innovativeness (see Appendix Table A4.4). Since finance and trade have especially strong linkages with the rest of the economy, their relatively low innovation intensity may be a future Achilles' heel for the Danish economy as a whole, if no action is taken to stimulate more innovative behaviour within these two sectors. However, a new survey of product innovation activities in the Danish private knowledge-intensive service sector indicates important ongoing changes. This new survey (2005) covers services within finance, communication and KIBS.[3] The overall conclusion is that these three service sectors fully match the most product-innovative manufacturing industries (electronics and chemicals) in terms of the rate of innovation. Seventy-nine per cent of the financial firms, 86 per cent of the firms within communication and 73 per cent within KIBS had developed new products/services in the period 2003–5 (ECON Analyse, 2005).

Despite the increasing awareness of innovation, it is interesting to note that different independent surveys all indicate that the share of product-innovative firms in Danish manufacturing has been decreasing over the last ten years.[4] Although the CIS asks non-innovative firms why they do not innovate, there are no conclusive answers to this question. The reasons given by the firms are rather broad, encompassing in the Danish case 'market conditions' (half of the firms), 'not necessary due to earlier innovations' (one out of seven), and 'other reasons' (one-third). Among reasons listed under 'other reasons', the most frequent was either 'lack of economic resources', 'innovation is done elsewhere' or 'restructuring'. Other surveys (Christensen et al., 2004; Drejer, 2006) find that 'lack of time' and 'no need' are by far the most frequently mentioned reasons. Interestingly, the statement that there is no need for innovation is equally distributed between high-tech and low-tech industries.

Tables A4.12 to A4.15 in the Appendix show the share of the firms' turnover in 2000 due to product innovations made during the period 1998–2000. Although some differences exist between the sectors, it is clear that the lion's share of the turnover is based on unchanged products. On average, 13 per cent of the turnover is based on products that are new or significantly changed from the perspective of the firm.[5] Genuine new products – i.e. products that are not only new to the firm but also new to the market – account for around 7 per cent of the turnover. Data prepared for the latest version of the European Innovation Scoreboard (2005) indicate a small decrease in the sales of new-to-the-market products but a substantial increase in the sales of new-to-the-firm not new-to-the market products from 18 per cent in 2000 to 26 per cent in 2002.[6] To the extent that these data sets provide a fair picture, Danish firms seem to be especially successful in introducing new-to-the-firm products.

In general, larger firms are more often innovative than smaller firms. However, there is no evidence that larger firms should generally be more innovative than smaller firms. On the contrary, evidence from innovation surveys in Denmark shows clearly that small firms tend to be relatively more innovative than large firms if the intensity is calculated as the number of product innovations per employee (Christensen et al., 2004). Moreover, when the share of turnover based on genuine new products is taken as indicator, there seems to be no significant difference between large and small firms.[7]

3.1 Non-technological Innovations

The CISs have hitherto mainly focused on technological innovation in relation to product and process innovations. However, the capability to introduce non-technological innovations in the form of organizational innovations, management techniques and industrial or pure aesthetic design are often crucial for firms' competitive advantages.

CIS3 opens up a tiny possibility to shed some light on the importance of non-technological innovations. Although the variance in the data suggests that there may be some problems with a common understanding of the underlying concepts, it is currently the only indicator available within CIS3. Half the Danish firms in the survey claim to have made one or more of these types of changes. Strategy, organizational and marketing changes are the ones listed by most firms, whereas aesthetic design and management changes are less frequent, according to the survey.

Compared to similar results from other countries, the results for Denmark are quite clear. In general, Denmark occupies the bottom rank in relation to Norway, Sweden, Finland and the Netherlands (Ireland not included). According to the data, these other countries have considerably higher shares of firms implementing non-technical innovations. According to a composite indicator summarizing the different non-technical innovations, Denmark is ranked as number 21 out of the 23 countries for which data are available (European Commission, 2004, p. 15).

It is interesting that the relatively low score also holds for the aesthetic design indicator. One explanation could be that Danish firms are still benefiting from design activities carried out (long) before the survey; see also Section 2. This would of course be true for products such as classical furniture and porcelain, but not, for instance, for consumer electronics. However, if the data are representative of the Danish firms, they surely indicate future challenges for Danish high-value-added firms to stay competitive, although recent data on for instance the number of new registered community designs per million inhabitants may give rise to less pessimism (see Section 4.1.1).

Recent work by Lorenz and Valeyre (2006) shows further interesting results concerning organizational forms and innovative performance. Based on European survey data on working conditions (conducted in 2000), these authors found wide differences in the importance of different forms of work organization across 15 European countries. They identified four basic forms (clusters) of work organizations: learning organization, lean production, Taylorism and simple organization. The learning form of work organization features the overrepresentation of problem-solving activities, learning new things in work, autonomy in setting work rate and work methods, and complexity of tasks. This form of organization seems to be most widely diffused in the Nordic countries, the Netherlands and to a lesser extent Germany and Austria. In the Danish case, learning organization accounts for 60 per cent of employment (EU-15 40 per cent), lean production 22 per cent (EU-15 28 per cent), Taylorism 7 per cent (EU-15 14 per cent) and simple organization 11 per cent (EU-15 19 per cent).

4 ACTIVITIES THAT INFLUENCE INNOVATION

The previous sections have briefly described the production structure and innovativeness of the Danish NSI. With this overall picture in mind, Section 4 analyses selected key elements and activities that constitute the Danish NSI.

4.1 Knowledge Inputs to Innovation

4.1.1 R&D activities

The R&D figures in Table 11.2 indicate a significant increase in the total R&D input for the Danish NSI. As can be seen, public R&D spending has not increased to the same extent as private, and for the moment, despite the decrease in 2004, it may seem more realistic that the private sector will reach the Barcelona goal of spending 2 per cent of GDP on R&D before 2010 than that the public sector reaches the minimum 1 per cent goal. In 2005 public R&D spending in Denmark even decreased to 0.75 per cent of GDP while most other countries increased such spending.

Going a bit below the surface of the aggregate numbers, a more differentiated picture emerges. Although the averages are level with, or even

Table 11.2 Denmark's R&D spending as a proportion of GDP, 1995–2005, %

Sector	1995	1997	1998	1999	2000	2001	2002	2003	2004	2005
Private	1.05	1.19	1.33	1.42	1.51	1.65	1.75	1.77	1.69	n.a.
Public	0.78	0.75	0.76	0.76	0.75	0.75	0.78	0.79	0.79	0.75[1]
Total	1.83	1.94	2.09	2.18	2.26	2.40	2.53	2.56	2.48	n.a

Notes: From 1998 there has been a change in the methods of data collection in order to include more R&D done by SMEs.
[1] Budget.

Source: Danish Institute for Studies in Research and Research Policy (2003, 2005).

above, EU and OECD averages, the distribution of R&D expenditures is skewed. Only 2 per cent of all firms conducted nearly 40 per cent of the total R&D in 2001 (Danish Institute for Studies in Research and Research Policy, 2004, p. 46), and in particular two major companies, Novo Nordisk and Sauer Danfoss, conducted a substantial part of the Danish R&D. Since the mid-1990s, though, the SMEs have increased their R&D effort, also as compared to the SME segment in the other Nordic countries.

The main part (in 2001, 80 per cent) of private R&D expenditure consists of in-house activities.[8] It is interesting, however, that nearly 60 per cent of external R&D is acquired from foreign firms and research organizations. Only 6 per cent of external R&D is purchased from Danish public research organizations.

The industries with the greatest research activities are the pharmaceutical industry and the information and telecommunication (ICT) industries. The research departments in these industries cooperate closely with Danish and foreign universities, but there is an increasing tendency for Danish firms to establish laboratories outside Denmark (Danish Institute for Studies in Research and Research Policy, 2003). Novo Nordisk, for instance, spends around half a billion euros per year on R&D and employs globally around 3000 people in research. Forty-four per cent of Novo Nordisk's 22460 employees are based outside Denmark (www.novonordisk.com). This trend is also prevalent in other countries hosting large research-intensive international companies, but especially for a country with only a relatively few R&D-intensive firms this may be a cause of concern.

Science and technology profile Patent statistics confirm the general picture of a Danish innovation mode characterized by incremental innovations and only a few radical breakthroughs (in technical terms), although sporadic changes in the general picture are visible. According to the 2005 European Innovation Scoreboard, Denmark has been moving ahead in terms of new European Patent Office (EPO) patents per million population, but is losing momentum with regard to new US Patent and Trademark Office (USPTO) patents and new triad (USA, Europe and Japan) patents, although both the EPO and the triad indicators are well above the EU-25 average. Regarding registered new community industrial designs per million inhabitants, Denmark held the leading position among the EU-25 in 2003 and 2004, supporting the hypothesis about prevailing Danish strongholds within industrial design.

Using the revealed technological advantage in patenting as an indicator of specific technological strongholds, Denmark has a relatively high share of patenting within biotechnology, pharmaceuticals, food and tobacco, agriculture, industrial process equipment, and glass, clay and cement (see Appendix Figure A3.2). Moreover, despite Denmark's relatively low total patenting performance compared to the leading European countries Sweden and Finland, it scores high on other output indicators. For example, Denmark has in general a high publication rate measured as number of publications in international journals per capita. Looking at the specific scientific areas with relatively high publication rates – here measured by the revealed comparative advantage in scientific publication output – the Danish strongholds are within ecology/environment, immunology, agricultural science, astrophysics, biology and biochemistry, molecular biology and genetics, and clinical medicine (see Appendix Figure A3.1). In that sense, there is a clear correspondence between several of the identified clusters of competence (see Section 2) and the technological and scientific strengths indicated by the specific areas of relative high patent activity and publication rates.

4.1.2 Competence building

Denmark spends a relatively high proportion of GDP on education as a whole – 8.5 per cent compared to the OECD average of 5 per cent. However, only 20 per cent of the expenditure on education is on higher education. Among other countries with high total education expenditure, the corresponding share is between 25 and 30 per cent (Ministry of Science, Technology and Innovation, 2003).

In the last two decades the number of higher education (HE) students has increased considerably. In 2005, 33.5 per cent of the Danish population between 25 and 64 years had a tertiary education. The equivalent figures for Norway, Sweden and Finland were 32.6 per cent, 29.2 per cent and 34.6 per cent, respectively, compared to the EU-25 average of 22.8 per cent (European Innovation Scoreboard, 2006). Even if the number of new graduates (per 1000 population aged 20–29) with training in science and engineering (S&E) has been growing, from 8.1 per cent in 1998 to 13.8 per cent in 2004, it is now just above the EU-25 average of 12.7 per cent and there is still a large gap compared to the leading EU countries Ireland and France with S&E graduates per 1000 inhabitants above 20 per cent.

Denmark has a long tradition of adult education and training. According to the European Innovation Scoreboard 2006, Denmark belongs together with Sweden, Switzerland, Ireland and Finland to the group of countries with the highest proportion of the population participating in lifelong learning activities. In 2005, 27.6 per cent of the Danish population aged 25–64 participated in such activities. In comparison, the equivalent share in Sweden was 34.7 per cent, and the EU-25 average was only 11.0 per cent.

Competence and competence building is linked to the labour market and labour mobility in various ways (Tomlinson and Miles, 1999; Tomlinson, 2004). First of all, mobility rates seem to increase during booms and decrease during recessions in overall economic activity. Second, as Graversen et al. (2003) found in their analysis of labour mobility trends in the Nordic countries, employees operating in different sectors and with different demographic characteristics have different propensities to shift jobs. Employees in high-tech sectors such as ICT seem to shift jobs more frequently. Moreover, employees with the highest human capital appear to be the most mobile, and older workers seem much less willing (or able) to move around than younger ones.

Even if mobility between firms is an important mechanism for knowledge dissemination, there are also drawbacks. If interfirm mobility becomes too high, it may reduce firms' incentive to invest in education and training activities or, conversely, increase the costs for these activities without the possibility of obtaining the long-term benefits. Furthermore, employees' capability and incentive to participate in innovation activities (including product, process and organizational changes) may be weakened.

4.2 Demand-side Factors

The CIS3 data provide a general picture of how firms perceive the importance of market pressure for their propensity to innovate. Nearly 90 per cent of the Danish innovative firms mention clients and customers as key sources of information for innovation, and only around 10 per cent of both the innovative and the non-innovating firms mention lack of consumer interest in new products and services as an important innovation barrier. Taking a longer-term perspective, it is obvious that 'demand factors' and market conditions are decisive for stimulating innovation activities, and in general factors such as high per capita income and a high degree of buyer sophistication influence the propensity to innovate positively. On both these indicators Denmark performs relatively well (European Commission, 2005). In the Danish case 'demand factors' in the form of public technology procurement and regulation have played (and still play) a key role in creating and maintaining important international strongholds.

4.2.1 Public technology procurement and regulation
As already mentioned in Section 2, public technology procurement, regulation and advanced public demand have been important driving forces in both the creation and the further development of several of the existing

Danish clusters of competence such as wind energy, water supply and wastewater treatment, hearing aids, disability equipment and the medical and pharmaceutical industry.

The development of both the Danish wind energy sector and the Danish hearing aid industry are in many ways illustrative examples of the systemic nature of innovation processes and its dependence on co-evolution and interaction between technological, economic, political and institutional elements. In both cases, public quantitative and qualitative demand and regulation have played a crucial role in innovation and performance of business activities. See Boxes 11.1 (wind energy) and 11.2 (hearing aids).

BOX 11.1 THE DANISH WIND ENERGY CASE

Since the late 1970s, wind power has played an increasingly important role in Danish energy production and consumption and over the same period the Danish wind turbine industry has obtained a leading world market position. In 2004, Danish firms accounted for around 40 per cent of the world production of wind turbines measured both in power (MW) and in market share.

The strong anti-nuclear power movement and the energy supply crises in the late 1970s spurred a growing interest in alternative sustainable energy technologies in Denmark. Most wind energy projects in the 1970s began as private projects, where technically interested people made experiments with scaled-down versions (10–15 kW) of the Gedser machine (Krohn, 1999). When the more 'professional' turbine manufacturers entered the scene in the late 1970s and the early 1980s, most of them came from a background in agricultural machinery (e.g. Vestas, Nordtank, Bonus, Nordex and later Micon), although one company, WindWorld, was founded on gearbox and marine technology (Krohn, 2000). The wind turbine companies illustrate how learning is cumulative and often based in the national production structure and at the same time 'accidental' or unplanned.

A mixed palette of policy instruments has been introduced to stimulate Danish wind power production. The obligation for utilities to buy wind power at 85 per cent of the market price level was crucial. Another key element was a 30 per cent subsidy of investments in new wind turbines. The investment subsidy was introduced in 1979, but was gradually reduced and was abandoned ten years later. Since 1985, the Danish government has ordered the utilities to install various amounts of wind power. Relatively high green taxes on all electricity – but with a partial refund for renewable energy including wind power – has been another measure to make wind power more attractive to the power companies.

The establishment of the public wind power test station at Risoe Research Centre in 1978 turned out to be crucial for the development of Danish wind power activities in relation to the knowledge requirements for production, distribution and regulation of wind power. To receive the public investment grants, a wind turbine type approval from the national laboratory was required. This approval process was an important mechanism for knowledge development and diffusion among both the wind turbine manufactures and the investors, and it stimulated an interactive learning process. The strict safety and performance requirements put a persistent pressure on manufacturers to upgrade their design and manufacturing skills, and today Risoe is among the leading international institutes when it comes to basic research in wind turbine technology and wind resource assessment.

Most wind turbine owners are organized in the Danish Wind Turbine Owners' Association, which publishes a monthly magazine with production figures and notes on technical failures for more than 1500 turbines. The statistical database, user groups and technical consulting services for members have been important instruments to secure a transparent market based on shared knowledge (Krohn, 2000). The wind industry manufacturers have their own organization too – the Danish Wind Industry Association. The organization carries out extensive information work, makes policy analyses, takes part in standardization activities, and is involved in national and international R&D activities. It seems fair to conclude that knowledge sharing and interactive learning among key players have been (and still are) important characteristics of the evolving Danish wind power innovation system. Hitherto, an 'open source strategy' seems to have prevailed for the benefit of the whole system, but new tendencies towards patenting and other forms of knowledge commodification may appear in the future.

Source: Gregersen and Segura (2003), www.windpower.org, www.dkvind.dk.

BOX 11.2   THE DANISH HEARING AID CASE

At the beginning of the 1950s, the Danish association of hearing-impaired people persuaded the Danish government to provide full public support for any Dane needing a hearing aid. This public-financed demand helped pave the way for the modern Danish hearing aid industry. It is worth noting that a similar decision on public support was taken in the UK, but without the same positive effects on the British hearing aid industry. According to Lotz (1997), this may have to do with the differences in the specification of the tenders. In the UK case, the tender was based on a specific design made by a medical research council, whereas the Danish tender only included a range of minimum specifications to be met.

An important part of the success story has also to do with a strong Danish knowledge base within audiology and acoustics – a knowledge base that beside hearing aids is applied in a vast range of other high-tech products with a Danish niche production like loudspeakers, room acoustics, advanced measurement techniques, and various forms of noise control. Furthermore, the Danish hearing aid industry has a long tradition of cooperation with Danish universities. Among the latest initiatives is the Centre for Applied Hearing Research, established in 2003 at DTU. The Centre is supported by various public and private funding including funds from the Danish hearing aid industry.

Today there are six global companies covering about 90 per cent of the world's hearing aid production: Oticon (William Demant Holding), GN ReSound, Phonat, Siemens, Starkey and Widex. Three of these are Danish:

1. Oticon (turnover: nearly DKK4 billion (2003), 3700 employees, of whom 1200 are located in Denmark, 95 per cent of turnover from abroad).
2. GN ReSound (turnover: approximately DKK3 billion (2003), approximately 3800 employees).
3. Widex (1400 employees worldwide, of whom 700 are located in Denmark, started 1956 as spin-off from Oticon).

Source: www.oticon.com, www.gnresound.com, www.widex.com; Lotz (1997).

4.3 Provision of Constituents

4.3.1 Provision of organizations

Renewal of existing organizations together with creation of new ones (firms, technological service institutes, regulatory bodies, etc.) reflects important dynamics of the NSI. By focusing on entrepreneurship, this section thus touches on one of the key factors creating dynamic changes within the system.

Formation of new organizations, here interpreted as formation of new firms, is most often either analysed by data from Global Entrepreneurship Monitor (GEM), which basically asks the adult, working-age population about their participation in start-up of new firms, or from statistics based upon establishment of new firms, for example using VAT numbers. The data show that Denmark is a little below the EU-15 average. The same goes for the total entrepreneurial activity (TEA) index from GEM. The TEA index for Denmark in 2004 was 5.32, meaning that one in every 19 Danish adults between the ages of 18 and 64 years had been directly involved in starting or running a new business. This was 10 per cent less than the year before. In comparison, the mean index for the 32 countries participating in the GEM survey in 2004 was 9.74 and for the 17 participating EU countries the mean TEA rate was 5.73 (GEM, 2004, p. 142). Taking the latest GEM surveys into account, they indicate a widening gap between Denmark and the rest of the world. While the Danish TEA rate decreased by 10 per cent in both 2003 and 2004, the world TEA increased. According to GEM (2004), Denmark has during the 2000–2004 period been consistently positioned in the middle among European countries, and below most of the developed countries outside Europe (ibid., p. 143).

In 2004, nearly half (48 per cent) of the new Danish businesses started within the framework of an existing firm (intrapreneurship). This is well above the 28.5 per cent mean for all countries and just below the top-ranked Hong Kong with an intrapreneurship rate of 51 per cent. Looking at the recent trends, there are indications of a decreasing interest among the Danish adult population to engage in intrapreneurial and entrepreneurial start-ups (GEM, 2004, p. 171).

Explanations for these differences across countries must refer to a combination of various factors including the country-specific production structure, unemployment rates, labour market institutions, culture, finance and supporting policies. In the Danish case, a low unemployment rate combined with an extended social benefit system implies that entrepreneurial activities simply motivated by lack of other income possibilities are low. Furthermore, the possibilities to start up and survive within traditional service and retail areas are not strong, due to a wide-reaching public service sector and a relatively concentrated retail market structure. Another explanation of the relatively low (and falling) rate of entrepreneurial activity may be that many Danish employees feel they have sufficient opportunities to influence their own working conditions and participate in learning and innovation activities at their existing workplaces. The previously mentioned study by Lorenz and Valeyre (2006) showing a relatively high share of learning organizations in the Danish case may support this hypothesis. Given the fact that the survival rate for new firms in general is low, there are – from a policy perspective – good reasons not to focus only on the creation of new firms, but also to pay attention to the crucial renewal process of existing firms.

4.3.2 Networking, interactive learning and knowledge integration

According to the CIS3 for Denmark, 40 per cent of innovating firms collaborated with one or more other partners in relation to their innovation activities in 1998–2000. As indicated in Table 11.3, nearly half of the firms (47 per cent) collaborated with their domestic (Danish) clients and customers, and 26 per cent had collaborations with their foreign clients and customers. From a globalization perspective, it is interesting to note that most of the foreign cooperation partners were located in the near EU/EFTA countries.

Table 11.3 Collaboration partners 1998–2000, % of firms with collaboration

Collaboration partner	Domestic (DK)	Foreign	Total CIS3-DK	Total CIS3-EU
Firms within same concern	42	29	65	42
Suppliers	48	23	60	61
Clients & customers	47	26	55	50
Competitors	22	14	34	29
Consultants	39	15	46	41
Private labs & R&D firms	17	7	22	22
Universities & other higher education	23	9	29	31
Public & private non-commercial research centre	19	3	21	20

Notes: CIS3-DK is based on an extended population including supplementary data from construction and selected service industries in order to fit the Danish R&D statistics. CIS3-EU is equivalent to the Eurostat dataset.

Source: Danish Institute for Studies in Research and Research Policy (2003), CIS3-DK, Table 15 and CIS3-EU, Table 32a.

Although suppliers and customers are the most frequent collaboration partners, it is interesting to note that around 30 per cent of the firms state that they have collaborated with universities. This indicates an important increase compared to earlier studies. A survey on the Danish NSI (the so-called DISKO project, 1996–99) found, for instance, that only 17 per cent of the Danish firms collaborated with universities in relation to product innovation, while 38 per cent collaborated with technological service institutes.[9] It is still too early to conclude to what extent the marked increase in the importance of universities as collaboration partners registered by the survey reflects an emerging transformation of the firms in general, i.e., whether they have begun to pay more attention to R&D or whether the data instead indicate that universities have become more engaged in consultancy and various technological service activities. The increasing private R&D (see Section 4.1) expenditures could support the former interpretation, and the growing pressure on universities to increase collaboration with private industry might support the latter.

4.3.3 Provision of institutions

An important role of institutions is to support or hamper processes of change for both individuals and organizations, making it easy or difficult for these actors to cope with changes in technology, organization, markets, procurement, occupation, income, residency, etc. Another focus point is how individuals and firms acquire and renew the skills and competences necessary for survival and success in a learning society. What makes people and firms 'abandon' and maybe forget existing competences and acquire new ones, and how is this accomplished? With this in mind, we shall concentrate in what follows on a few sub-sets of institutions that we think are crucial for the Danish NSI: educational institutions, research institutions and labour market institutions. In Section 7 we shall return to a discussion of some of the related governmental restructuring and policy initiatives.

Educational institutions Central and local governments are the main providers of education at primary and secondary levels. Generally speaking, curricula, study programmes, textbooks, etc., as well as most teachers, seem to acknowledge that supporting communicative and cooperative skills and 'social competence' is a major task of the school system. Informal institutions supporting interactive learning and innovation in the economy are provided by the school system, even if they are not often explicitly recognized. The recognition of the importance of social skills is, however, somewhat ambivalent. In the political debate on education, the proposition that too much time is devoted to social competence and too little to basic skills such as mathematics, language and hard facts about geography, history, etc. seems to have strong support.

University-level education is a state responsibility in Denmark. As a rule, the state supplies the education free of charge and is also responsible for a relatively generous system of loans and grants for the students. At the same time, it has been recognized that it is important for state-owned universities to cooperate with local interests and with labour market organizations. For engineering students, in particular, it has proved to be productive to connect their studies to concrete problems and tasks in firms and other organizations.

If we go back some ten years, the Danish investment in R&D and university training was at a lower level than in many of the other European countries. Some historical peculiarities in the training of engineers may have contributed to the difficulties as well. Until the beginning of the 1970s, Denmark's Technical University had a monopoly on the education of civil engineers, and it is important to note that it had a long tradition of giving priority to natural science rather than to solving technical problems in close collaboration with industry.

Research institutions The Danish NSI includes a wide grid of academic organizations: universities, governmental research institutes for specific sectors and areas (food, construction, environment, energy, social security, etc.), university hospitals, approved technical service institutes,[10] several centres of tertiary education (CVU in the Danish abbreviation) and business academies, science parks and innovation incubators. Within the college sector, the Ministry of Education has throughout the last ten years urged the organizations concerned with professional fields to merge, and these mergers have diminished the number of these types of organizations rather dramatically. Recently, further mergers have taken place within the university sector. Sixty per cent of public R&D investment is allocated to the universities, around 20 per cent to governmental research institutes, 15 per cent to university hospitals and the rest (around 5 per cent) to the remaining organizations.

Most of these organizations are public or semi-public. A large but decreasing share of funding for their activities comes from the yearly state budget. A smaller but increasing share of the institutional budget consists of funding for strategic purposes, programmes and projects, from both public and private sources. Private firms and funds pay for an increasing share of research, especially in the engineering sciences, which might reflect a growing orientation of the Danish NSI towards a more science-based and knowledge-intensive innovation mode. Institutional adaptations necessary to support this trend (new rules and regulations for contract research, publication, ownership of results, etc.) have been built up for a number of years.

Labour market institutions High participation rates, high mobility in terms of job changes, publicly organized and relatively generous unemployment support, considerable latitude for hiring and firing labour, and basic social security provided by a developed welfare state constitute, as already mentioned in Section 1, some of the most important institutional characteristics of the Danish labour market. This combination can be regarded as supporting interactive learning and innovation in the somewhat peculiar Danish village economy. The particular systemic characteristics of this institutional combination are not generally recognized in the Danish political system, however, and for several years now both the unemployment support regulations and the scope and generosity of some aspects of the welfare state have come under political pressure for reform. The Danish labour market still has institutional characteristics that support interactive learning and a specific Danish mode of innovation, but it cannot be taken for granted that they will survive the economic and political challenges and pressures connected to globalization and increasing competition.

National and international provision of institutions An increasing part of institutional change in Denmark has its roots abroad. This is a common tendency. Most countries are affected by new agreements within WTO and regional trade agreements (like NAFTA and CAFTA) also affect a great number of countries. For Denmark, the EU is a main provider of new formal institutions. Often, the international tendencies of institutional change take specific national shape through deliberate policy actions. One example is the international tendency towards extended and strengthened intellectual property rights (IPR), which has resulted in different regulations of university patenting in different countries.[11] In Denmark, a new patent law inspired by the Bayh–Dole principle has since 2000 (L347) given the universities an opportunity to take out patents, just as in a range of other European countries. It is too early to evaluate the long-term effects, but a recent evaluation (Ministry of Science, Technology and Innovation, 2004) confirms that setting up the necessary institutional infrastructure related to IPR is a costly, risky and lengthy learning process. Up to now it has not proved to be an important net income source for the universities.

4.4 Support Services for Innovating Firms

This section focuses on three types of support services for innovating firms: incubating activities, finance and consultancy services. At the general level such services are important ingredients in most countries' innovation policy, but the concrete manifestation typically varies according to the specific national context.

4.4.1 Incubating activities[12]

The Danish policy of business incubation has changed over time. Initially, the primary objective was to provide a supportive and conducive environment for SME start-ups by investing in incubators as an integral part of the business infrastructure. Today, incubation activities are mainly seen as an addition to the early-stage venture capital (VC) market, rather than an extension of the business support infrastructure. While the VC market in Denmark has developed considerably over the past few years, incubators still play an important role in providing support in the very early stages of new business development where traditional VC is sparse.

Incubation in Denmark is very much focused on establishing and supporting high-tech-oriented SMEs. Since the late 1990s, government support has been provided for technology incubators or 'innovation centres'. These centres give grants to entrepreneurs with innovative projects, thereby creating networks of universities, research institutes and other technology service institutes. The incubators amplify public funds with private sources of capital, such as VC firms and business angels, and they offer managerial and administrative services to start-ups as well as providing seed financing. A positive evaluation of these innovation incubators led to their continued funding through 2004, and in September 2004 they were granted another DKK400 million to secure future activities. At the beginning of 2005, most of these incubators were generating profit (Ministry of Science, Technology and Innovation, 2006).

Recently, there has been a growing focus on university incubators as a new organizational form to support students and young researchers in starting their own business. Although there are variations in both the content and the extent of the incubator activities among the 8 Danish universities, all 8 today have some form of incubator activity and all of them seem to intensify their efforts year by year. A recent survey conducted by the Ministry of Science, Technology and Innovation (2005b) shows that on average around 11 per cent of the Danish university students in 2005 were engaged in some form of concrete entrepreneurial activity, while around 30 per cent of the students indicated that they had considered starting their own business.[13] There thus seems to be an important potential for future knowledge-intensive entrepreneurial and intrapreneurial activities among the increasing number of graduates from higher education.

4.4.2 Financing

The Danish VC supply structure differs from that in European countries such as Germany, France and the Netherlands, where large private industrial firms constitute the main players on the supply side for VC. In Denmark, a large proportion of the VC supply originates instead from the state-owned Danish Investment Fund[14] and from pension funds – both these types of actors having a longer time horizon than many private firms, who often choose (or are forced) to concentrate on their core business areas when market conditions are tightened (Danish Venture Capital Association and Danish Investment Fund, 2005). These structural differences may be one of the reasons why, in the last four to five years (since 2000), Denmark has been able to maintain growth in VC investments while many other European countries have experienced a relative decline. This trend is also reflected in the European Innovation Scoreboard, where Denmark ranks third (after Sweden and Finland) among the EU-25 on the indicator for early-stage VC (measured as percentage of GDP).

The investment pattern among business angels differs slightly from that of VC investors. In general, business angels make fewer and smaller investments, and they have a broader investment focus than the Danish VC investors. Danish VC providers seem to prefer investments in high-tech companies – especially life sciences and ICT – located in the Copenhagen region, while business angels spread their investments to broader areas within manufacturing and services, and more than half of their investments take place outside the Copenhagen region (Danish Venture Captial Association and Danish Investment Fund, 2005).

4.4.3 Provision of consultancy services
The highly developed Danish technology consultancy system – including the Approved Technological Service Institutes (GTS institutes) and consultancy firms – functions as a bridging institution between research and firms. The GTS institutes are independent organizations selling their services on normal commercial terms. They do, though, also have non-profit objectives and the government co-funds some of their activities. Today, public funding provides on average only about 10 per cent of the costs, where this percentage was twice as high one decade ago. The GTS institutes employ approximately 3000 people and have a total turnover of DKK2.3 billion (about €310 million). Around 34 per cent of their revenue is generated from export of technological services, and this share is increasing (www.teknologiportalen.dk/EN). Concurrent with the commercialization of the technological service system, a comprehensive restructuring has taken place, and the number of approved institutes has been reduced from 16 to nine. The GTS institutes are encouraged to pay special attention to SMEs, and these firms receive a financial subsidy of their first-time use of one of the GTS institutes. In 2004, 60 per cent of the turnover came from SMEs. The institutes vary with respect to size and scientific/technical scope. Some are very R&D-intensive, while others emphasize scientific/technical activities such as consultancy, testing, standards, etc.[15]

As in most OECD countries, the Danish knowledge-intensive service (KIS) sector has grown rapidly during the 1990s. The KIS sector – here including finance, communication and business services (KIBS) – accounts for approximately one-third of the Danish private sector's value-added (2004). Forty per cent of total R&D expenditure within the private sector originates from KIS, and compared to most other OECD countries this is a relatively high share. Only Norway with 45 per cent shows a higher share (ECON Analyse, 2005). However, this might – as mentioned in Section 3 – rather reflect a low R&D level within the manufacturing sector in general than an extraordinarily innovative service sector in Denmark and Norway.

As in the manufacturing sector, the majority of the knowledge-intensive service firms collaborate with external partners when they innovate. Suppliers and private customers are ranked as the most frequent collaboration partners. However, it is interesting to note that the knowledge-intensive service sector collaborates less with universities and the technological service institutes than the manufacturing sector. On the other hand, affiliates, other consultancies, competitors and public sector customers are ranked as more important partners for service sector firms than for manufacturing firms. More than one-third of the KIS firms collaborate with the public sector when they develop new services (whereas the rate of public sector collaboration for manufacturing is only 10 per cent), and this makes the public sector an important driver of innovation within KIS.

4.5 Summary of the Main Activities Influencing Innovation

4.5.1 Knowledge inputs to innovation

The aggregated R&D expenditure as a proportion of GDP has steadily increased during the 1990s. The increase has taken place mainly in the private sector, while the public sector spending has stagnated, although the Danish government has stated recently that it will allocate more money to research in order to meet the Barcelona agreement of 1 per cent of GDP. However, compared to Sweden and Finland, this level of ambition will still place Denmark in the 'second division' concerning public R&D spending. On the supply side, the number of HE students has increased considerably in the last two decades, but the number of science and engineering graduates is still relatively low. On the positive side, Denmark has a long tradition of adult education and training. In 2005 more than 25 per cent of the Danish population aged 25–64 participated in various lifelong learning activities.

4.5.2 Demand-side Factors

According to the CIS3 data, nearly 90 per cent of the Danish innovative firms mention clients and customers as key sources of information. In this high-income country, the domestic market functions as an important 'test market' for more sophisticated consumer products – for instance, in relation to ICT.

At the same time, public technology procurement, regulation and advanced public demand have been crucial driving forces in creating several of the Danish clusters of competence such as wind energy, hearing aids, disability equipment, water supply and wastewater treatment. However, cuts in the public research activities within renewable energy may weaken the capability to stay in front.

4.5.3 Provision of constituents in the NSI

According to the Global Entrepreneurship Monitor 2004 (GEM, 2004), Denmark is positioned in the middle among European countries concerning entrepreneurship activities. This is the case even if the potential barriers in the form of costs and regulation are among the lowest in the world. The current low unemployment rate in combination with a long tradition of employee participation in innovation activities may be among the key explanatory factors for the relatively low level of entrepreneurial activities.

The majority of Danish firms innovate in collaboration with external partners – especially with suppliers and customers. Recently, more firms have started collaborations with universities and research institutes – a development that together with increasing private R&D investments may reflect a shift in the innovation mode toward an increasing focus on science and technology.

An important role of institutions is how they support or hamper processes of change for both individuals and organizations. Danish labour market institutions based on the so-called 'flexicurity model' have for many years been supportive of interactive learning and innovation, but it cannot be taken for granted that they will survive the prevailing political and economic challenges of the ongoing globalization process.

4.5.4 Support services for innovating firms

The Danish incubator activities emphasize high-tech SMEs. Since the late 1990s, government support has been provided for technology incubators or 'innovation centres'. These activities are supplemented by university incubators – a new organizational form to support students and young researchers in starting their own business.

A large proportion of the Danish VC supply is provided by the state-owned Danish Investment Fund and from pension funds. This basic funding opportunity has made the VC supply less vulnerable to shifting short-term private interests. Danish VC primarily invests in high-tech companies (especially life sciences and ICT), while business angels seem to spread their investments to broader areas within manufacturing.

The Danish technological consultancy system is widely ramified and functions to bridge the activities of research organizations and firms. The GTS institutes are independent organizations selling services on normal commercial terms. They pay special attention to SMEs but as certifying institutes they serve a vide variety of companies. Around 34 per cent of their revenue is generated from export. Furthermore, as in most OECD countries, the Danish knowledge-intensive service sector has grown rapidly during the 1990s and the sector today plays an increasing role in total private R&D activities. Although suppliers and private customers constitute the main external partners when firms in the knowledge-intensive service sector engage in innovation activities, collaboration with the public sector seems to be more important for this sector than for manufacturing.

In short, taking into account the various activities influencing innovation as well as the organizations and institutions related to these activities, it may be argued that the Danish NSI is gradually changing toward a more mixed mode of innovation combining an S&T-driven mode with the traditional mode of innovation dominated by doing, using and interacting.

5 CONSEQUENCES OF INNOVATION

As a first approximation it might be useful to distinguish consequences of innovation into two types: those that are expressed in terms of economic results, and those that are expressed in terms of broader social and cultural consequences. However, the focus here is limited to economic results.

When it comes to economic consequences of innovation in the narrower sense, it is important to make a further distinction between what is a potential consequence and what is a realized consequence. The institutional and organizational setting is of major importance for how a specific innovation is introduced. And this, in turn, will be reflected in the consequences that follow from the application. The many analyses pursued in the 1980s on the impact of information technology on the economy had to tackle this problem. For instance, a major result from a Danish study of the consequences of introducing new technologies in the manufacturing sector was that the impact on productivity of automation equipment was highly dependent on the efforts made in terms of organizational change and training of employees (Gjerding et al., 1990). Without such efforts, the new technologies seemed to contribute negatively to the productivity figures, and it took the innovating firms several years to compensate. This implies that productivity improvement is not only a consequence of technological and organizational change, among other things, but also of the interdependences between these factors. Since these interdependences appear to be quite strong, it is difficult to measure how much of the productivity growth may be a direct consequence of different kinds of innovation. We still do not know enough about how innovations are introduced, spread and utilized in the economy, and what effect they have in the short, medium and long run on production and productivity growth.

5.1 Macro-level Analysis

The Danish economic growth record does not deviate significantly from most other countries on about the same level of per capita income. The average growth rate of GDP per capita was just below 2 per cent in both the 1980s and the 1990s, and it followed the broad OECD pattern of slowing down after the oil crisis in 1973 (Kaitila, 2003). The Danish growth record during this period (1980–2000) followed the average OECD development closely except for the 1987–91 period, when it was almost 1 per cent below the OECD's.

The picture is about the same for aggregate labour productivity growth as measured by GDP per person employed or GDP per hour worked. During the 1990s the Danish productivity growth rate measured by GDP per hour was close to Sweden's and a little lower than Norway's and Finland's (Kaitila, 2003). However, Danish productivity growth rates are clearly below those of Singapore, Korea and Taiwan (ibid.). This means that the overall picture is that Denmark is no longer catching up with the USA but that South-East Asian countries are catching up with Denmark and similar high-income countries.

5.2 Micro-level Analysis

Moving from the macro- to the micro-level, there are as far as we know no clear research results on the relations between innovation and productivity growth. However, a report from the Danish Institute for Studies in Research and Research Policy concludes that value-added per employee in R&D-active firms was about 40 per cent higher than for R&D-inactive firms (Graversen and Mark, 2005).[16] The return was calculated at 34 per cent for R&D-active firms and only 11 per cent for firms on average. On the other hand, the R&D-active firms did not show a higher employment growth than other firms except for a small increase in personnel with higher education.

DISKO showed – on the basis of data for about 2000 Danish private manufacturing firms for the period 1993–97 – that firms that introduced product innovation in the period 1993–95 are more prone to create jobs than others. The data also show that much of the positive effect is located in the years after the innovation has taken place (Lundvall, 2002). This rules out possible dominance of the opposite causality – that employment growth stimulates product innovation.

These results should not come as a surprise. In most sectors, firms that introduce new products do so in order to attract demand from competitors, and if they succeed we should expect them to grow faster and to create more jobs than the ones losing market share. When it comes to the aggregate economy, a similar mechanism is at work. In all sectors producing tradables – goods and services exposed to international competition – the successful introduction of new products will conquer market shares from foreign firms competing in the global market.

By linking Danish labour market statistics (using the Integrated Database for Labour Market Research) to innovation activities at the firm level, it is possible to go into much more detail regarding the kinds of jobs created when firms introduce product innovations.[17] A general result is that unskilled workers, more than any other group of employees, seem to become highly dependent on their employer firms being able to respond dynamically to intensified competition.

6 GLOBALIZATION

Recently, outsourcing has played a central role in the public debate on globalization. In the Danish case the main focus has been on loss of labour-intensive production to Eastern Europe and East Asia. It is not a simple task to calculate the amount of job loss due to outsourcing, but some studies have estimated a net effect of currently around 5000 jobs every year (Danish Economic Council, 2004). In comparison, the Danish Economic Council estimates that the Danish export sector generates more than 500 000 jobs. Although outsourcing affects both unskilled jobs in manufacturing and high-skilled jobs in the ICT sector, the majority has hitherto consisted of unskilled jobs within manufacturing. This strong trend underlines the importance of new and additional policies aimed at stimulating an upgrading of the unskilled workforce.

Most developed countries have experienced a boost in foreign direct investment (FDI) since the liberalization of international capital flows in the late 1990s, and Denmark is no exception, with both inward and outward FDI near the EU average (see Appendix Table A2.6). The FDI openness of the Danish economy relates to both the inward and outward FDI stocks showing a balanced score. This also goes for the FDI share of services. However, in countries such as the Netherlands and Sweden, where multinational companies play a much more important role for the economy than in Denmark, FDI (both inward and outward) is also relatively higher.

The distribution of FDI by region of origin shows the majority of investment coming from Sweden, the USA and the Netherlands. These countries are also among the main recipients of Danish exports (along with Germany and the UK).

In many countries, R&D activities are less internationalized than production and services, but this is changing as more multinationals set up R&D laboratories abroad. OECD science, technology and industry data on the percentage of GERD (gross domestic expenditure on R&D) financed by foreign agencies shows a relatively significant reliance of the Danish research system on foreign linkages and international collaboration. Although there was a peak in these activities in the mid-1990s and it has fallen back recently, it is still one of the highest among the economies represented in this book – being outpaced only by Ireland (which had huge increases in FDI from a very small base throughout the 1990s, mainly from US firms).

Another indicator of how globalization affects R&D activities is the cross-border ownership of patents. As can be seen from Appendix Table A2.7, this indicator shows immense variation among the ten economies. In 2002, 41 per cent of the USPTO-granted patents based on inventions carried out in Denmark had foreign ownership, while only 6 per cent of the Danish patents were based on inventions made abroad. In comparison, the equivalent figures for Sweden, for instance, indicated that 22 per cent of the patents were based on Swedish inventions that had foreign ownership, while 14 per cent of the Swedish patents were based on foreign inventions. In Ireland, 71 per cent of the Irish patents had foreign ownership. On the one hand, a relatively high share of foreign-owned patents may indicate that Danish firms and research organizations are able to create interesting and important inventions in an international context. On the other hand, a high share of foreign-owned patents may also reflect a high vulnerability concerning the possibility of keeping and utilizing these inventions as a basis for creating domestic income and production in the long run.

7 STRENGTHS AND WEAKNESSES OF THE SYSTEM AND INNOVATION POLICIES

One way to approach strengths and weaknesses of a system of innovation (SI) is to compare certain variables that indicate strength and then to rank systems according to them. There are certain problems with such an approach that relate to what has been called naïve benchmarking. Actually aiming at reaching as high as possible in an international ranking on each single indicator disregards the systemic element in the SI (Lundvall and Tomlinson, 2002).

At the beginning of this chapter we emphasized some structural and institutional characteristics that have had major impacts on the form, content and rate of innovation activities. Some of these are outside the reach of innovation policy. That Denmark is a small country and located in the north of Europe is a geographical fact. The industrial structure – the low-tech specialization, and the large public sector, as well as the predominance of SMEs – has been shaped in a historical process and cannot be easily changed, not even in a long-term perspective, although public policies may of course influence the industrial structure.

Other apparent strengths and weaknesses discussed in Section 4 on framework conditions for innovation may be more or less easy to change through policy – but they, too, will reflect such historically rooted structural characteristics. For instance, the weak links between universities and the many SMEs in Denmark – although some changes have taken place recently – will reflect both historical and current weaknesses in demand from the industry side with roots in the industrial structure. Therefore policies aiming at bringing the national system 'to the very top' in this dimension might not necessarily strengthen the system as a whole. Given the importance of innovation in low-tech activities, other dimensions such as lifelong learning may be a much more important focus, even if Denmark already appears to be relatively strong in this respect.

Generally, indicators selected to rank SI and to define strengths and weaknesses are biased toward science-driven innovation and a science-based mode of innovation. Investments in R&D and in education are easier to quantify than the frequency of learning organizations and the quality of user feedback, and for this very reason the factors that support the more experienced-based mode of innovation tend to be relegated to secondary importance both in benchmarking exercises and on policy agendas. This is especially problematic when it comes to assessing the strengths and weaknesses of the Danish NSI, where innovation and learning in 'low-tech' activities are important for the overall performance of the economy.

7.1 Strengths and Weaknesses

7.1.1 Increasing the supply of human resources
Currently a lively debate is proceeding in Denmark under the heading of 'the ageing population and the crisis of the welfare state'. The major focus is on how to increase further the supply of labour in order to respond to

the ageing of the population. In this debate, the relatively low average number of hours per worker in Denmark (see Appendix Table A2.2) has been put forward as a problem, but the trade-off between long working hours and high rates of female participation in labour markets needs to be taken into account. It might not be a realistic strategy to increase all aspects of labour supply at the same time. It is already generally recognized among Danish politicians that families with children 'need more time for family life'.

A clear weakness in the Danish labour market is the low participation rates of workers without professional training and the even lower participation rates for workers with a non-Danish background. This poses a real dilemma for the future development of the Danish mode of innovation. The income distribution in Denmark is relatively egalitarian and the mode of innovation is quite participatory. This is one of the reasons why the system promotes interactive learning within and between organizations. But the egalitarian income distribution and the participatory mode of innovation tend to exclude those who have difficulties with engaging in interaction and informal communication.

This weakness is perhaps the most difficult to overcome. The high degree of 'social cohesion' sometimes referred to by defining Denmark as 'a village economy' is a major factor helping to explain the performance of the economy as a whole. Even if the proportion of the population belonging to ethnic minorities is small as compared to other countries, this has become a kind of Achilles' heel for the Danish model, including its innovation mode. The problem is exacerbated by discrimination, prejudice and populist xenophobic politics.

7.1.2 Incentives and the role of the public sector

To define what is a weakness and what is a strength always involves normative elements. This becomes especially clear when we discuss the role of government and taxes. There is a more or less implicit assumption in much of mainstream economic policy debate that high taxes and a large public sector constitute a weakness.

The fact that world rankings of 'competitiveness' tend to end with a very diverse group of national systems at the top of the overall ranking raises doubts about this perspective. In recent years, the US-market-dominated economy will typically appear in the top ten together with the Nordic welfare states. In this light, it is not self-evident that the high marginal tax should be seen as a weakness in Denmark. One might turn this around and argue that the 'exceptional willingness of Danish citizens to contribute to collective activities', documented in many different surveys, is a strength – especially as compared to Anglo-Saxon countries, where the negative attitude to the state gets in the way of collective investment in infrastructure and education.

But rather than arguing that a large public sector is either an advantage or a disadvantage, the focus should be on the specific role played by the public sector in the specific NSI. In such an area as labour market training this role is quite ambitious in Denmark. This reflects that the population of firms is dominated by small size and that there is high mobility in the labour market. In the absence of public initiative, there would be underinvestment in lifelong learning. Similar arguments are valid for public investment in research.

This perspective does not rule out that a high marginal tax might have a negative impact on, for instance, the start-up of new firms and the supply of labour. But it is important to take into account the trade-off between tax rate and the quality of public infrastructure. There might be direct positive effects on innovation and public expenditure as well. In Section 4.2 we gave some examples of how the public sector in Denmark has historically played a key role in creating the preconditions for major new technological strengths in relation to, for example, wind energy and hearing aid technology.

The fact that the public sector in Denmark is highly efficient should be brought in as another strength of the NSI. Over the last decade most public sector activities have been required to increase their productivity by 2.5 per cent per year. The low level of corruption is another positive feature that makes life easier for citizens and firms. This is a strength that is often neglected in economic analysis.

7.1.3 Entrepreneurship

In Denmark, the barriers to start up new firms are lower than in almost all other countries, and the fact that the rate of start-ups according to the TEA index remains low as compared to some other countries might be seen as a weakness in terms of lack of 'entrepreneurial culture'.

Entrepreneurship might be seen as a positive element in the process of competition and economic transformation. Some innovations can best be introduced through the establishment of new firms. New firms constitute a challenge for incumbent firms, stimulating those to engage in innovation. But, even so, it seems problematic to see the frequency of start-ups as a variable that ought to be maximized, independently of the wider systemic context.

Not all start-ups survive, and there are social costs connected to the birth and death of firms. Among those that survive, not all contribute to a dynamic economy. The fact that Argentina appears in the top category and Japan at the bottom in the international GEM surveys illustrates first that the frequency of start-ups may be a reflection mainly of high unemployment and lack of alternative options, and second that renewal of the economy can also come through other mechanisms, such as incumbent firms.

To start a new firm is not the only way to transform individual creativity into economic results. For the SI as a whole, it might actually be more important to have creative and entrepreneurial employees in existing organizations than to stimulate them to leave these organizations and start their own firms. This is another example of where benchmarking tends to operate more on a normative basis than on the basis of analytical results.

7.1.4 Activities related directly to innovation

Danish universities are mainly financed from the central state budget, and industry does not buy much research from the universities. Higher education has been exposed to the standard requirements of 2.5 per cent annual 'productivity growth' (reduced finance per student) for a decade. This policy has also undermined university research. In spite of this development, Danish scholars tend to be highly productive in terms of research publications, and their publications are widely cited.

In recent years, the focus has not been on high scientific productivity, however. The major debate has been on the commercialization of research and on the transfer of research results to industry. Most recent initiatives in research and innovation policy have been motivated by the need to increase the transfer and use of scientific results in the private sector. The starting point is that, according to benchmarking, the interaction between universities and industry has been less developed in Denmark than in many other NSI. This is the case both for industry procurement of research from universities and for the volume of cooperative projects organized and financed together by industry and university partners. However, according to the CIS data presented above (see Table 11.2), the gap between Denmark and other European systems has diminished when it comes to collaboration in innovation projects between firms and universities. Results from other surveys on research collaboration confirm this development (Christensen et al., 2004).

National differences in patterns of collaboration between firms and various knowledge institutions may reflect the different roles played by various knowledge institutions in different NSI. A broad system perspective including the firms, the universities and research institutions, as well as the technological service system, is needed if we want to decide what is the 'optimal' frequency of collaboration between firms and universities.

7.1.5 Strengths and weaknesses in relation to low-tech sectors

An intriguing question is how it has been possible to remain competitive in industries with low R&D intensities such as food, furniture and clothing, in spite of high wage rates. There are several specific explanations in addition to the general fact that such firms often have competitive competences even if they have only an indirect science base that is mediated through competences in design, flexible organization and marketing. First, Denmark is a small, homogeneous 'village economy' where communication (and transactions) can take place at low costs. The vertical communication between management and workers is also more open and efficient than in most other countries. This reflects the fact that Denmark, for historical reasons, is comparatively egalitarian both in terms of economic resources and in terms of cultural distance.

The large public sector has compensated the losers in industrial and technological change, and strong trade unions have also reduced the resistance to such change in wide portions of the population. This is reflected in the fact that Danish workers, in spite of not being well protected from redundancy, nevertheless feel more secure than their counterparts in, for instance, Germany and France.

There are, of course, many other characteristics worth mentioning, such as the long history of international trade and commercial efforts. In the relative absence of civil engineers, engineers trained at 'Teknikum', where skilled workers can get a degree in engineering, have contributed to strong links between the engineering side and workers.

In general, 'the social dimension' is crucial for understanding both the style and the specialization of the Danish NSI. Close interaction and easy communication make it possible to respond rapidly and effectively to new technological and market opportunities.

7.1.6 Strengths and weaknesses of the high-tech sectors

The relatively small – but increasing – high-tech sector primarily encompasses pharmaceuticals, mobile telecommunications and some other fields. Recently, also, biotechnology has grown to be as important as traditional areas of specialization such as textiles, furniture, and parts of the iron and metal industry. In high-tech fields, as in low-tech ones, we find elements of geographical concentration such as 'Medicon Valley' in the Ørestad region (around Copenhagen and Malmö) and 'Mobilecom Valley' in Northern Jutland.

Why is Danish industrial production so weak in high-tech products? There are several factors that need to be taken into account. First, there is the general small country handicap in most high-tech sectors. Part of the explanation is that high-tech activities are costly and difficult to build on a small scale. Finland and the example of Nokia (see Chapter 10, this volume) has been used as a kind of model for other countries, but it remains to be seen if the domination of one company also yields sustainable and stable economic growth in the long run.

There is a cumulative causation between the low investment in R&D and the historical specialization pattern, and not least the predominance of SMEs in Danish manufacturing. Firms' absorptive capacity in relation to formal codified knowledge of the kind that is produced at universities was limited and therefore, until recently, there was little 'demand pull' coming from the manufacturing sector.

7.2 Summary and Evaluation of the Innovation Policy Pursued

The Danish history of innovation policy is not very long. The first ambitious attempts to develop an innovation policy took place at the beginning of the 1980s under the heading of TUP (Danish acronym for Technology Development Programme). Information technology (IT) was seen as the most important challenge, and a number of initiatives were taken to stimulate its development. At the beginning of the 1990s, the emphasis on IT remained, but the focus changed toward technological infrastructure and the application side of ICT (Christensen, 2003).

At the beginning of the 1990s a broader approach was signalled as policies shifted from research and technology policies, often with technology programmes and subsidies as instruments, towards policies more directed towards innovation and framework conditions (Lundvall and Borrás, 2005). A similar change in policies took place in most other Western European countries (see for instance Larédo and Mustar, 2001). One initiative along these lines was the definition of 'development areas' covering almost all private economic activities in Denmark. The idea was to establish a dialogue with industry in specific broadly defined sectors aiming at policy development and optimizing framework conditions. After some enthusiasm and implementation of a number of ideas stemming from this dialogue, the initiative was mobilizing less and less interest.

In Denmark a policy area related to management, organization and competence building in firms became important at the end of the 1990s. The so-called LOK project (Danish acronym for Management, Organization and Competence) had interesting ambitions but ended up more as an academic exercise with an emphasis on management research and training.

With the outcome of the 2001 election, a liberal–conservative government took over and was re-elected for a second four-year period in February 2005. During these years, some important restructuring in the organization of the research and innovation policy has taken place. Innovation-related policies were moved from the Ministry of Economic and Business Affairs to a newly established Ministry of Science, Technology and Innovation. Similarly, innovation-related policies connected with trade and business services were transferred from the former

Ministry of Trade and Industry to the new Ministry of Science, Technology and Innovation. Further, the administration of the universities was moved from the Ministry of Education to the Ministry of Science, Technology and Innovation. The overall idea was to improve the coordination of the various research and innovation policies by centralizing the competences in one ministry. At the same time, however, the Ministry of Economic and Business Affairs became more focused on creating good general conditions for private firms, promoting entrepreneurship and supporting any kind of start-up firms, but especially high-tech and university spin-offs.

Although the new ministry has been labelled the Ministry of Science, Technology and Innovation, the restructuring has hitherto mainly focused on science-based sectors and 'high-technology research' in fields such as nanotechnology, IT and biotechnology. There has also been a stronger interest in understanding and developing the 'knowledge-based economy', but here too there has been a bias in the direction of formal knowledge and too little understanding of the importance of learning by doing, using and interacting (DUI learning).

The concrete initiatives have been focused on strengthening industry by making research more relevant and accessible. This orientation has resulted in a university reform that intends to bring universities closer to users in industry. Several programmes have thus aimed at strengthening the interaction between universities and the small minority of science-based firms. A new fund for 'high-technology research' using incomes from the sale of the North Sea oil rights has been established.

7.3 Future Innovation Policy

What will happen next with innovation policy in Denmark? In some respects, the direction is clear. The tendency to bring universities and their research efforts closer to industry will probably be pushed further ahead in the near future. The dominating focus on science-based innovation and on technical innovation – and the relative neglect of innovation in low-tech and service sectors – will remain for some time, even if there are some counteracting tendencies such as, for instance, a new initiative to stimulate 'user-driven innovation'.

There is still some dialogue between several interested parties, including academia, business and labour unions, regarding the development of innovation policy. The Ministry of Economic and Business Affairs and the Ministry of Science, Technology and Innovation have kept contact with innovation scholars in order better to understand developments in the field (Nyholm et al., 2001, p. 270). The SI perspective has certainly had a bearing on Danish policy and the way policy is discussed.

Even so, innovation policy has become strongly based on a benchmarking methodology that goes directly against the logic of the systems perspective. The terminology has changed, but policies are still implemented in an *ad hoc* fashion – even though the policy makers now speak of the coherence of the SI rather than market failure.

In order to enhance and maintain innovation capability of the Danish NSI, there is a need to transform innovation policy to take into consideration that the Danish NSI currently seems to be changing towards a more mixed picture combining the 'traditional' innovation mode based on learning by doing, using and interacting with a more science-based innovation mode. First, it is necessary to take into account the importance of the wider socioeconomic setting for the successful operation of the Danish model. Second, there is a need for a new and more ambitious policy for knowledge diffusion. The recent initiatives towards more university–industry interaction is one important part, but there is a need to include more actors than private industrial firms, for instance the public service and utility sector, the educational sector, the healthcare sector and KIS. Furthermore, there might be a need to support specific types of knowledge diffusion and innovation activities related to, for instance, environmental issues including energy. Third, the current focus on high-tech firms needs to be supplemented with policies that support the absorptive capacity of SMEs in traditional sectors. Fourth, although Denmark is doing relatively well compared to other European countries, there is a need to further develop policies that promote the diffusion of good organizational practices in terms of learning organizations and network formation.

NOTES

1. It should be noted that the use of the term 'clusters of competence' mainly expresses international specialization areas identified via trade statistics that do not necessarily qualify as clusters in the Porterian sense of the term, which emphasizes strong labour and commodity linkages within each cluster. However, each of the clusters of competence has strong linkages to related knowledge institutes (e.g. universities and technological services institutes).
2. According to the survey, over a five-year period companies that have increased their design investments experienced 40 per cent greater gross result gains than other companies (Danish Enterprise and Construction Authority, 2003).
3. Finance includes NACE codes 65, 66 and 67. Communication services include NACE codes 92.11, 92.2 and 80.42. Business services include NACE codes 72, 73 and 74.
4. The fact that CIS3 includes smaller firms (ten employees and above) than CIS2 (20 employees and above) may account for part of the decrease in shares of innovative firms. The development is consistent with what has been found in the manufacturing sector in several other European countries, and in this book, for example, in most of the countries for which we have statistics.

5. These numbers are consistent with other, independent surveys (Christensen et al., 2004; DISKO).
6. There are some smaller deviations between the data presented in the Appendix and the 2005 version of the European Innovation Scoreboard concerning the 2000 figures. According to EIS, the sales of new-to-the-market products (percentage of turnover) is 9 per cent compared to 7 per cent in the Appendix. New-to-the-firm products account for 18 per cent of total turnover in EIS and 13 per cent in the Appendix. However, these differences do not interfere with the overall picture.
7. The turnover of new-to-the-market products for larger firms accounts for 7.6 per cent of total sales, while around 5.7 per cent of the turnover for small firms is based on genuine new products (see Appendix Table A4.13). This difference is small and statistically insignificant. Moreover, the share of unchanged products new to the firm is the same.
8. Looking at the distribution between the 'R part' and the 'D part', the vast majority (around 70 per cent) is spent on development. Around 77 per cent is spent on product development.
9. Christensen et al. (2004) find that collaboration on product development has not increased between 1997 and 2004, but the pattern of collaboration partners among those who do collaborate has changed: firms have more partners of different types, and collaboration with knowledge institutions has increased significantly. In particular, collaboration with universities increased from 17 per cent to 29 per cent, much the same pattern as that found in CIS.
10. The GTS institutes function as a bridging institution between research and firms. We shall return to these institutions in Section 4.4.3.
11. At the same time, the trend is towards a more uniform European patent system, notably spurred by the recent implementation of the European patent.
12. Definitions of incubators differ. According to a study commissioned by the European Commission's Enterprise DG, a 'business incubator' is a broad term, embracing 'technology centres and science park incubators, business and innovation centres, incubators without walls, "new economy" incubators, and a variety of other models'. 'Incubators without walls' are organizations with no single physical location; instead they concentrate on managing a network of enterprise support services (European Commission, Centre for Strategy & Evaluation Services, 2002, p. 3).
13. According to the 2005 survey, students within humanities and social sciences seem to be a bit more entrepreneurial than students within engineering and natural sciences, while students within agriculture and health are less entrepreneurial. The three dominating single areas (2005) are IT services (27 per cent), consultancy (34 per cent) and public and private services such as entertainment, culture and sport (23 per cent). Around 4250 out of 111 000 Danish students participated in the email-based survey.
14. The Danish Investment Fund (in Danish 'Vaekstfonden') is a public fund under the supervision of the Danish Ministry of Economic and Business Affairs. The size of the base fund is around DKK2 billion and the main focus of its activities is investment in development and innovation in SMEs – especially within ICT and life sciences. More than half of the investments concern early-stage activities (seed and start-up capital) but the share allocated to growth (expansion) activities has recently been increasing.
15. The nine GTS institutes are: Danish Technological Institute, DHI Water and Environment, FORCE Technology, Danish Standards Association, Danish Institute of Fundamental Metrology, Danish Toxicology Centre, DELTA Danish Electronics, Light & Acoustics, Danish Institute of Fire and Security Technology, Bioneer.
16. The analysis is based on data for one year (2001) and covers around 2200 Danish firms with more than ten employees. R&D-active here simply indicates registered R&D activity.
17. It should of course be kept in mind that innovation destroys as well as creates jobs, depending on the type of innovation and other factors.

REFERENCES

Christensen, J.L. (2003), 'Changes in Danish innovation policy: responses to the challenges of a dynamic business environment', in P.S. Biegelbauer and S. Borrás (eds), *Innovation Policy in Europe and the USA*, Aldershot, UK: Ashgate, pp. 93–111.
Christensen, J.L., I. Drejer and A. Vinding (2004), 'Produktudvikling i dansk fremstillingsindustri', ACE Notat no. 8, Aalborg Universitet.
Danish Economic Council (2004), *Dansk økonomi, Efteråret 2004*.
Danish Enterprise and Construction Authority (2003), *Kortlægning af danske kompetenceklynger*.
Danish Institute for Studies in Research and Research Policy (2003), *Innovation i dansk erhvervsliv 2000*.
Danish Institute for Studies in Research and Research Policy (2003), *Erhvervslivets forskning og udviklingsarbejde. Forskningsstatistik 2001*.
Danish Institute for Studies in Research and Research Policy (2005), *Erhvervslivets forskning og udviklingsarbejde. Forskningsstatistik 2003*.
Danish Venture Capital Association and Danish Investment Fund (2005), *Det danske venturemarked, Investeringer og forventninger, Kvartalsanalyse, 4. Kvartal 2005*.
Drejer, I. (2006), *Et regionalt perspektiv på udviklingsaktiviteter inden for vidensintensive serviceerhverv, en sammenlignende analyse af udviklingen af nye serviceydelser i de nye regioner*, ACE, Aalborg University.
ECON Analyse (2005), *Innovation i videnservice*, Report 2005-105, Copenhagen.
European Commission, Centre for Strategy & Evaluation Services (2002), *Benchmarking of Business Incubators*.
European Commission (2004), *European trend Chart on Innovation. Annual Innovation Policy Trends and Appraisal Report, Denmark 2004–2005*.
European Commission (2005), *European Innovation Scoreboard 2005*.
European Commission (2006), *European Innovation Scoreboard 2006*.
Gjerding, A.N., B. Johnson, L. Kallehauge, B.-Å. Lundvall and P.T. Madsen (1990), *Den forsvundne produktivitet*, Copenhagen: Jurist og Økonomforbundets Forlag.
Global Entrepreneurship Monitor (2004), Part 2: M. Hancock, *Global Entrepreneurship Monitor – Denmark*.
Graversen, E. and M. Mark (2005), 'Forskning og Udviklingsarbejdes påvirkning af produktivitet og beskæftigelse', Rapport fra Dansk Center for Forskningsanalyse, 2005/1, Aarhus University.
Graversen, E. et al. (2003), 'Mobility of human capital – the Nordic countries, 1988–1998', Report 11/03, STEP, Norway.
Gregersen, B. and O. Segura (2003), 'A learning and innovation approach to social and ecological sustainability', paper presented at The First Globelics Conference, 'Innovation Systems and Development Strategies for the Third Millennium', Rio de Janeiro, 2–6 November.
Kaitila, V. (2003), 'GDP and productivity indicators for small European and Asian economies, and large reference countries', working papers provided for Eurocore Ten-country project.
Krohn, S. (1999), 'Danish wind turbines: an industrial success story', Danish Wind Industry Association.
Krohn, S. (2000), 'The wind turbine market in Denmark', Danish Wind Industry Association.

Larédo, P. and P. Mustar (eds) (2001), *Research and Innovation Policies in the New Global Economy: An International and Comparative Analysis*, Cheltenham, UK and Northampton, MA, USA: Edward Elgar.

Lorenz, E. and A Valeyre (2006), 'Organizational forms and innovative performance: a comparison of the EU-15', in E. Lorenz and B.-Å. Lundvall (eds) (2006), *How Europe's Economies Learn: Coordinating Competing Models*, Oxford: Oxford University Press.

Lotz, P. (1997), 'Supply and demand factors explaining exports in health care products: interdependencies between export specialization, health care expenditures, and research activities', paper presented at the DRUID Winter Conference, January, Copenhagen.

Lundvall, B.-Å. (2002), *Innovation, Growth and Social Cohesion: The Danish Model*, Cheltenham, UK and Northampton, MA, USA: Edward Elgar.

Lundvall, B.-Å. and S. Borrás (2005), 'Science, technology and innovation policy – old issues and new challenges', in J. Fagerberg, D. Mowery and R.R. Nelson (eds), *Handbook of Innovation*, Oxford: Oxford University Press, pp. 599–631.

Lundvall, B.-Å. and M. Tomlinson (2002), 'International benchmarking as a policy learning tool', in M.J. Rodriguez (ed.), *The New Knowledge Economy in Europe: A Strategy for International Competitiveness with Social Cohesion*, Cheltenham, UK and Northampton, MA, USA: Edward Elgar, pp. 199–226.

Madsen, P.K. (2006), 'How can it possibly fly? The paradox of a dynamic labour market in a Scandinavian welfare state', in J.L. Campbell, J.A. Hall and O.K. Pedersen (eds), *National Identity and the Varieties of Capitalism: The Danish Experience*, Copenhagen: DJØF Publishing, pp. 321–55.

Maskell, P. (2004), 'Learning in the village economy of Denmark. The role of institutions and policy in sustaining competitiveness', in P. Cooke, M. Heidenreich and H.J. Braczyk (eds), *Regional Innovation Systems*, 2nd edn, London: Routledge, pp. 154–85.

Ministry of Science, Technology and Innovation (2003), *Regeringens Vidensstrategi: Viden I Vækst*, Baggrundsrapport, Januar 2003, Videnskabsministeriet.

Ministry of Science, Technology and Innovation (2005a), *Danmark skal vinde på kreativitet: perspektiver for dansk uddannelse og forskning i oplevelsesøkonomien*, Copenhagen.

Ministry of Science, Technology and Innovation (2005b), *Universiteternes Iværksætterbarometer 2005*, Copenhagen.

Ministry of Science, Technology and Innovation (2006), *Evaluering af innovationsmiljøerne årene 2001–2004*, Copenhagen.

Nyholm, J., L. Normann, C. Frelle-Petersen, M. Riis and P. Torstensen (2001), 'Innovation policy in the knowledge based economy: can theory guide policy-making?', in D. Archibugi and B.-Å. Lundvall (eds) (2001), *Europe in the Globalising Learning Economy*, Oxford: Oxford University Press, pp. 253–72.

Tomlinson, M. (2004), 'Learning, knowledge and competence building at employee level in the UK', in J.L. Christensen and B.-Å. Lundvall (eds) (2004), *Product Innovation, Interactive Learning and Economic Performance*, Amsterdam and Oxford, Elsevier, pp. 211–28.

Tomlinson, M. and I. Miles (1999), 'The career trajectories of knowledge workers', in *Mobilising Human Resources for Innovation: Proceedings from the OECD Workshop on Science and Technology Labour Markets*, DSTI/STP/TIP(99)2/FINAL, pp. 152–67, OECD: Paris.

12. Globalization and innovation policy
Leif Hommen and Charles Edquist

1 INTRODUCTION

We argued in the introductory chapter that comparisons of the ten national systems of innovation (NSI) addressed in this volume are facilitated by an agreed-upon conceptual and theoretical framework as well as the use of a common table of contents in the country studies. Both the framework and the common table of contents were presented there.

In this chapter we will largely limit ourselves to addressing two issues on a comparative basis:

- The role of growth and globalization for small NSI
- Innovation policy, which will be dealt with more extensively.

Many other kinds of comparisons can be made on the basis of the case studies included in this anthology. However, making such comparisons will largely be left to the readers of this book, for whom the choice of issues for comparison, the selection of countries and the manner of comparison will naturally be governed by particular backgrounds, interests and objectives. By incorporating case studies of the ten NSI into this volume, we have simply provided a basis for pursuing these various kinds of comparisons.

Before dealing with the two issues listed above, we shall begin with a few general remarks on the activities-based framework presented in the introduction, concentrating on its usefulness for comparing NSI (Section 2). Subsequently, we shall address methodological issues regarding comparison between NSI (Section 3) and then present a classification of the ten NSI, identifying different contexts of growth and globalization (Section 4). Thereafter, we shall conduct an extended discussion of innovation policy issues, carrying out a comparative analysis of innovation policy in the ten NSI (Section 5). Finally, we shall present some 'telegraphic' conclusions (Section 6).

2 GENERAL REMARKS – THEORY AND POLICY

In the country chapters it has been shown that the ten national systems of innovation (NSI) are considerably different with regard to the propensity to innovate (or innovation intensity) – in terms of different kinds of innovations (product, process, radical, incremental) and the sectors in which these innovations occurred, etc. This is one possible comparison that, in its details, will not be spelled out in this chapter. Here, the main focus will instead be on innovation policy in the context of a globalizing learning economy.

In the introductory chapter we presented the conceptual and theoretical approach, which we called an activities-based framework for analysing and comparing systems of innovation. Before entering into the discussion of globalization and policy, we want to present a few general remarks related to issues raised in the introductory chapter.

The specification of a common set of key concepts and utilization of the same model table of contents in all country chapters means that they have all addressed the same issues and activities in similar ways – with, as previously noted, some interesting variations.[1] As a consequence, direct comparisons between the various cases are facilitated. And, as a further consequence, it should also be easier to draw conclusions from such comparisons and to generalize to other 'small country' NSI. Generating conclusions and generalizations on this basis might be thought of in terms of drawing lessons for theory and policy from the cases. But in that connection it is important to be cautious about the kinds of lessons that can be learned.

When it comes to drawing lessons for theory, it is essential to bear in mind that theories about systems and processes of innovation still remain very much at the level of what Nelson and Winter (1982) referred to some 25 years ago as appreciative theorizing. Such analyses are, to a large extent, context-based. Appreciative theorizing stays close to empirical matter, providing interpretation and guidelines for further exploration; it also has a rather partial focus, and develops causal arguments in a selective way (Edquist, 1997, p. 28). Both evolutionary economics and the systems of innovation (SI) approaches have been inspired by appreciative theorizing. SI approaches, in particular, rely to a large extent on grounding theory development in cases. To do so is not a matter of relating cases to law-like propositions of a formal theory, but rather one of remaining sensitive to how particular cases do not fit into larger theoretical assertions, paying close attention to contexts, and, in comparative analysis, 'constructing new contextual statements in order to link cases' (Ashford, 1992, p. 4).

The application of this approach in innovation studies has been described as 'building descriptive theory', by proceeding from careful observation, through classification into frameworks or typologies, to defining relationships between attributes of the phenomena and observed outcomes (Christensen, 2006, pp. 39–40). In this process, examining how phenomena vary across contexts is important for testing classifications and the hypotheses developed to explain them, since this is 'most often . . . done by exploring whether the same correlations exist between attributes and outcomes in a different set of data from which the hypothesized relationships were induced' – i.e., a different empirical context or type of case. Where such research reveals outcomes for which the theory cannot account, 'anomalies give researchers the opportunity . . . to define and measure the phenomena more precisely and less ambiguously, or to categorize the data better – so the anomaly and the prior association of attributes and outcomes can all be explained' (ibid., p. 40).

For such reasons, it is important to investigate contexts in a systematic way. In comparative case study research the analytic strategy of focused, controlled comparison is 'to formulate the idiosyncratic aspects for the explanation of each case in terms of general variables' in order to 'find ways of describing and explaining individual cases that render them comparable' (George, 1979, pp. 46–7). In testing hypotheses, the object is to apply general explanatory theories to relevant cases in order to produce 'typological' theory – generalizations relating to specific types of contexts. Such theory is especially valuable for policy makers because it 'enables them to make more discriminating diagnoses of emerging situations' (ibid., p. 59).

In the Ten Countries Project, we have engaged in building descriptive theory, as in Section 5 of this chapter. Although we have not reached the stage of hypothesis testing, we have nevertheless engaged in structured, controlled comparison by using the same concepts, the same theoretical approach, and the same format for describing ten different NSI. We have thus laid the foundations for policy-relevant typological theory about how innovation performance varies systematically according to differences in – or different types of – contexts.

The emphasis on context, including historical background, also leads to some important limitations concerning the possibilities of drawing lessons for policy. Perhaps the most important of these qualifications concerns the pitfalls of identifying widely transferable best-practice models. Direct imitation or copying of such models is rarely successful, particularly when crucial differences in context are not taken into account. Innovation policy, especially, can therefore not simply be reduced to institutional borrowing, i.e. copying institutions. It must instead remain essentially a matter of institutional learning, i.e. adapting to the national or local context (Lundvall and Borrás, 1997, p. 62).

This argument has been forcefully restated in the introduction to a recent volume on *Asia's Innovation Systems in Transition* (Lundvall et al., 2006). In their introduction, the editors reason that an SI perspective 'helps to avoid naïve borrowing of "best practice" policy across national borders' – an effort that often fails, since 'what seems to work well in one systemic context might not do so in another' (ibid., p. 16). They go on to claim, however, that although highly specific copying is seldom feasible, it is still possible to draw policy lessons of a more general character. There is much that can be learned, for example, about the importance of public intervention for promoting innovation and economic change, as well as the conditions under which it should (or should not) take place, the manner in which it ought (or ought not) to be implemented, the types of problems that policy should (or should not) address, and so forth (ibid., pp. 17–18). These are the kinds of conclusions that we attempt to draw from the comparative analysis of innovation policy carried out in the later sections of this chapter.

The activities-based conceptual and theoretical framework developed and used in this book differs from traditional SI approaches, which have focused on the constituents of systems of innovation (institutions, organizations, interactions). Instead, it develops a central focus on the activities in SI – i.e. the factors that influence the development and diffusion of innovations. By focusing on what 'happens' in the systems, an activities-based approach provides a more dynamic perspective than does focusing on the constituents. This does not mean, however, that constituents are ignored. In the case studies presented in this book, the ten activities outlined in the introductory chapter were used as points of entry into the analysis of NSI – but for each activity this analysis also addressed the relevant organizations, institutional frameworks and relations. In this respect, an additional virtue of the activities-based approach is that it enables us to capture the 'many ways countries may develop . . . to organise . . . "functionally equivalent" activities' (Kogut, 1993, p. 7).

The activities-based approach to innovation is certainly a broad one, and is, in this sense, more in line with the broad approaches to conceptualizing NSI advanced by Lundvall (1992) and Edquist (1997) than with the narrow approach propounded by Nelson (1993) – as discussed in the introductory chapter. Evidence from the ten countries investigated in this project also generally favours a broad perspective on NSI, rather than a narrow one. Most of the ten activities have been shown to matter for innovation processes in most of the ten countries.

One reason why an activities-based approach is advantageous is that it provides an explicitly multidimensional view of the determinants of innovation processes. This is important for innovation theory as well as for innovation policy – which are certainly related to each other. For example, explicitly addressing ten (albeit hypothetical) determinants excludes one-dimensional thinking such as that represented by the Barcelona Declaration of the EU.[2] Hence we argue that an activities-based approach is useful in securing a sufficiently broad perspective when it comes to determinants of the development and diffusion of innovations – and therefore facilitates an appropriate, i.e. multidimensional, perspective with regard to analysis as well as to policy.

Drawing attention to the multiplicity and variety of activities that influence processes of innovation within NSI naturally leads to a focus on the coordination of these activities. If we consider briefly two of the main frameworks that have been applied in recent comparative work on innovation – namely, varieties of capitalism (VoC) (Hall and Soskice, 2001) and business systems (Whitley, 2002) – we find that one of the main characteristics they have in common is a focus on the coordination of multiple kinds of activities.

In the case of the VoC approach, this focus is made explicit and underlined by the ideal-typical distinction drawn between coordinated market economies (CMEs) and liberal market economies (LMEs). As explained by the proponents of this approach, the central idea is that 'national political economies can be compared by reference to the way in which firms resolve the co-ordination problems they face' in five spheres: industrial relations, vocational education and training, corporate governance, interfirm relations, and employment relations (Hall and Soskice, 2001, p. 8). In LMEs, the dominant coordination mechanisms are 'corporate hierarchies and competitive market arrangements', whereas in CMEs 'firms depend more heavily on non-market relationships' (ibid.).

The business systems approach complements the VoC approach, sharing its fundamental emphasis on coordination, but pointing to important variations within the two broad VoC in both national institutional frameworks and firm strategies and structures (Whitley, 2002, p. 499). Further, proponents of this approach argue that the comparative analysis of innovation processes also needs to take into account 'variations in the organization and control of national public science systems [that] constitute an important part of the institutional environment explaining differences in prevailing patterns of technological development between countries' (ibid., p. 500).

As stressed by both of these approaches, coordination is governed by the institutional environment and achieved through reliance upon institutions as coordination mechanisms.[3] This emphasis on institutions accords well with SI thinking, which 'has emphasised the essentially context-bound nature of technological change . . . especially in terms of the relevance that the institutional set-up has for innovative performance' (Borrás, 2004, p. 427). However, the SI approaches also differ from these approaches in some important respects. Whereas both the VoC and business systems frameworks have a very wide scope, SI approaches have maintained a primary focus on innovation. Another distinguishing feature of SI approaches is that they remain more open-ended, due to their stronger commitment to appreciative theorizing.

Hollingsworth sums up the position taken by the SI approaches as follows: 'since we do not presently have an adequate theory on how institutions, firms and technologies co-evolve, we are not at a stage to test a set of hypotheses that flow from some well defined model' (Hollingsworth, 2000, p. 597). This statement certainly applies also to the current state of knowledge about how activities influencing innovation are coordinated within NSI; the present level of knowledge does not permit us to formulate a set of hypotheses that can be tested. The comparative analysis of innovation policies that we conduct in Section 5 of this chapter may be considered partly as a contribution to the further development of knowledge on this topic.

As indicated, we have regarded the ten activities outlined in the introductory chapter as constituting a set of factors hypothetically influencing innovation processes. In the NSI case studies the authors have tried to find out whether or not this perspective is valid – and, if so, to what degree – for the various activities in the specific countries. And, as we have commented above, the case studies have demonstrated, for most countries, that most of the ten activities have indeed exerted a significant influence on the extent and direction of innovation.

Of course, the case studies have also shown that the relative importance of specific activities varies widely across countries. For example, demand-side activities aimed at the formation of new markets have been very important for innovation in Singapore and a number of other countries, but in Hong Kong the stimulation of demand has been largely neglected as a policy measure, due to the continuing prevalence of a *laissez-faire* attitude. The case studies have also shown that the quantity of each activity and the efficiency with which it is performed vary considerably among NSI. Sweden, for instance, is an exceptionally strong performer with respect to investing in research and development (R&D) and education as knowledge inputs into innovation processes. However, the country's performance with respect to innovation outputs is low, given the amount of resources invested. In contrast, this pattern is almost completely reversed in Ireland, where high levels of growth and innovation are associated with low levels of R&D investment.

The case studies have further indicated that both the pattern of activities and the institutional and organizational arrangements – i.e. the set-up of organizations performing these activities and the set-up of institutions influencing those organizations – vary across NSI. In the comparative analysis carried out in this chapter, we shall focus on the diversity with respect to activities, organizations and institutions in NSI, along with the issue of coordination, which we have identified above as a key issue for comparative research on SI. The empirical focus and point of entry for this analysis is, as noted previously, innovation policy.

3 COMPARING NSI

In the literature on NSI, there has been an ongoing tension between, on one hand, demands for a more structured conceptual framework that would facilitate systematic comparisons and, on the other hand, insistence on recognizing the unique character of individual SI. The first position is well represented, for example, in various contributions on 'benchmarking' by Niosi and colleagues (Niosi, 2002; Niosi and Bellon, 1994; Niosi et al., 1993), and it has also been adopted by the OECD (1997). The second position has been championed by, among others, Miettinen (2002), who argues for a more contextually oriented approach to describing SI, based on the principles of historicity, industrial specificity and geographical specificity.

In the Ten Countries Project, we tried to strike a balance between these opposing views of how to study NSI. On one hand, we paid close attention to quantitative data (on, e.g., educational attainment of the labour force, industrial structure and globalization) and performance indicators for growth, scientific activity (publication), patenting and innovation. Community Innovation Survey 2 data were used extensively in descriptions of the European NSI, and parallel data sets were used for their Asian counterparts.[4]

As we have explained above and also in the introductory chapter, the common format developed for the country studies, and particularly its largest component, addressing the activities that influence innovation, represents the application of a structured conceptual and theoretical framework based on the activities-based approach developed in Edquist (2005) and Edquist and Chaminade (2006). On the other hand, however, we also required each national study to take the NSI's historical background into full account, mainly based on qualitative information. Further, we asked for an assessment of the NSI's particular strengths and weaknesses, as well as its past accomplishments and future challenges in innovation policy. Perhaps most importantly, we encouraged contributors to identify for each NSI a central issue, problem or paradox that illuminated its essential character – and these 'puzzles' have provided central themes in each of the country case studies.

Comparative research requires a sound rationale and clearly defined criteria for case selection. We have addressed these issues in the introductory chapter by comparing our sample of cases with the wider range of countries represented in Nelson's (1993) anthology on NSI. There we explained that the ten NSI included in the Ten Countries Project reflected a consistent rationale through our selection of countries by:

- a focus on (relatively) small, dynamic, high-income economies;
- the inclusion of both 'late industrializing countries' and 'newly industrializing countries' or NICs.

In the comparative analysis of innovation policy that we present in Sections 4 and 5 of this chapter, we implement a case-based approach to comparison as follows. First, before focusing on the identification of patterns in innovation policy, we systematically map national contexts. We begin with an overview of growth in gross domestic product (GDP) per capita over time. In this part of the analysis (Section 4) we address growth profiles, as well as the processes and mechanisms by which economic growth has been achieved. We also relate the growth patterns to issues of globalization – i.e. to another important contextual dimension. What results from this examination is a classification of the ten countries into two main groups and an elaboration of some key differences in 'political economy' between these two groups.

Subsequently, the discussion turns to innovation policy itself (Section 5). We elaborate a broad view of innovation policy, consistent with an equally broad view of innovation systems, and we therefore refer to policy areas such as labour market policy that are sometimes excluded from discussions of innovation policy. Our examination of national patterns in innovation policy and policy making is informed by a theoretical discussion that focuses the analysis by addressing general policy rationales and the specific issues of selectivity and coordination. For each of these themes, we relate relevant evidence from the country case studies to particular theoretical arguments. These arguments enable us to identify general patterns of innovation policy and to examine the distribution of these patterns across countries. Where these distributions correspond to our classification of the ten countries, we are able to develop a context-based explanation of the differences among them. In the final section (Section 6), we sum up the main conclusions of the analysis.

4 GROWTH AND GLOBALIZATION

The countries included in this book can, of course, be classified in a number of ways. To begin with, they can naturally be divided into European and

Asian countries. This classification would be similar to dividing the countries into earlier and more recent industrializers – or, again, similar to a division between countries that are trying to stay ahead and countries that are trying to catch up in economic growth. In recent years, though, Ireland would be an exception in the last two dimensions.

4.1 Fast versus Slow Growth Countries

The patterns of economic growth related to the ten NSI included in this book are shown in Figure 12.1. The figure, which depicts growth in GDP per capita over the 1950–2005 period, shows a clear separation between two groups of countries, with a marked shift in the growth patterns of each group occurring around 1970–75.

Our analysis is especially concerned with the patterns of growth for the countries during more recent decades, and it is therefore appropriate here to characterize the two groups of countries according to their growth patterns over the past 30 years – i.e. from 1970/75 to 2005. We can thus distinguish between a first group of five countries marked by 'slow growth' and a second group of five countries that have exhibited 'fast growth'. The first slow growth group includes Denmark, Finland, the Netherlands, Norway and Sweden. The second fast growth group includes Hong Kong, Ireland, Korea, Singapore and Taiwan.

Figure 12.1 also shows that the two groups had different growth profiles before 1970/75, when the general pattern was almost reversed. During that earlier period, the growth trajectories of the slow growth countries were generally quite strong, whereas, with the notable exception of Hong Kong, those of the fast growth countries were relatively flat compared to their later profiles.

This grouping of the ten countries is neither surprising nor controversial, corresponding as it does to the well-known distinction between 'catching-up' economies and those that are either 'falling behind' or risk doing so (Abramowitz, 1986; 1994). In fairness to the first group of slow growth countries, though, they generally appear to have been holding, and in some cases improving, their positions in recent years, rather than losing ground. Norway, in particular, has experienced remarkable gains in the past few years. On the whole, the difference between the countries in terms of GDP per capita has decreased during the period.

Finland provides a case in point of slow growth dynamism.[5] Since this country is usually identified as a highly successful example of innovation-based growth, it may be surprising to some that Figure 12.1 shows very similar levels of GDP per capita relative to the USA for Finland in both 1975 and 2005.[6] Therefore Finland is classified as a slow growth country.

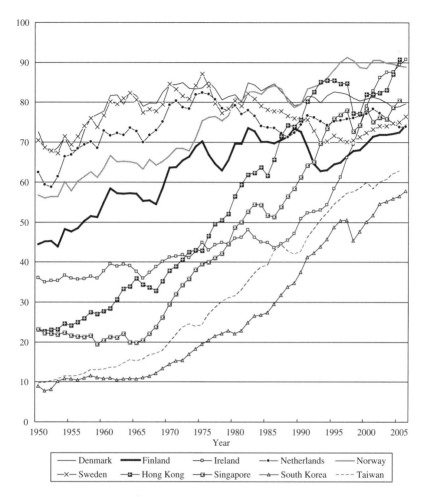

Source: Groningen Growth and Development Centre and The Conference Board, *Total Economy Database*, January 2007, available at: http://www.ggdc.net.

Figure 12.1 GDP per capita 1990 PPP US$, USA = 100, ten European and Asian countries

However, the historical circumstances must be taken into account – in particular the depression of the early 1990s. As discussed in the Finnish chapter (Chapter 10), this crisis was due to an overheated economy, a large current account deficit, and drastic decline of the Soviet/Russian export market – all of which made export of new products to Western markets imperative for recovery. Further, recession in Western countries and limited growth potential in mature industries meant that Finland's recovery would depend mainly on high-tech exports by its emerging electronics industry, which by 2000 contributed as much as 1.5 percentage points to the Finnish GDP growth rate.

More generally, the evidence from various country studies serves to highlight that growth has been closely related to globalization. Thus the reversal of growth patterns for the two groups of countries that is depicted in Figure 12.1 coincides with and reflects the significant increases in international flows of capital and technology that occurred during the later decades of the post-Second World War period (Fagerberg and Godinho, 2005, pp. 521–30). With regard to the newly industrialized fast growth countries, the evidence also shows that rapid economic growth has depended, for the most part, on successful entry into one and the same global growth sector – i.e. the electronics and information and communication technology (ICT) industries. However, as we shall discuss in the following subsection, their entry into this sector has been achieved by very different means and followed divergent paths. With regard to slow growth countries, the country studies show that they have been characterized by very different sectoral specializations and technological trajectories during different parts of the post-Second World War period. In all cases, however, the historical accounts indicate that these late industrializing countries in Europe originally caught up by targeting growth industries of the time and developing strong export orientations (cf. Fagerberg and Godinho, 2005, pp. 516–18).

4.2 Mechanisms and Instruments

The fast growth countries in this book can all be categorized as newly industrializing countries (NICs) – including the 'Celtic tiger', Ireland. Their rapid economic growth can largely be explained by reference to Lall's (2000) analysis of Asian NICs, particularly Korea, Singapore and Taiwan. In those three countries, and also in Ireland, the state played an important role in stimulating and supporting rapid growth based on entry into especially dynamic industrial sectors, particularly through policies for education (especially in engineering), targeting production for export and rewarding high export performers, and supporting R&D and innovation. Hong Kong is an important exception, though. There, the colonial state supported neither entry into new industrial sectors nor innovation based on R&D.

However, as also noted by Lall (2000) for earlier decades of the post-Second World War period, there were significant historical policy differences, as well as similarities, among three of the 'tigers'. Singapore, for example, relied more heavily on inward foreign direct investment (FDI) to industrialize than did either Taiwan or Korea, which both used protectionist measures to nurture their 'infant' indigenous electronics industries. Our country studies indicate that Ireland has followed much the same path as Singapore, though it first pursued protectionist policies, abandoning them in the late 1950s. In Hong Kong the colonial government was also eager to attract FDI, though not necessarily in order to support industrialization.

Country studies for the fast growth group also indicate that these catching-up economies have historically displayed considerable diversity with respect to industrial structure (in terms of size composition of firms in the country – as opposed to specialization among production sectors). Large, diversified business groups, or *chaebols*, were fundamental to industrialization in Korea, but the key actors in this process in Hong Kong and Taiwan were indigenous small and medium-sized enterprises (SMEs) of the family firm type. A further contrast is provided by the cases of Ireland and Singapore, where foreign multinational corporations (MNCs) were the main agents of industrialization.

The country studies also indicate that there have been important differences, too, with regard to specific instruments employed in state policies for economic growth. Countries' choices concerning policy instruments appear to have been partly determined by the differences with respect to size composition of firms. Thus Korea during the 1970s relied primarily on 'directed credit' (state credit rationing) to guide the *chaebols* – i.e. the instrument used was adapted to large firms. During the 1980s and 1990s, however, dynamic growth in Korea was more directly driven by large firms, and governmental efforts shifted to providing support for this process in the form of R&D and educational infrastructure. In contrast, both Taiwan and Hong Kong relied on so-called intermediate organizations with mixed public and private sector participation to nurture the development of small indigenous firms, although they accomplished this in very different ways. In Taiwan, especially, policy initiatives focused on the build-up of R&D infrastructure and the creation of state-owned firms to support nascent light manufacturing industries further downstream. Ireland and Singapore, which, like Hong Kong, adopted free trade policies beneficial to foreign MNCs, made themselves attractive locations for production units of foreign firms by pursuing competence building to develop skilled labour forces via public sector vocational education and training, polytechnic institutes and universities.

The five slow growth countries in this study are late industrializing Northern European countries that first caught up with the leading industrial economies during the late nineteenth and early twentieth centuries, following the example of Germany (Abramowitz, 1994). In both the

Netherlands and Denmark, early specialization in agriculture and services strongly influenced this process. Denmark developed a diversified export-oriented agro-industrial complex, and more recently has also created new clusters, such as within hearing aid technology, based not only on traditional industries but also on interaction with the provision of public services (e.g. healthcare). SMEs with extensive networks have been more important than large firms in Denmark. In the Netherlands, the adoption of technologies from abroad, dependence on foreign markets and historical involvement in international trade bred strong reliance on exports and large global companies of Dutch origin (such as Philips, Shell and Unilever). In this respect the Netherlands is similar to the Swedish case, with a considerable number of large and internationalizing firms of a national origin.

Sweden, Finland and Norway provide examples of late but rapid industrialization – in the second half of the nineteenth century – in forestry, metals and in engineering related to resource extraction and processing. In the latest decade there has been an increase of export specialization in R&D-intensive sectors in Sweden and Finland. In Sweden, state infrastructure and technology projects were used extensively to support industry development during the 1960s and 1970s. In Finland, the state participated actively in developing the industrial base, contributing to a late but rapid build-up of new technologies and industries aimed at export markets beginning in the late 1970s and continuing into the 1980s. A similar path was followed by Norway, but with a much stronger concentration on the extraction of natural resources and without a strong profile in high-tech industries. Norway used major state infrastructure projects, as in Sweden, as well as employing state-led finance, to support industrial development.

Concerning the main instruments of public policy in the slow growth countries, there was a considerable reliance upon state-owned firms, which led the development of key sectors in Finland during the early decades of the twentieth century and in Norway in the postwar industrial build-up and consolidation period, from 1945 until the 1970s. State-owned organizations, in the form of public utilities and public service providers in, for example, healthcare, were also important in Denmark and Sweden during most of the twentieth century. In these two countries, public technology procurement was (along with other demand-side measures such as regulation) an important means of creating new technological clusters – and, in Sweden, developing the technological capabilities of large industrial firms, such as ASEA/ABB and Ericsson. The Netherlands diverges from the Nordic countries in this respect. More generally, though, state-led economic development necessitated by late industrialization (Gershenkron,

1962) is clearly a common characteristic of all of the slow growth countries, particularly the Nordic ones.

4.3 National versus Sectoral Systems of Innovation

In the introductory chapter, we discussed the dichotomy between the broad and narrow versions of the SI approach, and we noted that different authors have different views on the appropriateness of a narrow versus a broad definition of NSI. Along another dimension Nelson and Lundvall also expressed fundamental differences of opinion concerning whether national systems will continue to be important even under conditions of increasing globalization and regionalization. These, of course, are trends that may challenge the coherence of national systems.[7]

According to Lundvall, 'both globalization and regionalization might be interpreted as processes which weaken the coherence and importance of national systems' (Lundvall, 1992, p. 3). However, he believed 'that national systems still play an important role in supporting and directing processes of innovation and learning' (ibid.). Hence, he argued that NSI would continue to pursue distinctive national trajectories, even under the homogenizing influence of globalization processes. One of his arguments for this is that interactive learning and innovation will be easier to develop 'when the parties involved originate in the same national environment – sharing its norms and culturally based system of interpretation' (ibid., pp. 3–4).

In contrast, Nelson and Rosenberg (1993), in the introduction to the Nelson book, expressed considerable scepticism about the overall coherence or consistency of NSI. They argued that the 'system of institutions supporting technical innovation' is very different between sectors of production, such as pharmaceuticals and aircraft. Moreover, they stressed that in many sectors, including the two mentioned, 'a number of the institutions are or act transnational' (Nelson and Rosenberg, 1993, p. 5). On this basis, Nelson and Rosenberg questioned whether the concept of a national system makes sense. One implication of this line of reasoning is that NSI may ultimately be largely reducible to ensembles of increasingly transnational sectoral SI (SSI) that would tend to look more and more the same in different countries. These two authors thus expressed fundamental differences of opinion with Lundvall concerning the coherence of NSI.

Malerba also addresses the relations between sectoral systems and national systems in reporting a major international project on SSI. He stresses that 'major differences exist between sectors', but also that 'differences in national systems matter, and they affect some of the features that a sectoral system may take on in a country' (Malerba, 2004, p. 34). The main theoretical issue, related to the different perspectives advanced by the national and sectoral schools, concerns the character and coherence of NSI – i.e. the question of whether or not they exhibit truly systemic properties at the national level, even in the context of increasing globalization.

Our country studies have shown that economic growth can be based on different patterns of sectoral specialization (composition of production across sectors) and trajectories of technology development. Significantly, most of our ten countries have moved (at least in some periods), or attempted to move, in the direction of greater specialization in rapidly expanding R&D-intensive (high-tech) sectors, in order to achieve or maintain high rates of economic growth.[8] However, the countries most successful in the attempt to enter high-tech growth sectors in recent decades have been the fast growth countries. These countries, moreover, have accomplished their respective entries into (in almost all cases) the same high-tech sector (i.e. electronics) in very different ways. Some, like Korea and Taiwan, have focused on developing domestic firms capable of competing globally; others, like Ireland and Singapore, have focused on attracting foreign MNCs and promoting 'innovation by invitation'. On the whole, the fast growth countries have been specialized to a larger extent than the slow growth countries in rapidly expanding high-tech sectors.

Further, the evidence on globalization effects reveals a growing diversity in the technological trajectories that both fast growth and slow growth countries have chosen to pursue, even within the same sectors. In ICT manufacturing, for example, Sweden has increasingly become a centre for R&D and design (and located production facilities abroad in many sectors of production). Taiwan has mastered and refined sophisticated production technologies. Hong Kong, on the other hand, has focused on 'brokering' between Mainland China and foreign firms through coordination. This means adding high-value-added services to regionally based international production networks in a wide range of production sectors. Both Denmark and the Netherlands have developed strengths in biotechnology, but while the former has benefited from international collaboration, the latter has been doing so to a lesser extent.

Rather than converging, then, the NSI in our study have established distinctive roles within an increasingly differentiated international division of labour. Moreover, these roles tend to be consistent across sectors, as demonstrated by the cases of countries as widely different as Sweden and Hong Kong. In these respects, our findings on how globalization has affected NSI by reinforcing and accelerating national patterns of specialization both within and across sectors corroborate – and are corroborated by – recent research on international performance within SSI (Coriat et al., 2004; Malerba, 2004). Thus there is considerable evidence to indicate that globalization does not erode NSI or render them incoherent. Rather, evidence from the ten countries investigated in this project generally favours the view that 'national systems still matter', indicating that national characteristics and strategies have been crucial in processes of globalization.

It could be added that the case studies in this book, as well as in Nelson (1993), indicate that there are sharp differences among various national systems with regard to propensity to innovate, the importance of different activities as determinants of the development and diffusion of innovations (e.g. in terms of institutional and organizational set-up), etc. In addition, public innovation policies are still mainly designed and implemented at the national level – and the policies differ greatly across countries with regard to objectives, characteristics, instruments and consequences for performance (Edquist, 1997, p. 12).

Hence we believe that NSI are not reducible to ensembles of SSI. However, with regard to the creation of diversity and dynamics in the national systems, the picture is different. In this respect, the emergence and development of new product areas and new SSI is an absolutely crucial element. We shall return to this issue in the next section on innovation policy issues. Here, our summary conclusion is that national systems as well as sectoral systems will remain crucial constructions for the foreseeable future – but for partly different 'purposes'.

Before proceeding to the next section, however, we shall briefly underline the importance of the sectoral perspective. For illustration, we refer to three quotations from the country chapters on Taiwan, Ireland and Korea:

- Taiwanese policy is 'actively promoting' FDI in high-tech industries (Balaguer et al., Chapter 2, Section 7.1).
- Irish-owned industry has 'grown relatively quickly by international standards, it has experienced significant upgrading in terms of skills and R&D performance, and it has generally had higher growth rates in high-tech sectors than in more traditional industries' (O'Malley et al., Chapter 5, Section 2.2).
- 'Korea's pattern of catching up can be described as "catching up by specializing in new industries" – that is, the ICT industries' (Lim, Chapter 4, Section 6).

5 INNOVATION POLICY ISSUES

The issue of innovation policy was briefly mentioned in Section 4 of the introductory chapter. In the country chapters, we have dealt with innovation policy in two (different) ways: policies pursued and policy proposals for the future. In what follows, we shall compare innovation policies across the various countries.

In order to lend greater depth to this comparison, and also to provide it with a thematic structure, we shall integrate our empirical account of differences and similarities in innovation policy with a theoretical discussion of innovation policy. In this discussion, we shall first address rationales for innovation policy, and then take up the specific issues of selectivity and coordination. For each of these three themes there will be a corresponding discussion of the empirical patterns of innovation policy found in the ten country chapters.

5.1 Rationales

A common basis for public policy intervention is the identification of a 'market failure' that is supposed to be corrected through intervention by the public sphere. As discussed elsewhere (e.g. Edquist, 1997, 2001), a market failure in mainstream economic theory implies a comparison between conditions in the real world and an ideal or optimal system. However, innovation processes are path-dependent over time, and it is not clear which path will be taken as they have evolutionary characteristics. We do not know whether the potentially optimal path is being exploited. Moreover, the system never achieves equilibrium. For these reasons, the notion of optimality is irrelevant in an innovation context. It follows that we cannot specify an ideal or optimal system of innovation and, therefore, comparisons between an existing system and an optimal system are not possible. As a corollary, it is not meaningful to talk about optimal policies and the notion of market failure loses its meaning and applicability (Edquist, 2001).

Instead of market failures, researchers and policy makers following a systemic approach often speak of 'systemic problems'. The criteria for policy intervention proposed here are, first, that there exists a systemic problem not spontaneously solved by private actors and market forces (i.e. private organizations fail to achieve the public policy objectives) and, second, that the public agencies must have the ability to solve or mitigate the problem (Edquist, 2001). This means that these systemic problems have to be identified, which can be done through empirical analyses comparing different systems of innovation with each other (Edquist and Chaminade, 2006).[9]

An important focus of the systemic approach is the complex interactions that take place among the different organizations and institutions that constitute the SI. From this perspective, policy makers need to intervene in those areas where the system is not functioning well. The rationale for innovation policy should therefore be based on systemic problems rather than on market failures. The policy discussion at each point should focus on changes in the division of labour between the private and the public spheres, and on changes in those activities already carried out by the public agencies. This includes adding new public policy activities as well as terminating others – and changing some. Terminating activities carried out by public organizations is not the least important (Chaminade and Edquist, 2006; Edquist, 2001).

As also argued elsewhere (Edquist and Chaminade, 2006), the activities-based framework for analysing SI, outlined in Section 2.3 of the introduction to this book, can fruitfully be used for innovation policy purposes. The activities that influence innovation processes in SI provide a useful point of entry into policy analysis. Thereafter, one can identify the organizations performing the ten activities and examine the relations among them as well as the institutions constituting constraints for the organizations when they pursue the innovation processes. When part of an activity is performed by a public organization, it is a matter of innovation policy – and most activities have a policy element.[10]

Often, there is not a one-to-one relation between organizations and activities. A certain kind of organization can perform more than one activity – for example, universities carry out research as well as teaching. Many activities can involve more than one category of organization, for example R&D is performed by universities, research institutes (public and private) as well as firms. With respect to innovation policy, we can analyse the division of labour between private and public organizations with regard to the performance of each of the activities in innovation systems and determine whether this division of labour is justified or not.

As noted above, one of the main differences between conventional (or mainstream economics) and systemic approaches to innovation policy is that the conventional approach begins by trying to define an ideal or optimal economic system, whereas the systemic approach compares existing systems with each other. These two approaches tend to define 'failures' or 'problems' very differently, and they also recommend different overall strategies for problem solving.

In all ten of the NSI covered by this study, policy makers have, without exception, proclaimed the adoption of the SI approach as a framework and guide for the design and implementation of future innovation policy. However, this wide acceptance might merely reflect a general tendency within the policy-making community to follow current 'trends'. Thus, what adoption of the SI approach might mean for the design and implementation of innovation policy within specific national contexts certainly remains far from clear – and policy makers often might not even know themselves.

There are widely differing views among and within national policy-making communities on what innovation policy consists of, and the same might also be said of the SI approach, which is often used 'more as a label than an analytical tool' (Edquist, 2005, p. 202). These points are illustrated below.

Norway currently faces 'the issue of deciding between a broad versus a more targeted approach' to innovation policy (Grønning et al., Chapter 8, Section 7.3). However, the broad approach taken in Norway remains vague as to specific policy measures, expresses an underlying philosophy of 'general upgrading', avoids setting priorities, and reverts at least partially to a linear view of the innovation process. All of this is highly incompatible with an SI approach, as demonstrated by the critical analysis developed in the Norwegian chapter, which proposes a much more 'targeted' kind of innovation policy for Norway.

In the Netherlands, the Ministry of Economic Affairs has explicitly adopted a systems perspective, initiated many activities inspired by SI approaches and the Porterian cluster concept, and seriously addressed the notion of so-called 'systemic failure'. At the same time, though, 'ideas from traditional economic analysis on market failure and how to address it continue to play a role in innovation policy' within the Ministry of Economic Affairs. In subsidy schemes and export promotion measures connected to innovation policy, 'competition policy has become an aim . . . and the relation between competition and innovation is now one of the focal points'. Further, 'it seems clear that too much emphasis on competition as a way to stimulate innovation may be counter-productive' (Verspagen, Chapter 9, Section 7.2).

In Denmark, too, there are indications of a similar confrontation between policy rationales. On one hand, current economic policy debates reflect a 'more or less implicit assumption in much of mainstream economic policy debate that high taxes and a large public sector constitute a weakness'. On the other hand, there has been considerable effort to improve certain aspects of the NSI, but very little appreciation of its real strengths. For instance, 'there has also been a stronger interest in understanding and developing the "knowledge-based economy", but . . . [with] . . . a bias in the direction of formal knowledge and too little understanding of the importance of learning by doing, using and interacting (DUI learning)' (Christensen et al., Chapter 11, Section 7).

Similar observations can be made about Sweden, Norway and even Finland. Thus the overall picture of innovation policy rationales that emerges for the slow growth countries is one of fragmentation, debate and a lack of consensus. As in Denmark, many policies in these countries seem to be motivated by an urge to 'follow the leaders' in a given area of activity, rather than by trying to determine whether or not there is actually either a market failure or a systemic failure – whatever these terms mean for different people.

For the fast growth countries, the picture is one of greater consensus, more pragmatism and less debate about rationales. Thus the Singaporean chapter does not directly discuss policy rationales, and neither does the Taiwanese chapter. However, they do present their own accounts of potential or actual system failures. The Taiwanese chapter also indicates that innovation policy has assumed an extremely important role in the Taiwanese NSI. It emphasizes 'the key role of policy leading the process of systemic upgrading' (Balaguer et al., Chapter 2, Section 1), and it adds, 'By the early 1980s, many policy makers in Taiwan had become conscious that the market price mechanisms were too slow to propel the kind of development that Taiwan needed' (ibid., Section 4.3.3).

The Hong Kong chapter notes that 'policy makers have explicitly integrated the SI approach as an aid to overall policy discussion and implementation', but it also states that 'Lateness in tackling and introducing innovation policy and subsequently weak implementation have left many initiatives fragmented and ineffectual' (Sharif and Baark, Chapter 6, Section 7). The Korean chapter also describes a situation of extreme fragmentation of innovation policy among separate policy fields, which it illustrates by explaining that 'The science and technology policy has mainly been driven by the perspectives of the science and engineering fields, and has ignored the economic and social aspects involved in the process of innovation' (Lim, Chapter 4, Section 7). Only in Ireland has innovation policy apparently been guided to any significant extent by an explicit analysis of systemic failures – as in the case made by Forfás for adopting systematic initiatives to strengthening innovation networks in Ireland (O'Malley et al., Chapter 5, Section 7).

With the partial exception of Ireland, policy makers in the fast growth countries, like their slow growth counterparts, display a general lack of clarity about the economic rationales for innovation policy interventions. Given the rather diffuse and uneven development of innovation policy in both sets of countries, though, it could certainly be argued, as in the Hong Kong chapter, that these and other NSI could 'benefit from an approach . . . whereby policy initiatives are better coordinated and understood in terms of a larger conceptual framework' (Sharif and Baark, Chapter 6, Section 7).

5.2 Selectivity

One of the main differences between conventional (or mainstream) and systemic approaches to innovation policy is reflected in current debates on non-selective versus selective innovation policies. In this respect, current understandings of innovation policy are divided into two main camps. A non-interventionist '*laissez-faire* version . . . [which] signals that the focus should be on "framework conditions" rather than specific sectors or technologies competes with a "systemic" version . . . [for which] a fundamental aspect . . . becomes . . . reviewing and redesigning . . . the linkages between the parts of the [innovation] system' (Lundvall and Borrás, 2005, p. 611). From the systemic perspective, innovation policy – like most other public policy – is naturally selective, since even policies that try to avoid 'picking winners' by addressing market operations in general tend in practice to favour certain sectors. As we shall see below, the existing power structure often tends to preserve 'the existing structure of production' and already established technological trajectories (Edquist, 2001, pp. 224–5). However, other parts of the power structure can, of course, also favour new and emerging sectors of production and the development of new technological trajectories. Some examples that point to selectivity are the following.

Any public policy that is intended to solve or mitigate a societal problem must focus on the nature of the problem and on its causes – and in this way be selective. R&D policy instruments involve public financing of research, which means allocating economic resources between different research fields. An increase in public funding of R&D with 1 billion (Swedish crowns or euros) necessitates a decision about which field of R&D the additional resources should be used in. Should they be used for electronics research or for research in the life sciences? Decisions are typically made in complex political and administrative institutional set-ups, on the understanding that those allocations will serve to stimulate and enhance levels of innovative and knowledge capacity of the economy, in areas where private investment has been insufficient. Another instrument of innovation policy, such as a tax deduction for R&D expenditures by private firms, tends to favour those firms that have (large) R&D expenditures, and industries with a high R&D intensity. It is, therefore, also a selective instrument (Borrás et al., 2008, forthcoming).

Another conventional instrument of innovation policy is public technology procurement, which focuses upon a certain function, such as air strikes, transportation of high-voltage electricity, or telecommunications exchange. Through this instrument, government or the public sector subsidizes the development of a system that can fulfil this function – i.e. a public agency pursues public innovative (or technology) procurement. Hence this instrument is highly selective. The list of examples could be made much longer, but it shows that public policies are normally selective in one sense or another. They may be selective with regard to problems, regions, sectors, products, firms, instruments, etc. (Borrás et al., 2008, forthcoming).

When uncertainty and risk are high, the danger that private actors will underperform relative to public policy objectives is particularly great. For example, private actors might underinvest in basic R&D (Arrow, 1962) or they might not invest at all in activities of great social return but low individual return (e.g. some drugs). High uncertainty might also prevent the emergence of innovations. Empirical evidence suggests that large-scale and radical technological shifts rarely take place without public intervention. Carlsson and Jacobsson (1997) have shown this for technological breakthroughs in electronics, semiconductors, and genetic engineering in the USA and Sweden. Mowery (2005) has shown that publicly funded R&D in combination with public technology procurement has played a crucial role in developing new high-tech sectors in the USA (and thereby in the world). Examples are computer hardware, computer software, large aircraft, biotechnology, and the Internet. Hence public intervention seems to be the rule in new fields and industries, i.e. for the early development of new SSI. The new fields mentioned above are also those where large-scale and radical technological shifts have taken place. Such shifts seem rarely to have taken place without public intervention (Carlsson and Jacobsson, 1997; Mowery, 2005). On the other hand, incremental innovations in established and mature sectors seem to take place mainly on the basis of private initiatives funded by private organizations. Examples could be gradual improvements of ballbearings during the past century or the improvement of integrated circuits during the last 30 years.

Many mainstream economic analysts claim that innovation policy 'should be' neutral and they sometimes even believe that policies are neutral. They seem even to pretend that some selective policies are neutral (such as tax breaks for R&D expenditures). We argue, however, that innovation (and other) policies are normally not neutral, because they naturally entail a selection. We also claim that they cannot possibly be neutral in any genuine sense if they are to solve or mitigate specific problems.

The acceptance that most innovation policies are selective makes it possible for us to transcend the sterile debate about whether or not policies are, or should be, neutral. A much more relevant and interesting issue to address is in what way, or in what sense (innovation) policies are or should be selective. This is just what has been done in countries such as Korea and Taiwan. Policy makers in these countries have accepted the idea that they are trying to 'pick winners' and avoid 'subsidizing losers'. This is very similar to what private firms are trying to do – i.e. to bet on winning products and concepts.

Globalization adds another dimension to the discussion on selectivity, particularly when resources invested in innovation might not generate externalities in the country or region but somewhere else. As Archibugi and Iammarino (1999) acknowledge, with increasing globalization, 'the choices of public actors are strongly limited by processes they are not entirely in control of' (ibid., p. 326). Should the government encourage (subsidize) foreign firms to establish R&D labs in their country or should they instead support R&D in domestic firms (that might later leave the country)? How can the government select those interventions that might have a large positive impact in their territory when innovation activities are becoming increasingly global (Borrás et al., 2008, forthcoming)?

The question then becomes in which direction or in which respects innovation policy is – or should be – selective. Most large firms, and sometimes also the labour unions, in established sectors of production pursue lobbying intended to make public actors support their own sectors and firms.[11] New and nascent sectors normally do not include strong actors and can therefore not pursue lobbying in an effective way. However, public innovation policy intervention is generally more justifiable in new sectors or in new operations in established sectors, since 'problems' that are not solved or mitigated by private organizations are more frequent in these contexts.[12] Such interventions are also more justified for radical innovations, as compared to incremental ones, and for early stages of innovation processes, rather than later stages in those processes (Borrás et al., 2008, forthcoming).

Because strong vested interests are often associated with mature industrial sectors and established technological trajectories, lobbyism often seems to work for an innovation policy that should not be pursued.[13] Instead the support should be channelled to operations and sectors where risk and uncertainty are greatest, where uncertainties are too large for private organizations to invest, i.e. where the public policy action would really be a complement to private actors and not a substitute for or duplicate of them. Innovation policy should play the role of a midwife – not provide support towards the end of life. This requires that policy makers and politicians have a sophisticated analysis at their disposal, as well as a high degree of integrity, to counterbalance private lobbyism and a considerable amount of power. There are seldom strong private lobbyists for the solutions of the future!

Let us now return to innovation policy in a more 'practical' context. As we have briefly indicated earlier, innovation policy, in the context of this book, may mean two different things:

- The policies that have historically been pursued in the ten countries during recent decades (how policy has been formulated, the content of policy, how policy has been implemented and what consequences it has had).
- Identifying either problems that are currently not solved or mitigated by private organizations (mainly firms) or opportunities currently

not exploited by private actors (since if such problems or unexploited opportunities can be identified, they should be subject to future innovation policy).[14, 15]

With reference to these issues, the evidence from the ten country chapters included in this book provides strong support for the view that innovation policies are normally selective rather than neutral. Let us now illustrate this by means of examples from the country case studies.

The findings from our case studies generally indicate that the main challenge facing NSI is that of creating diversity – i.e. escaping lock-in into well-established production specializations and technological trajectories by launching new alternatives. In contrast, the more familiar problems of providing adequate factor inputs and ensuring competitive framework conditions – which are typically emphasized by the supposedly neutral or non-selective policies proposed on the basis of a *laissez-faire* approach to innovation policy (Lundvall and Borrás, 2005, p. 611) – appear to be issues whose solution is much easier.

Ireland has excelled in ensuring positive framework conditions and abundant factor inputs such as human capital and finance well adjusted to the needs of innovative firms. Despite these successes, Irish policy makers are concerned to reduce Ireland's present degree of dependence 'on . . . the willingness of external companies to continue to transfer technology to Ireland' and strengthen Ireland's indigenous innovative capability (O'Malley et al., Chapter 5, Section 6). In this connection, building up basic R&D capabilities within the public sector and strengthening absorptive capacity among private sector SMEs through upgrading labour force skills are fairly simple tasks. The challenge of improving the innovative capability of domestic firms by increasing the degree of interactivity and networking within the NSI is a much more difficult task. As noted in the Irish chapter, policy makers lack previous experience in, and instruments for, addressing this problem, since until now 'no systematic initiative has been adopted for strengthening innovation networks in Ireland' (ibid., Section 7.2).

Turning to the slow growth countries, the evidence from both Sweden and the Netherlands supports the view that innovation policies are normally selective. Historically, Swedish innovation policies have reinforced the dominance of large firms and industries characterized by low innovation intensity, and have also supported high levels of investment in education and R&D. More recent policies have emphasized providing support to start-up firms in science-based sectors. However, other so-called neutral policies – for example, corporate taxation, labour market policies and a plethora of liberalization initiatives – have promoted the globalization of Sweden's major industrial firms while maintaining their dominant position within the national economy. New firms and new industries have therefore developed slowly, since terms favourable to incumbent actors have been created by policy. Herein lies a possible explanation for the rather slow diversification of Sweden's technology profile.

In the Netherlands, past and present policies, directed towards increased competition, on one hand, and higher levels of public–private interaction, on the other, seem too broadly framed to accomplish a fundamental reorientation of the NSI. The main question concerning interaction in the Netherlands' NSI, for example, is arguably not one of 'how much?' but rather 'what kind?' As argued in the chapter on the Netherlands, from an SI perspective 'it might make (more) sense to look at the interaction between the university system and the public research organizations' than to 'propose measures to stimulate interaction between university research and private firms' (Verspagen, Chapter 9, Section 7.2). The latter type of policy measure has been pursued for some time in the Netherlands, without any clear effect; in contrast, the former could address problems of coordination, gaps in funding, and a problematic focus of specialized public research institutes on 'relatively old specializations, such as civil water works and shipbuilding' (ibid.).

Strong support for the proposition that innovation policies are necessarily selective policies – and also clear indications of the problems entailed in developing and implementing selective policies – emerges from several of the fast growth countries. Korea exemplifies the limitations of policies addressing the market in general. Liberalization was a necessary response to the financial crisis of the 1990s, and involved the introduction of reforms that were both wide-ranging and comprehensive. However, Korean policy makers apparently underestimated the difficulties of implementation, as evidenced, for example, by the relative underdevelopment of new arrangements for financing innovation.

As stated in the Korean chapter: 'The financial system is new and it is not sufficiently well developed to channel financial resources to those firms that display good performance, because there is a limited pool of knowledge evaluating the credibility and performance of firms. Banks and other organizations are reluctant to take risks in making loans, and their hesitance is reducing firms' investments' (Lim, Chapter 4, Section 7.1). Further, continuing problems with the output and organization of both public education and public sector R&D indicate that the extent and pace of reform have not been sufficient and may need to be increased considerably in certain areas. Arguably, Korea requires strategic systemic initiatives for coordinating diversity creation efforts to counterbalance the strong selection pressures exerted by increased competition on both domestic and global markets. Thus the Korean chapter concludes as follows:

Finally, in the ever more globalized world, strengthening the capabilities of small firms, which are increasingly vulnerable to international competition, will be crucially important for future policy. . . . In order to resolve the chronic problems of small firms, future innovation policy needs to find strategic ways of enhancing the technological capabilities of small firms and the networking of small firms with domestic and international actors for knowledge and market access. The importance of strategic policies to ensure that the poor competences of small firms do not become a barrier to upgrading the competitiveness of the nation cannot be over-emphasized. (Lim, Chapter 4, Section 7.3)

The Taiwanese case illustrates the difficulty of designing selective policies. Historically, innovation policies implemented in Taiwan during the 1970s and 1980s succeeded in fostering competitive original equipment/design manufacturing firms in ICT manufacturing. Many aspects of the NSI have been geared to this effort – for example the public sector's role in building competences in strategic areas through a variety of mechanisms for technology diffusion and learning. However, the past achievements of Taiwanese innovation policy have also contributed to current problems of lock-in. To date, the change of Taiwan's technology profile has mainly been confined to advances in semiconductors and electronics in a more general sense. But even to consolidate and extend the gains made in this one sector, Taiwan currently faces the challenges of strengthening indigenous R&D and intellectual property rights (IPRs) of domestic firms as part of a more general effort to develop the broader and more comprehensive set of assets and capabilities now required for further progress. When it comes to addressing these problems through selective policy measures, though, the Taiwanese government has increasingly limited room for manoeuvre. Thus the Taiwanese chapter notes that:

> While, in the past, technology absorption, diffusion and the accumulation of capabilities in industry were core elements of economic and industrial policy at the highest level, at present Taiwan lacks a well-articulated policy approach where innovation could be promoted in a highly strategic fashion from a top executive level. (Balaguer et al., Chapter 2, Section 7.3)

The cases cited above are not exceptional, in so far as similar findings occur in other national studies. For instance, the Norwegian case supports the same conclusions as the studies of Sweden and the Netherlands, and the Singaporean case develops insights similar to those articulated in relation to Taiwan.

In Norway, past innovation policies were directed towards breaking out of the existing pattern of industrial and technological specialization by focusing on selected science-based and information-intensive industries, and corresponding research fields. However, these policies were not supported by accompanying reforms in areas such as taxation – a specialized R&D tax credit scheme targeting small firms was only introduced rather recently, in 2002 – and they were also undermined by an economic downturn during the early 1990s. As a consequence of this crisis, 'public R&D policy towards the private sector changed'. Thereafter, 'priority was no longer to be given to selected sectors. Although there were some main areas of focus . . . the new policy aimed at improving the general performance in all firms with innovation potential' (Grønning et al., Chapter 8, Section 7.2).

In the case of Singapore, innovation policies have evolved with the NSI, typically leading its development. But the achievements of past policies for 'MNC-leveraging economic development' – i.e. 'leveraging of foreign MNCs to jump-start local economic and technological growth' – may have reduced policy makers' scope of action in recent efforts to build up indigenous innovation capabilities' (Wong and Singh, Chapter 3, Section 7.2). Although the policy shift towards investment in R&D has had positive impact on R&D intensity and innovative performance, policies geared towards promoting high-tech entrepreneurship have not enjoyed similar success thus far.

To sum up, it is clear that in all ten countries there is strong evidence that the innovation policies that have been pursued have been selective. In different ways, all the various country studies point clearly to problems requiring selective policies, or what may also be referred to as opportunities for the development of such policies. However, there are also some significant differences between fast growth and slow growth NSI with respect to the extent to which they have developed and implemented selective policies.

When it comes to the actual performance of deliberately selective rather than so-called neutral or non-selective innovation policies, what distinguishes fast growth from slow growth countries is that the former have been much more active and also more successful in pursuing explicitly selective policies. This means that they have been selective in the 'right' direction.[16] It will be recalled from Section 4 that all these countries have developed successful policies directed towards entry into the ICT and electronic sectors. In contrast, the slow growth countries have been more weakly committed to new growth industries, and more attached to traditional ones. In at least some of the slow growth countries, moreover, there appears to be an ideological commitment to so-called neutral policies that has in effect ruled out the adoption of pursuing more selective policies in a conscious way – i.e. enhancing selectivity in the 'right' direction. The Netherlands and Norway provide very clear examples of this tendency. In other cases, such as Sweden, there may be a mix of both selective and supposedly neutral policies – in which, for Sweden, at least, the avowedly non-selective policies may counteract and even undermine the explicitly selective ones. Thus there appears to be greater resistance to selective policy measures in the slow growth countries, and greater scope for developing and applying such policies in the fast growth countries.

In trying to explain these differences, we can point to an obvious contextual difference between the two groups of countries. As NICs, the fast growth countries have fewer mature industries that would stand to benefit from so-called neutral policies favouring the existing structure of production and already established technological trajectories. Hence there are few particularly strong lobbies for so-called neutral policies (that actually support existing industries and actors in a selective way). The situation in the slow growth countries is, of course, just the opposite – i.e. established firms and industries have generally supported and benefited from so-called 'neutral' policies, rather than explicitly selective policies. Finland might be considered to constitute an exception in this regard. Selective policies to develop the ICT sector in Finland paid off during the late 1990s, but part of the explanation for their success lies in a severe economic crisis during the early 1990s that greatly diminished the lobbying power of traditional industrial sectors and created greater scope for policy actions favouring the ICT sector.

5.3 Coordination

Another key difference between conventional (or mainstream) and systemic approaches to innovation policy concerns the issue of coordination. As we pointed out in Section 2 above, coordination is a central theme in SI approaches, which lay great stress on how activities supporting innovation are coordinated by institutional frameworks. Coordination is also a vital concern for the systemic approach to innovation policy, since – as in the examples considered in Section 5.2, above – much research on innovation processes and systems points to 'tension or mismatch between different kinds of designed institutions that often represent different levels of policy-making' (Edquist et al., 1998, p. 38). Further, as explicitly recognized and underlined by the activities-based framework outlined in the introductory chapter, SI approaches generally recognize the importance of complementarity within systems and therefore emphasize the importance of policy coordination: for example, 'the coordination of support for R&D with support for . . . other kinds of learning, which operate through different mechanisms' (Edquist et al., 2001, p. 155). Thus one of the general policy implications of the SI approach is that it is important 'to integrate and co-ordinate policy areas like R&D policies, educational policies, regional policies, and even macro-economic policies when formulating innovation policies' (Edquist, 2001, p. 230).

Returning to the question of rationales for policy intervention, an emphasis on coordination can be regarded as fundamental for the systemic approach to innovation processes and innovation policies, whereas this is not the case for conventional (or mainstream) approaches. Thus Metcalfe (1995) has highlighted the issue of coordination in elaborating a comparison between conventional or optimizing approaches and evolutionary and systemic or adaptive approaches to innovation policy making. In the optimizing approach, which is informed by equilibrium economics, the 'favourite metaphor . . . is of the policy maker as a fully informed social planner who can identify and implement optima' for altering incentive schemes in order to change the behaviour of economic actors and thereby correct situations of market failure where 'social and private welfare [are] out of step' (ibid., p. 30). In contrast, the adaptive approach, based on evolutionary economics, does not presume 'that the policy maker has a superior understanding of market circumstances or technological information; rather what s/he does enjoy is a superior coordinating ability across a diverse range of institutions' (ibid., p. 31). For the adaptive policy maker, moreover, the central problem is not market failure but rather the 'evolutionary paradox that competitive selection consumes its own fuel, destroying the very variety which drives economic change' (ibid., p. 30). It follows that 'superior coordinating ability' must be harnessed to the cause of regenerating the diversity fundamental to economic progress by promoting and supporting 'experimental behaviour' on the part of economic actors.

Drawing upon the evolutionary economics tradition – as well as insights from institutionalist theories such as those that inform the VoC (Hall and Soskice, 2001) and business systems (Whitley, 2002) approaches discussed in Section 2 – Storper et al. (1998) have applied this reasoning to industrial policy for latecomer nations in the globalizing learning economy as follows. Orthodox economic theory proposes macroeconomic structural adjustment policies, accompanied by liberalization policies, as the alternative to traditional industrial policies, which cannot be practised in the context of increasingly open markets and internationally mobile capital. However, structural adjustment offers no solutions for the creation of new wealth through 'mobilization of local competitive specificities such as technology and know-how, trust and culture' (ibid., p. 3). Such solutions depend, instead, on 'coordination-for-learning', where learning is understood to involve multiple kinds of actors and forms of activity, and to depend on institutional frameworks and relations among actors – an approach in which 'the role of the state as a catalyst for starting and sustaining coordination becomes crucial' (ibid., p. 5). This approach to industrial policy is, of course, highly consistent with the systemic approach to innovation policy and with the institutionalist orientation and emphasis on interactive learning fundamental to systems of innovation approaches. As Storper et al. emphasize, 'learning involves coordination over shifting terrain, where the agents and institutions involved must be reflective . . . in order to alter their own parameters over time' (ibid., p. 6).

With these considerations in mind, we now turn to an examination of policy coordination in the ten countries represented in this book. Here, we will focus primarily on the consequences of past innovation policies, rather than examining how policy has been formulated, the content of policy, or how policy has been implemented. However, the analysis may also contribute to a more profound understanding of the existence of problems or opportunities as reasons for innovation policy intervention that were outlined in the preceding discussion of selectivity.

Sweden provides an instructive example of the problems of policy coordination in slow growth countries. Generally, Sweden's pattern of innovative activity reflects an ongoing imbalance between the supply of innovation inputs, particularly R&D, and innovation outputs. The 'Swedish paradox' refers essentially to low pay-off in terms of innovations from very large investments in R&D and innovation. This problem can be attributed to several causes: globalization resulting in commercialization of Swedish innovations abroad; ineffective technology transfer from research organizations to commercial application by firms; and a sectoral allocation of R&D investment favouring industries with low innovation intensity. The dominance of incumbent large firms (MNCs) is a common thread in all these lines of explanation.

Sweden's extensive support for innovating firms and entrepreneurial start-ups has resulted in only modest rates of new firm creation and only moderate success in strengthening specialization in fast growing high-tech industries. New firm creation and interfirm networking remain dominated by large firms, and institutional arrangements (in, e.g., labour markets and taxation) also sustain the dominance of large firms, many of them based in industries with low innovation intensity. There is considerable lack of coordination between policies related to the provision of constituents for SI and policies with regard to support services for innovation firms.[17] The ultimate beneficiaries of the latter appear to be those least in need. Thus the Swedish chapter points out that, despite a wide range of policies aimed at supporting the creation of new firms, the growth of these firms 'frequently seems to be enhanced by becoming part of a larger corporate structure through acquisition', and that despite numerous policies to support networking for small firms, innovation networks in Sweden are characterized by a 'rather high degree of vertical integration' (Bitard et al., Chapter 7, Section 4). The overall pattern of evolution in Sweden is one of gradual transition from an innovation system dominated by large mechanical engineering firms to one in which science-based and information-intensive sectors will feature more prominently, but large incumbent firms are unlikely to be easily displaced by new entrants.

Similar dynamics can be observed in two other slow growth countries. Norway exhibits little entrepreneurship in science-based and information-intensive sectors. Thus extensive provision of support for innovating firms has brought poor results, due to the restrictive investment climate and structural rigidity bred by the dominance of large firms in the resource extraction and transportation services sectors. Now that Norwegian policy makers have again shifted direction towards a more targeted approach, many questions remain as to how this new approach will actually be integrated with the earlier and ongoing emphasis on general upgrading to which it has been tied. As stated in the Norwegian chapter,

> In the case of general upgrading as the main and underlying philosophy, there is a need to further formulate the conditions under which general upgrading is to take place. In other words, if the rationale is that heterogeneity leads to increased output and that a subsequent core strategy is to foster heterogeneity by way of having broad and multiple targets, this must be stated in an explicit way. (Grønning et al., Chapter 8, Section 7.2)

In Finland, public sector support for innovation, networking arrangements and institutional reforms have been geared mainly to the successful development and internationalization of large firms such as Nokia. Entrepreneurial small firms have been much less well provided for – even though Tekes, an important R&D financing organization, requires that SMEs be included in the R&D networks they finance. For such reasons, the Finnish chapter points to the need for increased networking and collaboration between low-tech and high-tech firms, as well as between large and small firms, combined with measures developing the innovative capabilities of small firms, such as 'development of financing of especially young, small, growth-oriented, R&D-intensive and high-technology firms' (Kaitila and Kotilainen, Chapter 10, Section 7.3).

In the Netherlands, despite very high performance with respect to knowledge inputs, innovative activity has levelled off and private sector R&D has begun to decline, at least in relative terms, implying diminishing demand for these inputs. A pattern of sectoral specialization that de-emphasizes manufacturing, together with globalization effects, makes it difficult to improve system performance simply by increasing knowledge inputs. Instead, balancing supply and demand requires the Netherlands to gear inputs to emerging growth sectors.

The problems encountered in the Netherlands include low levels of entrepreneurship and inadequate interaction between universities and other public research organizations and private sector actors, reflected in low levels of knowledge transfer and research commercialization. The overall pattern is one of an impasse bred by lock-in to institutional and organizational arrangements that serve incumbent firms in declining (or de-industrializing) sectors better than new entrants in emerging sectors. The Netherlands NSI appears to be a dual system. The universities, most research institutes and many public research organizations cater primarily to large incumbent firms, well-established but declining industries, and relatively old technological fields. In contrast, only a few public organizations provide inputs to new firms, industries and technologies. Also, creation of new firms, industries and technological innovation platforms is often poorly coordinated with corresponding forms of support, some of which are inadequately funded. Thus the chapter on the Netherlands argues for a fundamental reorientation of several different kinds of innovation policies, in which coordination should be based on 'taking into account where the "new economy" will have its focus in terms of technological development' (Verspagen, Chapter 9, Section 7.1).

Denmark, like the Netherlands, has maintained essentially the same technological profile over the past two or more decades. Its current transition towards a more mixed mode of innovation combining a science-and-technology-driven mode with the traditional mode based on doing, using and interacting – i.e. 'a "mode of innovation" dominated by small and medium-sized firms continuously making incremental innovations based on learning by doing, learning by using, and learning by interacting especially with customers and suppliers' (Christensen et al., Chapter 11, Section 1) – has so far not entailed any major disruptions.

Against this background, though, recent policy appears to have led in a potentially disruptive direction that may actually become problematic, depending on the strength and focus of initiatives to bring about structural change through efforts to strengthen high-tech sectors by promoting university–industry interaction and the creation of science-based firms and industries. As the Danish chapter points out, measures aimed at stimulating activities in this area will probably require considerable adjustment to complementary or related sets of activities within the Danish NSI – through, for example, close coordination and even integration with existing forms of university–industry interaction. In Denmark, where low-tech activities predominate and a large population of SMEs has only weak networking linkages with the universities, 'policies aiming at bringing the national system "to the very top" in this dimension might not necessarily strengthen the system as a whole'. Other technology transfer strategies 'such as lifelong learning' might be more effective in strengthening the innovative capabilities of many firms and sectors (Christensen et al., Chapter 11, Section 7). More generally, policies aimed at promoting a transition to a 'science-based' mode of innovation are unlikely to succeed in Denmark without taking into account the wider socioeconomic setting and other factors contributing to the successful operation of the Danish model.

Among the fast growth countries, Taiwan can be regarded as a leading example of a country whose economic success has depended on effective policy coordination, but which now faces new coordination problems. Taiwan's rising rate of investment in R&D has been characterized by the fast growth of business expenditure on R&D and a dramatic increase in patenting. Similarly, the strong expansion of Taiwanese post-secondary education has increasingly focused on ICT-related scientific, engineering and technical skills. Balance between supply and demand has been maintained, since these developments have been driven by the upgrading of firms in high-tech manufacturing sectors. These firms excel in production for high-tech markets, drawing their competitive advantage from manufacturing and process innovation skills.

As it reaches the limits of factory automation and the adoption of advanced production techniques as a competitive strategy, though, Taiwan now confronts mismatches between policies related to activities concerning constituents for SI and policies related to activities concerning support services for innovating firms (see Box 1.2 in Chapter 1). These problems reflect the need of existing firms to diversify, through product innovation based on independent scientific and engineering capabilities. For Taiwanese firms to break out of the trajectory that they have established for themselves as technology followers and 'second movers', it has become necessary to reform existing institutions for protection of intellectual property and develop forms of R&D collaboration that facilitate appropriation of innovation.

Similar patterns emerge in the cases of Singapore and Hong Kong, both of which have recently diversified their pattern of innovative activity. In Singapore, recent increases in indigenous R&D within targeted fields of science and technology have not been matched by the commercialization of indigenous intellectual property in corresponding sectors – especially on the part of entrepreneurial high-tech start-up firms. Historically, Singapore has concentrated on developing labour force competences and skills, and its main R&D inputs have come from technology-intensive foreign MNCs with local operations.

Singapore has successfully reconfigured public and private innovation capabilities several times over the past decades, through institutional reforms and changes in the provision of support to innovating firms. However, there has been little recent private sector response to the current build-up of support services targeting entrepreneurial high-tech SMEs. Prospects for developing indigenous R&D and innovation capability remain uncertain, and technological entrepreneurship in strategic sectors remains low, due to cultural factors (risk aversion), gaps in institutions and organizations (a lack of mechanisms 'bridging' R&D and seed investment) and the conservatism natural to a small domestic market.

In Hong Kong, support for innovating firms largely reinforces the dominant producer services trajectory by developing consultancy services and financing ICT projects, but incubation targets other nascent industries. This divided focus indicates possible future tension and conflict between established and emerging industries. As in Taiwan and Singapore, path dependence makes diversification difficult.

Korea, like Taiwan, has a history of developing the technological capabilities of its own large firms. Similar to other 'catching-up' economies, Korea has matched supply- and demand-side activities by utilizing imported technology to support a strong specialization in rapidly growing export markets, especially for high-tech products. Increasing globalization of both production and R&D by major Korean firms has meant, however, that efforts to upgrade domestic knowledge inputs will have to be coordinated with initiatives to strengthen the absorptive capacity and innovative capabilities of small domestic suppliers to these large firms.

There has been an aggressive drive for liberalization and the promotion of a more entrepreneurial economy in Korea since the last 1990s. Despite regulatory reforms and the reorganization or creation of support functions suitable to new venture businesses, the economy continues to be dominated by large conglomerates. The *chaebols* have adapted poorly to liberalization and continue to constrain innovation networks, in particular. Thus a mismatch of organizations and institutions frustrates support for innovating firms. Generally, the Korean NSI is experiencing a difficult transition from large firm dominance and top-down government steering to entrepreneurship, open competition, and more interactive partnerships between government and industrial actors. In both the 'old' and the 'new' versions of the Korean NSI, small domestic firms have tended to be disadvantaged, but the latter benefits new venture businesses while threatening more traditional SMEs.

A similar problem arises in Ireland, albeit in relation to foreign, rather than domestic, MNCs. In activities concerning constituents for SI and activities concerning 'support services for innovating firms' (see Box 1.2 in Chapter 1), the Irish NSI has, on one hand, bolstered inward investment by embedding foreign MNCs within local or regional clusters of interrelated firms. On the other hand it has promoted the formation of new firms, the development of innovative capabilities, and effective innovation networks within indigenous industries. Success in the second type of effort appears to have depended greatly on the degree to which there has been a significant overlap with the first type. The predominance of foreign MNCs has ensured high overall consistency in Ireland's NSI, such that its main strengths are based on alignments of different kinds of activities that support this industrial order. Thus, for example, foreign MNCs based in Ireland constitute important sources of demand for indigenous firms, whose innovation and growth performance has been improved by vertical linkage to them. Small indigenous firms in traditional industries outside this virtuous circle remain the NSI's weakest components.

To sum up the evidence on policy coordination, we can observe very different patterns in our two groups of countries. For the slow growth countries the overall picture consistently drawn by the country case studies is one of a problematic lack of policy coordination. In all these countries, policies related to specific activities are often poorly aligned with one another and in some instances they even appear to be working at cross-purposes. The case studies repeatedly point to mismatches, imbalances, inconsistencies, impasses and even immanent conflicts among specific policies related to particular fields of activity.

For the fast growth countries, the overall picture is instead one of past successes and emerging problems in the coordination of innovation policy. In all cases, successful entry into technologically dynamic growth sectors, particularly the ICT and electronics sector, has depended on effective policy coordination – for example, the close alignment of education and training with other policies designed to encourage foreign or domestic investments in the development of technological capabilities. In general, these countries all provide highly instructive historical examples of 'coordination for learning'. However, they now face the problem of diversifying beyond established sectoral and technological specializations – and, in this respect, they can be seen to have begun experiencing coordination problems that are essentially similar in nature to those that have for some time characterized the slow growth countries.

To develop an explanation for these differences between fast growth and slow growth countries we can refer to the same basic set of contextual differences that was found in Section 5.2 to be relevant for differences with respect to selectivity. In the slow growth countries, policies designed to support new firms in emerging industries are to a large extent counteracted by policies that sustain established firms and industries. This is especially the case for Sweden, Norway and the Netherlands. There is also a parallel and overlapping tug-of-war in innovation policy between large firms and SMEs, and another between 'traditional' small firms and 'science-based' ones. In Finland, the first type of conflict is evident, while the second is becoming more pronounced in Denmark. For all of these countries, though, the basic problem is that established firms and industries exercise strong inertia, constituting a powerful lobby for the maintenance of policies that work to their benefit. Policy makers appear to have responded in many cases (especially Sweden, Norway and the Netherlands) by implementing completely different sets of policies for new firms and emerging industries – but the result is serious gaps in coordination.

Turning to the fast growth countries, it can be argued that their past successes in policy coordination were at least partly due to the greater freedom of manoeuvre afforded by the lack of strong vested interests in policies aimed at maintaining the existing sectoral composition of production. In these countries, there is less overt conflict among policies aimed respectively at old and new industries, since nearly all industries are relatively new, chronologically. Further, innovation policy has, as noted in Section 4, as well as earlier parts of this section, been concentrated to a much greater extent on the development of a much narrower range of high-tech growth industries. Thus there tend to be fewer problems related to the coordination of policy across sectors. Of course, the fast growth countries do provide some evidence of conflicting policy interests among firms, occurring mainly along the axis of traditional versus science-based small firms, as in Korea and Ireland. More generally, though, firms and governments in these countries appear to agree on the need to develop new innovative capabilities and 'rules of play'. In that respect, the fast growth countries may still have the advantage of greater scope for coordination.

Notwithstanding these considerations, we can also observe that the fast growth countries are now confronting new challenges with respect to the creation or regeneration of variety. As a consequence, they may therefore begin to experience coordination problems similar to those of the slow growth countries – as hinted at, for example, in the case of Hong Kong. Here, the relevant explanation from context does not concern the sectoral composition of production, the lobbying power of established versus emerging industries, or the relative age (or youth) of the institutional and organizational set-up. Rather, it concerns the countries' levels of technological development within their respective sectors of specialization – which have remained, at least until very recently, largely the same for the fast growth countries.

It has often been argued in the literature on innovation policy that it can be difficult to alter a country's specialization towards high-tech sectors 'because technological and economic uncertainty increases with the complexity of the required competence' (Archibugi and Iammarino, 1999, p. 329). The same argument, however, can also be applied to continued advance in such sectors. Many of the fast growth countries have mastered the production of electronics but now they require the development of new capabilities within these sectors – 'technological entrepreneurship', as in Singapore – or capabilities for product innovation based on independent scientific and engineering capabilities, as in Taiwan. To develop such capabilities, the fast growth countries can no longer concentrate on competence building and infrastructure development focused on the build-up of production. Thanks in no small part to globalization processes, they must now also address issues of IPR and domestic as well as global R&D collaboration. And, as highlighted in the case of Korea, they must also accomplish a transition towards more participatory policy making based on interactive partnerships between government and industrial actors. In many respects, therefore, the fast growth countries must now contend with a much higher level of complexity in the coordination of innovation policies. In this sense, they increasingly resemble their slow growth counterparts.

6 CONCLUDING REMARKS

We now formulate the main conclusions of the discussion in this chapter in terms of the following almost telegraphic statements.

- The performance with regard to the development and diffusion of innovations has been very different in the ten NSI studied in this book. There is no doubt that the ten countries are quite different with regard to propensity to innovate. In Section 2, for example, we contrasted the Swedish pattern of high innovation inputs versus low innovation outputs with the Irish pattern of low inputs versus high outputs. In other parts of this chapter, we have noted contrasting patterns of specialization – for example, Norway's low or medium-technology specialization versus Finland's growing specialization in high technology, or Hong Kong's focus on 'coordination' of production networks versus Taiwan's focus on production.
- The activities-based approach, presented in Section 2.3 in the introductory chapter of this book, addresses the activities or determinants that, hypothetically, are important for the development and diffusion of innovations. This concerns the issue of why the countries are different with regard to propensity to innovate. The activities-based approach to innovation is in line with the broad approach to conceptualizing NSI rather than with the narrow one – as also discussed in the introductory chapter. Evidence from the ten countries investigated in this project generally favours a broad perspective on NSI, rather than a narrow one. We have shown that this approach is useful in the sense that all the ten activities listed in the introductory chapter influence the innovation processes in most of the countries, and since it ensures a broad and multidimensional perspective and thereby avoids monocausality in analysis and policy. The activities-based approach also constitutes a dynamic perspective on

determinants of innovation rather than the more static one offered by the traditional approach focusing on constituents and components in SI. Further, the activities-based approach also makes it easier for the SI approach to draw lessons for theory and for policy. With respect to policy, for example, we have argued in Section 2 that drawing attention to the multiple activities that influence innovation naturally leads to a focus on the coordination of these activities, and in Section 5.3 that SI approaches generally recognize the importance of complementarity within systems and therefore emphasize the importance of policy coordination.

- In all ten of the NSI covered by this study, policy makers have, without exception, proclaimed the adoption of the SI approach as a framework and guide for designing future innovation policy. What that means, however, for the design and implementation of innovation policy is certainly not clear – and policy makers often do not even know themselves.
- We can distinguish between a first group of five countries marked by slow growth and a second group of five countries that have exhibited fast growth during the latest three decades. The first, slow growth group includes Denmark, Finland, the Netherlands, Norway and Sweden. The second, fast growth group includes Hong Kong, Ireland, Korea, Singapore and Taiwan.
- Innovation policy has been more important, at least in the sense of having had a higher profile, in some countries (Taiwan, Finland) than in others (Hong Kong and Norway). However, all the countries in this study have formulated and pursued innovation policies, and it is not clear what difference the degree of importance attached to innovation policy has made for the actual performance of innovation by private economic actors. For instance, innovation policy may have had a much higher profile in Sweden than in Denmark, but the Danish NSI might be considered to perform better than Sweden's.
- Innovation policies that have been practised have been largely selective in all countries. This is simply because very few public actions that influence innovation processes can be neutral or non-selective. The important issue is therefore not whether policies are selective or not, but in which direction they are selective! Do they, for example, support existing sectors and firms or are they engines for change in this respect? On this score, we have observed that explicitly and consciously selective innovation policies have tended to be those that support the development of new firms, industries and technologies. Such policies have been pursued to a much larger extent, and more successfully, in the fast growth countries than in the slow growth countries. One apparent reason for this difference is the relative lack

of strong lobbies in the fast growth countries for so-called non-selective or neutral policies – which could, in turn, be explained by the relative lack of established mature industries that would stand to benefit from neutral policies.
- With regard to the fast growth countries, the evidence also shows that rapid economic growth has depended, for the most part, on successful entry into one and the same global growth sector – i.e. the electronics and ICT industries. Policies for changing the sectoral composition of production have been crucial in these countries. On the whole, the fast growth countries have been specializing, to a larger extent than the slow growth countries, in rapidly expanding high-technology sectors. In Korea, Singapore, Taiwan and Ireland, the state played an important role in stimulating and supporting rapid growth based on entry into especially dynamic industrial sectors, particularly through policies for education (especially in engineering), targeting production for export and rewarding high export performers and supporting R&D and innovation. The slow growth countries have been slower and more rigid when it comes to going into new sectors. In this sense, the (selective) policies of the fast growth countries have been more successful than the (also selective) policies in the slow growth countries. They have simply been selective in a different 'sense' or 'direction'.
- Some fast growth countries, such as Korea and Taiwan, have attempted to enter the ICT industries by developing domestic firms capable of competing globally; others, such as Ireland and Singapore, have focused on attracting foreign MNCs and promoting 'innovation by invitation'.
- Arguments have been put forward that national SI still matter in spite of the homogenizing influence of increasing globalization and, alternatively, that SSI matter more. Rather than converging, the NSI in our study have established distinctive roles within an increasingly differentiated international division of labour. Moreover, these roles tend to be consistent across sectors. Thus there is considerable evidence to indicate that globalization does not erode NSI or render them incoherent. Hence evidence from the ten countries investigated in this project generally favours the national perspective.
- Hence NSI are not 'reducible to ensembles of SSI'. However, with regard to the creation of diversity and dynamics in the national systems, the picture is different. In this respect, the emergence and development of new product areas and new SSI is an absolutely crucial element. Therefore our summary conclusion is that national systems as well as sectoral systems will remain crucial constructions for the foreseeable future – but for partly different 'purposes'.

- As noted above, SI approaches generally recognize the importance of complementarity within systems and therefore emphasize the importance of policy coordination. With respect to this issue, we have observed that the slow growth countries commonly exhibit a problematic lack of policy coordination, whereas accounts of the fast growth countries attest to both past successes and emerging problems in the coordination of innovation policy. Our explanation for this difference has stressed that the historically superior performance of fast growth countries with respect to coordination was at least partly due to the greater freedom of manoeuvre afforded by the lack of strong vested interests in policies aimed at maintaining an existing sectoral composition of production. However, this reasoning is at best only plausible for the explanation of past successes. With respect to emerging problems, we therefore pointed to the increasing complexity of policy coordination required for further advance (i.e. movement up the value chain) within the sectors and technological trajectories that the fast growth countries have specialized in.
- In innovation policy, the difficult thing is to 'pick and support winners', but avoid 'supporting losers' (in terms of sectors, technologies and products). Despite arguments to the contrary, this is the challenge for innovation policy – as well as for firm strategy. At the same time innovation policy measures should be a complement to what private organizations do – and not duplicate or substitute what they can do. This often means new activities where uncertainty is very large. Innovation policy should play the role of a midwife – not provide support towards the end of life.

ACKNOWLEDGEMENTS

All coordinators of the national teams have provided valuable comments on this chapter, particularly on what is written about their own country. We therefore thank Chau-Chin Chang, Erik Baark, Birgitte Gregersen, Terje Grønning, Nola Hewitt-Dundas, Markku Kotilainen, Chaisung Lim, Wong Poh Kam and Bart Verspagen sincerely for their valuable input.

NOTES

1. See Section 4 of the introductory chapter of this book.
2. The Barcelona Declaration is a follow-up to the Lisbon Agenda, which has the objective of transforming the EU into the most dynamic knowledge-based economy in the world, by using and transforming knowledge in the interest of growth and employment. The Barcelona Declaration is one-dimensional in the sense that it focuses exclusively on the

R&D expenditures of the member countries – with the objective that these should be 3 per cent of GDP. The Barcelona Declaration may also be questioned, since it focuses on a measure of input into the innovation process without addressing the efficiency with which these resources are being used, i.e. what the resulting innovation output is. In addition, some countries already have a R&D intensity higher than 3 per cent of GDP.
3. These authors tend to include both rules of the game and players in the term 'institution'. Compare how this term is used by Freeman and by Lundvall, as addressed in Section 2.1 of the introductory chapter in this book.
4. Some of these quantitative data are presented in the Appendix.
5. The discussion in this paragraph is based on the Finnish chapter in this volume (Sections 5 and 7).
6. Of course, the absolute levels of GDP per capita for these years are quite different. In 1975, GDP per capita (in 1990 US dollars converted to 'Geary–Khamis' purchasing power parities) for Finland was $11 441 and in 2006 it was $23 191, having doubled over the 1975–2006 period. (*Source:* Groningen Growth and Development Centre and The Conference Board, *Total Economy Database*, January 2007, available at: http://www.ggdc.net.)
7. The two differences between the approaches of Nelson and Lundvall discussed here are treated as separate issues – although there are certainly relations between them.
8. Here, specialization is measured in terms of change in production structure, not in terms of patenting.
9. This means that the systemic approach consciously reduces the degree of rigour and formality. It specifies 'problems' on an empirical basis and in a pragmatic way – not by referring to a formal model.
10. This is why innovation policy is not considered to be a separate activity.
11. This led to public subsidies to old sectors such as textiles and shipyards in Sweden in the 1970s and 1980s. The support to the shipyard industry absorbed 0.5 per cent of the Swedish GDP over a ten-year period, but did not have any lasting results.
12. See the discussion of rationales in Section 5.1.
13. The support to the shipyard and textiles industries in Sweden is one example.
14. As already mentioned, such problems and unexploited opportunities cannot be identified by comparing empirically existing NSI with optimal or ideal ones (simply since no one can specify an optimal SI). Hence the main methodology for identifying such problems and opportunities is to compare empirically existing SIs with each other.
15. The rationale for dealing with these two aspects together is that they are partly related in the sense that an analysis of strengths and weaknesses of the NSI should be a basis for the formulation of future innovation policies. So should policies pursued earlier, and the evaluation of these policies.
16. Here, the 'right' direction refers to sectors with potential for growth based on further technology development. Naturally, these sectors may vary widely across countries. In Norway, some of the main sectors with high growth potential appear to be specialized supplier industries related to certain established and emerging sectors (including both high- and low-tech industries).
17. For the specification of these two categories, see Box 1.2 in the introductory chapter.

REFERENCES

Abramowitz, M. (1986), 'Catching up, forging ahead and falling behind', *Journal of Economic History*, **46**, 386–406.
Abramowitz, M. (1994), 'The origins of the post-war catch-up and convergence boom', in J. Fagerberg, B. Verspagen and N. von Tunzelmann (eds), *The Dynamics of Technology Trade and Growth*, Aldershot, UK: Edward Elgar, pp. 21–52.
Archibugi, D. and S. Iammarino (1999), 'The policy implications of the globalisation of innovations', *Research Policy*, **28**, 317–36.

Arrow, K. (1962), 'Economic welfare and the allocation of resources for invention', in R.R. Nelson (ed.), *The Rate and Direction of Inventive Activity*, Princeton, NJ: Princeton University Press, pp. 609–25.

Ashford, D.E. (ed.) (1992), *History and Context in Comparative Public Policy*, Pittsburgh, PA: University of Pittsburgh Press.

Borrás, S. (2004), 'Systems of innovation: theory and the European Union', *Science and Public Policy*, **31**(6), 425–43.

Borrás, S., C. Chaminade and C. Edquist (2008, forthcoming), 'The challenges of globalisation: strategic choices for innovation policy', in G. Marklund, N. Vorontas and C. Wessner (eds), *The Innovation Imperative: Globalisation and National Competitiveness*, Cheltenham, UK and Northampton, MA, USA: Edward Elgar.

Carlsson, B. and S. Jacobsson (1997), 'Diversity creation and technological systems: a technology policy perspective', in C. Edquist (ed.), *Systems of Innovation: Technologies, Institutions and Organisations*, London: Pinter/Cassell, pp. 266–94.

Christensen, C.M. (2006), 'The ongoing process of building a theory of disruption', *Journal of Product Innovation Management*, **23**, 39–55.

Coriat, B., F. Malerba and F. Montobbio (2004), 'The international performance of European sectoral systems', in F. Malerba (ed.), *Sectoral Systems of Innovation: Concepts, Issues and Analyses of Six Major Sectors in Europe*, Cambridge: Cambridge University Press, pp. 388–426.

Edquist, C. (1997), 'Systems of innovation approaches – their emergence and characteristics', in C. Edquist (ed.), *Systems of Innovation: Technologies, Organisations and Institutions*, London: Pinter, pp. 1–35.

Edquist, C. (2001), 'Innovation policy – a systemic approach', in D. Archibugi and B.-Å. Lundvall (eds), *The Globalising Learning Economy*, Oxford: Oxford University Press, pp. 219–38.

Edquist, C. (2005), 'Systems of innovation – perspectives and challenges', in J. Fagerberg, D.C. Mowery and R.R. Nelson (eds), *The Oxford Handbook of Innovation*, Oxford: Oxford University Press, pp. 181–208.

Edquist, C. and C. Chaminade (2006), 'Industrial policy from a systems-of-innovation perspective', *European Investment Bank Papers*, **11**(1), 108–33.

Edquist, C., L. Hommen and M. McKelvey (2001), *Innovation and Employment: Process versus Product Innovation*, Cheltenham, UK and Northampton, MA, USA: Edward Elgar.

Edquist, C., L. Hommen, B. Johnson, T. Lemola, F. Malerba, T. Reiss and K. Smith (1998), *The ISE Policy Statement: The Innovation Policy Implications of the Innovation Systems and European Integration (ISE) Research Project*, Linköping, Sweden: Linköping University, Department of Technology and Social Change.

Fagerberg, J. and M. Godinho (2005), 'Innovation and catching up', in J. Fagerberg, D.C. Mowery and R.R. Nelson (eds), *The Oxford Handbook of Innovation*, Oxford: Oxford University Press, pp. 514–42.

George, A.L. (1979), 'Case studies and theory development: the method of structured, focussed comparison', in P.G. Lauren (ed.), *Diplomacy: New Approaches in History, Theory and Policy*, New York: The Free Press, pp. 43–68.

Gershenkron, A. (1962), *Economic Backwardness in Historical Perspective*, Cambridge, MA: Belknap Press.

Hall, P.A. and D. Soskice (eds) (2001), *Varieties of Capitalism: The Institutional Foundations of Comparative Advantage*, Oxford: Oxford University Press.

Hollingsworth, J.R. (2000), 'Doing institutional analysis: implications for the study of innovations', *Review of International Political Economy*, **7**(4), 595–644.
Kogut, B. (1993), 'Introduction' in B. Kogut (ed.), *Country Competitiveness: Technology and the Organisation of Work*, Oxford: Oxford University Press, pp. 3–12.
Lall, S. (2000), 'Technological change and industrialization in the Asian newly industrializing economies: achievements and challenges', in L. Kim and R.R. Nelson (eds), *Technology, Learning and Innovation: Experiences of Newly Industrialising Economies*, Cambridge: Cambridge University Press, pp. 13–68.
Lundvall, B.-Å. (ed.) (1992), *National Systems of Innovation: Towards a Theory of Innovation and Interactive Learning*, London: Pinter.
Lundvall, B.-Å. and S. Borrás (1997), *The Globalizing Learning Economy: Implications for Innovation Policy*, Brussels: European Commission.
Lundvall, B.-Å. and S. Borrás (2005), 'Science, technology and innovation policy', in J. Fagerberg, D.C. Mowery and R.R. Nelson (eds), *The Oxford Handbook of Innovation*, Oxford: Oxford University Press, pp. 599–631.
Lundvall, B.-Å., P. Intarakaumnerd and J. Vang (2006), *Asia's Innovation Systems in Transition*, Cheltenham, UK and Northampton, MA, USA: Edward Elgar.
Malerba, F. (2004), 'Summing up and conclusions', in F. Malerba (ed.), *Sectoral Systems of Innovation: Concepts, Issues and Analyses of Six Major Sectors in Europe*, Cambridge: Cambridge University Press, pp. 465–507.
Metcalfe, J.S. (1995), 'Technology systems and technology policy in an evolutionary framework', *Cambridge Journal of Economics*, **19**(1), 25–46.
Miettinen, R. (2002), *National Innovation Systems: Scientific Concept or Political Rhetoric?*, Helsinki: Edita.
Mowery, D.C. (2005), 'National security and national innovation systems', paper presented at the PRIME/PREST workshop on 'Re-evaluating the role of defence and security R&D in the innovation system', University of Manchester, 19–21 September.
Nelson, R.R. (ed.) (1993), *National Systems of Innovation: A Comparative Study*, Oxford: Oxford University Press.
Nelson, R.R. and N. Rosenberg (1993), 'Technical innovation and national systems', in R.R. Nelson (ed.), *National Systems of Innovation: A Comparative Study*, Oxford: Oxford University Press, pp. 3–21.
Nelson, R.R. and S. Winter (1982), *An Evolutionary Theory of Economic Change*, Cambridge, MA: Harvard University Press.
Niosi, J. (2002), 'National systems of innovation are "x-efficient" (and x-effective): why some are slow learners', *Research Policy*, **31**(2), 291–302.
Niosi, J. and B. Bellon (1994), 'The global interdependence of national innovation systems', *Technology in Society*, **16**(2), 173–97.
Niosi, J., P.P. Saviotti, B. Bellon and M. Crow (1993), 'National systems of innovation: in search of a workable concept', *Technology in Society*, **15**(2), 207–27.
OECD (1997), *National Innovation Systems*, Paris: OECD.
Storper, M., S.B. Thomadakis and L. Tsipouri (eds) (1998), *Latecomers in the Global Economy*, London: Routledge.
Whitley, R. (2002), 'Developing innovative competences: the role of institutional frameworks', *Industrial and Corporate Change*, **11**(3), 497–528.

Appendix: statistical bases of comparison for ten 'small country' NSI

Pierre Bitard, Leif Hommen and Jekaterina Novikova

1 NATIONAL CHARACTERISTICS

1.1 Size and Population

Table A1.1   Surface area and population of the ten countries, 2004

Country	Surface area			Population		
	Total (sq.km)	Water (sq.km)	Land (sq.km)	Total	Median age (yrs)	Growth rate (%)
Denmark	43 094	700	42 394	5 413 392	39.2	0.35
Finland	338 145	33 672	304 473	5 214 512	40.7	0.18
Hong Kong	1 092	50	1 042	6 855 125	39.4	0.65
Ireland*	70 280	1 390	68 890	3 969 558	33.4	1.16
Korea**	98 480	290	98 190	48 598 175	33.7	0.62
Netherlands	41 526	7 643	33 883	16 318 199	38.7	0.57
Norway	324 220	16 360	307 860	4 574 560	37.9	0.41
Singapore	692.7	10	682.7	4 353 893	36.2	1.71
Sweden	449 964	39 030	410 934	8 986 400	40.3	0.18
Taiwan	35 980	3 720	32 260	22 749 838	33.7	0.64

Notes:
*   Excluding Northern Ireland.
** South Korea.

Source: CIA, *The World Factbook*, Washington, DC, 2004. Available at: http://www.cia.gov/cia/publications/factbook/index.html.

1.2 Human Development Indicators

Table A1.2 UNDP Human Development Index standing for the ten countries, 2004[1]

Country	Human development rankings and values					Indicators: life expectancy, education and GDP per capita		
	HDI rank 2002 (177 countries)	GDP per capita rank 2002 (177 countries)	GDP per capita (PPP US$) rank minus HDI*	HDI value 2002	GDP per capita value (PPP US$) 2002	Life expectancy at birth (years) 2002	Combined primary, secondary & tertiary gross enrolment ratio (%) 2001/2002[2]	GDP per capita (PPP US$) 2002
Denmark	17	5	−12	0.932	30 940	76.6	95.8	30 940
Finland	13	19	6	0.935	26 190	77.9	106.0	26 190
Hong Kong	23	17	−6	0.903	26 910	79.9	72.0	26 910
Ireland**	10	3	−7	0.936	36 360	76.9	90.1	36 360
Korea***	28	37	9	0.888	16 950	75.4	92.0	16 950
Netherlands	5	11	6	0.942	29 100	78.3	99.5	29 100
Norway	1	2	1	0.956	36 600	78.9	97.9	36 600
Singapore	25	22	−3	0.902	24 040	78.0	87.0	24 040
Sweden	2	21	19	0.946	26 050	80.0	114.0	26 050
Taiwan	n.a.	n.a.	n.a.	n.a.	23 400	77.06	n.a.	23 400

Notes:
* PPP US$ is the US dollars converted at 'Geary–Khamis' purchasing power parities. Higher means better on the Human Development Index (HDI).
** Excluding Northern Ireland.
*** South Korea.
1. The Human Development Index (HDI) is a composite index that measures the average achievements in a country in three basic dimensions: life expectancy at birth, adult literacy and the combined gross enrolment ratio for education at all levels, and GDP per capita. The HDI covers 175 countries, using data from international data collection agencies. Further information on construction of the index, specific indicators, and data sources is available at: http://hdr.undp.org/docs/statistics/indices/stat_feature_2.pdf.
2. In this column, some countries (Finland and Sweden) have scores of higher than 100 per cent. Such scores are possible due to the derivation of this particular indicator, which is defined as follows: 'The number of students enrolled in primary, secondary and tertiary levels of education, regardless of age, as a percentage of the population of official school age for the three levels.' The definition of this and other statistical terms is available at: http://hdr.undp.org/docs/statistics/understanding/definitions.pdf.

Sources: UNDP, *Human Development Report*, New York: United Nations Development Programme (country fact sheets), 2004. Available at: http://hdr.undp.org/statistics/. Figures for Taiwan are for 2003–4. They are taken from: CIA, *The World Factbook*, Washington, DC, 2004. Available at: http://www.cia.gov/cia/publications/factbook/index.html.

1.3 Literacy Indicators[1]

Table A1.3 Percentage of population aged 16–65 at each prose, document and quantitative literacy level in selected European countries, 1994–98

Types of literacy/ countries	Literacy levels[1]			
	Level 1	Level 2	Level 3	Levels 4 and 5
Prose				
Denmark	9.6 (0.6)	36.4 (0.9)	47.5 (1.0)	6.5 (0.4)
Finland	10.4 (0.4)	26.3 (0.7)	40.9 (0.7)	22.4 (0.6)
Ireland	22.6 (1.4)	29.8 (1.6)	34.1 (1.2)	13.5 (1.4)
Netherlands	10.5 (0.6)	30.1 (0.9)	44.1 (1.0)	15.3 (0.6)
Norway	8.5 (0.5)	24.7 (1.0)	49.2 (0.9)	17.6 (0.9)
Sweden	7.5 (0.5)	20.3 (0.6)	39.7 (0.9)	32.4 (0.5)
Document				
Denmark	7.8 (0.5)	24.2 (0.8)	42.6 (0.9)	25.4 (0.7)
Finland	12.6 (0.5)	24.1 (0.8)	38.1 (0.8)	25.1 (0.6)
Ireland	25.3 (1.7)	31.7 (1.2)	31.5 (1.3)	11.5 (1.2)
Netherlands	10.1 (0.7)	25.7 (0.8)	44.2 (0.9)	20.0 (0.8)
Norway	8.6 (0.5)	21.0 (1.0)	40.9 (1.0)	29.4 (1.2)
Sweden	6.2 (0.4)	18.9 (0.7)	39.4 (0.8)	35.5 (0.6)
Quantitative				
Denmark	6.2 (0.4)	21.5 (0.8)	43.9 (1.2)	28.4 (0.9)
Finland	11.0 (0.4)	27.2 (0.8)	42.1 (0.8)	19.7 (0.6)
Ireland	24.8 (1.5)	28.3 (0.8)	30.07 (1.0)	16.2 (1.6)
Netherlands	10.3 (0.7)	25.5 (0.9)	44.3 (1.0)	19.9 (0.8)
Norway	7.7 (0.5)	22.0 (1.0)	42.9 (1.3)	27.4 (1.2)
Sweden	6.6 (0.4)	18.6 (0.6)	39.0 (0.9)	35.8 (0.7)

Note: [1] Level 1 is lowest in terms of task variety, complexity and difficulty; level 5 is highest.

Source: OECD/Statistics Canada, *International Adult Literacy Survey, 1994–1998*, Paris/Ottawa: OECD/Minister of Industry, Canada, 2000, Table 2.3.

Table A1.4 Mean prose, document and quantitative scores on a scale with range 0–500 points by level of educational attainment, population aged 16–65, for selected European countries, 1994–98

Types of literacy/ countries	Levels of educational attainment		
	With less than upper secondary education	Completed upper secondary education	Completed tertiary education
Prose			
Denmark	252.8 (1.1)	278.1 (0.8)	298.5 (1.0)
Finland	261.6 (1.6)	295.9 (1.3)	316.9 (1.4)
Ireland	238.8 (2.8)	288.2 (2.7)	308.3 (2.6)
Netherlands	257.5 (1.2)	297.0 (1.3)	312.1 (1.4)
Norway	254.5 (2.8)	284.4 (1.2)	315.1 (1.0)
Sweden	275.4 (2.1)	302.3 (1.2)	329.1 (1.7)
Document			
Denmark	266.9 (1.5)	298.2 (1.0)	319.3 (1.5)
Finland	257.3 (1.7)	297.4 (1.2)	322.3 (1.7)
Ireland	231.5 (2.6)	280.5 (2.9)	303.5 (3.3)
Netherlands	262.6 (1.5)	302.3 (1.4)	311.2 (1.6)
Norway	257.0 (3.8)	293.1 (1.7)	326.7 (1.2)
Sweden	280.6 (2.4)	308.3 (1.0)	331.2 (2.0)
Quantitative			
Denmark	272.3 (1.4)	303.6 (1.1)	321.3 (1.4)
Finland	259.9 (1.6)	291.6 (1.3)	316.2 (1.6)
Ireland	236.8 (2.6)	285.6 (3.1)	310.5 (3.2)
Netherlands	263.7 (1.6)	300.2 (1.5)	316.2 (2.0)
Norway	262.2 (3.5)	291.6 (1.4)	326.6 (1.0)
Sweden	282.3 (2.1)	307.4 (1.1)	331.7 (2.0)

Source: OECD/Statistics Canada, *International Adult Literacy Survey, 1994–1998*, Paris/Ottawa: OECD/Minister of Industry, Canada, 2000, Table 2.4.

2 ECONOMIC STRUCTURE AND PERFORMANCE

2.1 Sectoral Composition of Production

Table A2.1 *Manufacturing structure by industry groups, 1980, 1990 and 2000*

	Total output			Value-added			Employment			Exports		
	Share	Change in percentage share		Share	Change in percentage share		Share	Change in percentage share		Share	Change in percentage share	
	1980	1990	2000	1980	1990	2000	1980	1990	2000	1980	1990	2000
Denmark												
high	4.8	2.2	4.3	6.8	2.7	5.0	6.6	1.7	0.9	8.6	4.6	7.5
med.–high	18.0	2.4	3.4	23.3	1.1	1.5	22.8	0.3	1.1	26.3	1.7	−0.1
med.–low	21.7	−0.3	0.7	22.0	0.9	0.3	22.2	−0.5	1.9	17.2	−2.0	−1.3
low	55.6	−4.3	−8.4	47.9	−4.7	−6.8	48.5	−1.5	−3.9	47.9	−4.3	−6.1
Finland												
high	2.5	2.7	16.6	3.3	4.4	15.9	3.9	2.3	6.9	3.6	5.2	18.5
med.–high	17.3	3.5	−1.6	20.4	3.6	−4.6	20.7	2.0	−0.3	17.3	9.8	−3.3
med.–low	26.1	−2.9	−1.2	22.7	−0.3	−3.3	20.2	1.1	2.6	21.2	−2.8	−1.0
low	54.1	−3.4	−13.7	53.6	−7.7	−8.0	55.2	−5.4	−9.2	57.9	−12.2	−14.2
Hong Kong												
high	21.6	−2.9	−6.5	16.4	−2.0	−2.6	15.5	−4.1	−1.0	n.a.	n.a.	n.a.
med.–high	6.1	1.8	10.4	5.6	1.6	13.8	5.9	0.7	2.7	n.a.	n.a.	n.a.
med.–low	17.9	−2.0	−0.8	20.4	−3.5	−4.3	20.1	−4.0	−3.1	n.a.	n.a.	n.a.
low	54.4	3.1	−3.1	57.6	3.9	−6.8	58.4	7.5	1.4	n.a.	n.a.	n.a.
Ireland (production benchmark share is 1991)												
high	19.0	n.a.	16.5	9.4	7.7	11.0	7.3	4.8	11.7	14.8	20.7	14.6

med.–high	18.4	n.a.	8.0	18.0	4.4	11.4	17.5	3.8	−0.2	21.8	−1.1	10.2
med.–low	8.9	n.a.	−3.8	14.1	−4.0	−4.1	17.9	−1.6	−0.8	9.1	−0.6	−5.5
low	53.7	n.a.	−20.8	58.4	−8.1	−18.3	57.3	−7.0	−10.7	54.3	−19.0	−19.3
Japan												
high	8.8	5.4	3.4	12.1	4.2	2.4	10.5	2.5	0.4	19.8	10.5	2.6
med.–high	29.0	4.1	0.1	28.4	2.1	−1.5	23.5	2.2	1.3	47.4	4.0	−0.6
med.–low	33.6	−10.0	−2.0	28.0	−4.9	−0.6	21.3	−2.4	−0.6	23.5	−11.6	−1.1
low	28.7	0.5	−1.5	31.5	−1.4	−0.3	44.7	−2.3	−1.2	9.2	−3.0	−0.9
Netherlands (production benchmark share is 1990)												
high	10.6	n.a.	0.9	10.9	−0.1	0.3	11.5	1.2	−1.5	10.0	6.1	16.5
med.–high	24.7	n.a.	1.9	22.6	2.8	−1.1	17.1	1.9	−0.1	29.4	2.2	−4.1
med.–low	23.7	n.a.	1.2	23.5	−0.1	−0.5	22.1	−1.6	1.4	31.9	−10.3	−4.4
low	41.0	n.a.	−4.0	43.0	−2.7	1.4	49.4	−1.5	0.3	28.9	2.1	−8.0
Norway												
high	3.1	1.5	2.0	4.8	1.9	1.2	4.6	1.6	0.7	4.5	3.4	2.4
med.–high	17.4	0.2	−1.4	21.3	−0.4	−2.6	18.4	−0.1	−0.9	25.0	−0.9	1.6
med.–low	34.8	−2.7	3.2	33.9	−4.0	2.5	29.4	−2.0	0.9	44.1	1.7	−5.0
low	44.7	1.0	−3.8	40.0	2.5	−1.2	47.6	0.6	−0.7	26.4	−4.2	0.9
South Korea (export benchmark is 1994)												
high	8.5	4.9	8.4	11.6	3.9	8.9	10.1	3.6	2.2	28.0	n.a.	9.1
med.–high	14.6	10.5	2.9	20.4	6.4	0.0	14.4	7.6	5.4	26.0	n.a.	0.7
med.–low	31.1	−3.2	1.2	26.1	2.5	−0.5	19.1	0.2	4.3	19.9	n.a.	1.1
low	45.9	−12.1	−12.5	41.9	−12.8	−8.3	56.4	−11.4	−11.9	26.1	n.a.	−10.9
Singapore (all benchmarks are 1991)												
high	43.0	−1.2	18.7	42.9	−1.2	17.6	39.2	−1.8	0.4	27.3	27.5	43.2
med.–high	14.6	−0.4	−4.4	16.7	0.0	0.1	15.9	0.2	0.3	10.4	2.5	3.0
med.–low	29.2	0.3	−7.5	25.1	−1.0	−10.4	21.8	−1.0	5.3	48.5	−27.6	−36.3
low	13.2	0.2	−6.9	15.4	0.5	−7.6	23.1	0.6	−6.3	13.9	−3.7	−9.9
Sweden												
high	6.6	2.6	11.6	10.5	1.1	5.3	9.7	0.3	3.1	10.7	5.3	12.8

Table A2.1 (continued)

	Total output			Value-added			Employment			Exports		
	Share	Change in percentage share		Share	Change in percentage share		Share	Change in percentage share		Share	Change in percentage share	
	1980	1990	2000	1980	1990	2000	1980	1990	2000	1980	1990	2000
med.–high	27.4	2.0	1.0	30.4	1.0	−0.7	25.9	3.0	1.2	34.0	2.8	−3.1
med.–low	26.4	−3.9	−3.9	25.3	−2.6	−2.3	25.5	−2.7	−1.1	25.1	−5.5	−3.1
low	39.6	−0.7	−0.7	33.8	0.5	−2.3	38.9	−0.6	−3.2	30.1	−2.6	−6.6
Taiwan (pharmaceuticals in med.–high)												
high	n.a.	n.a.	n.a.	9.5	2.2	12.3	12.8	1.1	5.8	n.a.	n.a.	n.a.
med.–high	n.a.	n.a.	n.a.	18.8	5.6	0.1	17.5	3.6	1.9	n.a.	n.a.	n.a.
med.–low	n.a.	n.a.	n.a.	28.6	1.8	−0.6	27.2	1.5	−0.1	n.a.	n.a.	n.a.
low	n.a.	n.a.	n.a.	43.1	−9.6	−11.9	42.5	−6.1	−7.5	n.a.	n.a.	n.a.
USA												
high	12.2	4.5	2.5	16.1	5.2	1.7	16.1	1.4	−1.7	23.4	9.3	5.7
med.–high	25.1	0.2	1.6	26.5	−1.3	0.9	23.2	−1.2	1.6	45.0	−5.7	−2.2
med.–low	29.7	−6.7	−0.8	25.2	−5.0	−0.1	22.3	−1.8	1.6	13.6	−2.2	−0.7
low	33.0	1.9	−3.3	32.2	1.1	−2.5	38.5	1.6	−1.5	18.0	−1.3	−2.8

Notes: Starting year for Ireland is 1991. Export data for Korea start in 1994. Ending years can vary. The classification of industries according to technology levels is based on OECD, *OECD Science, Technology and Industry Scoreboard*, Paris: OECD, 2001, pp. 137–8. High-tech industries include science-based industries (e.g. pharmaceuticals). Medium-high-tech industries include scale-intensive manufacturers (e.g. motor vehicles and trailers). Medium–low-tech industries include process industries (e.g. coke and petroleum products and rubber and plastic products). Typical low-tech industries are food, beverages and tobacco and wood and cork products.

Source: Tables in *Annex B*, OECD STAN database as of November 2003, complemented by the UNIDO database and national income accounts on the website of the Hong Kong Statistical Office. Compilation by Mark Knell. (Adapted from Table 2 in M. Knell, 'On measuring structural change in small European and Asian economies', ESF Working paper, TIK Centre: University of Oslo, 2003.)

2.2 Productivity

Table A2.2 Labour productivity, 2003

	2003			1995–2003
	GDP per capita[1]	Average number of hours worked	GDP per hour worked	Growth rates in productivity (=GDP per hour worked)
Denmark	78.0	1 481	94.8	1.8
Finland	70.1	1 588	86.6	2.8
Hong Kong[2]	79.3	2 287	67.4	1.4
Ireland	83.5	1 612	102.4	5.3
Netherlands	73.0	1 338	98.1	0.5
Norway	88.6	1 336	119.1	2.3
Singapore[2]	75.7	2 307	60.3	1.9
South Korea	53.7	2 402	43.3	3.9
Sweden	73.9	1 562	87.1	2.3
Taiwan[2]	58.8	2 174	57.3	4.6
USA	100	n.a.	100	100

Notes:
[1] The GDP data here are in 1990 PPP US$.
[2] Data for Hong Kong, Singapore and Taiwan are from 2002 apart from 'Average number of hours worked', which are from 2003. In column 5, the period covered for these countries is 1995–2002.

Source: Groningen Growth and Development Centre and The Conference Board, *Total Economy Database*, August 2004, available at: http://www.ggdc.net.

2.3 Growth[2]

Table A2.3 GDP per capita, 1990 PPP US$, ten European and Asian countries plus USA

	Denmark	Finland	Ireland	Netherlands	Norway	Sweden	Hong Kong	Singapore	South Korea	Taiwan	USA
1950	6943	4253	3453	5971	5430	6739	2218	2219	854	924	9561
1951	6936	4571	3544	6007	5670	6949	2295	2253	787	991	10116
1952	6955	4674	3642	6058	5814	6996	2377	2280	835	1063	10316
1953	7292	4652	3747	6515	5985	7145	2460	2314	1072	1140	10613
1954	7371	5002	3794	6878	6226	7402	2546	2320	1124	1193	10359
1955	7395	5197	3920	7295	6301	7566	2636	2358	1169	1250	10897
1956	7439	5295	3897	7467	6575	7797	2729	2333	1149	1270	10914
1957	7965	5490	3914	7582	6711	8092	2825	2318	1206	1314	10920
1958	8095	5474	3870	7451	6652	8083	2924	2295	1234	1382	10631
1959	8637	5754	4038	7704	6874	8288	3027	2186	1243	1462	11230
1960	8812	6230	4282	8252	7204	8688	3134	2310	1226	1492	11328
1961	9312	6658	4508	8167	7595	9137	3244	2422	1247	1551	11402
1962	9747	6819	4636	8603	7746	9469	3652	2520	1245	1632	11905
1963	9732	6994	4821	8795	7982	9917	4083	2701	1316	1804	12242
1964	10560	7307	4986	9398	8316	10515	4327	2541	1390	1977	12773
1965	10953	7670	5051	9757	8690	10815	4825	2667	1436	2056	13419
1966	11160	7824	5080	9895	8945	10936	4865	2891	1569	2205	14134
1967	11437	7947	5352	10298	9423	11219	4824	3163	1645	2395	14330
1968	11837	8093	5770	10849	9551	11561	4880	3540	1812	2539	14863
1969	12531	8878	6089	11414	9899	12055	5345	3965	2040	2706	15179
1970	12686	9577	6199	11923	10027	12716	5695	4439	2167	2980	15030
1971	12934	9765	6354	12301	10472	12748	5968	4904	2332	3324	15304

1972	13 538	10 448	6 663	12 554	10 922	13 002	6 473	5 460	2 456	3 767	15 944
1973	13 945	11 085	6 867	13 081	11 324	13 494	7 105	5 977	2 824	4 091	16 689
1974	13 751	11 361	7 042	13 508	11 726	13 885	7 091	6 276	3 015	3 942	16 491
1975	13 621	11 441	7 316	13 417	12 271	14 183	6 991	6 430	3 162	3 958	16 284
1976	14 466	11 358	7 302	13 937	12 930	14 282	7 906	6 797	3 476	4 566	16 975
1977	14 655	11 355	7 795	14 177	13 425	14 004	8 707	7 224	3 775	5 020	17 567
1978	14 826	11 559	8 250	14 424	13 840	14 207	9 277	7 752	4 064	5 542	18 373
1979	15 313	12 332	8 366	14 647	14 411	14 721	9 796	8 362	4 294	5 831	18 789
1980	15 227	12 949	8 541	14 705	15 076	14 937	10 503	9 058	4 114	5 869	18 577
1981	15 096	13 134	8 716	14 525	15 169	14 917	11 202	9 450	4 302	6 229	18 856
1982	15 563	13 485	8 821	14 292	15 145	15 058	11 333	9 654	4 557	6 446	18 325
1983	15 966	13 767	8 740	14 483	15 636	15 315	11 797	10 298	5 007	7 036	18 920
1984	16 676	14 107	9 056	14 900	16 513	15 908	12 846	10 938	5 375	7 790	20 123
1985	17 384	14 522	9 306	15 283	17 320	16 189	12 763	10 710	5 670	8 113	20 717
1986	17 993	14 819	9 265	15 617	17 882	16 505	13 960	10 900	6 263	9 088	21 236
1987	18 023	15 382	9 698	15 737	18 164	16 949	15 597	11 743	6 916	9 641	21 788
1988	18 224	16 088	10 234	16 043	18 059	17 232	16 716	12 718	7 621	9 623	22 499
1989	18 261	16 946	10 880	16 695	18 157	17 524	17 043	13 475	8 027	9 665	23 059
1990	18 452	16 866	11 818	17 262	18 466	17 609	17 541	14 220	8 704	9 886	23 201
1991	18 644	15 727	11 969	17 520	19 045	17 284	18 332	14 682	9 423	10 524	22 849
1992	18 949	15 058	12 275	17 747	19 561	16 980	19 255	15 172	9 866	11 244	23 298
1993	18 870	14 849	12 533	17 765	19 974	16 545	20 112	16 482	10 355	11 906	23 616
1994	19 847	15 316	13 191	18 055	20 905	17 070	20 775	17 853	11 117	12 662	24 279
1995	20 350	15 859	14 389	18 510	21 703	17 644	21 041	18 682	12 007	13 360	24 603
1996	20 810	16 395	15 450	19 055	22 728	17 852	21 359	19 425	12 717	14 092	25 230
1997	21 388	17 345	17 231	19 773	23 775	18 257	22 091	20 323	13 180	14 896	26 052
1998	21 771	18 200	18 668	20 425	24 251	18 915	20 729	19 513	12 167	15 434	26 824
1999	22 255	18 867	20 594	21 243	24 598	19 755	21 378	20 581	13 213	16 195	27 699
2000	22 966	19 773	22 448	21 933	25 133	20 580	23 299	22 258	14 220	17 007	28 403

Table A2.3 (continued)

	Denmark	Finland	Ireland	Netherlands	Norway	Sweden	Hong Kong	Singapore	South Korea	Taiwan	USA
2001	23 049	20 248	23 492	22 202	25 688	20 762	23 257	21 309	14 658	16 524	28 347
2002	23 075	20 530	24 626	22 074	25 855	21 142	23 513	21 761	15 591	17 110	28 535
2003	23 151	20 849	25 389	22 011	26 033	21 462	24 096	21 985	15 996	17 581	28 982
2004	23 493	21 532	26 166	22 310	26 723	22 310	25 995	23 479	16 675	18 527	29 839
2005	24 116	22 121	27 295	22 531	27 219	22 912	27 709	24 571	17 259	19 163	30 519
2006	24 871	23 191	28 360	23 094	27 774	23 870	n.a.	n.a.	18 041	n.a.	31 229

Source: Groningen Growth and Development Centre and The Conference Board, *Total Economy Database*, January 2007, available at: http://www.ggdc.net.

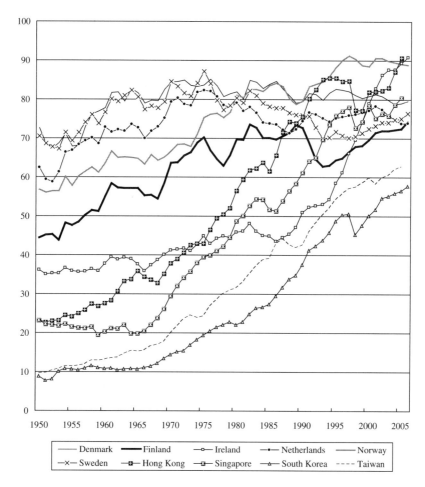

Source: As for Table A2.3.

Figure A2.1 GDP per capita, 1990 PPP US$, USA = 100, ten European and Asian countries

Table A2.4 GDP per person employed, in 1990 PPP US$, ten European and Asian countries plus USA

	Denmark	Finland	Ireland	Netherlands	Norway	Sweden	Singapore	South Korea	Taiwan	USA
1950	14 992	8 704	8 386	18 055	12 492	13 835	n.a.	n.a.	n.a.	23 615
1960	18 758	13 012	10 536	25 226	18 140	18 022	7 691	4 071	4 976	30 480
1973	28 867	23 575	19 778	38 225	26 578	28 349	16 393	8 846	11 924	40 727
1974	29 017	23 591	20 335	39 544	27 825	28 642	16 975	9 159	11 371	39 843
1975	29 554	24 380	21 910	39 611	28 746	28 648	17 455	9 541	11 559	40 967
1976	31 808	23 667	22 449	41 236	29 306	28 767	17 909	10 044	13 249	41 702
1977	32 130	24 227	23 551	41 777	29 753	28 232	18 583	10 735	14 091	42 045
1978	32 478	25 084	24 968	42 259	30 581	28 615	19 027	11 217	15 227	42 578
1979	32 126	26 160	24 483	42 626	31 629	29 251	19 522	11 849	15 821	42 790
1980	31 583	26 700	25 457	41 847	32 966	29 373	20 302	11 462	16 000	42 603
1981	32 637	26 907	26 373	40 797	32 776	29 418	20 763	11 879	16 970	43 190
1982	33 551	27 499	27 094	40 822	32 730	29 755	20 967	12 464	17 512	42 751
1983	34 180	28 088	27 246	42 022	33 890	30 241	22 138	13 776	18 715	43 965
1984	34 693	28 658	28 973	43 143	35 343	31 235	23 627	15 051	20 341	45 293
1985	35 249	29 330	30 112	43 631	36 343	31 511	23 847	15 457	21 117	46 107
1986	35 032	30 100	30 038	44 145	36 379	31 988	24 671	16 648	22 982	46 630
1987	34 923	31 438	31 238	39 977	36 461	32 888	25 901	17 601	23 746	47 048
1988	35 144	32 885	32 869	39 909	36 612	33 094	27 375	18 988	23 711	47 951
1989	35 911	34 184	34 780	40 875	38 142	33 352	28 592	19 405	23 651	48 606
1990	35 960	34 230	36 020	41 176	39 324	33 559	29 159	20 633	24 203	48 851
1991	36 723	33 837	36 843	40 961	40 940	33 677	30 450	21 823	25 550	49 053
1992	36 933	35 239	37 811	40 952	42 342	34 615	31 372	22 579	26 847	50 225
1993	37 815	37 117	38 201	41 295	43 500	35 837	34 999	23 455	28 362	50 810
1994	40 582	38 872	39 101	42 240	45 132	37 649	37 649	24 654	29 718	51 678

498

1995	40 757	39 490	40 917	41 886	45 861	38 423	39 399	26 161	31 256	52 291
1996	41 348	39 911	42 493	42 262	46 804	39 194	41 286	27 410	33 080	53 403
1997	41 760	41 578	45 239	42 506	47 620	40 469	42 799	28 390	34 873	54 543
1998	42 739	42 809	45 271	42 759	47 651	41 289	41 907	27 963	36 023	56 055
1999	43 429	43 129	46 734	43 148	47 805	42 127	43 986	30 568	37 589	57 460
2000	44 380	44 840	49 357	43 744	48 612	42 686	n.a.	n.a.	n.a.	59 083
2001	44 861	44 433	51 099	43 594	49 241	42 573	n.a.	n.a.	n.a.	59 805

Source: Groningen Growth and Development Centre and The Conference Board, *Total Economy Database*, August 2004, available at: http://www.ggdc.net.

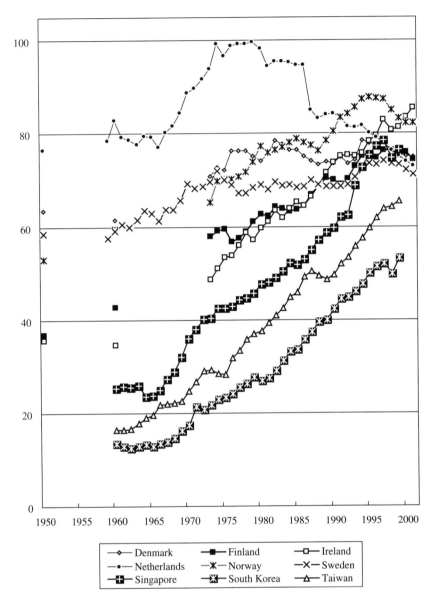

Source: As for Table A2.4.

Figure A2.2 GDP per person employed, in 1990 PPP US$, USA=100, nine European and Asian countries

Table A2.5 GDP per hour worked, in 1990 PPP US$, ten European and Asian countries plus USA

	Denmark	Finland	Ireland	Netherlands	Norway	Sweden	Singapore	South Korea	Taiwan	USA
1950	7.24	4.28	3.44	8.38	6.12	6.79	n.a.	n.a.	n.a.	10.90
1960	9.72	6.38	4.54	12.60	9.36	9.46	3.11	n.a.	1.80	15.49
1973	18.27	13.81	9.09	22.36	15.91	17.27	6.80	3.64	4.37	21.64
1979	20.58	14.61	11.86	27.01	20.90	19.29	8.21	4.64	5.95	23.19
1980	19.97	15.20	12.57	26.67	21.80	19.54	8.56	4.51	6.06	23.26
1981	20.37	15.45	13.28	26.26	21.82	19.68	8.78	4.70	6.60	23.79
1982	20.66	15.97	13.92	26.56	21.97	19.73	8.88	4.73	6.38	23.75
1983	20.78	16.32	14.27	27.81	22.83	19.93	9.40	5.27	7.19	24.32
1984	22.59	16.64	15.24	29.01	23.89	20.55	10.05	5.80	8.08	24.86
1985	22.70	17.09	15.81	29.71	24.67	20.68	10.17	5.96	8.50	25.27
1986	22.84	17.80	15.52	30.41	24.77	21.02	10.55	6.56	9.09	25.86
1987	23.07	18.34	16.24	27.74	25.27	21.47	11.10	6.75	9.39	26.06
1988	22.95	18.95	17.11	27.89	25.36	21.34	11.76	7.24	9.48	26.27
1989	23.81	19.94	18.03	28.76	26.48	21.52	12.31	7.55	9.58	26.57
1990	24.10	20.41	18.74	29.11	27.46	21.70	12.58	8.17	9.89	26.86
1991	24.75	20.43	19.47	29.21	28.68	21.97	13.17	8.69	10.56	27.14
1992	24.57	21.02	20.50	29.32	29.47	22.32	13.60	9.07	11.19	27.92
1993	25.74	22.44	20.85	29.74	30.33	22.86	15.17	9.44	11.83	28.00
1994	26.37	23.00	21.31	30.83	31.54	23.45	16.32	9.93	12.46	28.31
1995	27.15	23.42	22.30	30.69	32.43	23.82	17.08	10.48	13.20	28.42
1996	27.40	23.45	23.14	30.47	33.26	24.15	17.90	11.05	13.96	29.05
1997	27.47	24.56	25.17	30.80	34.03	24.92	18.55	11.59	14.72	29.51
1998	28.23	25.56	26.32	31.26	34.07	25.37	18.17	11.64	15.21	30.08
1999	28.18	25.68	27.49	32.03	34.27	25.78	19.07	12.18	15.87	30.70
2000	28.80	27.39	29.03	32.47	35.32	26.31	n.a.	12.88	n.a.	31.45
2001	29.11	27.14	30.06	32.36	35.78	26.24	n.a.	13.02	n.a.	32.02

Source: As for Table A2.4.

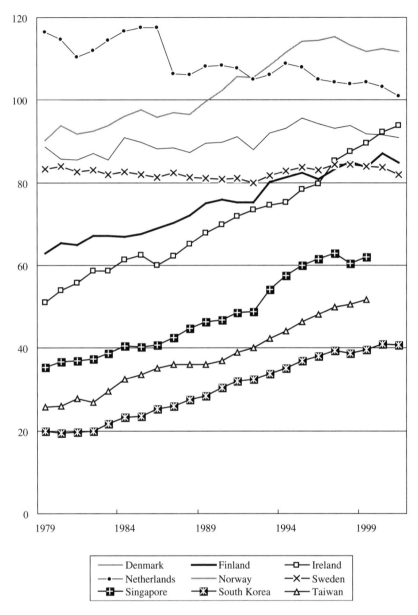

Figure A2.3 GDP per hour worked, in 1990 PPP US$, USA=100, nine European and Asian countries

2.4 Globalization

Table A2.6 *Selected criteria for openness and globalization of the economy*

	1980s	1990s	2000s
Denmark			
Import and export of goods and services/GDP ratio rank[1]	7	5	8
Exports of high-tech products as a share of total exports[2]	6.05	9.08	13.03
Inward FDI stocks as % of GDP	6.0	13.5	36.1
Outward FDI stocks as % of GDP	3.0	13.7	36.6
Inward FDI share of services	n.a.	78.9	87.1
Outward FDI share of services	n.a.	75.2	86.3
Inward FDI performance index ranking	55	58	40
Outward FDI performance index ranking	17	22	12
Finland			
Import and export of goods and services/GDP ratio rank[1]	14	8	7
Exports of high-tech products as a share of total exports[2]	2.55	9.45	20.37
Inward FDI stocks as % of GDP	2.5	6.5	36.1
Outward FDI stocks as % of GDP	3.4	11.5	42.4
Inward FDI share of services	n.a.	59.9	89.5
Outward FDI share of services	n.a.	34.5	28.9
Inward FDI performance index ranking	65	81	43
Outward FDI performance index ranking	12	17	23
Hong Kong			
Import and export of goods and services/GDP ratio rank[1]	n.a.	n.a.	n.a.
Exports of high-tech products as a share of total exports[2]	n.a.	n.a.	n.a.
Inward FDI stocks as % of GDP	525.5	160.6	236.5
Outward FDI stocks as % of GDP	6.7	55.6	211.9
Inward FDI share of services	n.a.	84.1	96.4
Outward FDI share of services	n.a.	79.1	87.6
Inward FDI performance index ranking	3	14	9
Outward FDI performance index ranking	5	8	6
Ireland			
Import and export of goods and services/GDP ratio rank[1]	5	1	1
Exports of high-tech products as a share of total exports[2]	15.69	28.95	38.92
Inward FDI stocks as % of GDP	157.7	60.2	129.7
Outward FDI stocks as % of GDP	42.2	19.9	22.5
Inward FDI share of services	n.a.	60.9	50.7
Outward FDI share of services	n.a.	n.a.	26.4*
Inward FDI performance index ranking	51	40	4
Outward FDI performance index ranking	13	33	20

504   *Small country innovation systems*

Table A2.6   (continued)

	1980s	1990s	2000s
Netherlands			
Import and export of goods and services/GDP ratio rank[1]	4	6	4
Exports of high-tech products as a share of total exports[2]	6.57	14.32	19.46
Inward FDI stocks as % of GDP	18.8	28.0	65.6
Outward FDI stocks as % of GDP	36.1	41.6	75.0
Inward FDI share of services	n.a.	57.1	67.7**
Outward FDI share of services	n.a.	63.7	69.2**
Inward FDI performance index ranking	13	35	16
Outward FDI performance index ranking	3	11	4
Norway			
Import and export of goods and services/GDP ratio rank[1]	6	12	13
Exports of high-tech products as a share of total exports[2]	1.57	3.12	3.91
Inward FDI stocks as % of GDP	11.7	12.7	20.4
Outward FDI stocks as % of GDP	1.7	15.2	18.4
Inward FDI share of services	n.a.	40.0	85.6
Outward FDI share of services	n.a.	38.1	33.3
Inward FDI performance index ranking	50	60	108
Outward FDI performance index ranking	19	18	33
Singapore			
Import and export of goods and services/GDP ratio rank[1]	n.a.	n.a.	n.a.
Exports of high-tech products as a share of total exports[2]	19.2	47.6	53.9
Inward FDI stocks as % of GDP	73.6	78.2	161.3
Outward FDI stocks as % of GDP	24.8	41.8	99.5
Inward FDI share of services	51.9	70.7	58.1
Outward FDI share of services	n.a.	79.9	99.5
Inward FDI performance index ranking	1	3	6
Outward FDI performance index ranking	9	10	3
South Korea			
Import and export of goods and services/GDP ratio rank[1]	n.a.	n.a.	n.a.
Exports of high-tech products as a share of total exports[2]	n.a.	n.a.	n.a.
Inward FDI stocks as % of GDP	2.3	1.8	7.8
Outward FDI stocks as % of GDP	0.5	2.0	5.7
Inward FDI share of services	n.a.	41.9	44.2
Outward FDI share of services	n.a.	45.7	51.7
Inward FDI performance index ranking	81	121	120
Outward FDI performance index ranking	32	37	47
Sweden			
Import and export of goods and services/GDP ratio rank[1]	11	4	6
Exports of high-tech products as a share of total exports[2]	7.68	14.47	20.71
Inward FDI stocks as % of GDP	4.2	12.5	47.5

Table A2.6 (continued)

	1980s	1990s	2000s
Outward FDI stocks as % of GDP	10.4	29.5	62.7
Inward FDI share of services	n.a.	21.7	43.7
Outward FDI share of services	n.a.	23.6	58.8
Inward FDI performance index ranking	52	29	42
Outward FDI performance index ranking	2	15	7
Taiwan			
Import and export of goods and services/GDP ratio rank[1]	n.a.	n.a.	n.a.
Exports of high-tech products as a share of total exports[2]	n.a.	n.a.	n.a.
Inward FDI stocks as % of GDP	4.7	5.9	11.9
Outward FDI stocks as % of GDP	0.3	9.5	22.8
Inward FDI share of services	n.a.	44.4	59.5
Outward FDI share of services	n.a.	66.9	69.8
Inward FDI performance index ranking	49	100	82
Outward FDI performance index ranking	6	29	24

Notes:
* 2001–2 average.
** 2000–2001 average.

Import and export of goods and services/GDP ratio rank is a ranking among 23 selected OECD countries. Export of high-tech products as a share of total exports is based on data for the EU-15 plus Japan, Switzerland, USA and Australia. Measuring years are for import and export of goods 1985, 1995 and 2000; for exports of high-tech products 1981, 1993 and 1999; for inward and outward FDI stocks 1985, 1995 and 2003; for inward and outward FDI share of services 1995–99 and 2000–2002; for inward and outward FDI performance ranking 1988–90, 1994–96 and 2001–3.

Sources: [1] Table 3 in Torben Andersen and Tryggvi Herbertsson, *Measuring Globalisation*, Bonn: IZA Discussion Paper No. 817, July 2003 (selected countries); [2] Table 1 in Pontus Braunerhjelm and Per Thulin, *Can Countries Create Comparative Advantages?*, Stockholm: Svenska Nätverket För Europaforskning i Ekonomi Working Paper, 2004 (EU-15 plus Japan, Switzerland, USA and Australia); and UNCTAD, *World Investment Report 2004: The Shift Towards Services*, New York and Geneva: United Nations Conference on Trade and Development, 2004.

Table A2.7 Cross-border ownership of patents, 2002

	Foreign ownership (%) of domestic inventions (2002)[1]	Domestic ownership (%) of inventions made abroad (1980–2002)[2]
Denmark	41	6
Finland	15	18
Hong Kong	n.a.	n.a.
Ireland	71	43
Netherlands	44	44
Norway	42	13
Singapore	51	8
South Korea	11	3
Sweden	22	14
Taiwan	44	9

Notes:
[1] Share of patents granted by the US Patent Office (USPTO) owned by foreign residents in total patents invented domestically.
[2] Share of patents granted by the USPTO invented abroad in total patents owned by country residents.

Source: C. Lim, *Progress Report*, Korea: Korea Christian University, 2003. Derived from the USPTO data base at www.uspto.gov, 3–9 March 2003.

3 SCIENCE AND TECHNOLOGY PROFILES

3.1 Science Profiles

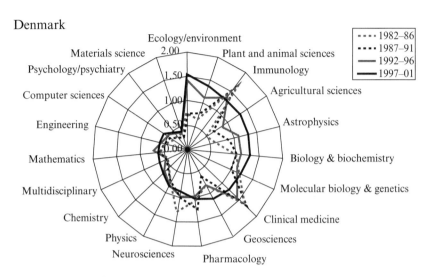

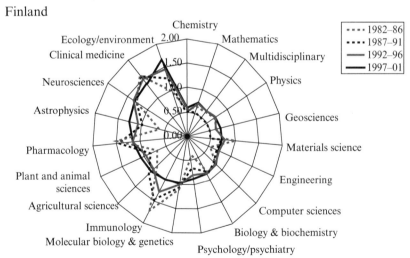

Source: K. Wang, M.-T. Tsai, Y.-L. Luo, A. Balaguer, S.-C. Hung, F.-S. Wu, M.-Y. Hsu and Y.-Y. Chu, *Intensities of Scientific Performance: Publication and Citation at a Macro and Sectoral Level of Nine Countries*, ESF Working paper, Science and Technology Information Centre – National Science Council, Taipei, Republic of China, 2003.

Figure A3.1   Historical changes in revealed comparative advantage (RCA) in scientific publication output in eight nations, 1982–2001[1]

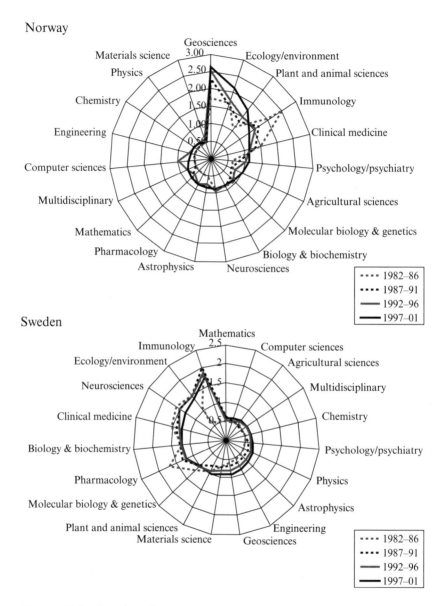

Figure A3.1 (continued)

Appendix 509

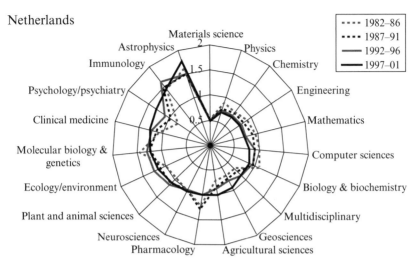

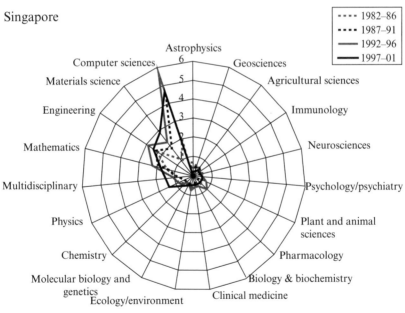

510 *Small country innovation systems*

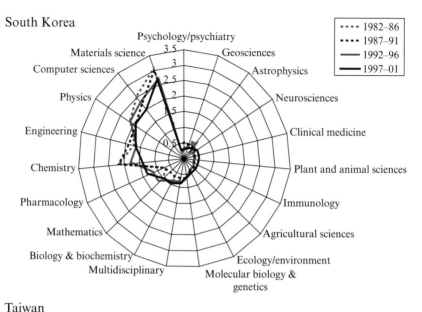

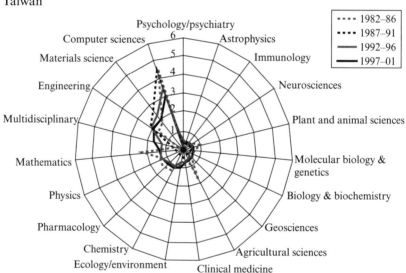

Figure A3.1   (continued)

3.2 Technology Profiles

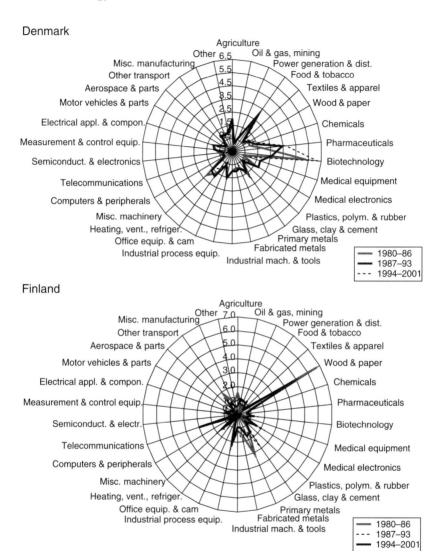

Source: K. Wang, M.-T. Tsai, Y.-L. Luo, A. Balaguer, S.-C. Hung, F.-S. Wu, M.-Y. Hsu and Y.-Y. Chu, 'Intensities of technological performance: Patenting at a macro and sectoral level for eight countries', Science and Technology Information Centre – National Science Council, Taipei, Republic of China, 2004.

Figure A3.2   Historical changes in revealed technological advantage (RTA) in patenting in eight nations, 1982–2001

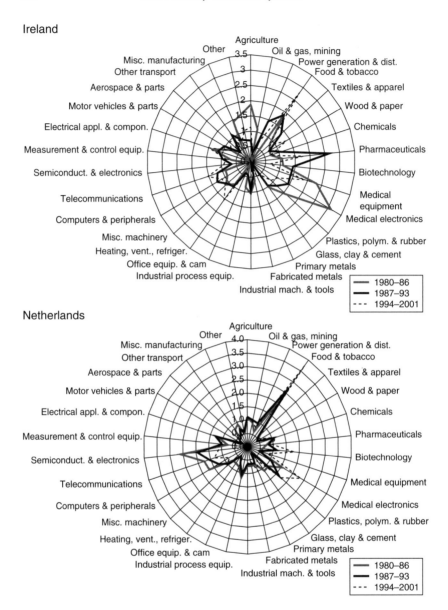

Figure A3.2 (continued)

Appendix

Norway

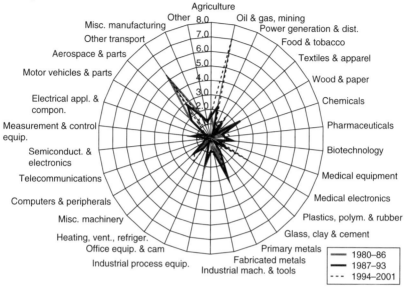

South Korea

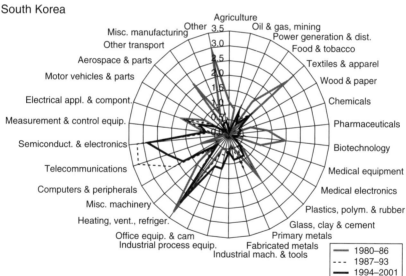

Sweden

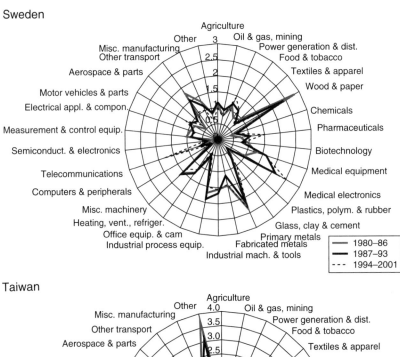

Taiwan

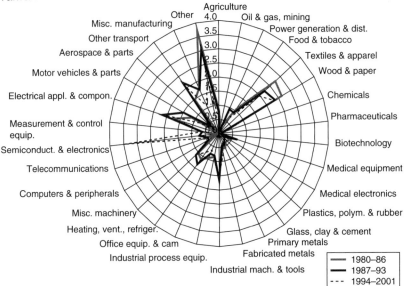

Figure A3.2 (continued)

3.3 Patenting

Table A3.1 Patenting estimates for 2000, according to inventors' residence

	EPO			USPTO			Triadic				
	Number of patent applications to the EPO	Share of countries in total EPO patent applications, %	EPO patent applications over GDP[1]	EPO patent applications per million population	Number of patents granted at the USPTO[2]	Share of countries in total USPTO patent grants,[2] %	Number of triadic patent families[2]	Share of countries/ economies in triadic patent families,[2] %	Triadic patent families[3] over GDP[1]	Triadic patent families[3] per million population	Triadic patent families[3] over industry-financed R&D[1]
Total	109 609	100.00	n.a.	n.a.	179 658	100.00	43 664	100.00	n.a.	n.a.	n.a.
Denmark	886	0.81	6.61	160.63	554	0.3	254	0.58	1.90	47.65	0.151
Finland	1 353	1.23	11.04	261.35	992	0.6	489	1.12	3.99	94.53	0.194
Hong Kong	45	0.04	n.a.	n.a.	243	0.1	28	0.06	n.a.	n.a.	n.a.
Ireland	196	0.18	1.90	38.99	151	0.1	45	0.10	0.44	11.91	0.063
Netherlands	3 351	3.06	8.34	210.47	1 431	0.8	857	1.96	2.13	53.83	0.220
Norway	384	0.35	2.69	67.54	294	0.2	109	0.25	0.76	24.19	0.107
Singapore	141	0.13	1.64	25.76	476	0.3	93	0.21	1.08	23.17	0.116
South Korea	1 211	1.10	1.85	35.12	4 100	2.3	478	1.09	0.73	10.17	0.046
Sweden	2 227	2.03	10.10	51.01	1 594	0.9	811	1.86	3.68	91.40	0.155
Taiwan	n.a.	n.a.	n.a.	n.a.	n.a.	n.a.	n.a.	n.a.	n.a.	n.a.	n.a.

Table A3.1 (continued)

	EPO				USPTO				Triadic			
	Number of patent applications to the EPO	Share of countries in total EPO patent applications, %	EPO patent applications over GDP[1]	EPO patent applications per million population	Number of patents granted at the USPTO[2]	Share of countries in total USPTO patent grants,[2] %	Number of triadic patent families[2]	Share of countries/economies in triadic patent families,[2] %	Triadic patent families[3] over GDP[1]	Triadic patent families[3] per million population	Triadic patent families[3] over industry-financed R&D[1]
USA	29 207	26.65	3.26	96.91	95 140	53.0	14 985	34.32	1.67	53.11	0.098
EU	49 668	45.31	5.84	150.90	28 007	15.6	13 699	31.37	1.61	36.24	0.161
Japan	20 388	18.60	6.76	166.06	36 781	20.5	11 757	26.93	3.90	92.63	0.188

Notes:
1. Gross domestic product (GDP), billion 1995 PPP US$.
2. Patents all applied for at the EPO, USPTO and JPO. Figures for 1999 and 2000 are estimates.
3. Patents all applied for at the EPO, USPTO and JPO. Figures for 2000 are estimates.

Source: OECD, Patent Database, September 2004, available at: http://www.oecd.org/document/10/0,2340,en_2825_497105_1901066_1_1_1,00.html. Compilation by J. Novikova.

4 INNOVATION

4.1 CIS Data[3]

Table A4.1 Innovation intensity and R&D intensity, all firms

Country	(1994–96)		(1998–2000)
	Innovation intensity[1]	R&D intensity[2]	Innovation intensity[3]
Denmark	0.0482	0.0229	0.0054
Finland	0.0408	0.0222	0.0250
Ireland	0.0300	0.0221	n.a.
Netherlands	0.0328	0.0182	0.0150
Norway	0.0299	0.0143	0.0122
Sweden	0.0668	0.0403	0.1207
Average incl. Ireland	0.0414	0.0233	
Average excl. Ireland	0.0437		0.0357
Rankings			
Denmark	2	2	5
Finland	3	3	2
Ireland	5	4	n.a.
Netherlands	4	5	3
Norway	6	6	4
Sweden	1	1	1

Notes:
[1] 'Innovation intensity' measures all expenditures made in order to innovate, divided by firms' turnover (all firms in the category).
[2] 'R&D intensity' measures the R&D expenses designed to increase the firm's stock of knowledge systematically, by means of creative activities (performed by the firm itself or bought from another organization) divided by firms' turnover (all firms in the category).
[3] The figures for 1998–2000 should, if possible, be compared to other data sources. We regard these data – e.g. the 0.1207 figure for Sweden and the 0.0054 figure for Denmark – as highly uncertain. Since R&D intensity is a component of innovation intensity, we judge that the figure for Sweden is high, though perhaps not as high as indicated, and that the figure for Denmark should not be as low as indicated.

Source: Calculations by P. Bitard from Eurostat NewCronos 2 and 3, reference years: 1996 and 2000.

Table A4.2 Innovation intensity and R&D intensity, size decompositions

Country	(1994–96)		
	SME	LE	
	Innovation intensity	Innovation intensity	R&D intensity
Denmark	0.0489	0.0478	0.0254
Finland	0.0194	0.0472	0.0244
Ireland	0.0269	0.0353	0.0149
Netherlands	0.0222	0.0378	0.0208
Norway	0.0226	0.0365	0.0168
Sweden	0.0270	0.0802	0.0475
Average	0.0278	0.0474	0.0250
Rankings			
Denmark	1	2	2
Finland	6	3	3
Ireland	3	6	6
Netherlands	5	4	4
Norway	4	5	5
Sweden	2	1	1

Notes: Internal and external R&D expenditures are unavailable for SMEs in CIS2 (1994–96). Neither total innovation expenditures nor internal and external R&D expenditures are available for Sweden in the CIS3 database (1998–2000).

Source: Calculations by P. Bitard from Eurostat NewCronos 2.

Table A4.3 Innovation intensity and R&D intensity, manufacturing sector

Country	1994–96		1998–2000	
	Innovation intensity	R&D intensity	Innovation intensity	R&D intensity
Denmark	0.0484	0.0194	0.0095	0.0055
Finland	0.0434	0.0230	0.0391	0.0267
Ireland	0.0333	0.0127	n.a.	n.a.
Netherlands	0.0379	0.0197	0.0307	0.0244
Norway	0.0272	0.0125	0.0206	0.0155
Sweden	0.0704	0.0415	0.0642	0.0559
Average	0.0434	0.0215	0.0328	0.0250
Rankings				
Denmark	2	4	5	5
Finland	3	2	2	2
Ireland	5	5	–	–
Netherlands	4	3	3	3
Norway	6	6	4	4
Sweden	1	1	1	1

Notes: Figures for R&D intensity for 1998–2000 are dubious and therefore should be treated with caution.

Source: Calculations by P. Bitard from Eurostat NewCronos 2 and 3 (percentages and rankings).

Table A4.4 Proportions of innovating firms, macro-sectoral and size decompositions[1]

Country	SME		LE		All		Manu.		KIBS		Finance		Trade	
	1996	2000	1996	2000	1996	2000	1996	2000	1996	2000	1996	2000	1996	2000
Denmark	0.4306	0.4057	0.8211	0.6492	0.4548	0.4166	0.7061	0.5020	0.5675	0.5069	0.4760	0.3837	0.2650	0.3505
Finland	0.2707	0.3887	0.6810	0.6644	0.3027	0.4050	0.3637	0.4387	0.4423	0.5109	0.2846	n.a.	0.1487	0.4074
Ireland	0.6261	n.a.	0.8618	n.a.	0.6362	n.a.	0.7345	n.a.	0.7466	n.a.	0.6746	n.a.	0.5152	n.a.
Netherlands	0.4381	0.4085	0.7830	0.7273	0.4587	0.4229	0.6219	0.5121	0.5837	0.5221	0.3984	0.4407	0.3597	0.3743
Norway	0.2977	0.3135	0.6478	0.5923	0.3153	0.3253	0.4809	0.3578	0.4173	0.4894	0.4399	0.4238	0.1807	0.2849
Sweden	0.3912	0.3855	0.6753	0.6362	0.4095	0.3965	0.5421	0.4012	0.5014	0.5156	0.5645	0.4481	0.2902	0.4599
Average (excl. Ireland)	0.3656	0.3804	0.7217	0.6539	0.3882	0.3933	0.5429	0.4424	0.5024	0.5090	0.4327	0.4241	0.2489	0.3754
Average (incl. Ireland)	0.4091		0.7450		0.4295		0.5749		0.5431		0.4730		0.2933	
Rankings														
Denmark	3	2	2	3	3	2	2	2	3	4	3	4	4	4
Finland	6	3	4	2	6	3	6	3	5	3	6	–	6	2
Ireland	1	–	1	–	1	–	1	–	1	–	1	–	1	–
Netherlands	2	1	3	1	2	1	3	1	2	1	5	2	2	3
Norway	5	5	6	5	5	5	5	5	6	5	4	3	5	5
Sweden	4	4	5	4	4	4	4	4	4	2	2	1	3	1

Note: [1] In the CIS3 'Core Questionnaire', an 'innovating firm' is defined as a firm that, during the reporting period (1998–2000), engaged in innovation activities by implementing new or significantly improved products (in either goods or services) or introducing within the enterprise new or significantly improved processes based on science, technology or other knowledge areas. The indicator 'proportion of innovating firms' is the ratio 'number of firms having introduced either a new product and/or a new process' divided by 'total number of firms' according to the selected size classes and sector decompositions of firms.

Source: Calculations by P. Bitard from Eurostat NewCronos 2 and 3 (percentages and rankings), reference years: 1996 and 2000.

Table A4.5 *Proportions of firms that have innovated (1) in products and (2) in processes*[1]

Country	1994–96		1998–2000	
	Ranking of the proportions of new-to-the-firm product innovators[2]	Ranking of the proportions of process innovators	Ranking of the proportions of new-to-the-firm product innovators	Ranking of the proportions of process innovators
Denmark	2	2	2	1
Finland	5	6	3	3
Ireland	1	1	–	–
Netherlands	3	3	1	2
Norway	6	5	5	4
Sweden	4	4	4	5

Notes:
[1] In the CIS3 'Core Questionnaire', 'product innovation' is defined as a good or service that is either new or significantly improved with respect to its fundamental characteristics, technical specifications, incorporated software or other immaterial components, intended uses or user-friendliness. 'Process innovation' includes new and significantly improved production technology, new and significantly improved methods of supplying services and of delivering products. The outcome should be significant with respect to the level of output, quality of products (goods/services) or costs of production and distribution.
[2] In the CIS3 'Core Questionnaire', 'new-to-the-firm' product innovation should be new to the enterprise, but it does not necessarily have to be new to the market. It does not matter whether the innovation was developed by the enterprise or by another enterprise. Changes of a solely aesthetic nature and simply selling innovations wholly produced and developed by other enterprises should not be included.

Source: As for Table A4.4.

Table A4.6 *Proportions of firms that have innovated (1) in products and (2) in processes, all firms*

Country	Proportions of new-to-the-firm product innovators	Proportions of process innovators
Denmark	0.3655	0.2589
Finland	0.3499	0.2346
Netherlands	0.3765	0.2554
Norway	0.2971	0.2226
Sweden	0.3240	0.2013
Average	0.3426	0.2346
Rankings		
Denmark	2	1
Finland	3	3
Netherlands	1	2
Norway	5	4
Sweden	4	5

Source: Calculations by P. Bitard from Eurostat NewCronos 3 (percentages and rankings), 1998–2000.

Table A4.7 *Proportions of firms that have innovated in product (DUCT) and proportions of firms that have innovated in process (CESS), SMEs and large firms*[1]

Country	SME				LE			
	1994–96		1998–2000		1994–96		1998–2000	
	DUCT	CESS	DUCT	CESS	DUCT	CESS	DUCT	CESS
Denmark	0.5524	0.4906	0.3543	0.2482	0.8022	0.7253	0.6040	0.4884
Finland	0.2441	0.2105	0.3347	0.2185	0.6923	0.5692	0.5910	0.4905
Ireland	0.6469	0.5290	n.a.	n.a.	0.7820	0.7218	n.a.	n.a.
Netherlands	0.5350	0.4344	0.3629	0.2404	0.8010	0.7416	0.6628	0.5722
Norway	0.3193	0.3824	0.2862	0.2113	0.6798	0.6629	0.5436	0.4769
Sweden	0.4540	0.3487	0.3146	0.1895	0.7132	0.6103	0.5288	0.4595
Average (excl. Ireland)	0.4210	0.3733	0.3306	0.2216	0.7377	0.6619	0.5860	0.4975
Average (incl. Ireland)	0.4586	0.3993			0.7451	0.6718		
Rankings								
Denmark	2	2	2	1	1	2	2	3
Finland	6	6	3	3	5	6	3	2
Ireland	1	1	–	–	3	3	–	–
Netherlands	3	3	1	2	2	1	1	1
Norway	5	4	5	4	6	4	4	4
Sweden	4	5	4	5	4	5	5	5

Note: [1] To convey the content of the indicators more clearly, 'proportion of firms that have innovated in product' (DUCT) and 'proportion of firms that have innovated in process' (CESS) have been provisionally renamed. Originally, DUCT was identified as NPDT, and CESS was identified as INPCS. For a list of the indicators as they were originally identified, see P. Bitard, *The Use of the CIS Data in the Nine Countries Study – A Short Introduction*, Working paper, Division of Innovation, Lund University, prepared for the NSI9-ESF meeting in Taipei, 17 November 2003.

Source: Calculations by P. Bitard from Eurostat NewCronos 2 and 3 (percentages and rankings), reference years: 1996 and 2000.

Table A4.8 DUCT and CESS, macro-sectoral decompositions

Country	Manufacturing				KIBS				Finance				Trade			
	1996–2000		1998–2000		1998–2000				1998–2000				1998–2000			
	DUCT	CESS	DUCT	CESS	DUCT	CESS			DUCT	CESS			DUCT	CESS		
Denmark	0.5749	0.5115	0.4484	0.3208	0.4656	0.2439			0.2854	0.3445			0.3118	0.2026		
Finland	0.2945	0.2512	0.3674	0.2810	0.4900	0.2281			n.a.	n.a.			0.3919	0.1668		
Ireland	0.6565	0.5427	n.a.	n.a.	n.a.	n.a.			n.a.	n.a.			n.a.	n.a.		
Netherlands	0.5590	0.4621	0.4462	0.3609	0.5102	0.2781			0.3661	0.2732			0.3423	0.1622		
Norway	0.3468	0.4038	0.3274	0.2865	0.4628	0.2737			0.3920	0.3340			0.2654	0.1349		
Sweden	0.4816	0.3763	0.3113	0.2294	0.4335	0.2246			0.3431	0.2914			0.4192	0.1641		
Average (excl. Ireland)	0.4514	0.4010	0.3802	0.2958	0.4724	0.2497			0.3466	0.3108			0.3461	0.1662		
Average (incl. Ireland)	0.4856	0.4246														
Rankings																
Denmark	2	2	1	2	3	3			4	1			4	1		
Finland	6	6	3	4	2	4			–	–			2	2		
Ireland	1	1	–	–	–	–			–	–			–	–		
Netherlands	3	3	2	1	1	1			2	4			3	4		
Norway	5	4	4	3	4	2			1	2			5	5		
Sweden	4	5	5	5	5	5			3	3			1	3		

Source: As for Table A4.7.

Table A4.9 *Proportions of firms that have innovated in products that were new for the firm and proportions of firms that have innovated in products that were new on the market, all firms*[1]

Country	Proportion of new-to-the-firm product innovators	Proportion of new-to-the-market product innovators
Denmark	0.3655	0.2254
Finland	0.3499	0.2803
Netherlands	0.3765	0.1851
Norway	0.2971	0.1399
Sweden	0.3240	0.1728
Average	0.3426	0.2007
Rankings		
Denmark	2	2
Finland	3	1
Netherlands	1	3
Norway	5	5
Sweden	4	4

Note: [1] According to the CIS3 'Core Questionnaire', a 'new-to-the-market' product innovation is one that is both new to the enterprise and new to the market.

Source: Calculations by P. Bitard from Eurostat NewCronos 3 (percentages and rankings), reference year: 2000.

Table A4.10 DUCT and proportions of firms that have innovated in products that were new to the market (MARK), size class decompositions[1]

Country	SME		LE	
	DUCT	MARK	DUCT	MARK
Denmark	0.3543	0.2148	0.6040	0.4507
Finland	0.3347	0.2682	0.5910	0.4728
Netherlands	0.3629	0.1762	0.6628	0.3733
Norway	0.2862	0.1343	0.5436	0.2667
Sweden	0.3146	0.1664	0.5288	0.3138
Average	0.3306	0.1920	0.5860	0.3755
Rankings				
Denmark	2	2	2	2
Finland	3	1	3	1
Netherlands	1	3	1	3
Norway	5	5	4	5
Sweden	4	4	5	4

Note: [1] To convey the content of the indicators more clearly, 'proportion of firms that have innovated in products that were new to the market' (MARK) have been provisionally renamed. Originally, MARK was identified as NTTM. For a list of the indicators as they were originally identified, see P. Bitard, *The Use of the CIS Data in the Nine Countries Study – A Short Introduction*, Working paper, Division of Innovation, Lund University, prepared for the NSI9-ESF meeting in Taipei, 17 November 2003.

Source: As for Table A4.9.

Table A4.11 DUCT and MARK, macro-sectoral decompositions

Country	Manufacturing		KIBS		Finance		Trade	
	DUCT	MARK	DUCT	MARK	DUCT	MARK	DUCT	MARK
Denmark	0.4484	0.2552	0.4656	0.2953	0.2854	0.1837	0.3118	0.2614
Finland	0.3674	0.2882	0.4900	0.3814	n.a.	n.a.	0.3919	0.3334
Netherlands	0.4462	0.2397	0.5102	0.2760	0.3661	0.0027	0.3423	0.1742
Norway	0.3274	0.1509	0.4628	0.2641	0.3920	0.1285	0.2654	0.1167
Sweden	0.3113	0.1566	0.4335	0.2640	0.3431	0.1659	0.4192	0.2238
Average	0.3802	0.2181	0.4724	0.2962	0.3466	0.1202	0.3461	0.2219
Rankings								
Denmark	1	2	3	2	4	1	4	2
Finland	3	1	2	1	–	–	2	1
Netherlands	2	3	1	3	2	4	3	4
Norway	4	5	4	4	1	3	5	5
Sweden	5	4	5	5	3	2	1	3

Source: As for Table A4.9.

Table A4.12 Average proportions of turnover due to products that were new to the firm and average proportions of turnover due to products that were new to the market, all firms

Country	Turnover due to new-to-the-firm product	Turnover due to new-to-the-market product
Denmark	0.1348	0.0665
Finland	0.1751	0.1446
Netherlands	0.1212	n.a.
Norway	0.0703	0.0191
Sweden	0.2137	0.0457
Average	0.1430	0.0690
Rankings		
Denmark	3	2
Finland	2	1
Netherlands	4	–
Norway	5	4
Sweden	1	3

Source: As for Table A4.9.

Table A4.13 *Average proportions of turnover due to products that were new to the firm and average proportions of turnover due to products that were new to the market, size class decompositions*

Country	SME		LE	
	Turnover due to new-to-the-firm product	Turnover due to new-to-the-market product	Turnover due to new-to-the-firm product	Turnover due to new-to-the-market product
Denmark	0.1366	0.0565	0.1331	0.0755
Finland	0.0839	0.0480	0.2166	0.1885
Netherlands	0.0926	n.a.	0.1423	n.a.
Norway	0.0719	0.0212	0.0689	0.0174
Sweden	0.1545	0.0663	0.2519	0.0325
Average	0.1079	0.0480	0.1626	0.0785
Rankings				
Denmark	2	2	4	2
Finland	4	3	2	1
Netherlands	3	–	3	–
Norway	5	4	5	4
Sweden	1	1	1	3

Source: As for Table A4.9.

Table A4.14 Average proportions of turnover due to products that were new to the firm and average proportions of turnover due to products that were new to the market, macro-sectoral decompositions

Country	Manufacturing		KIBS		Finance		Trade	
	Turnover due to new-to-the-firm product	Turnover due to new-to-the-market product	Turnover due to new-to-the-firm product	Turnover due to new-to-the-market product	Turnover due to new-to-the-firm product	Turnover due to new-to-the-market product	Turnover due to new-to-the-firm product	Turnover due to new-to-the-market product
Denmark	0.1936	0.1140	0.2433	0.1242	0.1102	0.0301	0.1076	0.0477
Finland	0.2728	0.2387	0.2085	0.1233	n.a.	n.a.	0.0288	0.0169
Netherlands	0.1988	n.a.	0.1204	n.a.	0.1149	n.a.	0.0595	0.0001
Norway	0.1251	0.0313	0.2317	0.0784	0.0509	0.0042	0.0451	0.0081
Sweden	0.2764	0.0303	0.1790	0.0878	n.a.	n.a.	0.2267	0.0906
Average	0.2133	0.1036	0.1966	0.1034	0.0920	0.0171	0.0936	0.0327
Rankings								
Denmark	4	2	1	1	2	1	2	2
Finland	2	1	3	2	–	–	5	3
Netherlands	3	–	5	–	1	–	3	5
Norway	5	3	2	4	3	2	4	4
Sweden	1	4	4	3	–	–	1	1

Source: As for Table A4.9.

Table A4.15 *Average proportions of turnover due to products that were new to the firm and average proportions of turnover due to products that were new to the market, manufacturing sector*

Country	1996		2000	
	Turnover due to new-to-the-firm product	Turnover due to new-to-the-market product	Turnover due to new-to-the-firm product	Turnover due to new-to-the-market product
Denmark	0.0703	0.0512	0.1936	0.1140
Finland	0.0915	0.0730	0.2728	0.2387
Ireland	0.1693	0.0845	n.a.	n.a.
Netherlands	0.0731	0.0656	0.1988	n.a.
Norway	0.1026	0.0408	0.1251	0.0313
Sweden	0.1371	0.0694	0.2764	0.0303
Average (incl. Ireland)	0.1073	0.0641	0.2133	0.1036
Average (excl. Ireland)	0.0949	0.0600		
Rankings				
Denmark	6	5	4	2
Finland	4	2	2	1
Ireland	1	1	–	–
Netherlands	5	4	3	–
Norway	3	6	5	3
Sweden	2	3	1	4

Source: Calculations by P. Bitard from Eurostat NewCronos 2 and 3 (percentages and rankings), reference years: 1996 and 2000.

NOTES

1. Prose literacy is defined as the ability to understand and use information contained in various kinds of text. Document literacy refers to the knowledge and skills needed to process information contained in materials such as schedules, charts, graphs, tables, maps and forms. Quantitative literacy entails the performance of arithmetic operations.
2. All GDP data reported in this subsection are from University of Groningen and The Conference Board, GGDC Total Economy Database, 2002. Available at: http://www.eco.rug.nl/ggdc/homeggdc.html. The term 1990 PPP US$ refers to '1990 US dollars converted at "Geary–Khamis" purchasing power parities'.
3. The data in this subsection are drawn from the third Community Information Survey (CIS3), which covers only European countries. In some tables, the names of indicators have been altered. These changes are noted in relation to the specific tables where they first occur, as are explanations of terminology. For a list of the indicators as they were originally identified, see P. Bitard *The Use of the CIS Data in the Nine Countries Study – A Short Introduction*, Working paper, Division of Innovation, Lund University, prepared for the NSI9-ESF meeting in Taipei, 17 November 2003.

Index

Titles of publications are in *italics*.

A*STAR (Agency for Science, Technology and Research), Singapore 90, 97
Aalsmeer flower auction 335
academic entrepreneurship, Sweden 251–2; *see also* university research
Action Community for Entrepreneurship (ACE) 93–4
activities-based approach 9–11, 445–6, 459, 478–9
activities influencing innovation 8–9
   Denmark 411–27
   Finland 363–85
   Hong Kong 202–22
   Ireland 161–81
   Korea 119–41
   The Netherlands 328–42
   Norway 292–305
   Singapore 79–103
   Sweden 243–57
   Taiwan 37–57
adaptive approach 470
adult education and training, Denmark 413
agricultural institutes, The Netherlands 335
agro-industry, Denmark 405
Ali-Yrkkö, J. 378, 381
Amsden, A.H. 52
Andersson, T. 251
appreciative theorizing 443
Approved Technological Service Institutes (GTS), Denmark 424, 427
Archibugi, D. 463, 477
ARF venture capital fund, Hong Kong 219
Ashford, D.E. 443

Asia's Innovation Systems in Transition 445
ASML 339–40

Barcelona Declaration 446
Benner, M. 266, 305
BERD, *see* Business Expenditure on R&D
Bio-Medical Research Council (BMRC), Singapore 97
Biopolis, Singapore 100
biotech hub policies, Taiwan 63
biotechnology sector, The Netherlands 337–8
BIT programme, Norway 296
Borch, O.J. 303
Borrás, S. 446, 462
Brennan, N. 176
Buijink, C. 339
building descriptive theory 443–4
business angel investment, Singapore 101–2
business expenditure on R&D (BERD) 40
   Hong Kong 202
   Ireland 163–4
   Taiwan 37, 39
business growth, *see* firm-level growth
business incubators, Korea 138
business sector R&D
   Finland 364
   The Netherlands 329, 346, 349
   Sweden 245
business survival, *see* survival rates
business systems approach 446

Carlsson, B. 248, 463
Cassidy, M. 181, 183
chaebols, Korea 115–16, 133–4, 146

531

532 *Index*

Challenge 2008 National Development Plan, Taiwan 53, 63
Cheung Kong Group 212–13
China
  relationship with Hong Kong 196–7, 225
  relationship with Taiwan 33–4, 59–60
China Productivity Centre, Taiwan 56
Chinese family businesses, Taiwan 52
Christensen, C.M. 444
Christensen, J.L. 439
Chu, W.-W. 52
CITB, Hong Kong 216
CK Life Sciences 213, 215
Clancy, P. 169
clusters of competence, Denmark 405–6
Cogan, D. 178
collaboration
  Denmark 419–20, 425, 426, 434
  Finland 377–9
  Ireland 173–5
  Korea 134–5
  The Netherlands 336–7
  Norway 297–8
  Sweden 249–51
Collins, P. 179
Commitment to Research White Paper, Norway 299, 311–12
competence building
  Denmark 413–14
  Finland 369–71
  Hong Kong 206–8
  Ireland 166–8
  Korea 125–8
  The Netherlands 331–3
  Norway 294–5
  Singapore 84–8
  Sweden 245–7
  Taiwan 43–7, 61
competition, Norway 299
consequences of innovation
  Denmark 427–9
  Finland 385–9
  Hong Kong 222–4
  Ireland 181–3
  Korea 141–2
  The Netherlands 342–5
  Norway 305–7
  Singapore 103–5

Sweden 257–62
Taiwan 57–9
constituents provision
  Denmark 418–22, 426
  Finland 374–80
  Hong Kong 211–17
  Ireland 171–7
  Korea 131–7
  The Netherlands 335–8
  Norway 296–301
  Singapore 91–9
  Sweden 248–52, 266–7
  Taiwan 49–55
consultancy services
  Denmark 424–5, 427
  Finland 382–4
  Hong Kong 221–2
  Ireland 179–80
  Korea 140
  The Netherlands 341
  Norway 303–4
  Singapore 102–3
  Sweden 255–6
  Taiwan 56–7
consumer associations and new market formation, The Netherlands 333
Consumer Council, Hong Kong 211
Cooper, C. 186
cooperation, *see* collaboration
coordination and innovation policy approach 469–78, 481
corporate governance
  Ireland 176
  Norway 300
creation versus diffusion of innovations 8
Cyberport project, Hong Kong 217–18

Danish Crown 405
Daveri, F. 183
demand-side factors
  Denmark 414–17, 425–6
  Finland 371, 373–4
  Hong Kong 208–11
  Ireland 168–71, 181
  Korea 129–31
  The Netherlands 333–4
  Norway 295–6
  Singapore 88–91

Sweden 247–8, 265–6, 270
Taiwan 47–9
DEMO 2000 programme, Norway 296
Denmark 24, 403–38
   activities influencing innovation
      411–27
   consequences of innovation 427–9
   future innovation policy 437–8
   globalization 429–30
   historical trends 404–8
   innovation intensity 408–11
   innovation policy 436–7, 460
   innovation policy coordination 473–4
   science profile 507
   strengths and weaknesses of
      innovation policy 430–36
   technology profile 511
Design for Environment in SMEs
   programme, Sweden 248
design innovations, Denmark 408,
   410
diffusion of new technologies,
   Singapore 91
Digital 21 Strategy, Hong Kong 210
dynamism of study countries 12

e-Norway programme 296
East Asian Miracle, The 58
Economic Development Board,
   Singapore 90
economic growth, *see* growth
Economic Society, Finland 356
Edquist, C. 238, 460, 462, 469
education
   Denmark 413, 420–21
   entrepreneurial, Singapore 92
   Finland 369–71, 398
   Hong Kong 206–8
   Ireland 166–8, 180–81
   Korea 125–8
   The Netherlands 331–3
   Norway 294–5
   Singapore 84, 86–7, 95
   Sweden 245–6
   Taiwan 43–7
   see also higher education; universities
education expenditure
   Finland 370
   Hong Kong 207
   Korea 125

Sweden 245–6
Taiwan 43, 45
EISC report, Singapore 92–4
electronics sector FDI, Singapore 88–9
Enright, M.J. 196
Enterprise Ireland 179–80
entrepreneurship
   Denmark 433–4
   education, Singapore 92
   Singapore 106–7
environmental policy, Taiwan 66
environmental regulation
   Korea 130–31
   Hong Kong 210–11
Esprit programme participation,
   Ireland 179
Estevão, M.M. 389
EU membership
   Ireland 168
   Sweden 239, 247–8, 252
EU Structural Funds R&D support,
   Ireland 178–9
expenditure, *see* education expenditure;
   innovation expenditure; R&D
   expenditure
export policy, Korea 115
extrapreneurship, The Netherlands
   339–40

fair trade policy, Korea 136
fast growth countries
   growth mechanisms and instruments
      452–3
   growth patterns 450–52
   innovation policy coordination 476,
      477
   see also Hong Kong; Ireland; Korea;
      Singapore; Taiwan
FDI, *see* foreign direct investment
Felisberto, C. 182–3
finance sector innovation
   Denmark 408
   The Netherlands 325–6
financing innovating firms
   Denmark 423–4
   Finland 380–82
   Hong Kong 219–21
   Ireland 178–9
   Korea 133, 138–40
   The Netherlands 340

Norway 300–301, 302–3, 308
Singapore 100–102
Sweden 253–5, 267–8
Taiwan 55–6
Finland 23–4, 355–99
　activities influencing innovation 363–85
　consequences of innovations 385–9
　future innovation policy 396–9
　globalization 389–93
　growth dynamism 450–52
　historical trends 356–8
　innovation intensity 358–63
　innovation policies 395–6
　innovation policy coordination 472
　policy instruments 454
　science profile 507
　strengths and weaknesses of NSI 394–5
　technology profile 511
firm-level growth
　Denmark 428–9
　Ireland 181–2
　Norway 306
　Sweden 257–8
firm size and innovation
　Denmark 409
　Hong Kong 200
　The Netherlands 326–7
　Norway 286
firm size and R&D activities
　Finland 364
　Korea 120–21
Fitzsimmons, P. 177
Florida, R. 13–14
flower auction, Aalsmeer 335
foreign direct investment
　Denmark 429–30
　Finland 389, 391
　Ireland 157–8, 168–9, 183–4
　Korea 136, 144
　Norway 308
　Singapore 89–90, 105
foreign-owned enterprises
　Ireland 169–71
　　and patent activity 164, 166
　R&D, Finland 366, 391–3
　Singapore 75, 88–90
　　R&D 82
　Taiwan 59

Forfás, Ireland 171–2
Foundation for Finnish Inventions (FII) 381
Freeman, C. 4, 64, 145
Fund for Research and Innovation (FRI), Norway 300
funding, *see* financing innovating firms
future innovation policy
　Denmark 437–8
　Finland 396–9
　Hong Kong 229–31
　Ireland 186–9
　Korea 148–9
　The Netherlands 350–52
　Norway 312–14
　Singapore 108–10
　Sweden 268–72
　Taiwan 64–6

GDP growth and innovation, Denmark 428
GDP per hour worked 501–2
GDP per person employed 498–500
George, A.L. 444
Georghiou, L. 383, 398
GERD, *see* gross expenditure on R&D
Ginarte, J.C. 379
globalization 16–17
　Denmark 429–30
　effect on NSIs 456–7
　Finland 385, 389–93, 399
　and growth 449–57
　Hong Kong 224–6
　Ireland 183–4
　Korea 143–6
　The Netherlands 345–8
　Norway 307–8
　and selectivity 463–4
　Singapore 105
　statistics 503–6
　Sweden 262
　Taiwan 59–61, 63
government
　as lead user of new technology, Singapore 91
　role in creating networks, Taiwan 52–3
government sector expenditure on R&D (GOVERD)
　Hong Kong 202, 203

Taiwan 39
government sector R&D
  Korea 121
  The Netherlands 330–31
  see also public R&D
Graversen, E. 308, 414
Grimes, S. 179
GRIs (government research institutes), Korea 121, 123, 146
gross expenditure on R&D (GERD)
  Ireland 161
  Singapore 79–81
  Taiwan 37
growth
  and globalization 452
  high-tech sectors 456, 480
  Ireland 158, 182–3
  Korea 113
  patterns 450–52
  policies 452–5
  Singapore 72
  statistics 494–502
growth effects of innovation
  Denmark 427–9
  Ireland 181–3
  Korea 141–2
  Norway 305–7
  Singapore 103–5
  Taiwan 57–9
  see also productivity
GTS Institutes, Denmark 424, 427

Hall, P.A. 446
Hämäläinen, T. 383, 397
Hannan, D. 166
He, Z.L. 107
hearing aid industry, Denmark 417
heavy and chemical industries (HCI), Korea 115–16
Henrekson, M. 246, 251
HERD, see Higher Education R&D
Hermans, R. 378
Hewitt-Dundas, N. 170
high-tech sectors
  Denmark 435–6
  and economic growth 456, 480
higher education
  Denmark 413, 421
  Finland 370
  Hong Kong 206–8
  Ireland 167–8
  Korea 125–6
  The Netherlands 331–3
  Norway 294–5
  Sweden 246–7
  Taiwan
  see also universities
higher education R&D (HERD)
  Hong Kong 202, 203, 205, 222
  Ireland 163
  The Netherlands 329–30
  Norway 299
  Sweden 245
Ho, Y.P. 104
holistic innovation policy, Norway 311
Hollingsworth, J.R. 447
Hong Kong 22, 194–231
  activities influencing innovation 202–22
  consequences of innovation 222–4
  future innovation policy 229–31
  globalization 224–6
  historical trends 195–7
  innovation intensity 197–201
  innovation policy approach 228–9, 461
  innovation policy coordination 475
  innovation system strengths and weaknesses 226–8
  relationship with China 196–7, 225
Hong Kong Industrial Technology Centre Corporation (HKITCC) 217
Hong Kong Institute of Biotechology Ltd (HKIB) 217
Hong Kong Productivity Council (HKPC) 221
Hong Kong Safety Institute Limited (HKSI) 211
HOTSpots (Hub Of Technopreneurs), Singapore 100
Hou, C.-M. 58
Howie, D.I.D. 186
Hsieh, C.T. 104
Hub Of Technopreneurs (HOTSpots), Singapore 100
human development indicators 486–7
human resources
  Denmark 431–2
  from abroad

The Netherlands 346, 348
Singapore 87–8
Hyytinen, A. 381

Iammarino, S. 463, 477
ICT sector
  and FDI, Finland 391
  and growth, Ireland 183
  Hong Kong 210
  Korea 145
  The Netherlands 333–4
  R&D, Korea 123–4
III (Institute for Information
    Industry), Taiwan 47, 57
incentives, Denmark 432–3
incubating activities
  Denmark 423, 426
  Finland 380
  Hong Kong 217–18
  Ireland 177–8
  Korea 137–8
  The Netherlands 339–40
  Norway 301–2
  Singapore 100
  Sweden 252–3
  Taiwan 55
Industrial Development Bureau (IDB),
    Taiwan 53
industrial relations, Taiwan 54
Industrial Technology Research
    Institute (ITRI), Taiwan 47, 57
industrialization
  Finland 356–7
  Ireland 157–8
  Korea 114
  The Netherlands 320–22
  Norway 282–5
  Sweden 238–9
industry–university links, see
    university–industry collaboration
Infocomm Development Authority
    (IDA), Singapore 86–7
infrastructure investment and market
    creation
  Hong Kong 209–10
  Korea 129
Innovation Bridging Foundations,
    Sweden 267–8
innovation centres
  Ireland 180

The Netherlands 337
innovation consequences, see
    consequences of innovations
innovation collaboration, see
    collaboration
innovation expenditure
  The Netherlands 324–8
  Norway 286
  Sweden 241, 263
  see also R&D expenditure
innovation intensity 15–26
  Denmark 408–11
  Finland 358–63
  Hong Kong 197–201
  Ireland 158–61
  Korea 117–19
  The Netherlands 323–8
  Norway 285–92
  Singapore 76–9
  statistics 517–19
  Sweden 240–43
  Taiwan 34–7
innovation networks, see networking
Innovation Norway 302, 303
Innovation Platform, The Netherlands
    338
innovation policy
  coordination 469–78
  definition 9
  Denmark 436–7
  Finland 395–6
  future, see future innovation policy
  Hong Kong 228–9
  Ireland 186
  Korea 174–5
  Norway 310–12
  The Netherlands 349–50
  rationales 458–61
  selectivity 461–9, 479–80
  Singapore 107–8
  Sweden 265–8
  Taiwan 63–4
innovation propensity, see innovation
    intensity
Innovation Research Programmes, The
    Netherlands 336–7
innovation statistics 517–30
Innovation and Technology
    Commission (ITC), Hong Kong
    206

Innovation and Technology Fund, Hong Kong 205–6
innovations, definition 8
Institute for Information Industry (III), Taiwan 47, 57
institutions
    definition 9
    Denmark 420–22, 426
    Finland 379–80
    Hong Kong 216–17
    Ireland 175–7
    Korea 135–7
    The Netherlands 337–8
    Norway 298–301
    Singapore 96–9
    Sweden 251–2
    Taiwan 53–5
instruments for economic growth 453–5
    selectivity 462
intellectual property rights
    Denmark 422
    Finland 379–80
    Hong Kong 216–17
    Ireland 177
    Korea 136–7
    The Netherlands 338
    Taiwan 54–5, 62
    see also patenting activity
interactive learning
    Hong Kong 214
    Korea 133–5
    Singapore 94–6
International Enterprise (IE), Singapore 90
international R&D
    Norway 293, 298
    Singapore 109–10
    Taiwan 59
inward FDI
    Norway 308
    Singapore 105
inward technology transfer, Ireland 184–5
Ireland 21, 156–89
    activities influencing innovation 161–81
    consequences of innovation 181–3
    future innovation policy 186–7

globalization 183–4
historical trends 157–8
innovation intensity 158–61
innovation policy approach 461
innovation policy coordination 475–6
innovation policy selectivity 465
innovation system strengths and weaknesses 184–5
technology profile 512
ITF funding, Hong Kong 219
ITRI (Industrial Technology Research Institute), Taiwan 47, 57

Jacobson, D. 170
Jacobsson, S. 248, 463

Kaitila, V. 58
Kaukonen, E. 378
Kearns, A. 181–2
Kenniswijk project, The Netherlands 334
KIBS (knowledge-intensive business service) sector
    Denmark 425
    Finland 397
    Korea 140
    Norway 303
    Singapore 76–9
    Sweden 255–6
    Taiwan 65–6
KIBS sector innovation
    Finland 359, 362
    The Netherlands 324
    Singapore 76–7
    Sweden 256
Kim, L. 115
Kleinknecht, A.H. 324
Knell, M. 345
knowledge inputs to innovation
    Denmark 411–14, 425
    Finland 363–72
    Hong Kong 202–8
    Ireland 161–8
    Korea 119–28
    The Netherlands 328–33
    Norway 292–5, 298–9, 304
    Singapore 79–88
    Sweden 243–7, 265
    Taiwan 37–47

knowledge-intensive business services, see KIBS sector
Kogut, B. 445
Korea 21, 113–50
    activities influencing innovation 119–41
    consequences of innovations 141–2
    economic growth 113
    future innovation policy 148–9
    globalization 143–6
    historical trends 114–17
    innovation intensity 117–19
    innovation policy approach 147–8, 461
    innovation policy coordination 475
    innovation policy selectivity 466–7
    innovation system strengths and weaknesses 146–9
    science profile 510
    support services for innovating firms 137–40
    technology profile 513
Korea Fair Trade Commission (KFTC) 136
Krugman, P. 58, 141

labour market institutions, Denmark 422
labour mobility
    Denmark 414
    The Netherlands 346, 348
    Singapore 87–8
    Taiwan 45, 47
labour productivity
    Denmark 428
    Ireland 183
    Korea 142
    statistics 493
    Sweden 258–9
    Taiwan 57–8
labour supply, Denmark 431–2
Lall, S. 452
land supply policies, Hong Kong 209
Larédo, P. 5–6
lead-user role
    government, Singapore 91
    public sector, Finland 373
Lee, K. 140
Leiponen, A. 379
Lemola, T. 357
Li, K. 223

Li & Fung 227–8
life-science sector, Singapore 89–90, 97
Lin, C.-Y. 32
literacy indicators 488–9
Local Industry Upgrading Programme, Singapore 94
long-term labour productivity, Taiwan 57–8
Lorenz, E. 410, 419
Love, J.H. 175, 179, 181
low-tech sectors, Denmark 434–5
Lundvall, B.-Å. 5, 12, 238, 455, 462

Malerba, F. 455
Mansfield, E. 58
manufacturing sector innovation
    The Netherlands 324
    Singapore 79, 82
    Taiwan 34–7
market formation, see demand-side factors
Maskell, P. 404
Matson, E. 297
McDevitt, J. 178
Metcalfe, J.S. 470
Miettinen, R. 448
MNCs, see multinational corporations
MNEs, see multinational enterprises
mobility, see labour mobility; student mobility
Mottiar, Z. 170
Mowery, D.C. 463
multi-agency coordination, Singapore 99
multinational corporations (MNCs)
    Singapore 75, 82
    Taiwan 59
multinational enterprises (MNEs)
    Ireland 157–8, 169–71
    Sweden 262, 270–71
Mustar, P. 5–6

National Linkage Programme, Ireland 169–70
national systems of innovation (NSI) 1
    compared with sectoral systems 445–7
    comparison 448–9
    definitions 4–6
    see also individual countries

National Technology Plans, Singapore 76, 96
Nelson, R.R. 4–5, 12, 25, 443, 455
Netherlands 23, 319–52
  activities influencing innovation 328–42
  consequences of innovation 342–5
  future innovation policy 350–52
  globalization 345–8
  historical trends 320–22
  innovation intensity 323–8
  innovation policy 349–50, 460
  innovation policy coordination 472–3
  innovation policy selectivity 466
  innovation system strengths and weaknesses 348–9
  science profile 509
  technology profile 512
Netherlands Organization for Applied Scientific Research (TNO) 330–31
networking
  Denmark 419–20
  Finland 377–9
  Hong Kong 214–15
  Ireland 173–5
  Korea 133–5
  The Netherlands 336–7
  Norway 297–8
  Singapore 94–6, 109–10
  Sweden 249–51
  Taiwan 52–3
new firm startups
  Norway 296–7
  Sweden 248–9
  Taiwan 49, 52
new technology-based firms (NTBFs), Korea 139
Next Lap, The 96
Nieminen, M. 378
Niosi, J. 1, 448
Nokia 365–6
non-technological innovation
  Denmark 410–11
  Hong Kong 200
  The Netherlands 327–8
Norway 23, 281–314
  activities influencing innovation 292–305
  consequences of innovations 305–7

  future innovation policy 312–14
  globalization 307–8
  historical trends 282–5
  innovation intensity 285–92
  innovation policies strengths and weaknesses 309–10
  innovation policy 310–12, 460
  innovation policy coordination 472
  innovation policy selectivity 467–8
  science profile 508
  technology profile 513
Novo Nordisk 412
NSE employment, KIBS sector, Sweden 255–6
NSI, *see* national systems of innovation
Nutek 248, 267

OEM agreements, Taiwan 36–7, 53
Office of Science and Technology (OST), Ireland 171
OG21 project, Norway 296
oil and gas industry, Norway 283
One North project, Singapore 92, 100
optimizing approach 470
organizational forms 410–11
organizations
  definition 9
  Denmark 418–19
  Finland 374–7
  Hong Kong 211–14
  Ireland 172–3
  Korea 131–3
  The Netherlands 335–6
  Norway 296–7
  Singapore 91–4
  Sweden 248–9
  Taiwan 49–52
Ørstavik, F. 306
outsourcing R&D, Hong Kong 225
outward FDI
  Norway 308
  Singapore 105

Pajarinen, M. 381, 391
Palmberg, C. 373, 377
Park, W.G. 379
Pasteur quadrant research, Singapore 108–9
patent system, Hong Kong 216–17

patenting activity 515–16
  cross-border 506
  Denmark 412–13, 430
  Finland 368
  Hong Kong 203
  Ireland 164, 166
  Korea 118, 136–7, 142, 144
  Norway 294
  Singapore 84
  Taiwan 41–3, 62
pharmaceutical industry, Singapore 89–90
Philips Electronics 339–40
Plan to Construct a National Innovation System, Korea 148
plastics industry and innovation, Taiwan 34, 36
poldermodel, The Netherlands 321–2
policy, *see* innovation policy
policy instruments for economic growth 453–5
population of study countries 12, 485
PRICs (public research institutes/centres), Singapore 81, 86, 96
private sector and technology diffusion, Taiwan 48
private sector R&D
  Korea 120
  Norway 292
  Singapore 80–82
process innovation 8
  Hong Kong 199
  Sweden 241
  Taiwan 36–7
product innovation 8
  Denmark 409
  Hong Kong 197, 199
  Norway 306
  Sweden 241–2
production network coordination, Taiwan 49
productivity, effects of innovation
  Denmark 427–9
  Finland 385–9
  Hong Kong 222–4
  Korea 142
  The Netherlands 342–4
  Sweden 258–9
  Taiwan 57–9

productivity statistics 493
propensity to innovate, *see* innovation intensity
property sector and demand, Hong Kong 209
provision of constituents, *see* constituents provision
Public Industry Organizations, The Netherlands 321
public policy and NSI development, Singapore 76
public R&D
  Denmark 411
  Finland 364, 369
  Korea 121
  Singapore 75–6, 81, 96–9
public sector role in NSI
  Denmark 432–3
  Singapore 76
public technology procurement (PTP)
  Denmark 414–15, 454
  Finland 373
  Norway 295
  and selectivity 462
  Sweden 247, 454
publications, *see* scientific publications
public–private partnerships, Sweden 248
public–private sector cooperation, The Netherlands 336

quality requirements and demand, Hong Kong 210–11; *see also* standards

R&D activities
  Denmark 411–13
  Finland 363–9
  Hong Kong 202–6
  Ireland 161–6
  Korea 119–25
  The Netherlands 328–31
  Norway 292–4
  Singapore 79–84
  Sweden 243–5
  Taiwan 32–3, 37–43
R&D collaboration
  Hong Kong 214
  Korea 134–5

Norway 298
Taiwan 62
R&D, effect of globalization
  Hong Kong 225
  The Netherlands 346
R&D expenditure
  Denmark 411–12
  Finland 363–4, 386–7, 395–6
  Hong Kong 202–3, 205
  Ireland 161, 163–4, 178, 180
  Korea 119–25
  Norway 292–3
  Singapore 79–82, 108
  Sweden 243–4, 245, 265
  Taiwan 37–8
R&D financing
  Hong Kong 219–21
  Ireland 178–9
R&D intensity
  Korea 119–20, 142
  The Netherlands 328–9
  Singapore 79, 108
  statistics 517–19
  Sweden 240
  Taiwan 37
rationales for innovation policy 458–61
regulation as driver of demand
  Denmark 414–15
  Hong Kong 210–11
  Korea 130–31
  The Netherlands 333–4
  Sweden 248
Remøe, S.O. 310, 311
research councils, The Netherlands 329–30
research institutes
  Denmark 421
  Finland 366
  Korea 121, 123, 146
  Norway 292–3
  Taiwan 47
Roper, S. 175, 179, 181
Rosenberg, N. 5, 25, 246, 251, 455
Ruane, F. 181–2

Samsung Electronics 130
San, G. 58
Schein, E. 97
Schienstock, G. 383, 397
Schot, J.W. 333

Science and Engineering Research Council, Singapore 97
Science Foundation Ireland 163
Science and Irish Economic Development 186
science parks
  Singapore 100
  Sweden 253
science profiles 507–10
Science and Technology Advisory Group, Taiwan 54
science and technology (S&T) policies
  Finland 357–8
  Singapore 109
  Taiwan 54
scientific publications
  Denmark 413
  Finland 368
  Norway 293
  Sweden 244
  Taiwan 39, 41
sectoral effects of innovation, Sweden 258
sectoral innovation intensity
  Denmark 408–9
  Finland 359–63
  The Netherlands 324–6, 327
  Norway 286
sectoral policies, Sweden 270
sectoral production statistics 490–92
sectoral specialization 456
sectoral systems of innovation 455–7
SEEDS (Startup Enterprise Development Scheme), Singapore 102
selectivity, innovation policy 461–9, 479–80
service sector
  Finland 397
  Ireland 188
  Korea 147
  R&D, Singapore 82
  see also KIBS sector
Shin, T. 133
SI, *see* systems of innovation
Singapore 20, 71–110
  activities influencing innovation 79–103
  consequence of innovations 103–5
  future innovation policy 108–10

globalization 105
historical trends 71–6
innovation intensity 76–9
innovation policy coordination 474–5
innovation policy selectivity 468
science profile 509
strengths and weaknesses of NSI 106–7
Singh, N. 58
slow growth countries
growth mechanisms and instruments 453–5
growth patterns 450–52
innovation policy coordination 476
see also Denmark; Finland; The Netherlands; Norway; Sweden
SMEs
and innovation, Hong Kong 200, 211–12
and networking, Taiwan 52
Norway 285
Smyth, E. 166
Soskice, D. 446
South Korea, *see* Korea
spending, *see* education expenditure; innovation expenditure; R&D expenditure
Stambøl, L.S. 303
standards as drivers of demand
Finland 373
Hong Kong 210–211
Korea 129
The Netherlands 334
Sweden 248
Taiwan 49
Startup Enterprise Development Scheme (SEEDS), Singapore 102
Statoil 295–6
Storper, M. 470–71
Strategic Economic Plan, Singapore 96
strengths and weaknesses of innovation system
Denmark 430–36
Finland 394–5
Hong Kong 226–8
Ireland 184–5
Korea 146–9
The Netherlands 348–9
Norway 309–10

Singapore 106–7
Sweden 263–5
student mobility
Finland 370
Hong Kong 206–7
Korea 126, 128
Norway 308
Taiwan 45
subsidies for innovation, The Netherlands 337
Sun, C.-H. 58
support services for innovating firms
Denmark 422–5
Finland 380–84
Hong Kong 217–22
Ireland 177–80
Korea 137–40
The Netherlands 339–41
Norway 300–304, 308
Singapore 100–103
Sweden 252–6, 267–8
Taiwan 55–7
survival rates
Ireland, effect of innovation 181–2
Sweden 249
sustainability, Taiwan 66
Sweden 22–3, 237–72
activities influencing innovation 243–57
consequences of innovation 257–62
future innovation policy 268–72
globalization 262
historical trends 238–40
innovation intensity 240–43
innovation policies 265–8
innovation policies strengths and weaknesses 263–5
innovation policy coordination 471–2
innovation policy selectivity 465–6
policy instruments 454
science profile 508
technology profile 514
Swedish Agency for Economic and Regional Growth (NUTEK) 248, 267
Swedish Agency for Innovation Systems (VINNOVA) 267
Swedish paradox 237–8, 240–42, 256–7
Synthens, The Netherlands 337

systemic approach for SI analysis 458–9
  and selectivity of innovation policy 462
systems, definition 6
systems of innovation (SI) approaches 4–6, 446–7
  definitions 6–8
  effect on innovation policy 458–60

Taiwan 20, 31–66
  activities influencing innovation 37–57
  consequences of innovation 57–9
  future innovation policy 64–6
  globalization 59–61
  historical trends 32–4
  innovation intensity 34–7
  innovation policy approach 461
  innovation policy coordination 474
  innovation policy selectivity 467
  relationship with China 33–4, 59–60
  science profile 510
  strengths and weaknesses of innovation system 61–3
  technology profile 514
Taiwan Technology Innovation Survey (TTIS) 34
tax incentives
  The Netherlands 337
  Norway 301
TEA, *see* total entrepreneurial activity
Technological Institute, Norway 304
technological trajectory perspective, Norway 287
technology entrepreneurship, Singapore 106–7
technology profiles 511–14
Technology Top Institutes (TTIs) 336
Technology Transfer Initiative, Ireland 180
TechnoPartner programme, The Netherlands 339
Technopreneurship 21 (T21), Singapore 92
Technopreneurship Fund, Singapore 101
Tekes 358, 379, 381, 383
tertiary education, *see* higher education

TFP, *see* total factor productivity
3TU agreement, The Netherlands 339
Tijssen, R. 330, 336
Tinagli, I. 14
Toivanen, O. 381
total entrepreneurial activity (TEA)
  Denmark 418
  Ireland 172
total factor productivity (TFP)
  Hong Kong 223–4
  Ireland 182–3
  Singapore 103–4
  Taiwan 58–9
trade associations, Finland 373
trade liberalization
  Finland 371, 373
  Korea 136
trade sector innovation
  Denmark 408
  The Netherlands 326
training, Singapore 84–8; *see also* vocational training
transport infrastructure and market formation, Hong Kong 209–10
Trieu, H. 58
Tsai, K.-H. 59
TSE (Taiwan Stock Exchange) 56
turnover, effects of innovation
  Denmark 409
  Sweden 257–8
24SJU (24SEVEN) project, Sweden 247–8
Two Trillion, Twin Star industry strategy, Taiwan 63

uncertainty, effect on innovation 463
universities
  business incubators
    Hong Kong 218
    Korea 138
  consultancy services, Ireland 180
  cooperation with business, Hong Kong 214
  Denmark 421, 434
  entrepreneurship, The Netherlands 339
  Finland 370
  Hong Kong 206–7, 222
  innovation support, Sweden 253
  Ireland 172, 287

patenting, The Netherlands 338
  Taiwan 64–5
university–industry collaboration
  Denmark 434
  Hong Kong 214–15
  Singapore 95
  Sweden 250, 267
  Taiwan 63–4, 64–5
university research
  Denmark 434
  Korea 123
  The Netherlands 329–30
  Norway 299
  Sweden 251–2
  Taiwan 57
Urban Knowledge Area project (Kenniswijk), The Netherlands 334

Valeyre, A. 410, 419
Van Beers, C. 392
Van Riel, A. 320, 321
Van Zanden, J.L. 320, 321
varieties of capitalism (VoC) 446
venture businesses, Korea 131–3
venture capital
  Denmark 423–4, 426–7
  Finland 381
  Hong Kong 219
  Korea 139–40
  The Netherlands 340
  Norway 302
  Singapore 101
  Sweden 253–5, 268
  Taiwan 55–6
VINNOVA (Swedish Agency for Innovation Systems) 267
vocational training
  Hong Kong 206
  Korea 128
  Singapore 84, 86
Volvo Korea 130
VTech 212

Wang, J.-C. 59
Wang, K. 330
Whelan, N. 186
Whitley, R. 446
Wicken, O. 310
wind energy sector, Denmark 415–16
Winter, S. 443
Wong, P.K. 72, 94, 99, 107
Woo, C. 126
World Bank Report, *The East Asian Miracle* 58

Yearley, S. 179
Ylä-Anttila, P. 377, 391
Young, A. 103–4

Zucker, L.G. 187

HC415.T4 S63 2008

Small country innovation
   systems : globalization,
      c2008.